International Conference on Computer Simulation in Physics and Beyond 2015

Journal of Physics: Conference Series Volume 681

Moscow, Russia
6 – 10 September 2015

Editors:

Lev N. Shchur
Serge A. Krashakov

ISBN: 978-1-5108-2091-3
ISSN: 1742-6588

Printed from e-media with permission by:

Curran Associates, Inc.
57 Morehouse Lane
Red Hook, NY 12571

Some format issues inherent in the e-media version may also appear in this print version.

Printed by Curran Associates, Inc. (2016)

For permission requests, please contact the Institute of Physics
at the address below.

Institute of Physics
Dirac House, Temple Back
Bristol BS1 6BE UK

Phone: 44 1 17 929 7481
Fax: 44 1 17 920 0979

techtracking@iop.org

Additional copies of this publication are available from:

Curran Associates, Inc.
57 Morehouse Lane
Red Hook, NY 12571 USA
Phone: 845-758-0400
Fax: 845-758-2634
Email: curran@proceedings.com
Web: www.proceedings.com

Table of contents

Volume 681

International Conference on Computer Simulation in Physics and Beyond 2015
6–10 September 2015, Moscow, Russia

Accepted papers received: 10 December 2015
Published online: 3 February 2016

Statistical Mechanics

Quantum Computing

Space Research

Dynamical Systems

Earth Physics

Computational Physics

Information technologies in research

Journal of Physics: Conference Series **681** (2016) 011001 doi:10.1088/1742-6596/681/1/011001

New trends in Computer Simulations in Physics and not only in physics

Lev N Shchur[1,2,3] and Serge A Krashakov[2,3]

[1] National Research University Higher School of Economics, 101000 Moscow, Russia

[2] Landau Institute for Theoretical Physics, 142432 Chernogolovka, Russia

[3] Science Center in Chernogolovka, 142432 Chernogolovka, Russia

In this volume we have collected papers based on the presentations given at the International Conference on Computer Simulations in Physics and beyond (CSP2015), held in Moscow, September 6-10, 2015. We hope that this volume will be helpful and scientifically interesting for readers.

The Conference was organized for the first time with the common efforts of the Moscow Institute for Electronics and Mathematics (MIEM) of the National Research University Higher School of Economics, the Landau Institute for Theoretical Physics, and the Science Center in Chernogolovka. The name of the Conference emphasizes the multidisciplinary nature of computational physics. Its methods are applied to the broad range of current research in science and society. The choice of venue was motivated by the multidisciplinary character of the MIEM. It is a former independent university, which has recently become the part of the National Research University Higher School of Economics.

The Conference Computer Simulations in Physics and beyond (CSP) is planned to be organized biannually. This year's Conference featured 99 presentations, including 21 plenary and invited talks ranging from the analysis of Irish myths with recent methods of statistical physics, to computing with novel quantum computers D-Wave and D-Wave2.

This volume covers various areas of computational physics and emerging subjects within the computational physics community. Each section was preceded by invited talks presenting the latest algorithms and methods in computational physics, as well as new scientific results. Both parallel and poster sessions paid special attention to numerical methods, applications and results.

For all the abstracts presented at the conference please follow the link http://csp2015.ac.ru/files/book5x.pdf

Journal of Physics: Conference Series **681** (2016) 011002 doi:10.1088/1742-6596/681/1/011002

International conference on Computer Simulation in Physics and beyond (CSP2015): Preface

The International conference on Computer Simulations in Physics and beyond (CSP2015) was held from 6–10 September 2015 at the campus of the Moscow Institute for Electronics and Mathematics (MIEM), National Research University Higher School of Economics, Moscow. Computer simulations are in increasingly popular tool for scientific research, supplementing experimental and analytical research. The main goal of the conference is contributing to the development of methods and algorithms which take into account trends in hardware development, which may help with intensive research. The conference also allowed senior scientists and students to have the opportunity to speak each other and exchange ideas and views on the developments in the area of high-performance computing in science.

We would like to take this opportunity to thank our sponsors: the Russian Foundation for Basic Research, Federal Agency of Scientific Organizations, and Higher School of Economics.

CSP2015 Committees

Conference Chair:

Prof. Lev Shchur (National Research University Higher School of Economics, Science Center in Chernogolovka, and Landau Institute for Theoretical Physics)

Members of local organizing committee:

Prof. Alexander Belov
Prof. Saygid Uvaysov
Dr. Boris Linetskiy
Dr. Serge Krashakov
Svetlana Shikota
Nadezhda Titkova

Journal of Physics: Conference Series **681** (2016) 011002 doi:10.1088/1742-6596/681/1/011002

Ilya Ivanov

Advisory Board:

Prof. Alexander Tikhonov
Prof. Yuri Evtushenko
Prof. Lev Zelenyi
Prof. Viktor Ivannikov
Prof. Alexander Litvak
Prof. Boris Chetverushkin

Program Committee:

Joan Adler - Technion, Israel
Wanda Andreoni - Technical University of Lozanna, Switzerland
Marat Biktimirov - Federal Research Center for Informatics and Control, Russia
Nithaya Chetty - University of Pretoria, South Africa
Giovanni Ciccotti - University of Rome, Italy
Alex Hansen - University of Trondheim, Norway
Wolfhard Janke - Leipzig University, Germany
Janos Kertesz - Budapest University, Hungary
Vladimir Lebedev - Landau Institute for Theoretical Physics, Russia
Hai-Qing Lin - Beijing Computational Science Research Center, China
Ravil Nazirov - Institute of Space Research, Russia
Mark Novotny - Mississippi State University, USA
Lev Shchur - Science Center in Chernogolovka, Russia
Alexei Starobinsky - Landau Institute, Russia
Hideaki Takabe - Osaka University, Japan
He Xiantu - Institute of Applied Physics and Computational Mathematics, China

List of Plenary Talks

1. Norbert Attig (Forshungzentrum Juelich, Germany) – "Impacts of Current Hardware and Software Developments on Simulation Sciences"
2. Bertrand Berche, Universite de Lorraine, Nancy, France – "Academic research groups: evaluation of their quality, quality of their evaluation"
3. Boris Chetverushkin (Keldysh Institute, Russia) – "Logically simple algorithms - the way to success of exaflops calculation"
4. Lin Hai-Qing (Beijing, Computational Science Research Center, China) – "Study of Superconductivity in Polycyclic Aromatic Hydrocarbon"
5. Hans Herrmann (Technical University of Zurich, Switzerland) – "Discontinuous Percolation"

6. Nail Inogamov (Landau Institute, Russia) – "Surface 3D nanostructuring by tightly focused laser pulse: lagrangian codes and molecular dynamics"
7. Wolfhard Janke (Universitat Leipzig, Germany) – "Computer Simulation Studies of Polymer Adsorption and Aggregation"
8. Helmut Katzgraber (Texas A&M University, USA) – "Beyond Moore's Law? Seeking Quantum Speedup Through Spin Glasses"
9. Ralph Kenna Coventry University, (United Kingdom) – "Maths Meets Myths - Quantitative Investigations of Ancient Narratives"
10. Yuri Lozovik (National Research University Higher School of Economics, Russia) – "Puzzle of supersolid: history and current state"
11. Mark Novotny (Mississippi State University, USA) – "Spanning Trees, Continents, and the Quantum/Classical Divide on D-Wave 2 machines"
12. Arkady Satanin, (Nizhny Novgorod State University, Russia) – "Generation of entangled microwave photons in superconducting circuits"
13. Vladimir Shchur (Sanger Institute, United Kingdom) – "On modern problems and methods for data analysis in human genomics"
14. Sauro Succi (Istituto Applicazioni Calcolo "Mauro Picone", C.N.R., Italy and Harvard University, USA) – "Lattice Boltzmann simulations of flowing matter across scales: classical, quantum and relativistic"

List of Invited Talks

1. Alexander Bogdanov (Saint Petersburg University, Russia) – "Solution of Financial Mathematics Real-Time Problems by Virtual Supercomputer"
2. Chin-Kun Hu (Institute of Physics, Taiwan) – "Protein aggregation might not be related to protein misfolding"
3. Nobuyasu Ito (University of Tokyo, Japan) – "Social sciences with supercomputer"
4. Oleg Druzhinin (Institute of Applied Physics, Russia) – "Numerical simulation of small-scale mixing processes in the upper ocean and atmospheric boundary layer"
5. Igor Petrov (Moscow Institute for Physics and Technology, Russia) – "Computational problems in Arctic Research"
6. Martin Weigel (Coventry University, United Kingdom) – "Connectivity properties of random-cluster model"
7. Alexandre Zagoskin (Loughborough University, United Kingdom) – "The grand challenge of quantum computing: bridging the capacity gap"

Journal of Physics: Conference Series **681** (2016) 011002 doi:10.1088/1742-6596/681/1/011002

Best Poster Prize

The Best Poster Prize commission of CSP2015 (Prof. A. Satanin, Prof. A. Bogdanov, and Prof. O. Druzhinin) decided to present the "Best Poster Award" and a 5000 rubles prize to Liliya Ziganurova for the presented poster "Virtual Time Profile Modeling in Parallel Discrete Event Simulation". Liliya Ziganurova graduated in 2015 from the Moscow Institute of Electronics and Mathematics Higher School of Economics (MIEM HSE, Moscow, Russia) with a master's degree in Applied Mathematics.

Peer review statement

All papers published in this volume of *Journal of Physics: Conference Series* have been peer reviewed through processes administered by the proceedings Editors. Reviews were conducted by expert referees to the professional and scientific standards expected of a proceedings journal published by IOP Publishing.

Journal of Physics: Conference Series **681** (2016) 012001 doi:10.1088/1742-6596/681/1/012001

Surface 3D nanostructuring by tightly focused laser pulse: simulations by Lagrangian code and molecular dynamics

Nail A. Inogamov[1] and Vasily V. Zhakhovsky[2]

E-mail: nailinogamov@gmail.com
[1]Landau Institute for Theoretical Physics, RAS, Chernogolovka, Russia
[2]Dukhov Research Institute of Automatics (VNIIA), Moscow, Russia

Abstract.

There are many important applications in which the ultrashort diffraction-limited and therefore tightly focused laser pulses irradiates metal films mounted on dielectric substrate. Here we present the detailed picture of laser peeling and 3D structure formation of the thin (relative to a depth of a heat affected zone in the bulk targets) gold films on glass substrate. The underlying physics of such diffraction-limited laser peeling was not well understood previously.

Our approach is based on a physical model which takes into consideration the new calculations of the two-temperature (2T) equation of state (2T EoS) and the two-temperature transport coefficients together with the coupling parameter between electron and ion subsystems. The usage of the 2T EoS and the kinetic coefficients is required because absorption of an ultrashort pulse with duration of 10–1000 fs excites electron subsystem of metal and transfers substance into the 2T state with hot electrons (typical electron temperatures 1-3 eV) and much colder ions. It is shown that formation of submicrometer-sized 3D structures is a result of the electron-ion energy transfer, melting, and delamination of film from substrate under combined action of electron and ion pressures, capillary deceleration of the delaminated liquid metal or semiconductor, and ultrafast freezing of molten material. We found that the freezing is going in non-equilibrium regime with strongly overcooled liquid phase. In this case the Stefan approximation is non-applicable because the solidification front speed is limited by the diffusion rate of atoms in the molten material. To solve the problem we have developed the 2T Lagrangian code including all this reach physics in. We also used the high-performance combined Monte-Carlo and molecular dynamics code for simulation of surface 3D nanostructuring at later times after completion of electron-ion relaxation.

1. Introduction

Figure 1 presents typical 2D and 3D structures produced in experiments [1–6] with the thin gold films. Let us call "nanobumps" all such structures. The film "f" and the underlying substrate "s" in Fig. 1 (a) are in solid state before a single laser impact. The final (i.e. long time after impact) nanobumps also are solid. Formation of a nipple in the tip of nanobump in Fig. 1 (c), or the jet in Fig. 1 (d-g) means that gold inside the heated spot have passed the melting stage and have been solidified after melting. The sequence of nanobumps in Fig. 1 (b-g) is shown in ascending order as the central fluence F_c of spatial distribution $F(r) = F_c \exp(-r^2/R_L^2)$ increases; here r is a distance from the axis of laser beam (we suppose the normal incidence) and the spatial scale R_L characterizes a focused laser beam "L" shown in Fig. 1 (a). Pulse duration should be

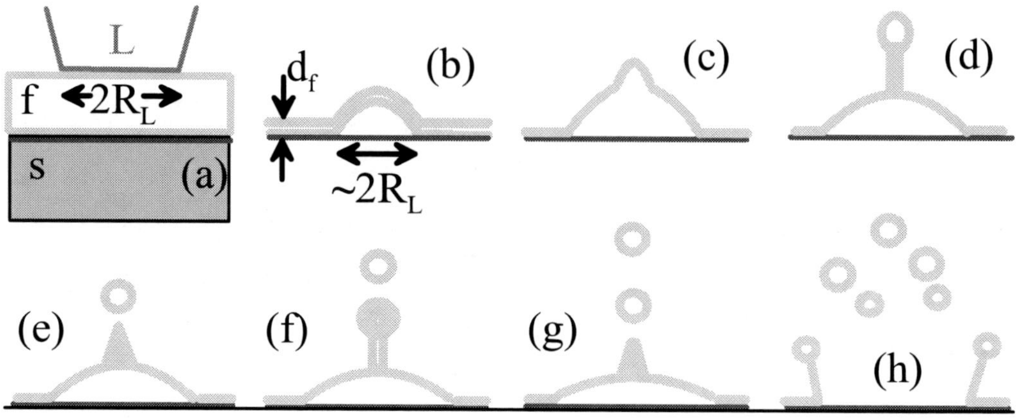

Figure 1. The sequence (b-g) of the nanobumps shown in ascending order as energy of a laser pulse increases while a radius R_L of a laser beam is fixed. Increase of energy at a fixed radius R_L means that the central fluence F_c of a distribution $F = F_c \exp(-r^2/R_L^2)$ increases. A pulse acts onto a layered target composed from a film "f" mounted above a substrate "s". Further increase of F_c above the case (g) results in formation of a hole in a tip region of a nanobump. A radius of the hole (h) increases while length of a delaminated remnants decreases as the F_c increases. In 2D we do not see formation of the long jets. Instead a droplet like condensation in the tip appears. In the 2D/3D cases there are the planar/circular laser beams "L" (a) and the nanobumps are slit type or rounded respectively.

of the order of, or shorter, than an acoustic time scale $t_s = d_f/c_s$ to lead to a *thermomechanical* delamination of film from substrate produced by a stress field generated by fast heating; here d_f is a thickness of film (see Figs. 1 a, b), c_s is sound speed in gold.

Experiments [7] with a femtosecond pulse and a thin $d_f \sim 40$ nm gold film, $t_s \sim 10$ ps demonstrate that the pulse duration τ_L doesn't influence results up to durations at least ≈ 8 ps, see Fig. 2 in [7]. Let us mention that for the thin films the acoustic scale t_s is comparable with the duration of two-temperature electron-ion relaxation t_{eq}; $t_{eq} \approx 5-7$ ps for gold. In a sense of delamination dynamics of separation of a film from a substrate the comparison τ_L versus t_{eq} is insignificant relative to the comparison τ_L versus t_s. In the both mentioned cases the condition $\tau_L < t_s$ ("supersonic" laser pulse) assures that the strong thermomechanical stresses drive the delamination irrespective to the origin of the stress – is it mainly electronic in the subcase $\tau_L < t_s < t_{eq}$, or is it electronic at the two-temperature stage and ionic later in the subcase $\tau_L < t_{eq} < t_s$. There is some quantitative difference [8] between these two subcases because the electronic Gruneisen parameter ≈ 1 is smaller than the ionic Gruneisen parameter ≈ 2. But the situation changes qualitatively in the case of the "subsonic" pulses $\tau_L > t_s$ when amplitude of the mechanical stress drops down as the ratio τ_L/t_s increases above 1.

A fluence distribution $F(r)$ along a surface with a pronounced central $r = 0$ maximum F_c produces a velocity distribution $v(r)$ with a maximum. This is a normal (relative to surface) component v of velocity of a local piece of a film *after* separation of this piece from a substrate (this piece is at a distance r). In experiments a film is thin in the sense that its thickness d_f is much less $d_f \ll R_L$ than the spot size R_L, see Fig. 1 (a); the typical values are: $d_f \sim 20-100$ nm, the radius is $R_L \sim 1$ micron for the cases with diffraction limited focusing and optical lasers. The lateral velocities are small relative to the normal one as a result of this shallowness. The maximum velocity v locates in the center $r = 0$, thus the dome like shapes of the nanobumps shown in Fig. 1 are formed. Duration of an acoustic stage t_s is enough long to homogenize

Journal of Physics: Conference Series **681** (2016) 012001 doi:10.1088/1742-6596/681/1/012001

distribution of locally absorbed laser energy across a thickness d_f of a film[1]; but this duration is very short for the temperature equalizing at the distances $\sim R_L \gg d_f$ along a hot spot in Fig. 1. Thus a distribution of energy $E(r)$ per unit of volume appears at the beginning of an acoustic stage. The pressure distribution $p(r)$ acting to delaminate a film [10] is $p(r) \propto E(r)$; because variation of Gruneisen parameter and volume expansion at the beginning of the acoustic stage both are small. If we neglect the adhesion between gold and silica glass then the velocity distribution along a film is $v(r) \propto p(r) \propto E(r) \propto F(r)$.

Firstly, there is a small value of the gold/silica adhesion and, secondly, absorbed energy decays strongly $\propto \exp(-r^2/R_L^2)$ outside the spot $\sim R_L$. Therefore in the external area $r > r_{delam} \sim R_L$ the film remains in mechanical contact with the film as shown in Fig. 1. We will see below that the mechanical contact connecting the internal $r < r_{delam}$ and the external $r > r_{delam}$ parts is important for deceleration of the delaminated film while the thermal contact between the hot dome and the outside cold film is necessary for cooling and freezing of the dome. Increase of the scale r_{delam} with increase of pulse energy $E_{total} \propto F_c$ (if R_L is fixed) has been studied in paper [11], see their plots in the $\ln E, r_{delam}^2$ plane in their Fig. 4. As a result of adhesion between gold and glass the velocity distribution $v(r)$ is equal to zero for the distances $r > r_{delam}$. Thus the velocity distribution $v(r)$ follows the fluence distribution $v(r) \propto F(r)$ everywhere except the margins $r \approx r_{delam}$. Experimentally an interesting situation with a ring type fluence distribution $F(r)$ has been considered [12], see Fig. 8 in this paper. Then complicated motion from the ring to the center is observed.

The droplets shown in Fig. 1 lose the thermal contact and therefore for a long time remain in a liquid state. Estimates of the radiation loses $t_{rad} \sim cTV/(\sigma_{SB}T^4S) \sim 0.6\,(R_{droplt}/100\text{nm})/(T/2000\text{K})^2$ millisec give large values for the radiative cooling time t_{rad}; here $c \approx 3nk_B$ is heat capacity, $n \approx 6 \cdot 10^{22}$ cm^{-3} is concentration of atoms, k_B is the Boltzmann constant, T is temperature of a droplet after separation from a jet, V, S, R_{droplt} are volume, surface, and radius of a droplet, σ_{SB} is the Stefan-Boltzmann constant. But in paper [13] a few orders of magnitude faster freezing of droplets seems has been observed; the frozen droplets reflect from a receiver while the molten - stick to it. May be this is because temperature of molten gold in the droplets produced by nanobumping is close to the melting temperature. Our molecular dynamics simulations presented below show that liquid in a dome and a jet shown in Fig. 1 are significantly overcooled below the melting temperature. Distances L_{dr} between a receiver and a donor plate in [13] were $L_{dr} \sim 10$ microns much less than $L_{dr} = 2$ mm in [14]. In [13] the time of flight $t_{frez} \sim 1 \cdot [(L_{dr}/10\text{um})/(v_{droplt}/10\text{m/s})]$ μs (to pass distance $L_{dr} \approx 10$ microns) divides molten and solid nanodroplets. In [14] nanodroplets were hotter, evaporative cooling enhances temperature decrease, and according to authors of paper [14] nanodroplets are still liquid after flying time 2[mm]/100[m/s]=20 μs.

The cooling time $t_{cond} = [R_L/(1\text{um})]^2/\ [\chi/(1\text{cm}^2/\text{s})] = 10$ ns due to conduction along high conductance metallic film is much shorter than the radiative and evaporative cooling times. This is why such a cooling is responsible for retention of the form of transient nanobump. Conduction along a film transfers heat from a hot spot $r < r_{delam}$ to the outside area $r > r_{delam}$, see Fig. 1. Thus this area serves as an infinitely capacious thermal sink. This means that we can influence the cooling of a nanobump by decreasing area or volume of cold metal external relative to the hot spot $r < r_{delam}$. The outside area may be done as a disk. We have to balance mechanical and thermal abilities of the disk varying its radius $R_{dsk} > r_{delam}$. From the one side the disk has to absorb mechanical momentum from a flying dome and to transfer it into a substrate, and from the other side the disk has to limit a cooling rate of a dome. Choice of material for substrate

[1] The direction of an illumination (from the vacuum side as in Fig. 1 a or through the transparent substrate) doesn't influence the subsequent dynamics if the energy homogenization across thickness of a film takes place prior to the onset of significant motion of a film. This is the case of the thin films, thinner than depth of a heat affected zone $d_T \approx 140$ nm for Au of an ultrashort laser pulse [9].

(low conductive glass versus high conductive crystallin silicon or diamond) also influence cooling ability. Using a film made of metallic alloys, like in the experiment [11] where Au/Pd alloy was used, with a low conductive component or pure metals like Pt, Fe with decreased heat diffusion coefficient χ is the another way to regulate thermal rate.

2. Resources required for molecular dynamics simulation

Number of atoms which might be involved in MD simulation can be determined by two geometrical parameters: radius R_L and thickness d_f of the metal film – see Fig. 1. One of the largest ($2r_{delam} \approx 20$ μm) and very accurate nanobumps with rich fine substructures have been obtained in the Livermore's experiments [6]. Typical diameters $2r_{delam}$ are ~ 5 μm [11–13, 16, 17]. There are also examples with smaller scales $2r_{delam} \sim 1$ μm [3, 12, 15–17]. Technically available diameters $2R_L$, see Fig. 1 (a), of laser beams are limited from the smallest scales by the diffraction limit $2R_L \sim \lambda$. From the opposite side, the largest diameters are defined by pulse energy and usually are as large as tens and hundreds microns. The minimal submicron values of $2r_{delam}$ may be achieved using ultraviolet lasers or higher harmonics.

Typical thicknesses of the thin metal films varies from 20 to 100 nm. There are $\sim 10 \times 10^9$ atoms within a cylinder of $\pi\,[r_{delam}/(1\mu\text{m})]^2\,[d_f/(50\text{nm})]$ covering the delaminated region. We have to triple this number to include the minimally thick glass substrate able to exclude the influence of the back reflection from the rear-side of the supporting glass layer onto process of delamination from substrate. This process lasts during an acoustic time scale t_s. It is necessary to follow this huge system along very extended time interval of the order of 100 ns and more. Typical time step of molecular dynamics (MD) simulation is ~ 1 fs. Thus $\sim 10^8$ steps are required. Taking into account that about 1 second is required to perform a step with MD system of 200,000 atoms on a single processor core, MD simulation of several billions of atoms during 10^8 steps would require hundreds of years on a cluster with 1000 cores.

This difficulty is usual for MD applications - the spatial-temporal size of application is much larger than available algorithmic/computer resources. We have to use physics of nanobump phenomenon in an attempt to overcome this difficulty. Two steps to decrease the computational demands were used. Separation of the delamination process from the process of slow inflation of a nanobump is the first step. The second step is based on the self-similarity of the problem relative to the capillary and cooling parameters. The second step is presented in next Section. The splitting of the delamination and inflation processes (the first step) is a consequence of very large (three or more decimal orders) difference between durations of the delamination stage and the inflation and capillary deceleration stage. This large difference arises from the difference in scales and velocities. The size $R_L \sim 1-5$

Table 1. The double scaling map covered by the MD-MC runs, see also Fig. 2. Number of atoms N_{atom} involved in MD-MC simulation is given in 10^6 units. The simulation box is a square $L_b \times L_b$ in the plane of a contact. The sizes L_b and initial thickness d_f of a film are given in [nm]. The last column presents temperature T_{margn} in [K] of a thermostat acting in the margin area between the square $L_b \times L_b$ and a circle where a film separates from substrate, see Figs. 5 and 6.

#	N_{atom}	L_b	d_f	$v_{0\sigma}$	$v_{0\chi}$	[K]
20	6.0	140	5.1	2.6	6.3	500
23	6.0	140	5.1	2.6	3.1	500
30	44	270	10	3.6	6	500
3	192	620	8.3	2.6	∞	500
21	6.0	140	5.1	5.2	12.5	500
22	6.0	140	5.1	10	25	500
24	6.0	140	5.1	5.2	6.3	500
25	6.0	140	5.1	5.2	6.3	500
26	2.1	140	1.8	6.2	12.5	500
27	2.1	140	1.8	4	8.1	500
28	6.0	140	5.1	10	12.5	500
29	6.0	140	5.1	10	12.5	300
29a	6.0	140	5.1	10	12.5	300*
31	44	270	10	3.8	6.3	500

μm is $F_x = 10 - 100$ times larger than film thickness $d_f = 20 - 100$ nm. The delamination is connected with rarefaction and compression waves, their interactions, and reflections from a glass/metal contact and a vacuum boundary of a film. The waves propagates distance d_f with elastic longitudinal sonic speed $c_s \sqrt{(K + 4G/3)/\rho} \approx 3.3$ km/s (for Au); here K, G are bulk and shear modulus. The speed c_s is $F_v = 30 - 100$ times higher than velocities 20-100 m/s of a film after its separation from a substrate. The large cumulative factor $F = F_x \times F_v$ means that we can with high accuracy split the delamination and the inflation of a nanobump, see [10]. We neglect curvature of film during delamination (this is one side) and the interaction of the delaminated film with substrate is absent after delamination (this the other side). Thus the inflation flight is simulated by MD without substrate. The substrate in a MD run serves only as a sink of momentum in the region near the point r_{delam} of attachment of a film to a substrate, i.e. near the lateral margins, see Fig. 1.

The delamination process is simulated with two-temperature one-dimensional hydrodynamics code [10]. This code is accurate, it includes full two-temperature physics, and it is inexpensive in the sense of the computer resources. The code is a very valuable addition to the MD code. Indeed, it is difficult to describe full two-temperature physics together with an electron pressure effects inside the MD subroutines. The hydrodynamics code supplies the MD simulation with the initial (for the MD code) temperature and velocity distributions along a film. This is the velocity distribution after delamination from substrate.

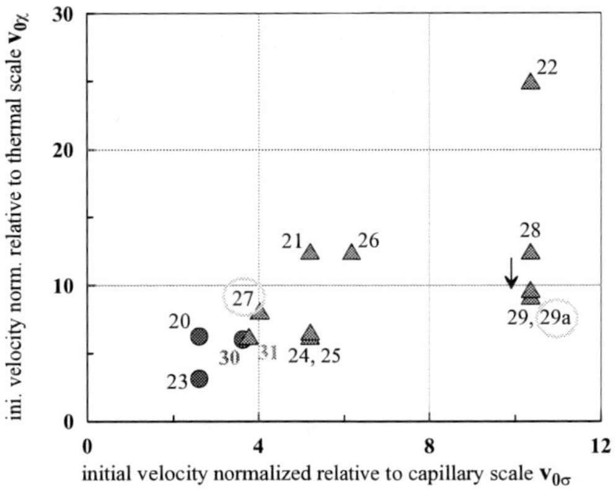

Figure 2. Performed MD-MC runs in plane of parameters. We vary ratio v_σ/v_χ in the range ~ 1 and vary ratio of initial velocity $v_c(0)$ to these velocities in a wider range. The blue circles corresponds to the first three lines in Table 1 presenting the non-ejecting (all matter of a film remains on substrate) cases. The numbers 27 and 29a with the circles around correspond to the cases shown in Figs. 3, 5 (#27) and 4, 6 (#29a).

3. Self-similarity thanks to capillary-cooling relation

3.1. Capillary dynamics. Dynamics of molten film inside its circle $r < r_{delam}$ of separation from a substrate is determined by inertia and surface tension. The swelling film is clamped to the immovable circular arc with a radius $r = r_{delam}$ placed on an immovable substrate. Therefore a capillary force acting onto the attachment point $r = r_{delam}$ doesn't produce a work. As a result a sum of potential energy of surface tension $2\sigma S$ and kinetic energy

$$E_{kin} = (1/2) \int_0^{r.delam} \rho(r,t) d(r,t) \vec{v}^2(r,t) 2\pi r dr$$

is conserved in time, here σ is a coefficient of surface tension, S is a surface area of a film inside the circle $r < r_{delam}$, the area is taken twice (coefficient "2") because there are two sides of a film; the integral is written for the 3D case with a circular illumination spot in Fig. 1, similar expression exists in the 2D case, r is a distance from a beam axis in Fig. 1, $\rho(r,t)$ is density of liquid gold. We neglect temperature and therefore density differences across thickness of a thin film: $\rho(\vec{r}, t) = \rho(r,t)$. Value $d(r,t)$ is thickness of a film along a normal to the substrate surface. Surface tension acts as a spring decelerating and returning back a moving point mass attached to the end of a spring.

The film, clamped to the circular arc with a radius $r = r_{delam}$, oscillates in time above and below the plane of a contact if we imagine that the substrate disappears inside the circle

$r < r_{delam}$. In case of small initial velocities $v \ll v_\sigma$ a period of oscillation is $3.7 r_{delam}/v_\sigma$ [10]. The capillary velocity scale

$$v_\sigma = 2\sqrt{\sigma/(\rho d_f)} = 46\,[\sigma/(1000\text{dyne/cm})]^{1/2}\,[\rho/(19.3\text{g/cm}^3)]^{-1/2}\,[d_f/(100\text{nm})]^{-1/2}\ \ [\text{m/s}] \quad (1)$$

is defined from expression $2\sigma = (1/2)\rho d_f v_\sigma^2$. Let's approximate the shape of a nanobump by a cone and a velocity distribution by a monotonous function $|\vec{v}| = v_c[\cos(\pi r/(2r_{delam}))]^2$ having the maximum value v_c in the tip and going to zero in the attachment point r_{delam}. Using an expression for a lateral area of a cone and integrating kinetic energy, we rewrite an energy conservation in a form

$$2\pi r_{delam}\left(\sqrt{h_c^2(t) + r_{delam}^2} - r_{delam}\right)\sigma + (1/2)\,2\pi\,(3/16 - 1/\pi^2)\,\rho_0\,d_f\,r_{delam}^2\,[v_c^2(t) - v_c^2(0)] = 0, \quad (2)$$

where $h_c(t), v_c(t)$ are the height of the tip above the initial contact plane and velocity of the tip. Energy equation (2) relates the current height h_c and velocity v_c : velocity decreases as the height and hence surface area and surface energy increase. The maximum height h_{max}, achieved in a stopping point $v_c(t_{stop}) = 0$, is

$$h_{max}/r_{delam} = v_{0\sigma}\sqrt{\alpha\,(2 + \alpha\, v_{0\sigma}^2)}, \quad \alpha = 0.172, \quad v_{0\sigma} = v_c(0)/v_\sigma. \quad (3)$$

In (3) the non-dimensional initial velocity $v_{0\sigma}$ is normalized relative to the capillary scale (1), α is a numerical coefficient. For the moderate values of velocity $v_{0\sigma} \le 1$, the turning instant

$$t_{stop} = 20\,[r_{delam}/(1\,\mu\text{m})]\,\sqrt{[d_f/(100\,\text{nm})][\rho/(19.3\,\text{g/cm}^3)]/[\sigma/(1000\,\text{dyne/cm})]}\quad [\text{ns}]$$

weakly depends on velocity $v_{0\sigma}$ and is approximately a quarter of a period $3.7 r_{delam}/v_\sigma$.

We see that it should be $v_{0\sigma} \sim 1$ to lift up the tip of a nanobump to the height $h_{max} \sim r_{delam}$. To take the integral of kinetic energy, we suppose that (i) $\rho(r,t) \approx \rho_0$ (19.3 g/cm^3 for Au), that (ii) $d_f(r,t) \approx d_f$, and (iii) we neglect dispersion of velocities in the vicinity $\sim d_f$ of the point $\vec{r}$ of the film, here d_f is initial thickness of a film. The assumption about density is approximately valid because our simulations show that in the nanobumping regimes the temperatures of a film shortly (at a $\sim$ 100 ps time scale) after irradiation are below 2 kK and thermal expansion is relatively low: density at the boiling curve is $\rho_2 = \rho_{bin}(T = 2\text{kK}) \approx 16.8$ g/cm^3 according to a wide-range equation of state for Au [18, 19], see also [20], thus $(\rho_0 - \rho_2)/(\rho_0 + \rho_2) = 0.07$. Energy conservation $2\sigma S + E_{kin}$ =const is valid but the assumption (ii) $d_f(r,t) \approx d_f$ is only approximately accurate as long as the stretching of a film is small. The stretching is small if velocity is small $v_{0\sigma} \ll 1$. There is significant stretching for the larger velocities $v_{0\sigma} \sim 1$ and higher (nonlinear regimes).

It is important that the stretching becomes very inhomogeneous in the nonlinear regimes. This is a consequence of a strong mass redistribution along a film. Central part of a film forms a droplet (in 2D [10, 20]) or a jet [or a jet plus droplet(s) in 3D] thus doing stretching very inhomogeneous - there are condensations (droplet or jet), from the one side, and strong stretching and thinning of the film remaining outside relative to the condensations, from the other side. Extremely long jet (many lengths r_{delam}) may be developed. The rate per unit of mass of transfer of kinetic energy into the surface energy decreases in a jet because surface to volume ratio decreases from film to jet. Nevertheless, surface tension will return fluid back for arbitrary large initial velocity $v_{0\sigma}$, because surface area grows with stretching while kinetic energy causing stretching is limited. Of course, this is true in the case of infinite stretchability of a film and a jet - there is no breaking of surface of a film or decay of a jet into droplets. In reality the jet formation process, stretching and thinning of a film, and rupture of a film or a jet together with interplay with a recrystallization wave make up a set of rather complicated regimes considered below in Chapter 5.

3.2. Cooling rate. Hot spot created by an ultrashort laser impact onto a metal film cools down as a result of conductive heat transfer along a film from the hot spot to the external area of a film, see Section 1. The temperature decrease causes crystallization of a film and a jet. Transition to a solid state sharply changes mechanical properties - polycrystalline solid strongly resists to stretching, while liquid resists weakly, only through surface tension. This jump of stretchability is a reason of dynamic phenomena near the solid-liquid crystallization zone, e.g., breaking of a liquid film surrounding growing crystal grains, or rupture of a very much thinned solid film under action of significant inertial forces stretching solid, see Sections 5 and 6 below.

Cooling in a 3D case is described by a heat conduction equation $T_t = (\chi/r)(r\,T_r)_r$, here $T_t \equiv \partial T/\partial t$, $T_r \equiv \partial T/\partial r$, r is a cylindrical radius, $\chi = \kappa/c$ is a heat diffusion coefficient, κ, c are heat conduction coefficient and heat capacity per unit of volume. Transmission of heat along distances $\sim R_L$ becomes significant much later after finishing of a two-temperature stage, therefore usual well-known values of the coefficients χ, κ, c have to be used. For simplicity the moderate dependencies of these coefficients on temperature are not included for our temperature range in the expressions below. A heat equation has a fundamental solution (thermal Green function)

$$T(r,\tau) = \frac{E}{\pi\, c\, d_f\, (4\,\chi\,\tau)} \exp\left(-\frac{r^2}{4\,\chi\,\tau}\right), \qquad (4)$$

describing diffusion of heat along a film after impulsive absorption of energy E [erg] in the point $r = 0$ in the instant $\tau = 0$.

Ultrashort pulse laser illumination shown in Fig. 1 (a) creates similar to (4) Gaussian one-temperature $T_e = T_i$ distribution in a short time scale 5-7 ps. Short duration of this stage prevents lateral expansion of heat during the first few tens of picoseconds. The Gaussian temperature profile made by laser is

$$T(r, t=0) = \frac{E}{\pi\, c\, d_f\, R_L^2} \exp\left(-\frac{r^2}{R_L^2}\right). \qquad (5)$$

Diffusive spreading of the profile (5) outside on the distances $\sim R_L$ takes tens of nanoseconds for $R_L \sim 1$ micron and usual values of thermal diffusivity $\chi \sim 1$ cm^2/s. In this regard it is written $t = 0$ in expression (5): ps is much shorter than ns. Around the instant $t = 0$ laser action takes place. There is a time lag between the instants $\tau = 0$ and $t = 0$. There is the *same* absorbed energy E in expressions (4) and (5).

Comparing expressions for the Green function (4) and for the initial deposition (5) we write solution of a Cauchy problem for heat conduction equation with initial data given by expression (5). The solution

$$T(r,t) = \frac{E}{\pi\, c\, d_f\, (R_L^2 + 4\,\chi\, t)} \exp\left(-\frac{r^2}{R_L^2 + 4\,\chi\, t}\right), \quad T(r=0,t) = \frac{E}{\pi\, c\, d_f\, (R_L^2 + 4\,\chi\, t)} = \frac{T_c}{1 + 4\chi\, t/R_L^2} \qquad (6)$$

describes evolution of a temperature field after laser action $t > 0$. Dependance presenting decrease of temperature in the tip point $T(0,t)$ is also given in (6).

There are the delamination r_{delam} and the melting r_m circles corresponding to the delamination F_{delam} and melting F_m thresholds on an absorbed fluence distribution $F = F_c \exp(-r^2/R_L^2)$; namely this absorption produces the initial temperature field (5); $F_c = E/(\pi\, R_L^2)$. The delamination F_{delam} threshold is below the melting threshold F_m in the case of weak adhesion between gold and a glass substrate, i.e., $r_m < r_{delam}$. We are studying formation of a nanobump with some substructure in the tip, see Fig. 1 (c-h). This is possible if a film in a hot spot transits to a liquid stage. Thus we are interested in the cases $F_c > F_m$.

A solid film weakly departs from a substrate surface. Thus the radius r_m should be regarded as a "true" or significant delamination radius. It marks position where an enhanced separation of a film from a substrate begins. A nanobump inflates mainly inside a circle $0 < r < r_m$. This means that these nanobumps after laser action have

$$T(r=0,t=0) \equiv T_c(t=0) = E/(c\,d_f\,\pi\,R_L^2) = F_c/(c\,d_f) > T_m,$$

i.e., their tip temperatures are above the melting threshold, see (6). Let's estimate duration t_{sol} necessary for total solidification: $T_c(t_{sol}) = T_m$, where T_m is melting temperature. From equation (6) we obtain

$$t_{sol} = [(T_c(t=0) - T_m)/T_m]\,R_L^2/(4\chi). \tag{7}$$

Expression (7) shows how the solidification time increases above a melting threshold. Estimates and simulations show that a degree above the melting threshold $T_c(0)/T_m$, where nanobumps shown in Fig. 1 (c-g) exist, is limited by a value $T_c(0)/T_m \approx 2-3$. Above this limiting value a hole shown in Fig. 1 (h) appears. Initial local temperature T of a film depends on local absorbed fluence $T = F/(d_f\,c)$. The initial (later in time whole nanobump will cool down and freezes) radius r_m is defined by the degree $T_c(0)/T_m$:

$$T_m(r_m, t=0) = T_c(0)\exp(-r_m^2/R_L^2), \qquad r_m = R_L\sqrt{\ln(T_c/T_m)}.$$

Let's define cooling velocity v_χ and initial velocity normalized relative to v_χ as this is done in expressions (1) and (3). We have

$$v_\chi = \chi/R_L = 100\,[\chi/(1\,\mathrm{cm}^2/\mathrm{s})]\,/\,[R_L/(1\,\mu\mathrm{m})]\ \ [\mathrm{m/s}], \qquad v_{0\chi} = v_c(t=0)/v_\chi. \tag{8}$$

3.3. Double scaling and reduction of demands on computer resources. Now we have two non-dimensional parameters $v_{0\sigma}$ (3) and $v_{0\chi}$ (8). In next Section the Monte-Carlo method used in the molecular dynamics algorithm MPD^3 [21] is presented. It allows to include an electron heat conduction into simulations. It is possible to vary a heat conduction coefficient κ in the Monte-Carlo method. This freedom together with the two non-dimensional parameters (3) and (8) give us ability to simulate a huge experimental object with a smaller molecular dynamics (MD) object, see Chapter 2.

Let's illustrate this with an example. The experimental object is: $R_L = 1$ micron, $v_c(t=0) = 100$ m/s, $\chi = 1$ cm^2/s, $\sigma = 1000$ dyne/cm, $d_f = 60$ nm, $T_c(t=0) = 2000$ K. Then the parameters (3) and (8) are: $v_{0\sigma} = 1.68$, $v_{0\chi} = 1$.

Let's choose the simulation object having the same temperature distribution and values of $v_{0\sigma}$ and $v_{0\chi}$ but ten times smaller geometrical values R_L and d_f; thus it is three order of magnitudes smaller in number of atoms. To do this we have to adjust initial velocity $v_c(0)$ and χ; heat capacity is the same in experiment and in MD, therefore changing κ is equivalent to the changing of χ. We have to take into account that in the used EAM (embedded atom method) interatomic potential of Au [22] the coefficient σ is 540 dyne/cm [10].

Calculating, we find parameters for our equivalent MD object: $v_c(0)|_{MD} = 229$ m/s and $\chi|_{MD} = 0.1$ cm^2/s.

4. Numerical simulation of conductive cooling

Approach used here combines multi-processor molecular dynamics (MD) algorithm MPD^3 [21] and Monte-Carlo method necessary to describe heat conduction. In the general case of a laser-matter interaction the energy of a laser pump is absorbed by electrons in a skin layer, and then

Journal of Physics: Conference Series **681** (2016) 012001 doi:10.1088/1742-6596/681/1/012001

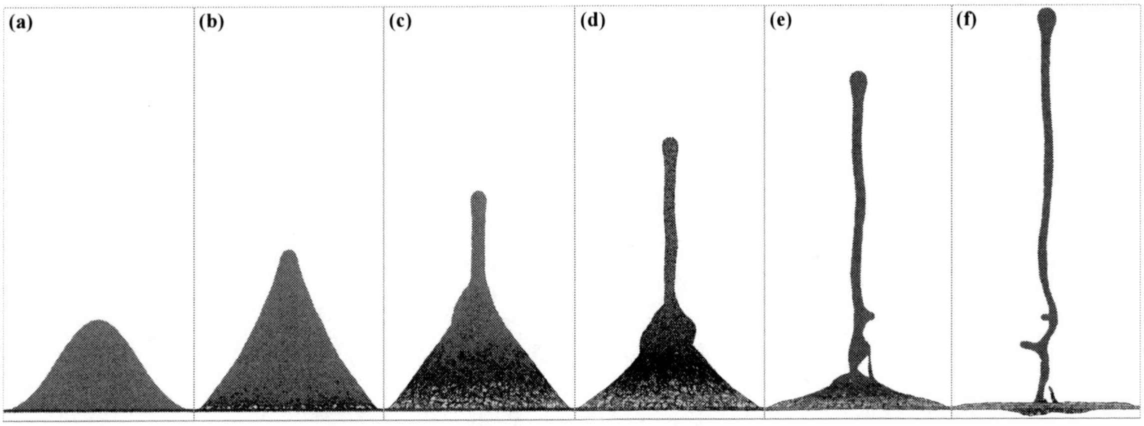

Figure 3. Evolution of nanobump in the case #27, see Table 1. Capillary forces decelerate inflation of the nanobump shell. Therefore the tip of nanobump (not the tip of jet) stops rising and turns back. It impacts the surface of substrate, compare with experimental pictures [6, 7]. This is obvious because the shell is seen below the plane of substrate in the last frame. In simulation we do not put a barrier in the plane of initial gold-glass contact. Thus the material can move through this plane in simulation. Contraction of surface area of solidified shell (the height of the cone decreases) will inevitably lead to formation of the radial folds seen in experiments, e.g., [6, 7]. The folds accept the excess of solid surface area of the thin shell. The shape of nanobump changes during its motion from convex to conical one. This is a result of pumping of mass into jet. Thus the region of attachment of the shell to the bottom of the jet is the main region resisting to the back motion of the shell because inertia of a massive jet is large relative to inertia of a strongly thinned shell - a jet strains a shell giving a conical shape to the shell. The lateral subjet like protrusions results from breaking of a moving, stressed, and solidifying liquid-solid shell (green is solid, red is liquid). Evolution of the protrusions covers the frames (c-f), see also Fig. 5 below. Shown frames correspond to the times: (a) $t = 72$ ps, (b) 144 ps, (c) 216 ps, (d) 290 ps, (e) 380 ps, (f) 475 ps.

the electron-ion coupling and electron heat transport are described by Monte-Carlo method (MC). In this paper MD-MC code is employed to follow one-temperature dynamics and cooling of the nanobump long after the short stages of absorption of laser energy and two-temperature relaxation.

To combine MC with MD the light Monte-Carlo particles, treated as classical "electrons", were added to atom subsystem. Each host atom/ion has one "electron", which has its own velocity but its position is always allocated on its host atom. It guarantees the charge neutrality in the combined electron-ion system in any conditions. The neighboring atoms exchange their electrons with some frequency determined from known experimentally thermal conductivity of liquid gold at the melting point. The pair of atoms, where electron-electron exchange takes place, is chosen randomly from the Verlet list of neighbors within a cutoff distance.

In addition to the frequency of electron-electron exchange between atoms, we adjusted the mass of "electron", in order to obtain the characteristic time of electron-ion energy exchange in e-i collisions. The electron-ion scattering depends on the velocity vectors of electron with mass m_e and ion with mass m_i according to equations:

$$\begin{aligned} \mathbf{v}'_e &= (m_e\mathbf{v}_e + m_i\mathbf{v}_i + m_i v\mathbf{n})/(m_e + m_i) \\ \mathbf{v}'_i &= (m_e\mathbf{v}_e + m_i\mathbf{v}_i - m_e v\mathbf{n})/(m_e + m_i) \end{aligned} \quad (9)$$

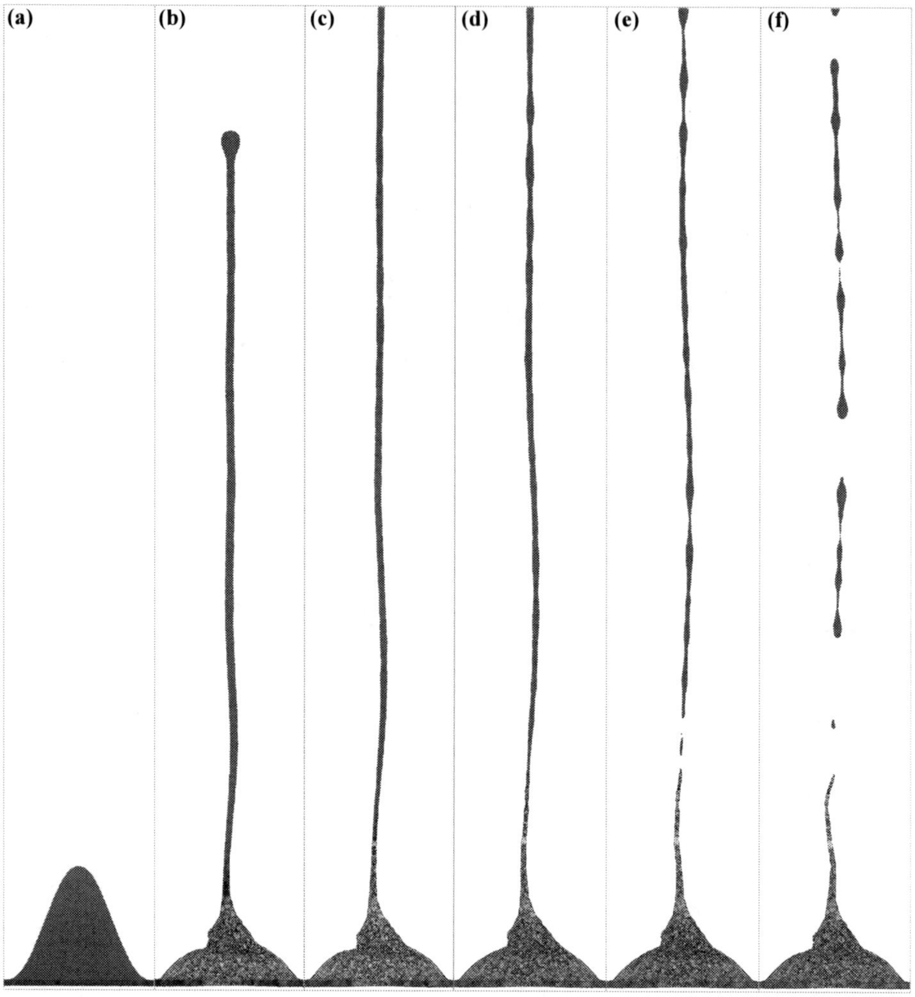

Figure 4. The case #29a, see Table 1. (a) $t = 72$ ps, (b) 650 ps, (c) 860 ps, (d) 1.1 ns, (e) 1.3 ns, (f) 1.5 ns.

where v is an electron-ion relative speed, and $\mathbf{v}'_e$ and $\mathbf{v}'_i$ are the velocities of electron and ion in the laboratory system after collision. The unknown unit vector $\mathbf{n}$ is assumed to be distributed isotropically and can be provided by a generator of random directions. Equations (9) conserve the total momentum and energy of electron and ion subsystems.

The described Monte-Carlo method has free parameters of effective electron mass m_{ef}, electron jump rate ν_D, and e-i-collision rate ν_{ei}. For simplicity the jump and e-i-collision rates are assumed to be equal $\nu = \nu_D = \nu_{ei}$ in this Monte-Carlo model. Thus, both electron diffusion and e-i energy exchange depend on the jump/collision rate ν. Furthermore, the e-i energy exchange rate depends on effective electron mass. Because electron thermal conductivity λ_e is only guided by ν and independent on m_{ef} in equilibrium systems where local condition $T_e = T_i$ is hold everywhere, the time-independent solution of heat equation is used first to fit ν with the known experimental thermal conductivity.

To find ν the steady MC-MD solutions of the Fourier law for the heat flux $q = -\lambda\, dT/dx$ is tested in a domain bounded by two subdomains, where the different temperatures 1300K and 1200K are supported by two electron thermostats separated by a layer of 100 nm thick.

Journal of Physics: Conference Series **681** (2016) 012001 doi:10.1088/1742-6596/681/1/012001

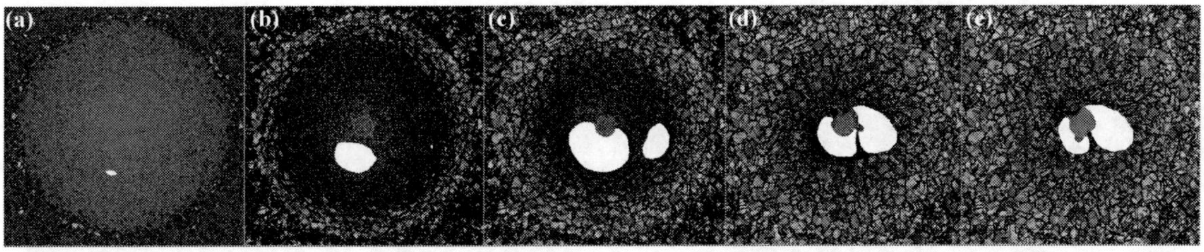

Figure 5. The run # 27. This is a top view from the upper point onto the plane of a film, see also the side view for the same run in Fig. 3. The instants (a) t = 144 ps, (b) 200 ps, (c) 270 ps, (d) 380 ps, and (e) 475 ps are shown. A map of the symmetry parameter is given: solid is green, liquid is red. We see definite difference between the margins and the central circle where a film delaminates from a substrate. The thermostat is working in the marginal area covering space from the sides of the square to the circle arc path. The thermostat absorbs momentum coming through the surface tension force and absorbs heat flux flowing from the hotter central part. The film was initially very thin in the runs 26 and 27, see Table 1. Of course, in a thinner film the decay processes proceed more easy. Thus the MD-MC simulation reveals appearance/disappearance of the atomic size holes in a very thin film as a result of thermal fluctuations. One of them crosses a metastable barrier and transforms into the viable expanding nucleus, see frame (a). Later another one appears, see frame (b). It is interesting, that appearance of the holes doesn't mean automatically termination of the inflation/deceleration development of a nanobump or immediate total rupture. A hole expands with moderate velocity, this conclusion follows from the calculations performed in [10].

After attainment of steady linear temperature profile with constant temperature gradient $dT/dx = 1 \times 10^9$ [K/m] within the layer, the time-averaged heat flux was calculated during 300 ps of MD-MC simulation. To provide the experimental flux the electron jump rate $\nu = 4.286 \times 10^{15}$ [1/s] was found to have been used in MC part of our combined MC-MD simulation of gold.

After determining the jump rate, the effective mass of "electron" was fitted to the relaxation time between electrons and ions obtained from the 2T model equations for evolution of electron T_e and ion T_i temperatures. In the case of a sample with uniform density and temperature where material is staying at rest, the 2T hydrodynamics model equations transform to the system of two coupled kinetic equations for temperatures $T_e(t)$ and $T_i(t)$ given by

$$c_e dT_e/dt = -\alpha\,(T_e - T_i), \qquad c_i dT_i/dt = +\alpha\,(T_e - T_i),$$

where c_e and c_i are electron and ion heat capacities, respectively, and α is an electron-ion coupling coefficient. Solutions of above equations gives the relaxation time of 5 ps. The same time can be achieved in MC-MD simulation if $m_{ef} = 6m_e$ of standard electron mass is used.

To simulate atomic motion in gold under high strain-rate deformations we use the well-fitted EAM potential [22], which was successfully utilized in [3, 9]. This EAM potential reproduces mechanical response of gold to strong compression and stretching, and the melting point as well.

5. Dynamics of inflation, deceleration and decay of nanobump in liquid or solid states

Using a series of two-temperature 1D hydrodynamics and MD-MC runs we have investigated the parameter space of a nanobump problem relative to the capillary $v_{0\sigma}$ (3) and thermal $v_{0\chi}$ (8) numbers. A set of the MD-MC runs is listed in Table 1 and in Fig. 2. The small initial

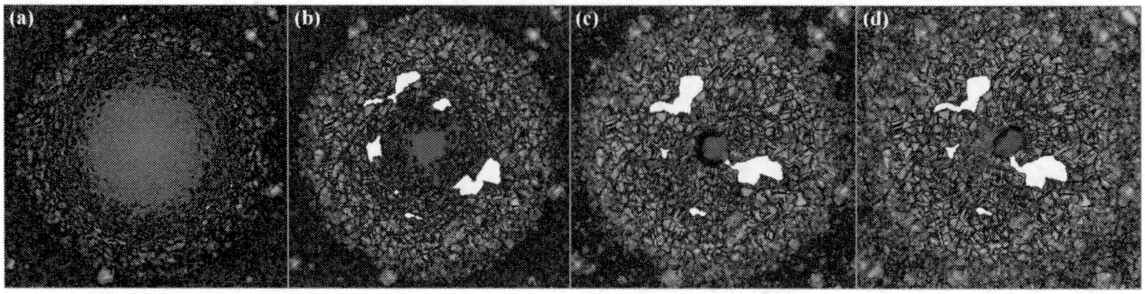

Figure 6. The top view onto dynamical and thermal processes going in the moving shell in the run #29a. Fluid is strongly overcooled and the two free surfaces are near to the overcooled volume (crystallization of a thin film) therefore crystallites nucleates inside molten gold. At the early stage (frame a), when area of a liquid shell constitutes the significant portion of a bump, there is a wide transition belt around the liquid portion of the shell. The crystal grains in the belt seems elongated in the azimuthal direction - this is a result of projection of the inclined lateral side of a dome onto plane of view. (a) $t = 230$ ps, (b) 360 ps, (c) 720 ps, (d) 1.8 ns.

velocity $v_c(t = 0)$ cases are located near the origin point in Fig. 2. They have been discussed in Subsections 3.1 and 3.2 above. In those cases a rupture from substrate and strong mass redistribution along a film are absent. At larger velocities $v_c(0)$ a nanobumps with a nipple type apex shown in Fig. 1 (c) are formed (the cases 23, 30 in Fig. 2). A rather short jet with a frozen droplet at the upper point of the jet is formed in the case 20. Length of this rather short jet is slightly above the radius of a nanobump at the contact plane. The group of the relatively low velocity runs is highlighted in Fig. 2 by blue color (the blue circles). The group 20, 23, 30 is separated from the other runs because in the cases 20, 23, 30 the whole of metal remains on the substrate.

The runs selected by the red triangles in Fig. 2 corresponds to the cases when part of a mass of a film loses its contact with a film and flies away. Transition between these ejecting and non-ejecting regimes is abrupt. This follows from a comparison of the runs 30 and 31. Small 5% increase of initial velocity $v_c(0)$ provokes sharp changes from a compact nanobump 30 with a moderate mass redistribution and a nipple apex to the case 31 with a strongly thinned shell of a nanobump and a massive very long jet.

To start the MD-MC runs an initial state has been prepared. The preparation involves heating and melting of a gold film by a Langevin thermostat. The thermostat creates a distribution of temperatures with a maximum temperature 2000 K in the central point. Temperatures decrease from 2000 K in the center to 1500 K in the margin. After that a hydrodynamic velocity distribution having only a normal component relative to the surface is imposed. The maximum velocity is velocity of a tip $v_c(0)$. Velocities decrease down to zero in the margin. After beginning of a run, the thermostat is removed everywhere except the margin. In the margin the thermostat holds a film at the margin and supports the margin temperature T_{margn} at a given level below a melting point. This temperature is given in the last column of a Table 1. It was $T_{margn} = 500$ K for all runs listed in Table 1 except the runs #29 and #29a; $T_{margn} = 300$ K for these runs. Temperature difference between the central part of a nanobump and the margin supports the cooling flow and cools down the central part, see Subsection 3.2. Gradually a recrystallization zone moves up from the margin to the central part - solidification of a nanobump proceeds, see Figs. 3-6 where the green color corresponds to solid, while the red color highlights molten gold.

Decreasing of the margin temperature T_{margn} intensifies the cooling flux, thus increasing rate of solidification. This essentially increases the thermal velocity scale v_χ (8) and decreases normalized velocity $v_{0\chi}$ (8). Estimates based on a thermal equation discussed in Subsection 3.2 give 25% reduction of the parameter $v_{0\chi}$ as a result of changing $T_{margn} = 500$ K $\rightarrow$ 300 K. This reduction is marked by the arrow in Fig. 2. In all other respects the runs 28, 29 and 29a are identical, see Table 1.

Another fine difference was tested in the runs 29 and 29a. Working of a thermostat in the margin area prevents freezing of fluid in this area because a stronger thermostat faster destroys atom-atom correlations directed to nucleation of lattice seeds thus impeding the crystallization, see Figs. 5 and 6. To check an influence on results, we increase 1.5 times the characteristic time of thermostat from 0.8 ps for all runs to 1.2 ps in the particular run 29a. An influence on dynamics of a nanobump is small.

Figs. 3 (#27) and 4 (#29a) show the side view of the process. These runs are in ejecting regime. In this regime large amount of mass is pumped from the nanobump shell into the jet. As a result there is strong thinning of a shell and very long jets develop. The thinning proceeds while a shell is in a liquid state or, when solidification begins, the thinning takes place in the liquid part of a film. The runs 27 and 29a have approximately similar cooling rates but the capillarity effects are weaker in the case 29a, comp. the marked points 27 and 29a in Fig. 2. This follows from the difference in the value $v_{0\sigma}$ (3). Therefore the jet 29a is longer. It begins to break into separate subjets and later decays into droplets as a consequence of elongation and Rayleigh instability.

Excess of kinetic energy above a binding initial surface energy defines a measure of kinetic deformations of a bump - larger excess - then longer jets, and then the decay processes become more active, see Subsection 3.1 about conservation of energy and Fig. 1. Comparing Figs. 3 and 4, we see how this excess works. The jet in Fig. 4 becomes thinner in its bottom part as a result of weakening and termination of supply from the shell when the crystallization process gradually covers whole shell. This decreasing feed together with elongation of a stretching jet and tearing off the subjet form the narrow peaked frozen jet in the apex of a bump. Separated hot droplets slowly cool down, see Chapter 1. Frozen shell is made from a dense set of the very small size nanocrystallites, comp. with [3]; colorizing presents distribution of a symmetry parameter: green is solid, red is liquid. The set is formed in the process of crystallization of strongly overcooled liquid gold. Thin overcooled liquid shell has two free surfaces, therefore the solid grains begin to grow in liquid at some distance from the continuous solid film, see Fig. 6 below. The green shell in Fig. 4 has holes, see also Fig. 6 below. Such holes are often observed in experiments, see, e.g., [7].

Why experimentally the frozen bumps are observed only in the cases with tight focusing when a radius of a nanobump $\sim R_L$ is small? This is a result of interplay between the value of a heat conduction coefficient κ and the size R_L limited from the small sizes by an optical wavelength $\sim$ 1 micron. Relative thickness d_f/R_L is also significant. Because for the large spots when R_L is tens and hundreds microns the relative thickness is small. Relative thickness d_f/R_L usually cannot be larger than d_T/R_L, where d_T is thickness of a heat affected zone. This limit will meet us if we attempt to increase film thickness and overcome the value d_T. Typical values are $d_T = 30 - 130$ nm for different metals. It should be emphasized that in the case $d_f > d_T$ the bump shell may spall out not at a contact but inside a film [8]. The shell made from the spallation plate has significant thickness inhomogeneities because this is the case of nucleation inside a film relative to the case of a spall from a very smooth contact with substrate. Of course, the inhomogeneities accelerate decay of the bump shell into cloud of the spatially separated droplets [14]. Flight of a hollow liquid bump or a dome or a cupola [23] created from the *bulk* targets is observed beginning from the first paper [24] thanks to microinterferometric pump-probe technique.

Duration of flight usually is 1-10 ns, and flight distance is $\sim$ 10% of $R_L \sim 100$ microns in these experiments. Initial velocity usually is rather large $v_c(0) \sim 0.1 - 1$ km/s, while the capillary scale (3) $v_\sigma \sim 10 - 100$ m/s is smaller if the shell is thicker. Thus it is difficult to observe the capillary deceleration of larger cupola: inflation velocity $v_c(t)$ should decrease per one unit of v_σ for every one unit of a large flight distance $\sim R_L$. Therefore the jet formation and transformation of shell shape from a convex parabolic cupola to a cone (see Fig. 3) do not have time to realize before the shell rupture.

The conduction coefficients κ and χ of metals are fixed (usually $\kappa = 30 - 300$ W/m/K for solid metals). Therefore the thermal velocity scale $v_\chi = \chi/R_L$ (8) strongly decreases for large spots. Corresponding points have similar values $v_{0\sigma}$ as shown in Fig. 2, but the values $v_{0\chi}$ are much higher. In these conditions only a frozen ring around a delamination circle remains. This ring is observed in all experiments with the large spots. The ring presents the solidified remnants of marginal part of the entire shell. The central part of shell is torn off from the margin part and flies away. This situation is similar to the case of so called "nano-crown" shown in Fig. 1 (h) but now at much larger scale R_L.

6. Non-equilibrium crystallization of thin film in conditions of strong and fast continuous stretching

Cooling and propagation of the recrystallization zone over a shell surface are shown in Figs. 5 (#27) and 6 (#29a). They correspond to the side view of the dynamic and thermal evolution presented in Figs. 3 (#27) and 4 (#29a). This is a non-equilibrium phase transition with a wide transformation belt separating pure liquid and pure solid. The belt is filled with a two-phase solid-liquid mixture. The newborn crystallites seems elongated perpendicularly to the radial direction because the lateral side of a shell is oblique relative to the plane of projection in Figs. 5 and 6.

The runs #27 $d_f = 1.8$ nm and #29a $d_f = 5.1$ nm have different thickness. It is easier to break a hole through a film in the case of a thinner film. In the run #27 shown in Figs. 3 and 5 the holes appear in the pure liquid. While in the run #29a (Figs. 4, 6) there are a liquid films surrounding the crystal grains. Stretching of a belt of a two-phase solid-liquid mixture stretches the liquid film in between the grains remaining the grains much less stretched. Therefore decrease of thickness of liquid is enhanced in the intergrain space. This process leads to formation of the holes penetrating through the pieces of the liquid film in the intergrain area, see Fig. 6 (b). The holes gradually grow and crystallize inside the solid bump.

7. Conclusion

We used Lagrangian hydrodynamics and molecular dynamics combined with Monte-Carlo codes to shed light onto the 3D processes initiated by an ultrafast laser action on the films deposited on substrates. New results (relative to [10]) about physics of formation of the bumps, including description of the thermal part of the problem discussed in Subsection 3.2., were obtained. Thermal problem should be solved in conjunction with the inertia/capillarity dynamical part of the problem. We use two non-dimensional parameters and scaling (see Section 3, Table 1, and Fig. 2), bypassing difficulties connected with the huge size for MD simulation of the considered experimental situation, see Section 2.

Acknowledgments

Authors acknowledge support from Russian Science Foundation (RSCF) (project No. 14-19-01599).

References

[1] F. Korte, J. Koch, and B. N. Chichkov, Appl. Phys. A **79**, p. 879 (2004)
[2] Y. Nakata, T. Okada, and M. Maeda, Jpn. J. Appl. Phys. **42**, L1452 (2003)

[3] D. Ivanov, A. Kuznetsov, V. Lipp, B. Rethfeld, B. Chichkov, M. Garcia, and W. Schulz, Appl. Phys. A. **111**, p. 675–687 (2013)
[4] A. Kuchmizhak, A. Ionin, S. Kudryashov, S. Makarov, A. Rudenko, Y. Kulchin, O. Vitrik, and T. Efimov, Optics Letters **40**, p. 1687–1690 (2015)
[5] M. Domke, S. Rapp, and H. Huber, Physics Procedia **39**, p. 717–725 (2012)
[6] M. Armstrong, *The evolution of ultrafast compression,* SIMES Seminar, (2015) https://simes.stanford.edu/events/mike-armstrong-simes-seminar
[7] D. Wortmann, J. Koch, M. Reininghaus, C. Unger, C. Hulverscheidt, D. Ivanov, and B.N. Chichkov J. Laser Appl. **24,** 042017 (2012)
[8] N.A. Inogamov, V.V. Zhakhovsky, V.A. Khokhlov, K.V. Khishchenko, Y.V. Petrov, D.K. Il'nitsky, K.P. Migdal, www.chemphys.edu.ru/pdf/2014-11-29-005.pdf (in Russian)
[9] B.J. Demaske, V.V. Zhakhovsky, N.A. Inogamov, I.I. Oleynik, Phys. Rev. B **82,** 064113 (2010)
[10] N. A. Inogamov, V. V. Zhakhovskii, and V. A. Khokhlov, JETP **120**, p. 15–48 (2015)
[11] V.I. Emelyanov, D.A. Zayarniy, A.A. Ionin, I.V. Kiseleva, S.I. Kudryashov, S.V. Makarov, T.H.T. Nguyen, A.A. Rudenko, JETP Lett. **99,** No. 9, p. 518–522 (2014)
[12] A.I. Kuznetsov, J. Koch, B.N. Chichkov, Appl. Phys. A **94,** p. 221–230 (2009)
[13] A.I. Kuznetsov, J. Koch, B.N. Chichkov, Optics Express **17,** No. 21, p. 18820–18825 (2009)
[14] C.M. Rouleau, C.-Y. Shih, C. Wu, L.V. Zhigilei, A.A. Puretzky, and D.B. Geohegan, Appl. Phys. Lett. **104,** 193106 (2014)
[15] P.A. Danilov, E.A. Drozdova, A.A. Ionin, S.I. Kudryashov, S.B. Odinokov, A.A. Rudenko, V.I. Yurovskikh, D.A. Zayarny, Appl. Phys. A **117,** 981–985 (2014)
[16] C. J. Koch,L. Overmeyer, and B. Chichkov, Optics Express **20**, p. 24864 (2012)
[17] A.I. Kuznetsov, C. Unger, J. Koch, B.N. Chichkov, Appl. Phys. A **106,** 479–487 (2012)
[18] A.V. Bushman, G.I. Kanel', A.L. Ni, and V.E. Fortov, *Intense Dynamic Loading of Condensed Matter* (Taylor & Francis, London, 1993).
[19] http://teos.ficp.ac.ru/rusbank/
[20] N.A. Inogamov and V.V. Zhakhovskii, JETP Lett. **100,** No. 1, p. 4–10 (2014)
[21] V. Zhakhovskii, K. Nishihara, Y. Fukuda, S. Shimojo, T. Akiyama, S. Miyanaga, H. Sone, H. Kobayashi, E. Ito, Y. Seo, M. Tamura, and Y. Ueshima, IEEE Proceeding of the 5th International Symposium on Cluster Computing and Grid (CCGrid 2005), May 9-12, 2005, (Cardiff, UK), Vol. 2, p. 848–854 (2005)
[22] V.V. Zhakhovskii, N.A. Inogamov, Yu.V. Petrov, S.I. Ashitkov, K. Nishihara, Appl. Surf. Sci. **255,** No. 24, p. 9592–9596 (2009)
[23] N.A. Inogamov, Yu.V. Petrov, S.I. Anisimov, A.M. Oparin, N.V. Shaposhnikov, D. von der inde, J. Meyer-ter-Vehn, JETP Lett. **69,** No. 4, p. 310–316 (1999)
[24] K. Sokolowski-Tinten, J. Bialkowski, A. Cavalleri, D. von der Linde, A.M. Oparin, J. Meyer-Ter-Vehn, S.I. Anisimov, Phys. Rev. Lett. **81,** 224–228 (1998)

Journal of Physics: Conference Series **681** (2016) 012002 doi:10.1088/1742-6596/681/1/012002

Maths Meets Myths: Network Investigations of Ancient Narratives

Ralph Kenna[1,3] **and Pádraig Mac Carron**[2]

[1] Applied Mathematics Research Centre, Coventry University, CV1 5FB, England
[2] Social and Evolutionary Neuroscience Research Group, Department of Experimental Psychology,University of Oxford, OX1 3UD, England
[3] The Doctoral College for the Statistical Physics of Complex Systems, Leipzig-Lorraine-Lviv-Coventry ($\mathbb{L}^4$).

E-mail: `r.kenna@coventry.ac.uk`; `padraig.maccarron@psy.ox.ac.uk`

Abstract. Three years ago, we initiated a programme of research in which ideas and tools from statistical physics and network theory were applied to the field of comparative mythology. The eclecticism of the work, together with the perspectives it delivered, led to widespread media coverage and academic discussion. Here we review some aspects of the project, contextualised with a brief history of the long relationship between science and the humanities. We focus in particular on an Irish epic, summarising some of the outcomes of our quantitative investigation. We also describe the emergence of a new sub-discipline and our hopes for its future.

1. Introduction

In 2012 and 2013 we published two papers [1, 2] which generated significant amounts of interest within and beyond academia. This resulted in invited articles in science magazines, newspaper reports, radio interviews and a host of coverage on blogs and online media. For us, this was an unprecedented amount of media attention and illustrated the potential that interdisciplinary research has for impact generation. So what led to such levels of interest?

Although the distinctions between them are not always sharp, myths differ from legends and folktales. Mythology entails a plethora of characters and timeless narratives outside documented history. Legends are couched in a definite historical timeframe and folktales are intentionally fictional. Although each of these is broadly accessible, and forms important components of our cultural identities, certain works of mythology are the most ethereal. Part of the reason for widespread interest in our work was that it involved the treatment of such ancient corpora in a very modern way – as complex systems.

The aim of our research was (and remains) to study networks of characters appearing in different mythologies to facilitate their classification and comparison. For example, the *Táin Bó Cúailnge* (Cattle Raid of Cooley) is the most famous epic of Irish mythology. The tale describes the invasion of Ulster (the northern part of Ireland) by the armies of queen Medb of Connacht (the western part) and its defence by the warrior Cúchulainn. Before it was committed to writing by medieval Irish scholars, the *Táin* had an extensive oral tradition. The historicity of the *Táin* has long been debated. Some argue that such narratives corroborate Greek and Roman accounts of the Celts while others object that such tales have no historical basis.

Medb's royal residence, with king Ailill, was at Cruachan, the ancient capital of Connacht. This is identified as a complex of archaeological sites at Rathcroghan (Ráth Cruachan, the Ringfort of Cruachan), in County Roscommon, about 50 km north-west of Athlone. In 1864 Samuel Ferguson explored nearby Oweynagat cave and found two Ogham stones there (monoliths inscribed in an early Irish alphabet). On returning to Dublin he discovered one letter of his record of these inscriptions to be indistinct. He immediately took the night train back to Roscommon and was in the cave by the following noon. He translated the inscription as `FRAICCI MAQI MEDFFI`, meaning "Fraic, son of Medf." One can imagine his excitement, having found what appeared to him to be evidence for a historical Medb [3]. Nowadays Medb is believed by some to be an amalgam of several entities and antecedents. Indeed, in Section 5 we will present evidence which may support this suggestion.

Mythology continues to provoke excitement today. Comparative mythology is a well developed subject within the humanities. Joseph Campbell maintained that mythological narratives from a variety of cultures around the world essentially share the same fundamental structure, called the *monomyth* [4]. The monomyth is supposed to represent the universality of the human condition; the idea is to explain and contextualise our world, different societies have constructed different mythologies. The similarities in the structures of these narratives, as Campbell saw it, reflect common or universal human concerns. However, the concept of the monomyth has been criticised and mainstream scholarship of comparative mythology values differences as well as similarities [5, 6]. Nonetheless, the idea of universality (or lack of it) overlaps with a fundamental concept in the statistical physics of critical phenomena.

Epic mythology also differs from folktales and legend in that it often entails large casts of characters. The number of interactions between these characters is vaster still. We know from statistical physics in general, and sociophysics in particular, the value of treating such medium-to-large data sets as complex systems. Therefore it is tempting to seek to apply statistical physics techniques – or at least viewpoints – to comparative mythology to see what comes out. The focus on the collection of interactions in large sets of characters is a novel approach to comparative mythology and, indeed, to literature generally, because many traditional methods tend to zoom in on the individual. We view the method as complementary to traditional approaches – it certainly cannot replace them, but can perhaps supplement them.

In the next section we briefly outline why physicists like to investigate subjects beyond the physical world and why they should be allowed to do so. In Section 3 we provide the context for our work by summarising some aspects of Irish mythology. In Section 4 we introduce our methodology and we apply it to three iconic European epics in Section 5. We conclude in Section 6.

2. Why do Scientists want to do Research in the Humanities?

Why do physicists wish to delve into subject areas covered traditionally by other disciplines and in which they have little or no expertise? Physicists, and statistical physicists in particular, are interested in *complex systems* – how properties can emerge in an aggregate system which are absent from the microscopic agents which comprise it. The classic example from a statistical physicist's point of view is phase transitions; water turns to ice or steam at sufficiently low or high temperatures. Yet its constituent parts, the molecules of H_2O from which it is constructed, do not themselves manifest different phases. Another example is that of flowing water. It can exhibit turbulence, but an individual molecule cannot. Similarly, the characteristics of a population are not simple aggregates of the behaviours of individuals. Examples include flocking, language, culture, and mythology itself. Each of these is a feature of the collective rather than the individual. In recent decades, simple models, many inspired by statistical physics, have been developed to account for such social phenomena. Disciplines with names such as *complexity science*, *sociophysics* and *econophysics* have emerged and are hotbeds of activity.

Journal of Physics: Conference Series **681** (2016) 012002 doi:10.1088/1742-6596/681/1/012002

Thus physicists are not only interested in the physical and, indeed, physics does not only concern the physical. There are many physicists and their motivations are varied. Some are interested in the practical and that which can be applied. Others are motivated by the challenge of tackling problems which are difficult intellectually. Still others are driven by curiosity. It is often the latter type that engages in cross-disciplinary pursuits. With curiosity as a map in one hand, and holding the passport that is academic freedom in the other, curious physicists are happy to make forays into anything that interests them.

Some of this activity is interdisciplinary involving cooperation with sociologists, psychologists and economists as well as humanities scholars. The relationship between science and humanities has sometimes been perceived to be a difficult one. *The Two Cultures* was the title of part of an influential Lecture by the English physical chemist and novelist Charles Snow in 1959 [7]. The gist of his talk was that our society, its intellectual life, priorities, and systems of education, is split into two cultures – humanities and the arts on one hand, and science on the other. Snow argued that difficulties in communication between the two is a significant impediment to solving many of the world's problems. Examples of these problems include informing and influencing politicians, developing science and technology strategies, deciding policies for the future of education and issues of cultural fragmentation. Snow's diagnosis provoked an enormous amount of controversy and is still debated to this day.

The *Two Cultures* dichotomy is unfortunate in our view, as science and the humanities are perhaps not so far apart as it pretends. Indeed, the spirit of curiosity that drives many physicists is the same one that motivates many activities in the humanities. These are often not financially-driven or even applications-driven pursuits. Such research activity may be removed from everyday life but, in pushing at theoretical and methodological boundaries, they can launch new and unpredictable developments. Statistical physics and the humanities have a common cousin in the social sciences. In the 18th century people noticed regularities in numbers of events such births, deaths, etc. This was surprising because individuals themselves are unpredictable. This partly motivated the development of statistical approaches to many-body systems [8].

The word *sociophysics* is related to the older term *social physics*. Both refer to the study of social systems from the perspective of physics or using concepts or tools of physics. The term *sociology* itself was first used by Auguste Comte, a founder of the discipline, as an alternative to the term social physics. It is interesting to compare the following two quotes in this context:

> "Now that the human mind has grasped celestial and terrestrial physics, — mechanical and chemical; organic physics, both vegetable and animal, — there remains one science to fill up the series of sciences of observation, – social physics. This is what men have now most need of."

> "I think the next century will be the century of complexity. We have already discovered the basic laws that govern matter and understand all the normal situations. We dont know how the laws fit together, and what happens under extreme conditions."

The first of these is from Compte in the 1850's [9]. The second is atributed to Steven Hawking in 2000 [10]. Given the long-standing links between statistical physics and sociology, and the current interest in complexity, it is perhaps not such a great leap to investigate aspects of humanities (literature, mythology) which may be viewed as many-body, interacting systems.

3. Aspects of Irish Epic Mythology

Besides being the location of the opening scenes of the *Táin Bó Cúailnge*, Cruachan features in *Táin Bó Flidhais* and *Táin Bó Aingen* (also known as *Echtra Neraí*). Through Oweynagat, it is also associated with *Samhain*, the Gaelic seasonal festival which gave rise to the modern Halloween. Archaeological evidence supports the suggestion that the site was an important location for ritualistic gatherings and a significant cemetery, although there is little archaeological evidence to support the assertion that it was a royal residence as often described in the mythological literature [3, 11].

To perform a broad investigation from a non-material perspective, we used social-network analysis. Here we report on these investigations, with a particular focus on the *Táin Bó Cúailnge*. This belongs to the Ulster Cycle, one of the four great groupings of Irish mythology. Early Irish literature presents ideal material for a social-network approach because of the preponderance of characters they contain, as well as the interactions between them. The publisher and Celticist Alfred Nutt estimated that known literature of the Ulster Cycle alone would occupy at least 2,000 pages of a modern volume if all repetitions were edited out [12, 13]. The question we asked, is how these mythological networks compare to other social networks; what are their universal features, if any, and what are the distinguishing ones?

We started with the *Táin Bó Cúailnge* because it is the most famous epic of Irish mythology, frequently compared to Homer's *Iliad* and the Anglo-Saxon *Beowulf*. Related to the *Táin* itself are a number of pre-tales and tangential tales (remscéla) which give the backgrounds and exploits of the main characters. The *Táin* has come down to us in three recensions. The first has been reconstructed from partial texts contained in *Lebor na hUidre* (the Book of the Dun Cow, dating from the 11th or 12th century and compiled at the monastery at Clonmacnoise) and *Lebor Buide Lecáin* (the Yellow Book of Lecan, a 14th century manuscript) and other sources. The second, later recension is found in *Lebor Laignech* (the Book of Leinster, a 12th century manuscript formerly known as the *Lebor na Nuachongbála* or Book of Nuachongbáil, after a monastic site at Oughaval in County Laois). A third recension comes from fragments of later manuscripts and is incomplete. Two popular English translations of the *Táin* [14, 15] are mainly based on the first recension, although they each include some passages from the second. We base our analysis on [14] which therefore, for the purposes of this analysis, serves as a proxy for what is commonly understood as the *Táin Bó Cúailnge*.

Before they were committed to writing, the *Táin* and other mythological narratives had extensive oral histories. The *Táin* was dated by medieval Irish scholars to the first centuries BC but this may have been an attempt by Christian monks to artificially synchronise oral traditions with biblical and classical history. It has been speculated that Christian scribes may have sought to de-deify Pagan gods to the level of (super-) human heroes to dilute Pagan influences [16]. Similar reasons may have led to Rathcroghan mound, for example, being depicted as a residence rather than a religious site [17]. It has been argued that heroic literature, written down in Christian times, contains attempts to fit to contemporary notions in order to make it more acceptable to Christian society [18, 19]. Indeed, the Book of Leinster version of the *Táin* ends with scribal notes in Latin and Irish which state [13]

> "A blessing be upon all such as shall faithfully keep the *Táin* in memory as it stands here and shall not add any other form to it.
> "I, however, who have copied this history, or more truly legend, give no credence to various incidents narrated in this history or legend. For, some things herein are the feats of jugglery of demons, sundry others poetic figments, a few are probable, others improbable, and even more invented for the delectation of fools."

Undoubtedly much material, which has come down to us today as mythology, has been influenced by religion and politics. Jeffrey Gantz has described the early Irish tales as verging "upon wishful thinking, if not outright propaganda" [20]. For example, the *Dindshenchas* texts on Tara (onomastic texts giving the origins of place-names and traditions associated with places) are believed to have been compiled to bolster the claims of the southern Uí Néill (from the middle Kingdom of Mide) over those of Brian Boru in Munster (Ireland's southern province) [21]. Myth could be used to explain why things are as they are and to justify the ambitions of dynasties [22].

The *Dindshenchas* and other texts in turn influenced Geoffrey Keating's famous seventeenth-century *History of Ireland* [23]. In the words of Keating (referring to the later Fenian Cycle of Irish mythology),

> "And whoever should say that Fionn and the Fian never existed, would not be stating truth... For it has been delivered to us from mouth to mouth that Fionn and the Fian existed; and, moreover, there are numerous documents that testify to this."

Through the 18th century, the historicity of Irish mythology was hotly debated. On the one hand, writers such as Edward Ledwich rejected the idea of a sophisticated civilization in ancient Ireland; on the other hand, Charles O'Connor and Sylvester O'Halloran defended the notion of a heroic past [21]. The historicity of the *Táin* continued to be contested in the 20th century. For example, while Thomas O'Rahilly [24] objected that such tales have no historical basis whatsoever, Myles Dillon [25] believed that heroic sagas give "a picture of pre-Christian Ireland which seems genuine." Kenneth Jackson [26] argued that while such narratives corroborate Greek and Roman accounts of the Celts and offer us a "window on the Iron Age", "the characters Conchobar and Cúchulainn, Ailill and Medb and the rest, and the events of the Cattle Raid of Cooley, are themselves entirely legendary and purely un-historical. But this does not mean that the traditional background, the setting, in which the Ulster cycle was built up is bogus." Gantz [20] concludes that "the Irish society represented by the Ulster Cycle is ... very similar to the Gaulish civilization depicted by Caesar." So may such old myths contain kernels of truth or are they entirely fanciful?

Currently there remains little direct material evidence for the events and the society in the *Táin*. However, there is also "no definite evidence against the medieval Irish belief that Conchobar and his court flourished around the time of Christ" [22]. Myles Dillon and Nora Chadwick [27] have argued that while the *Lebor Gabála Érenn* (Book of Invasions) which forms a major part of the Mythological Cycle, has no value as history, the *Táin* gives a more authentic picture of Iron-age Irish society. Using material from annals and genealogies to interpret ancient Irish epics, Francis Byrne examined the political landscape of early Irish kings and high-kings and suggested that the wars between Connacht and Ulster may indeed have occurred at any time before the fifth century. He suggests that such heroic tales usually contain "a kernel of historical fact" [22]. There is also some evidence that characteristics of the landscape are reflected in the story [28]. Thus, the extent to which Irish mythological narratives may or may not reflect traces of a historical, albeit distorted, version of an ancient reality continues to fascinate and generates an enormous amount of public interest [16, 29].

There are obvious caveats and limitations regarding the material to be analysed. These include enormous political and religious influences as discussed above, the obvious lapses in time between the creations of the medieval tracts and the societies and events they purport to describe and the use of modern texts (in English) as proxies for the originals. The reliability of any statistical analysis is dependent on the sample size, the larger the better. Ideally we would prefer extensive narratives with a large numbers of characters and plenty of interactions between them for increased statistical accuracy. Nonetheless, we can only analyse the material which has come down to us and we can only apply the techniques which we possess. The reader should keep in mind that we cannot offer proofs only evidence and comparisons from new perspectives. It is up to the experts in archaeology and mythology to inform further and ultimately it is up to the readers to make informed judgements of their own. Our purpose is to help inform such judgements from a wholly new perspective.

4. Complex Networks

Networks can be used to quantitatively characterise a large range of structures, such as the internet, power grids, transport networks and food-webs. In a social network, the nodes are

Journal of Physics: Conference Series **681** (2016) 012002 doi:10.1088/1742-6596/681/1/012002

representations of people and the links represent interactions or acquaintanceships between those people. A number of statistical tools have been introduced to describe and quantify various properties of networks. Here we briefly describe some of these tools in a non-technical manner, referring the reader to [1] for details.

There is a famous notion in sociology of *six degrees of separation*. This is the idea that, despite the worlds population of about seven billion, everyone is, on average, only about six steps away from any other person. In this case the number six is an example of the notion of average path length. This is a measure of the connectivity of the network and tends to be very short in social networks. A second mathematical measure of connectivity is the *clustering coefficient*, which gives an indication as to how cliqued a network is. In social networks the clustering coefficient tends therefore to be very high compared, for example to a random network. These measures of connectivity illustrate that social networks have different characteristics to other types of complex networks. The highly connected nature of social networks is a property which is known as *small world* [30].

The *degree* of a node in a network is the number of links it has. In the narratives we deal with, the degree of a character is the number of other characters he or she interacts with. The degree distribution in social networks tends to be well approximated by a power law and is often found to be scale free [31]. Removing the most connected characters in networks with scale-free degree distributions causes them to break down and become disconnected. However, removing characters at random does not strongly affect networks as they are robust to random attacks but not to targeted attacks.

In real-life social networks, people tend to be friends with other people who are similar to themselves. For example, popular people are more likely to have popular friends. We say that a network in which most nodes have this property is *assortative* [32]. Networks that do not have this property are *disassortative*. It turns out that assortativity is a very important property which distinguishes social networks from other networks. To investigate the assortativity properties of a network, one simply compares the degree of a node (acquaintances of a character) to the degrees of its neighbours.

In the past decades, a large array of social networks has been studied. Examples include networks of company directors [33], jazz musicians [34], movie actors [31], users of online-forums [35], and scientific co-authors [36]. Catalogues of such social networks have been assembled and they are normally small world, highly clustered, assortative and scale free [37]. Thus the general properties of real social networks are well established and well documented. On the other extreme, physicists have also examined the social networks underlying the panoply of characters that have appeared down the years in Marvel comics [38, 39]. Analysis of the network found that it does not have all the features of a real social network. Although it was found to be highly connected, it is also highly disassortative. All of the very high degree characters were too highly connected, making the network appear artificial. In the academic literature on networks, there are few if any other analyses of artificial or fictional networks.

For this reason, we also constructed and analysed the networks of several works of fiction. Clearly a small sample is not meant to embody all of the network features of the entire corpus of world fiction. Rather, they are examples of clear works of fiction — fiction involving fabulous characters, which do not attempt to emulate reality. It is hoped that, in the future, other researchers will analyse other works of fiction, so that a more extensive list of typical features will be built. For the moment, we found that each of the fictional networks we analysed were small world, just like real social networks. However, none of their degree distributions appeared to follow a clear power law and they were all disassortative. Moreover, unlike real social networks, the fictional ones were very robust upon targeted removal of the most connected nodes.

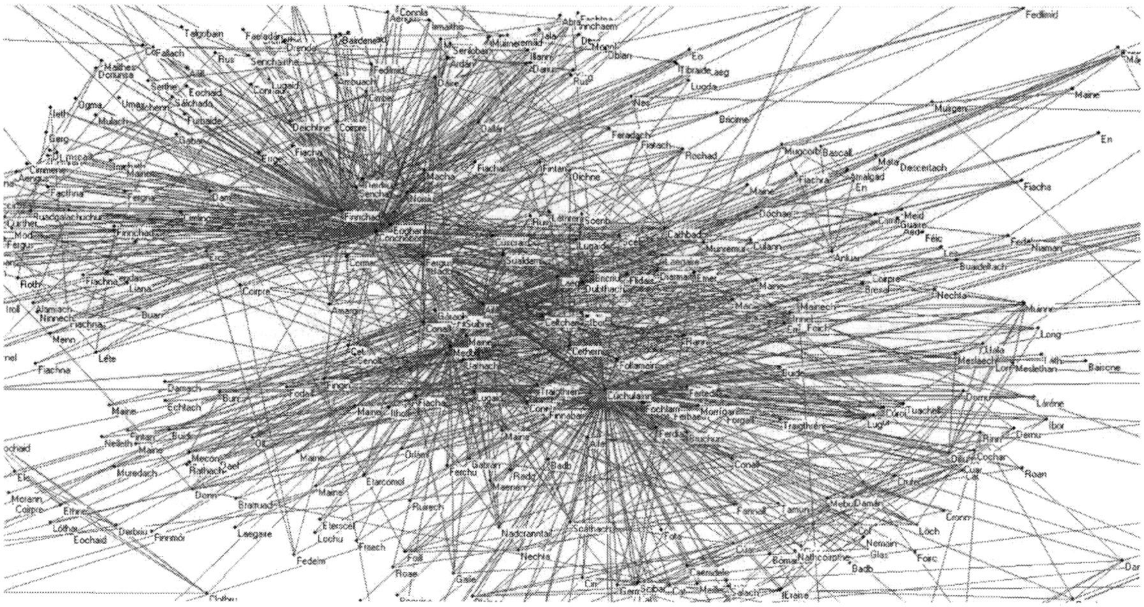

Figure 1. The *Táin* network. Green links are positive ("friendly") and red ones are negative ("hostile"). Two main clusters are identifiable by eye. The upper left is that of Ulster, with Conchobor at its centre. The lower right is that of Connacht, centred on Medb and Ailill. Between them is Fergus mac Roich and other warriors exiled from Ulster. Cúchulainn himself, although an Ulster warrior, is embedded amongst the Connacht foe as the majority of his links are conflictual.

5. Network Comparison of the *Táin Bó Cúailnge*, the *Iliad* and *Beowulf*

The *Táin Bó Cúailnge* is frequently compared to other European epics, in particular to *Beowulf* and the *Iliad*. *Beowulf* is an Old English heroic epic, set in Scandinavia. A single codex survives which has been estimated to date from between the 8th and 11th centuries. The story relates the coming of Beowulf, a Gaetish hero, to the assistance of Hrothgar, king of the Danes. After slaying two monsters, Beowulf returns to Sweden to become king of the Geats and, following another fabulous encounter many years later, is fatally wounded. Although the poem is embellished by obvious fictional elements, such as monsters and a dragon, archaeological excavations in Denmark and Sweden offer support for the historicity associated with some of the human characters [40]. Nonetheless, the character Beowulf himself is mostly believed not to have existed in reality [41, 42].

The *Iliad* is an epic poem attributed to Homer and is dated to the 8th century BC. It is set during the final year of the war between the Trojans and a coalition of besieging Greek forces. It relates a quarrel between Agamemnon, king of Mycenae and leader of the Greeks, and Achilles, their greatest hero and counterpart to Cúchulainn. Also much debated throughout the years [43], some historians and archaeologists maintained that the *Iliad* is entirely fictional [44], while recent evidence suggests that the story may be based on a historical conflict around the 12th century BC interwoven with elements of fiction [45, 46, 47].

The data for the networks were harvested by carefully reading each of the three narratives and entering all characters' names into databases, meticulously listing the characters they interact with. We defined two distinct types of interactions: links were designated as *positive* or "friendly" when two characters know each other, are related, speak to one another, or appear in a small congregation together. Links were deemed *negative* or "hostile" if two characters meet in combat. We also assigned a weight to the links between characters, based on how often they meet each other. In total, Kinsella's *Táin* has 404 characters and 1,233 links. *Beowulf* and the *Iliad* were found to have 74 and 716 unique characters and 165 and 2,650 links between them, respectively. The *Táin* network is depicted in Fig.5.

All three epics were found to be small world; they have low average path lengths and high clustering coefficients. All the networks became disconnected when the most highly linked 5% of characters were removed in a targeted manner. But they were all robust when 5% of characters were removed at random. These are characteristics of real social networks. However, of the three narratives, only the *Iliad* turned out to be clearly assortative. *Beowulf* was mildly disassortative and the *Táin* was highly disassortative.

Thus the social-network properties of the *Iliad* are similar to those of real social network and dissimilar to the fictitious social networks listed above. This is what one might expect if the narrative is a reasonably accurate portrayal of a real society. Obviously it does not categorically prove that the *Iliad* is based on real events, but it may be interpreted as evidence to support the case for some degree of historicity [45, 46, 47].

By the same token, the disassortativity of *Beowulf* may be a signal of its artificiality. However, recall that archaeological evidence suggests that the society in *Beowulf* may be based on reality but the eponymous character is believed fictitious. This clue prompted us to remove the character Beowulf from the network and to reanalyse it. Interestingly, the resulting network is assortative!

Thus, if assortativity and disassortativity are good markers of reality (or at least realism) and fiction respectively, our analyses of the *Iliad* and *Beowulf* and the *Táin* reflect some of the recent archaeological evidence and opinion; the *Iliad* appears realistic. *Beowulf* appears fictional at first sight, but on closer inspection, the artificiality is seen to reside in the lead character (as well as in other, more obvious, fabulous elements); *Beowulf* the society, without Beowulf the character, appears realistic. However the social network in the *Táin* is strongly disassortative. By the above criteria, it appears fictional. We ask wherein resides the artificiality of the *Táin*. Does the entire society appear fictional or is its artificiality localised in a few characters, as *Beowulf*'s artificiality is localised in Beowulf himself?

To investigate further, we next turned our attention to the degree distributions of the networks. We found that the distributions for all three epics are compatible with power laws, just like real social networks. Moreover, there is a striking similarity between the degree distributions of *Beowulf* and the *Táin* for all but the top six most connected characters in the larger set – see Fig.2(a). The solid line in Fig.2(a) is a curve which is best fitted to the data points which correspond to *Beowulf*. The six rightmost data points, which represent the six most connected characters in the *Táin*, are offset relative to the solid *Beowulf* line. This means that the degrees of these characters are too large relative to the characters of *Beowulf*, and, indeed, relative to the other characters in the *Táin*. The dashed line is a best fit to the *Táin* data. The anomalous six data points on the right pull the line upwards in that region of the plot. This hints that the network structure of the *Táin* may be similar to that of *Beowulf*, but for the six anomalous characters in the Irish narrative. This therefore might be an indication of where the artificiality of the *Táin*, is located.

The six anomalous characters in the *Táin* network are Conchobor mac Nessa, Cúchulainn, Finnchad Fer Benn (Conchobor' s son), Ailill mac Mágach, Medb and Fergus mac Roich. Four of these are those identified by Jackson as unhistorical [26]. To match the *Táin* line with the *Beowulf* one, the degrees of these six characters would have to be reduced. In terms of the narrative, this means that the number of interactions they have with other characters would

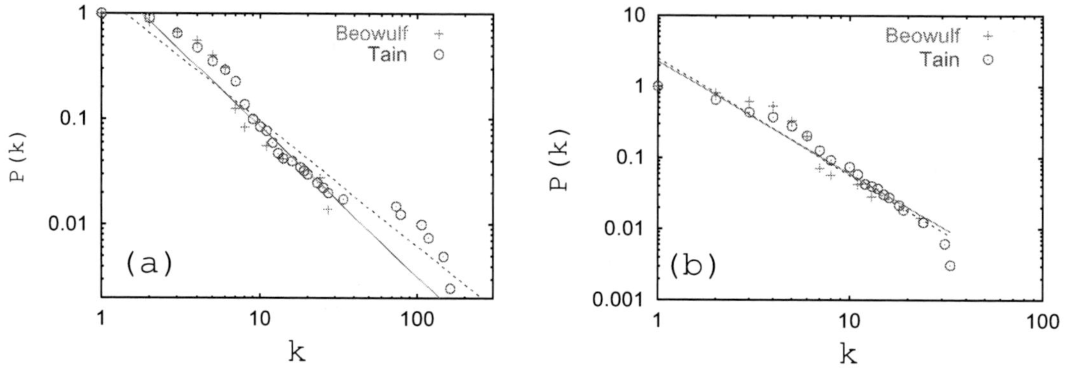

Figure 2. (a) the degree distributions for *Beowulf* and the *Táin* overlap except for the six most highly connected characters in the Irish epic. (b) Removing the weakest links of those characters renders the network similar to that of *Beowulf* with the eponymous protagonist removed. Both adjusted networks are assortative.

have to be lowered. To do this in a systematic manner, we define a weak link as one that occurs when two characters meet only once in the entire narrative. We then speculate that such weak links are proxy interactions in the tale. For example, the single encounter between Medb and one of her warriors chosen to fight Cúchulainn may rather represent an encounter between Medb's proxy and that individual. Removing the weak links from the top six *Táin* characters reduces their respective degrees. Note what we have done here is to remove weak links not to remove the characters themselves! Fig.2(b) shows the degree distributions for the adjusted networks. Now the top six points of the *Táin* no longer offset the best-fit line and the data points fall closer to this line for the high degree characters.

We next re-examined the assortativity of the *Táin*. Indeed, the removal of the weakest links of the six anomalous characters has the effect of rendering the entire network assortative. This supports the suggestion that the apparent artificiality of the *Táin* network is associated with just these top six characters. It also supports the notion that these characters may each be amalgams of several entities, as suggested in the Introduction. Now all of the *Táin* network properties which we investigated are similar to those of real social networks. We may conclude that, on the basis of network analysis alone, the adjusted *Táin* network is about as realistic (assortative) as that of *Beowulf*.

6. Conclusions

It has been claimed that archaeological evidence supports the historicity of some of the events of the *Iliad*. Manfred Korfmann argues that the onus lies now with those who deny any historical association between events in Late Bronze Age Troy and the *Iliad* to defend their positions, rather than the other way around [46]. There is also material evidence for the existence of some of the characters of *Beowulf*. However, the eponymous protagonist is not believed to have been a real person. There is little material evidence to support the historicity of the society or events depicted in the *Táin Bó Cúailnge*, but there is no definite evidence to the contrary either [22].

While we cannot comment on the purported events described in these narratives, here we have used ideas from modern statistical physics to analyse the relationships that underpin the societies depicted in them. We can draw on the experience and knowledge garnered in the complex-network and sociophysics community over recent years. This indicates, in particular, that real social networks have certain characteristics and assortativity appears to be most important amongst these.

On the spectrum from the assortative to the disassortative, the *Iliad* has properties most similar to those of real social networks, *Beowulf* is mildly disassortative and the *Táin* is strongly so. The beauty of applying maths to myths is that, unlike material artefacts, we can adjust the data set to address the question of what it would require to render these networks realistic. In the case of *Beowulf*, we follow established opinion and remove the protagonist to find the remaining network assortative. Thus, while the entire network is not realistic, we may suggest that an assortative subset has properties akin to real social networks, and this subset has corroborative evidence of historicity.

For the *Táin*, the social network of the full narrative is disassortative. On this basis, the society it describes appears to have "no credence" as the scribe noted at the end of the Book of Leinster. However, from a network-theoretic perspective, the artificiality is related to the top six most connected characters. Similar to the superheroes of the Marvel Universe, for example, they are too well connected. Speculation that these characters may be based on amalgams of a number of entities and proxies, suggests removal of their associated weak social ties. Indeed, the resulting network is assortative, similar to the *Iliad* and to other real social networks and very different to the society of the Marvel Universe and other works of intended fiction we examined. On this purely theoretical basis, we suggest that if the society in the *Táin* is to be believed, each of the top six characters is likely to have been formed by amalgamations of characters, as the narrative was passed orally throughout the generations.

In the years since our initial publication[1], we have become aware of excellent work taking place around the world. This includes traditional approaches to comparative mythology, folktales and to epic literature. Significant amounts of data have been gathered and we think that these are amenable to mathematical, statistical and computational methodologies. Indeed, fascinating approaches include conceptual mapping in a large folklore corpus [48]; analyses of virtual worlds [49]; studies of cognitive constraints in literature [50]; principal component analyses of distributions of folklore motifs [51]; and phylogenetic approaches to folk tales [52, 53]. To deepen and broaden quantitative investigations into mythology, annals and folktales, more collaboration between the humanities and sciences is needed. We have hosted interdisciplinary workshops which include physicists, applied mathematicians, complexity theorists, computer scientists, anthropologists, psychologists, authors, film makers, artists, historians, medievalists, mythologists and other humanities scholars. Resulting from these, an edited volume is under preparation and is scheduled to be published in 2016 [54]. We hope that these activities will spur further collaborative work and help break down the barriers that are perceived to exist between our two cultures.

Acknowledgments
We would like to thanks Peter Berresford Ellis for helpful comments on the text. We also thank staff at the Rathcroghan Visitor Centre for hospitality at a conference at Cruachan Aí.

References
[1] Mac Carron P and Kenna R 2012 Universal properties of mythological networks *EPL* **99** 28002
[2] Mac Carron P and Kenna R 2013 Network analysis of the Íslendinga sögur the Sagas of Icelanders *Eur. Phys. J.* B **86** 407
[3] Waddell J 1983 Rathcroghan A Royal Site in Connacht, *Journal of Irish Archaeology* **1** 21
[4] Campbell J 1949 *The Hero with a Thousand Faces* (Princeton: Princeton University Press)
[5] Cosentino D J 1998 African Oral Narrative Traditions in John Miles Foley (ed.) *Teaching Oral Traditions* Modern Language Association, New York, pp. 174188.
[6] Northup L 2006 Myth-Placed Priorities: Religion and the Study of Myth *Religious Studies Review* **32** 5-10
[7] Snow C P 1959 *The Two Cultures*, (London: Cambridge University Press)

[8] Fortunato S, Macy M and Redner S 2013 Editorial: Statistical Mechanics and Social Sciences *J. Stat. Phys.* **151** 1-8
[9] *The Positive Philosopy of Auguste Compte*, Freeley translated and condensed by Harriet Martineau Volume I, D. Appleton and Co. New York 1853.
[10] San Jose Mercury News, Morning Final Edition, January 23 (2000).
[11] Waddell J, Fenwick J. and Barton K 2009 *Rathcroghan, Co. Roscommon: Archaeological and Geophysical Survey in a Ritual Landscape* (Co. Wicklow: Wordwell)
[12] Nutt A 1900 *Cúchulainn, the Irish Achilles* (London: David Nutt)
[13] Dunn J 1914 *The ancient Irish epic tale Táin Bó Cúailnge* (London: David Nutt)
[14] Kinsella T 1969 *The Táin* (Oxford: Oxford University Press)
[15] Carson C (2007) *The Táin: a new translation of the Táin Bó Cúailnge* (London: Penguin)
[16] Green M J 1992 *Dictionary of Celtic Myth and Legend*, Thames and Hudson, London.
[17] Waddell, J. 2012 *Crúachain and Emain Macha: Archaeology and myth*, talk at Rathcroghan Archaeology Conference, Tulsk, Co. Roscommon, 13-15 April 2012 (available through the Rathcroghan Legacy Project, Rathcroghan Visitor Centre).
[18] Berresford Ellis P 1998 *The ancient world of the Celts*, Constable, London.
[19] Wallace B 2011 *Warriors and Warfare: Ideal and Reality in Early Insular Texts*, PhD thesis, The University of Edinburgh.
[20] Gantz J 1981 *Early Irish myths and sagas*, Penguin, London.
[21] Waddell J 2005 *Foundation Myths: The Beginnings of Irish Archaeology*, Wordwell, Co. Wicklow.
[22] Byrne F J 1973 *Irish Kings and High Kings* B.T. Batsford, London.
[23] Cunningham B 2000 *The world of Geoffrey Keating: History, Myth and Religion in Seventeenth-Century Ireland*, Four Courts Press, Dublin.
[24] O'Rahilly T F 1946 *Early Irish History and Mythology*, Dublin Inst. Adv. Studies, Dublin.
[25] Dillon M 1948 *Early Irish Literature*, University of Chicago Press, Chicago.
[26] Jackson K H 1964 *The oldest Irish Tradition: A Window on the Iron Age*, Cambridge University Press, Cambridge.
[27] Dillon M and Chadwick N 1967 *The Celtic Realms*, Phoenix Press, London.
[28] Lynn C 2003 *Navan Fort: Archaeology and Myth*, Wordwell, Co. Wicklow.
[29] Edel D Myth Versus Reality: Queen Medb of Connacht and her Critics, Ancient and Modern, in: Edel. D. (2001) *The Celtic West and Europe: Studies in Celtic Literature and the Early Irish Church*, Four Courts Press, Dublin.
[30] Watts D J and Strogatz S H 1998 Collective dynamics of "small-world" networks *Nature* **393** 440-442.
[31] Amaral L A N, Scala A, Barthélémy M and Stanley H E 2000 Classes of small-world networks *Proceedings of the National Academy of Sciences of the USA* **97** 11149-11152.
[32] Newman, M E J 2002 Assortative mixing in networks *Phys. Rev. Lett.* **89** 208701.
[33] Davis G F, Yoo M and Baker W E 2003 The Small World of the American Corporate Elite, 1982-2001 *Strategic Organization* **1** 301-326.
[34] Gleiser P M and Danon L 2003 Community Structure in Jazz, *Advances in Complex Systems* **6** 565-573.
[35] Kujawski B and Abell P 2011 Virtual communities?: the Middle East Revolutions at the Guardian Forum: Comment is Free, *European Physics Journal* B **83** 525-529.
[36] Newman M E J 2001 The structure of scientific collaboration networks *Proceedings of the National Academy of Sciences of the U.S.A.* **98** 404-409.
[37] Newman, M E J and Park J 2003 Why Social Networks are different from other types of networks, *Phys. Rev.* E **68** 036122.
[38] Alberich R, Miro-Julia J, and Roselló F 2002 Marvel Universe looks almost like a Real Social Network *Preprint* cond-mat/0202174.
[39] Gleiser P M 2007 How to become a superhero *Journal of Statistical Mechanics: Theory and Experiment* P09020.
[40] Anderson C E 1999 Formation and Resolution of Ideological Contrast in the Early History of Scandinavia, PhD. Thesis, the University of Cambridge.
[41] Klaeber F 1950 *Beowulf and the Fight at Finnsburg*, D C Heath, Massachusetts.
[42] Chambers R W 1959 *Beowulf: An Introduction*, Cambridge University press, Cambridge.
[43] Wood M 1998 *In Search of the Trojan War*, University of California Press, Berkeley.
[44] Finley M I 1954 *The World of Odysseus*, Viking Press, New York.
[45] Kraft J C, Rapp G, Kayan I and Luce J V 2003 Harbor Areas at Ancient Troy: Sedimentology and Geomorphology Complement Homer's Iliad *Geology* **31** 163-166.
[46] Korfmann M 2004 Was There a Trojan War? *Archaeology* **57** 36-38.
[47] Papamarinopoulos St P, Preka-Papadema P, Antonopoulos P, Mitropetrou H, Tsironi A and Mitropetros

P 2012 A New Astronomical Dating of Odysseus' Return to Ithaca *Mediterranean Archaeology and Archaeometry* **12** 117-128.
[48] Tangherlini T R (2013) The Folklore Macroscope: Challenges for a Computational Folkloristics: The 34th Archer Taylor Memorial Lecture *Western Folklore* **72** 7-27.
[49] Mryglod O, Fuchs B, Szell M, Holovatch Y, Thurner S (2014) Interevent time distributions of human multi-level activity in a virtual world *Physica A* **419** 681-690.
[50] Stiller J, Nettle D, and Dunbar R I M (2004) The small world of Shakespeare's plays *Human Nature* **14** 397-408.
[51] Berezkin Y E. (2010) Selecting separate episodes of the peopling of the New World: BeringianSubarcticEastern North American folklore links *Anthropological Papers of the University of Alaska* **5** 257-276.
[52] d'Huy J (2013) A phylogenetic approach to mythology and its archaeological consequences *Rock Art Research* **30** 115118.
[53] Tehrani J J (2013) The Phylogeny of Little Red Riding Hood *PLoS ONE* **8** e78871.
[54] Kenna R, Mac Carron M and Mac Carron P, *Maths Meets Myths: Complexity-science approaches to folktales, myths, sagas, and histories*, to appear in 2016.

Discontinuous percolation

H J Herrmann

Computational Physics, IfB, ETH Zürich, Switzerland

E-mail: hjherrmann@ethz.ch

Abstract. I present recent progress in understanding abrupt connectivity transitions, i.e. percolation-type models exhibiting first order phase transitions. The introduction of bridge bonds and their tricritical scaling allows obtaining exact expressions for the appearance of a jump in the order parameter. Another exact result can be obtained through the Potts model using Gliozzi's scheme. Finally also hybrid transitions are discussed.

Percolation is one of the simplest and oldest models in Statistical Physics in which sites or bonds of a lattice are randomly occupied with a probability p. It is known for its extremely robust continuous phase transition towards a spanning phase. But there exist many connectivity problems in nature and society that are clearly abrupt, like the explosion of a volcano, the breaking of a dam, the collapse of the banking system or the outbreak of pandemics. Thus it is not surprising that the search for a connectivity model exhibiting a first order transition represents an old challenge in statistical physics.

The oldest known example exhibiting a discontinuous transition is bootstrap percolation [1] in which, starting from a classical percolation configuration, all sites are removed that have less than a certain number z_c of occupied neighbors. This procedure, called "culling", is repeated, until every site has at least z_c occupied neighbors. Evidently the result depends on the type of lattice and on z_c and there are certain cases where the transition is shifted to $p = 1$ at which the order parameter P, i.e. the fraction of sites in the largest cluster, jumps to unity. This occurs because the culling mechanism pushes the curve P(p) towards larger p values, making it steeper and steeper. The disadvantage of bootstrap percolation is that culling cannot be described by a statistical weight and the number of iteration depends on the initial configuration and can become very long close to the transition.

In 2009 Achlioptas et al presented in Science a new model they called "product rule", in which instead of occupying one bond, two bonds are chosen and the one is occupied, which creates the smaller cluster. Numerically this model on the completely connected graph showed a sharp jump in the order parameter P, but strangely the cluster size distribution followed a power law. Soon after three different groups [3, 4, 5] were able to show convincingly that in the thermodynamic limit the jump should vanish, but that this would only be observable numerically for systems of at least 10^{23} sites, explaining why the original work saw effectively an abrupt transition. What happens in finite dimensions remains unclear.

Achlioptas' model was generalized to choose not between two, but between m bonds, giving rise to the so called "best of m model" [6,7]. Numerically clearly the jump in the order parameter becomes steeper the larger m and the smaller the spatial dimension. By interpolating on a square lattice between $m = 1$ and $m = 10$ it is even possible to identify a tricritical point, a clear indication that the transition for $m = 10$ is of first order [6]. Interestingly thc exponents of transport on the inicipient spanning cluster, i.e. the fractal dimension of the backbone, the cutting bonds and the shortest path as well as the electrical conductivity, depend on m, showing anomalies around $m = 4$ [7]. The situation of the best of m model is however still not settled.

A much clearer case is the "largest cluster model" in which bonds are occupied randomly as in usual percolation, except if the bond is going to generate a cluster larger than any cluster in the system. In this case the bond is only occupied with the probability

$$q = \exp\left[-\left(\frac{s-<s>}{<s>}\right)^2\right] \qquad (1)$$

where s is the size of the new cluster and <s> the average cluster size [8]. In this way the formation of a large cluster is exponentially suppressed and all clusters are roughly of the same size. At the transition the order parameter P(p) is discontinuous and displays a bimodal cluster size distribution. The clusters at the transition are compact but interestingly their surface is fractal with dimension 1.22 in 2d and 2.49 in 3d [9]. These values belong to a very large universality class, which will be discussed in the following.

The suppressing of spanning, that was a crucial ingredient of the models introduced hereto, can be studied best through what has been called "bridge percolation" [10]. A bridge is a bond, which if occupied would create a spanning cluster and in bridge percolation bonds are occupied randomly as in usual percolation, except if they are a bridge, so that bridges are never occupied. Therefore the system never spans and the percolation transition is pushed to $p = 1$. Since in 2d bridge bonds are dual to cutting bonds the fractal dimension of the set of bridge bonds N_{BB} at the threshold of classical percolation p_c is $1/\nu = 3/4$, where nu is the critical exponent of the correlation length. Above p_c, however, the fractal dimension of N_{BB} is 1.22, precisely the value found for the fractal dimension of the surface of the clusters of the largest cluster model. The crossover between the two dimensions at p_c is similar to a theta point, exhibiting a tricritical scaling law:

$$N_{BB}(p,L) = L^{\frac{1}{\nu}}\mathcal{F}\left[(p-p_c)L^{\theta}\right] \qquad (2)$$

At $p = 1$ the set N_{BB} becomes a continuous line, which is actually identical to the watershed of a random uncorrelated landscape [11]. Simulating large systems and making a careful correction to scaling analysis, its fractal dimension d_{BB} was determined with high precision, showing its universality and the equivalence of various models [12]. Amongst others the following models also belong to the same universality class: the shortest path of loop-less percolation [13], the optimal path crack [14], the touching line of two identical fluids invading a porous medium and the crack of fuse networks with infinite disorder [15]. It has also been shown that the line also follows the properties of the Schramm Loewner Evolution (SLE), i.e. is conformally invariant [16].

The scaling law for N_{BB} enabled proving that another model, namely the "spanning cluster model" exhibits a first order transition [17]. In this model m bonds are chosen and any of them, which is not a bridge, is occupied. Only if all m chosen bonds are bridges, one of them must be occupied and the system spans, exhibiting a jump in the order parameter. If m is above a certain value m_c the location of the transition p_c increases with system size towards $p = 1$, while it decreases towards the p_c of classical percolation when m is below m_c. It can be shown using Eq.(2) that

$$m_c(d) = \frac{d}{d - d_{BB}} \qquad (3)$$

where d is the spatial dimension and d_{BB} the fractal dimension of the bridge bonds. Above the critical dimension $d = 6$ the set of bridge bonds is dense, i.e. $d_{BB} = d$, and thus the transition is continuous in the thermodynamic limit for all values of m. For m above m_c the order parameter jumps to unity at $p = 1$, the transition being thus discontinuous. A similar phase diagram is expected for the product rule, but this is still disputed [7, 10].

An even more rigorous proof for a first order percolation transition came through the Potts model for a metallic breakdown model [18]. Imagine an insulating substrate adsorbing metallic particles on which an electric field is applied parallel to the substrate. The adsorption probability W should depend on the local potential drop ΔV as

$$W = \frac{p}{q}\left[1-(1-q)\left(1-\left(\frac{\Delta V}{V_0}\right)^{\gamma}\right)\right] \tag{4}$$

where p, q and γ are parameters of the model. Classical percolation is retrieved, when γ is infinite. More astonishingly the case $\gamma = 0$ corresponds to the clusters of the q-state Potts model as shown in a little known paper by Gliozzi [19]. In his Monte Carlo algorithm one choses a homogeneously distributed random number $x \in [0,1)$ and occupies the bond if $x < p/q$, leaves it empty if $x > p$ and if $p/q < x < p$ one occupies the bond only if it is an internal bond, i.e. which connects sites that belong to the same cluster. Since it is known that in 2d the transition of the Potts model changes from second to first order at $q = 4$, our metallic breakdown model will also exhibit a first discontinuous transition in 2d when $q > 4$ and $\gamma = 0$.

A more exotic type of transition is found in the restricted Erdös-Renyi model [20]. Here one connects preferentially the most disconnected sites. Concretely, bonds between two nodes are randomly placed with the condition that one of the nodes must belong to a cluster, which belongs to the fraction g of the smallest clusters. Considering this model on the infinitely connected graph, it becomes possible to prove the existence of a hybrid transition with a continuously varying exponent [21], which up to now is the only known example of such a transition. In other words one finds a jump in the order parameter and after that a power-law for the cluster size distribution, where the exponent τ continuously varies with g according to the following transcendental equation:

$$\frac{\varsigma(\tau)}{\varsigma(\tau-1)} = \frac{1}{g} - \frac{1}{g+1}\ln\left(\varsigma(\tau-1)\left(\frac{g+1}{2}\right)^{-\left(1+\frac{1}{g}\right)}\right) \tag{5}$$

where ς is Riemann's zeta function.

An applied example of how a global condition induces a discontinuous transition can be found in epidemic spreading. Consider a kind of Susceptible-Infected-Susceptible (SIS) model [22], in which individuals are infected by sick neighbors with an infection rate τ and will be cured the next time step, only using a certain amount c of money, which comes from a global budget, that is refilled by fixed contributions coming from the healthy individuals [23]. As for the usual SIS model one observes at a certain infection rate a percolation-like transition at which the fraction of infected individuals continuously becomes finite. However in this model one finds also a second transition at a higher pandemic infection rate τ^*, at which after some time the budget suddenly is exhausted and everybody gets infected. At this rate τ^*, which depends on the parameters of the budget, the order parameter jumps discontinuously.

Finally it is worth mentioning that first order connectivity transitions have also been found in the robustness of coupled networks [24], following a cascading dynamics that is similar to the culling in bootstrap percolation. The transition can be made continuous, i.e. less dangerous, by the introduction of autonomous nodes [25].

Summarizing, there are at least two different ways of making percolation transitions discontinuous. On one hand one can compress the p-axis around p_c by some culling or cascading mechanism, thus making the order parameter increasingly steep at the transition point. On the other hand one can introduce an occupation rule that requires global knowledge, like the spanning condition or the cluster sizes. Here the strategy consists in delaying spanning, which can be done either by suppressing bonds that foster spanning, or by favoring bonds that do not contribute to spanning. Such bonds can either be internal bonds, i.e. those that already belong to the same cluster or bonds that preferentially merge the smallest clusters. In this way clusters become more compact and all of a similar size, i.e. exhibit a narrow size distribution. Achlioptas' product rule certainly goes in the right direction but the condition is just not strong enough.

Acknowledgements
I thank my collaborators José Soares Andrade Jr., Nuno Araújo, Lucas Böttcher, Young-Sul Cho, Peter Grassberger, Dirk Helbing, Lucas Oliveira, Julian Schrenk and Bob Ziff for the fun and excitement we had disclosing the mechanisms described in this short review. I acknowledge financial support from the European Research Council (ERC) Advanced Grant 319968 FlowCCS.

References

[1] Adler J 1991 *Physica A* **171** 453
[2] Achlioptas D, D'Souza R M and Spencer J 2009 *Science* **323** 1453
[3] Nagler J, Levina A and Timme T 2010 *Nature Phys.* **7** 2645
[4] Riordan O and Warnke L 2011 *Science* **333** 322
[5] da Costa R A, Dorogovtsev S N, Goltsev A V and Mendes J F F 2010 *Phys. Rev. Lett.* **105** 255701
[6] Araújo N A M, Andrade Jr. J S, Ziff R M and Herrmann H J 2011 *Phys. Rev. Lett.* **106** 095703
[7] Andrade Jr. J S , Herrmann H J, Moreira A A and Oliveira C L N 2011 *Phys. Rev. E* **83** 031133
[8] Araújo N A M and Herrmann H J 2010 *Phys. Rev. Lett.* **105** 035701
[9] Schrenk K J, Araújo N A M and Herrmann H J 2011 *Phys. Rev. E* **84** 041136
[10] Schrenk K J, Araújo N A M, Andrade Jr. J S and Herrmann H J 2012 *Sci. Rep.* **2** 348
[11] Fehr E, Andrade Jr. J S, da Cunha S D, da Silva L R, Herrmann H J, Kadau D, Moukarzel C F and Oliveira E A 2009 *J. Stat. Mech.* P09007
[12] Fehr E, Schrenk K J, Araújo N A M, Kadau D, Grassberger P, Andrade Jr. J S and Herrmann H J 2012 *Phys. Rev. E* **86** 011117
[13] Porto P M, Havlin S, Schwarzer S and Bunde A 1997 *Phys. Rev.Lett.* **79** 4060
[14] Andrade Jr. J S, Oliveira E, Moreira A A and Herrmann H J 2009 *Phys. Rev. Lett.* **103** 225503
[15] Moreira A A, Oliveira C L N, Hansen A, Araújo N A M, Herrmann H J and Andrade Jr. J S 2012 *Phys. Rev. Lett.* **109** 255701
[16] Daryaei E, Araújo N A M, Schrenk K J, Rouhani S and Herrmann H J 2012 *Phys. Rev. Lett.* **109** 218701
[17] Cho Y S, Hwang S, Herrmann H J and Kahng B 2013 *Science* **339** 1185
[18] Oliveira C L N, Araújo N A M, Andrade Jr. J S and Herrmann H J 2014 *Phys. Rev. Lett.* **113** 155701
[19] Gliozzi F 2002 *Phys. Rev. E* **66** 016115
[20] Panagiotou K, Spöel R, Steger A and Thomas H 2011 *Elec. Notes in Discret. Math.* **38** 699
[21] Cho Y S, Lee J S, Herrmann H J and Kahng B 2015 Hybrid phase transition when connecting the disconnected *Preprint*
[22] Hethcote H W 1989 *Applied Mathematical Ecology* (Biomathematics vol 18) eds. Levin S A, Hallam T G and Gross L J (Heidelberg: Springer) pp 119-144
[23] Böttcher L, Wooley-Meza O, Araújo N A M, Herrmann H J and Helbing D 2015 *Sci.Rep.* in print arXiv:1508.05484
[24] Parshani R, Buldyrev S V and Havlin S 2010 *Phys. Rev. Lett.* **105** 048701
[25] Schneider C, Yazdani N, Araújo N A M, Havlin S and Herrmann H J 2013 *Sci. Rep.* **3** 1969

Journal of Physics: Conference Series **681** (2016) 012004 doi:10.1088/1742-6596/681/1/012004

Academic research groups: evaluation of their quality and quality of their evaluation

Bertrand Berche[1,4], **Yuri Holovatch**[2,4], **Ralph Kenna**[3,4] **and Olesya Mryglod**[2,4]

[1] Université de Lorraine, Statistical Physics Group, IJL, UMR CNRS 7198, Campus de Nancy, B.P. 70239, 54506 Vandœuvre lès Nancy Cedex, France
[2] Institute for Condensed Matter Physics of the National Academy of Sciences of Ukraine, 1 Svientsitskii Str., 79011 Lviv, Ukraine
[3] Applied Mathematics Research Centre, Coventry University, CV1 5FB, England
[4] The Doctoral College for the Statistical Physics of Complex Systems, Leipzig-Lorraine-Lviv-Coventry ($\mathbb{L}^4$).

E-mail: bertrand.berche@univ-lorraine.fr; hol@icmp.lviv.ua; r.kenna@coventry.ac.uk; olesya@icmp.lviv.ua

Abstract. In recent years, evaluation of the quality of academic research has become an increasingly important and influential business. It determines, often to a large extent, the amount of research funding flowing into universities and similar institutes from governmental agencies and it impacts upon academic careers. Policy makers are becoming increasingly reliant upon, and influenced by, the outcomes of such evaluations. In response, university managers are increasingly attracted to simple metrics as guides to the dynamics of the positions of their various institutions in league tables. However, these league tables are invariably drawn up by inexpert bodies such as newspapers and magazines, using arbitrary measures and criteria. Terms such as "critical mass" and "h-index" are bandied about without understanding of what they actually mean. Rather than accepting the rise and fall of universities, departments and individuals on a turbulent sea of arbitrary measures, we suggest it is incumbent upon the scientific community itself to clarify their nature. Here we report on recent attempts to do that by properly defining critical mass and showing how group size influences research quality. We also examine currently predominant metrics and show that these fail as reliable indicators of group research quality.

1. Introduction

Academia is undergoing profound change. All over the world, positions of leadership and senior management within universities are becoming professionalised. Whereas in times gone by, such positions were typically occupied by senior academics, having proved themselves through long and successful careers in research and education, nowadays professional managers – who lack such experience – are increasingly empowered. In the never-ending drive for increased efficiency, policy makers and university managers seek to commoditise all aspects of academic activity including research. In this context, a plethora of indicators has emerged to measure the previously unmeasurable. As a result, the academic sector is becoming ever more business-like and league tables encourage competition over cooperation.

One of the simplest indicator is the size of a research department or group. Funding bodies, governments and universities frequently claim that these must be above a certain minimum to be viable. The term "critical mass" has been borrowed from nuclear physics to capture this notion. An implication of the term, as it is still all too widely understood, is that a group or department which is subcritical does not have the ability to sustain quality research activity. It suggests (erroneously as we shall see) the existence of a quantum jump in quality once the critical threshold is passed – a first-order phase transition, to use the parlance of statistical physics. Extending the notion, it is often claimed that bigger is always better in research and that funding should therefore be focused into small numbers of large, elite institutions with smaller universities relegated to teaching-only roles.

Here we show that, despite their widespread uses, these notions are manifestly incorrect. They are based upon unfounded and conflated analogies and not on scientific rigour. A proper, measurable definition of critical mass in the research context has only emerged in the last five years [1, 2, 3]. It does *not* entail a quantum leap or first-order transition from low to high quality. Instead, the critical-mass phase transition is higher order – barely noticeable, in fact. There is also a second, measurable phase transition as group size grows. This is triggered by a *Ringelmann-type effect* and marked by a *Dunbar number*. The former is due to a slow-down in the rate of increase of quality with quantity due to coordination losses. The latter expresses the reasons for these losses, namely cognitive limits to the numbers of interactions individuals can sustain. In Section 2 we explain these terms in more detail and review their connection to critical mass.

A second indicator which has recently emerged is the h-index. Again this is very attractive to policy makers and managers as a simple, zero-dimensional metric which is supposed to measure quality on an individual level. The notion has recently been extended to research groups in Ref.[4]. Competing measures include the normalised citation index (NCI) [5, 6, 7]. While the h-index itself was developed by a physicist, the NCI is an invention of a private company. In the second part of this paper, we compare both of these against expert peer review measures of the quality of research groups. We show that the h-index is the better indicator of group quality but that its role is strongly dependent on group size. We also show that neither of these metrics is a reliable substitute for expert peer review estimates of quality.

Academics fear the increased use of metrics to measure research activity. The fear is that the increasingly professionalised management class lacks detailed knowledge of, and appreciation for, academic subjects and seeks to base judgment and decisions on automated metrics rather than expertise and academic experience. This, accompanied by a top-down management approach, bolstered by metrics, and in pursuit of league-table rankings, may impinge upon academic freedom as researchers are forced to chase arbitrary measures rather than follow where their curiosity leads. Here we show that these fears are well founded; despite the rising tide of metrics, they are a poor measure of quality.

2. Evaluation of Quality and Critical Mass

The notion of critical mass in research has been around for a very long time. The basic idea is that sub-critical research groups tend to produce research of poor quality. Once the critical threshold is passed, research quality becomes of an acceptable level. This idea has been extended to, and far beyond, its logical conclusion to the idea that "bigger is always better" so that benefit always accrues through increasing group size. However, multiple analyses based on citation counts have produced no evidence for such a concept of critical mass. Despite that, policies based upon this belief have been developed and implemented and these have serious consequences.

Mark Harrison has studied how government officials allocated funding amongst research projects in the former USSR. In Ref.[8], he describes how policies swung between phases of competition and phases based on critical mass. Harrison sees remarkably similar parallels in the UK's funding system today. He predicts that the current focus on concentrating research funding in pursuit of critical mass will eventually give way to competition, but only after the

Journal of Physics: Conference Series **681** (2016) 012004 doi:10.1088/1742-6596/681/1/012004

follies of the policy are declared. By then, he fears, emerging research groups in promising universities will have lost their funding; small but excellent research centres will have closed and individual careers curtailed. Amazingly, the policies in which these dangers are inherent are based upon concepts of critical mass which have no foundation.

Five years ago, we developed a theory for critical mass and the dependency of research quality on group size [1]. Our model was successfully tested against empirical data coming from peer-review based national research assessment exercises in the UK and France [2]. To date, this model remains the only theory in existence for the relationship between research quality and group size. A significant amount of literature has been developed since Ref.[1] and all quantitative evidence supports the theory. There is no quantitative evidence invalidating the model we are about to describe. Here we recall the model developed in Ref.[1] and how it was statistically tested against empirical evidence.

It is important to realise that what we are about to develop here is a theory of averages (a "mean-field theory" in the language of statistical physics). There will always be deviations from such averages; the brilliant lone researcher, beavering away on a profound topic for years - still exists, especially in an area like pure mathematics [9]. (We will show, in fact, that critical mass for that subject is very small – only one or two and perhaps it is unmeasurable.) Also, although the model predicts higher research success for large groups than small, there are exceptions: some smaller groups are as good as, or better than some large ones. We could overlay our model of averages with a random distribution to simulate deviations that always exist, but we feel such complications would only serve to obfuscate the trends we are trying to capture. With this in mind, then, we proceed to develop our theory of averages.

A naive assumption (but one that was/is widespread, especially amongst policy makers, funding bodies and research managers) would be that the strength S of a research group or department is a simple sum of the strengths of the individuals comprising it. The idea here is that excellent researchers will be attracted to larger research groups, and the best therefore become bigger. With this causation arrow, the driving mechanism is quality determining size: better becomes bigger. If this were the only relation between quality and size, one would expect to see an unbounded *Matthew effect.* The Matthew effect, or cumulative advantage, is a phenomenon whereby "the rich get richer and the poor get poorer". (The name comes from the Gospel according to St Matthew, which states: "For unto every one that hath shall be given, and he shall have abundance: but from him that hath not shall be taken away even that which he hath.") However, as we shall see, the empirical data does not support this scenario as the dominant driving mechanism.

Of course, individual strength is itself a function of many factors, including innate calibre, teaching and administrative loads, management support, the extent of interdisciplinarity, the availability of suitable equipment, whether the work is mainly experimental, theoretical or computational, the methodologies and traditions of the field, library facilities, journal access, external collaboration, grants held, confidence supplied by previous successes, prestige of the institute and many other factors. We denote the mean strength of the individuals in a group, resulting from all these factors and more, by a. If the group has N members, then, the strength of the group would simply be $S = aN$.

However, research groups are complex systems and we have to take interactions between individuals into account. We suppose that direct two-way interaction between individuals produces an added effect and we denote the mean strength of that effect by b. (We consider three-way interactions, and so on, as collections of two-way interactions, so we need not take these into account separately.) Since there are $N(N-1)/2$ possible two-way communication links in the group, the strength becomes $S = aN + bN(N-1)/2$. Note that certain institutions, publications and league tables refer to "research power" rather than "strength" [10]. We define research *quality* s as the strength or power per head, and write $s = a_1 + b_1 N$.

Journal of Physics: Conference Series **681** (2016) 012004 doi:10.1088/1742-6596/681/1/012004

In fact this is not the full story for there has to be a limit to the number of colleagues with whom a given researcher can effectively communicate. In anthropology, this limit is known as the *Dunbar number* [11]. This was proposed in the 1990's by Robin Dunbar who noticed a correlation between the average sizes of the neocortex in species of primates and the average sizes of their social groups. Dunbar suggested that it represents a cognitive limit to the number of individuals with whom one can maintain stable relationships. Extrapolating that correlation to human brain sizes, Dunbar predicts that we can comfortably maintain social relationships, on average, with about 150 people. Again, this is an average only - the number can, of course, vary; it has been proposed to typically lie between 100 and 250. There is a large body of evidence in support of Dunbar's theory: the estimated average size of a Neolithic farming village is 150, as is the fragmentation point of Hutterite settlements. It is also the size of basic units in professional armies in Roman antiquity as well as in modern times since the 16th century. It is the average village size in the Domesday Book.

In our original publications on the topic, we referred to this limit as the upper critical mass and denoted it N_c. But it is a Dunbar number for academic researchers and it is discipline dependent. Once the group exceeds N_c, groups tends to fragment into subgroups. We suppose that the average number of such subgroups is $N/(\alpha N_c)$, so that they have average size αN_c. The strength of the group is then ascribed to the accumulation of individual strengths together with the strength intra-subgroups interactions, $S = aN + bN(\alpha N_c - 1)/2$. Of course, each of the $N/(\alpha N_c)$ subgroups can themselves interact with another subgroup. If the strength of such interactions is c, say, the total strength of an average large ($N > N_c$) group is expected to be $S = aN + bN(\alpha N_c - 1)/2 + cN/(\alpha N_c)[N/(\alpha N_c) - 1]/2$. Gathering terms of the same order in N, we thus arrive at

$$s = \begin{cases} a_1 + b_1 N & \text{if } N \leq N_c \\ a_2 + b_2 N & \text{if } N \geq N_c, \end{cases} \tag{1}$$

where a_i and b_i are functions of the strength parameters a, b, c as well as αN_c. In particular $b_2 \propto 1/N_c^2$ so that slope of the s versus N curve to the right of the Dunbar point should be small for disciplines with large N_c values.

To test the above theory, we used empirical data from the UK's last Research Assessment Exercise (RAE). The RAE is a peer-review based assessment of the quality of university research which the Government then uses to decide how funds should be allocated between different universities, institutions and departments. Not surprisingly, it is taken very seriously by university managers as well as by researchers themselves. The exercise took place about every five years up to 2008. In 2014 it was replaced by a somewhat different exercise wherein, besides quality of research, the impact which that research had beyond academia played an important role. The 2014 event was called the Research Excellence Framework or REF.

While for the RAE research was categorised into 67 academic disciplines, in the 2014 REF there were only 36 units of assessment (UoA). Applied Mathematics, for example, which includes some theoretical physics, was a stand-alone unit UoA at RAE2008 but it was merged with Pure Mathematics, Statistics and Operational Research to form a Mathematical Sciences UoA for REF2014. The 2008 was therefore a more "fine-grained" exercise and for that reason the data we report on here comes from the RAE.

In the UK system, research was scrutinised by experts from each discipline to determine the proportions carried out at each of five levels: 4* (world-leading research); 3* (internationally excellent research); 2* (research that is internationally recognised); 1* (research recognised at a national level) and unclassified research. Following RAE, a formula is used to determine how funding is distributed to higher education institutes for the subsequent years. The original formula used by the Higher Education Funding Council for England (HEFCE) valued 4* and 3* research seven and three times more highly than 2* research and allocated no funding to lower quality research. We can use that formula for the relative values of the various categories of

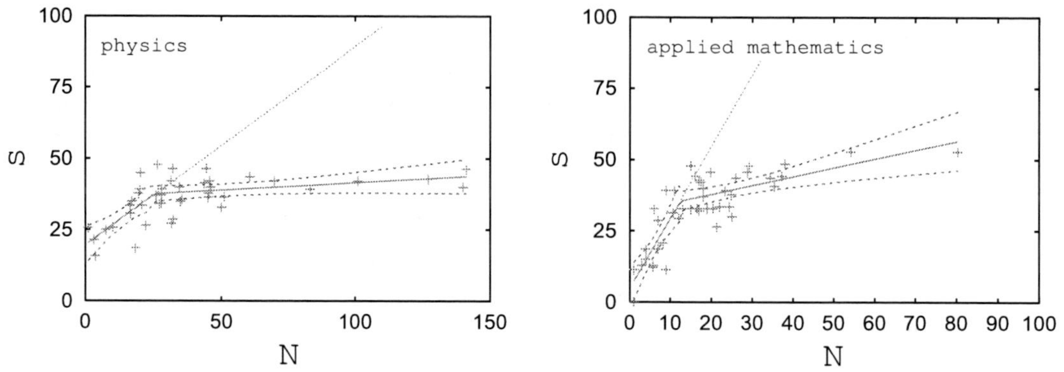

Figure 1. The relationship between quality and quantity of research groups submitted to the UK's last Research Assessment Exercise in Physics and Applied Mathematics. For Physics (which is dominated by experimental physics), the Dunbar number marking the breakpoint in the fitted red curve is $N_c = 25 \pm 5$. For Applied Mathematics (which includes some theoretical physics.) it is 13 ± 2. The critical mass is half the Dunbar number. The blue dashed lines are error boundaries. The purple dotted lines are extensions of the sub-breakpoint fits and capture some of the top performers.

research as a proxy for quality. If p_{n*} represents the percentage of a team's research which was rated $n*$, then, the team's *quality* is

$$s = p_{4*} + \frac{3}{7}p_{3*} + \frac{1}{7}p_{2*} \,. \tag{2}$$

With Eq.(2) to hand, it is a straightforward exercise to translate the RAE results to quality scores and plot them against group size.[1] Two such plots are given in Fig. 1: one for Physics and one for Applied Mathematics. Almost all experimental physics submitted to the RAE were assigned to the "Physics" UoA. Some theoretical also came under the "Physics" banner but others were entered into "Applied Mathematics". For our purposes then, we may consider "Physics" as meaning experimental physics and "Applied Mathematics" as being both applied mathematics and theoretical physics. Fitting to the formula (2) delivered the estimates $N_c = 13 \pm 2$ for Applied Mathematics and $N_c = 25 \pm 5$ for physics. Note how, as expected, the slopes of the s versus N curves reduce to the right of the Dunbar number. Note also that the right slope is smaller for the discipline with the higher N_c-value, as expected.

We have produced similar plots to those in Fig.1 for a variety of academic disciplines based on data from RAE2008. the reader is referred to the original literature for details [2]. We reproduce the N_c values in Table 1 for convenience. Three different candidate Dunbar numbers

[1] While generally throughout this paper, the word 'group' means a collection of researchers at a given university active in a common discipline, in the UK context it means those who were submitted to RAE2008 in a given unit of assessment. Usually these are permanent staff but they may include postdocs. A group is not, therefore, synonymous with a department because not all department members may be research active and, therefore, not all may have been submitted to RAE. Alternatively, a submission may draw from researchers interacting across different departments. The word 'group' in this sense is also not synonymous with research centre, as such entities may be involved more than one department. Indeed, at RAE, research centres may have been involved in submissions across more than one subject area. While "group" in this context captures how research is conducted at universities in countries such as Austria, France, Ireland, Ukraine and the UK, it misses the more hierarchical structure of, say, German universities, where research groups are more focused around individual senior professors.

Journal of Physics: Conference Series **681** (2016) 012004 doi:10.1088/1742-6596/681/1/012004

were detected for Computer Science. We speculate that this discipline comprises a number of very different approaches, from the formal and theoretical to the empirical. Similarly in Archaeology, two candidates for N_c were determined. In Pure Mathematics no breakpoint was found and since the smallest group submitted had $N = 4$, we believe the N_c value for that discipline is that number or less.

We have stated that N_c is the Dunbar number or upper critical mass. However, this is *not* analogous the old (unfounded) notions of critical mass as a threshold below which high quality research is impossible. (Clearly there is no such threshold.) Nonetheless, we can come closer to the old concept of critical mass [8] with the following considerations.

Suppose funding for a new academic were to become available. We ask whether it is more beneficial for society as a whole to allocate the new researcher to a group with $N > N_c$ or to one with $N < N_c$. The decision is made by considering the rate of change of *strength* $S = sN$ with N. We find that dS/dN is bigger for $N < N_c$ groups provided that $N > N_c/2$. In other

Table 1. The values of the Dunbar numbers or upper critical masses for a variety of academic disciplines.

Research discipline	N_c
Applied mathematics	13 ± 2
Statistics & operational research	17 ± 6
Physics	25 ± 5
Geography, environment & Earth studies	30 ± 3
Biology	21 ± 4
Chemistry	36 ± 13
Agriculture, veterinary & food sciences	10 ± 3
Law	31 ± 4
Architecture, the build environment, town & country planning	14 ± 3
French, German, Dutch & Scandinavian languages	6 ± 1
English language & literature	32 ± 3
Pure mathematics	≤ 4
Medical sciences	41 ± 8
Nursing, midwifery, allied health professions & studies	18 ± 5
Computer sciences 1	11 ± 5
Computer sciences 2	33 ± 9
Computer sciences 3	49 ± 10
Archaeology 1	17 ± 3
Archaeology 2	25 ± 4
Economics and econometrics	11 ± 3
Business and management studies	48 ± 8
Politics and international studies	25 ± 5
Sociology	14 ± 4
Education	29 ± 5
History	25 ± 5
Philosophy and theology	19 ± 3
Art & design	25 ± 8
History of art, performing arts, communication and music	9 ± 2

words, to maximally benefit society as a whole, additional researchers should be given to the smaller groups provided they are not too small. The cutoff is at

$$N_k = \frac{N_c}{2}. \tag{3}$$

We refer to this number as the *critical mass*. (In our earlier papers we called it *lower critical mass*. In other words, we have established a scaling relation between the critical mass and the Dunbar number.

We can now divide research groups into three different categories: small ($N < N_k$); medium ($N_k < N < N_c$); and large ($N > N_c$). These categories are discipline-dependent. For example, while a team of five pure mathematicians may be considered as "large", a similar number of experimental physicists is "small". This accords with experience.

The overall shapes of the curves in Fig.1 may be ascribed to a Ringelmann-type effect. The original Ringelmann effect, in psychology and in sports science, is the tendency for individual members of a group to become less productive as the group size increases [12]. The effect is due to an increase in inefficiency as the numbers grow. In psychology, this is often ascribed to motivational losses (also known as social loafing). But coordination losses may also account for the phenomenon. Here, as in Dunbar's theory, we also believe that the phenomenon is due to coordination losses. However, rather than the decrease in the average individual performance (which would be the actual Ringelmann effect), our model manifests a reduction in the *rate of change of quality* per unit staff member. In other words, it is the quality-versus-quantity gradient rather than in the quality itself that reduces as N increases. The challenge to managers, therefore, is to counteract or minimise this effect. Can we find point the direction to looking for a solution? We believe we can do so as follows.

The dotted line in Fig.1 is an extrapolation of the left fit for small and medium groups into the supercritical region where N exceeds N_c. We refer to such extrapolated lines as "overshoots". In the absence of a transition point N_c, if the mechanism which governs research quality for small and medium universities applied also to the large ones, then the research quality to the right of N_c would also be expected to follow this line. In other words, the naive model outlined earlier, and the one upon which many critical mass policies are currently based, would predict an unbounded Matthew effect along this line. In this case, a policy of continued concentration of resources would indeed drive up research quality. However, the figure clearly indicates that this is not the case. Similar figures for other research disciplines may be found in Ref.[3].

Instead, as the figure clearly demonstrates, large research teams tend to have a different interaction pattern than small and medium ones, as predicted in Eq.(1). With a large value of $N_c \approx 25$ for Physics, research quality is saturated to the right of the N_c as Dunbar's theory predicts. This suggests that the concentration of more staff into these teams only leads to a linear increase in research volume and *not* to an increase in research quality.

However, it is interesting to note that some of the best performing research teams, which appear as outliers to the overall fit, are bunched slightly above N_c but are well described by this overshoot. We suggest that this overshoot may be caused by a greater than usual degree of cohesiveness in these highly successful research teams. Somehow, two-way communication links are sustained despite their group sizes exceeding the Dunbar numbers for their disciplines. These are the research groups which are best managed: those in which communication is key.

We have shown that, it is most beneficial to society if new researchers are added to medium-sized groups. This is because the new researcher brings new links to existing academics in such groups, helping to drive them to the Dunbar number. However, joining a medium-sized group is *not* the best course of action from the new individual's point of view. For them, it is better to join a large group. The reason is that a medium group (unless it is on the border of becoming large) only provides the new researcher with a number of links which is below their cognitive limit. If they join a large group, on the other hand, they can enjoy maximal connectivity. (Of course establishing such connections takes time as the intra-group cooperative network becomes reconfigured.) This returns us to the causation mechanism mentioned at the start of Section 2. Of course high-quality individual researchers are attracted to high-quality, large groups. However, that is not how quality grows. Adding more and more researchers to large groups only increases the volume of high-quality outputs. It cannot increase average quality significantly because that is already at saturation level. The fastest way to drive up research quality is to help medium-sized groups attain the nirvana that is the Dunbar number.

3. Quality of Evaluation: Quantitative Indicators

So far, we addressed evaluation of the quality of research and the roles played by Dunbar's numbers and critical masses. Now we turn to the second main theme announced in the title of our paper: the quality of research evaluation through metrics.

Nowadays, many different measures exist claiming to assess academic research excellence or impact and these are subject to ongoing discussion and debate within the academic, scientometric, university-management and policy-making communities internationally. A topic of prime importance is the extent to which citation-based indicators compare with peer-review-based evaluation. Although flawed in many obvious respects, peer-review remains the only approach that is broadly accepted by the scientific community. But peer-review exercises such as the RAE and REF are expensive, both in terms of the labour involved and the time lost by academics who could otherwise have been engaged in research activity. For example, it has been estimated that UK universities spent about £4000 for each researcher submitted to REF2014 [13]. The total cost to the UK of running the national evaluation exercise in 2014 is estimated to have been £246 million [13]. This is a significant fraction of the £1.6 billion distributed by HEFCE for quality-related research funding in 2015-2016. (That figure is based on submission of 52 077 individuals, so the funding allocated to universities is over £30,000 per submitted researcher in that year.)

For this reason, it is desirable for governments, funders and policy makers to introduce, if it exists, a simple, cheap and reliable way to measure scientific excellence. In the absence of expert subject knowledge, professional research managers are also keen on such an approach to serve as a basis for regularly measuring the performances of research groups and individuals. However, while citation-based indicators and metrics are readily accessible, they are far from being universally accepted as a way to inform evaluation processes. They are even less accepted by academics as a way to replace evaluations based on peer review such as the RAE and REF.

In Refs.[14, 15], we considered a citation-based measurement of research at an amalgamated, institutional or group level, from the natural to social sciences and humanities. In particular, we examined how that measure correlates with the results of RAE2008. We found that the citation-based indicator is very highly correlated with peer-evaluated measures of group strength S for some disciplines. But it is poorly correlated with group quality $s = S/N$ for all disciplines. This means that, almost paradoxically, indicators could possibly form a basis for deciding on how to fund research institutions, especially for the so-called hard sciences, but they should not be used as a basis for ranking or comparison of research groups. Moreover, the correlation between peer-evaluated and citation-based scores is weaker for soft sciences.

The citation-based measure which we compared against peer review in Refs.[14, 15] was one provided by *Thomson Reuters Research Analytics.* This company has developed the so-called *normalised citation impact* (NCI) i as a measure of a department's citation performance in a given discipline. The NCI is calculated using data from the *Web of Knowledge* databases and by comparing outcomes to a mean or expected citation rate. The measure is determined for an entire group or department and then normalised by group size. It is therefore a *specific* (per-head) measure. A useful feature of the NCI is that it attempts to take account of different

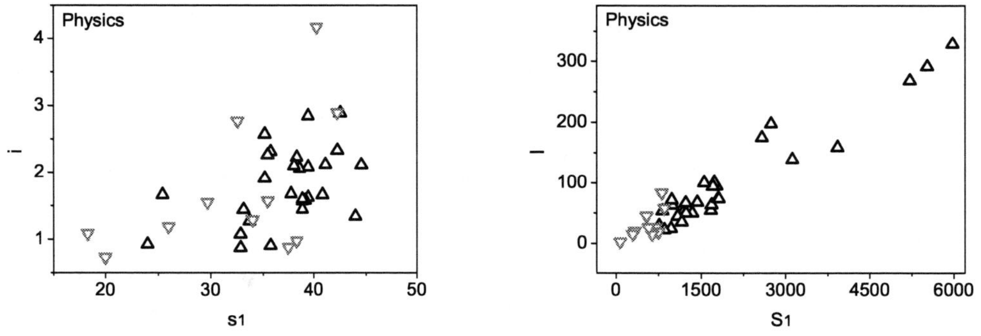

Figure 2. Left panel: The correlation between the normalised citation impact i and peer-review measures of group research quality s for Physics. Right panel: The correlation between the absolute impact $I = iN$ and research strength/power $S = sN$. The black and red symbols represent large and medium/small groups respectively.

citation rates in different disciplines. To achieve this, the total citation count for each paper is "rebased" to an average number of citations per paper for the year of publication and either the field or journal in which the paper was published. Here we denote the NCI specific measure by i. Scaled up to the size of a group or department, the corresponding *absolute* measure is denoted by I where $I = iN$.

The relationship between i and s for Physics research groups submitted to RAE2008 is given in Fig.2 (left panel). Clearly the correlation between the two measures is very weak. In fact the Pearson correlation coefficient is only 0.48. In Refs.[14, 15] we also analysed the correspondence between i and s for biology; chemistry; mechanical, aeronautical and manufacturing engineering; geography and environmental studies; sociology; and history. The maximum value for Pearson's coefficient was 0.60 (for biology and chemistry). These weak correlations indicate that the NCI cannot be use to estimate research quality - it cannot be used as a proxy for peer review.

We also ranked the various universities according to their i and s values to see if the NCI could, at least, serve as a guide to league tables. However Spearman's coefficient was of similar weakness to Pearson's so this is also ruled out.

Figure 2 (right panel) depicts the relationship between the absolute measure $I = iN$ and strength (or power) $S = sN$. The correlations are very good. In fact, for physics, Pearson's correlation coefficient for these measures is a remarkable 0.96. It is also in the nineties for the other scientific disciplines (biology and chemistry). However it is lower for sociology and history (0.88 in each case). The multiplication by N when replacing specific measures by their absolute counterparts stretches the data sets and the correlations between the rescaling leads to improved correlations. Moreover, when one partitions the various teams into small; medium; and large (as explained in the previous section), the correlation tends to be best for large groups. Nonetheless the conclusion is clear: citation-based measures may perhaps inform, or even serve as a proxy, for peer-review measures of the strengths of research groups. But they cannot be used for measures of quality.

In the UK system, quality-related governmental funding for research is strength based; the amount of money delivered to universities for a particular research discipline is a function of their S-scores at RAE or REF. The above results indicate that the use of citation-based alternatives may offer a far cheaper, and less intrusive alternative to the current system. The scale of the savings that could be made in the UK alone is a quarter of a million pounds per exercise. Such a proxy would work best for the hard sciences but would be less reliable for the social sciences and humanities. Also, it is crucial to realise that such citation-based indicators should not be used to compare or rank the qualities of research groups.

However, we see a grave danger in introducing any form of citation-based metric to evaluation exercises. The *Hawthorne effect* (also known as the *observer effect*) is the name given to reactions of individuals when an aspect of their behavior is observed or measured; they naturally seek to modify that behaviour. This would be disastrous in the academic context; university managers would force researchers to chase citations instead of curiosity. This would shift the entire landscape of academic research as everyone chases the same "hot topics" in a bid to increase their citation counts. It is well known that many profound and important scientific breakthroughs have come about through mixtures of curiosity, serendipity, accident and chance. Academic freedom is the cornerstone on which these features can thrive. Combined with the professionalisation of research management and "businessification" (indeed, moves towards privatisation) of many third-level institutions, citation-based league tables would pose a serious threat to academic freedom.

In 2014, Dorothy Bishop arrived at a similar conclusion to us. She used a different metric - the so-called *departmental h-index* instead of the NCI. In Ref.[16] we showed that Bishop's metric has an even better correlation with the RAE-measured strength index S than has the scaled-up NCI. The meaning of a departmental h-index of, say, n is that n papers, authored by researchers from a given department in a given discipline were cited at least n times over a given time period (we used the periods of assesment for RAE2008 and REF2014). This index takes into account, in principle, *all* researchers (not only those submitted to RAE or REF). However, in practice it can be dominated by a single, extremely strong individual. Also, unlike for REF where authors' addresses at the census date determine their affiliations for assessment purposes, author address at the time of publication determines to which university a given output is allocated for the departmental h-index.

Nonetheless, the departmental h-index is indeed better correlated with research quality; the Pearson and Spearman correlation coefficients for between h and s for Physics at RAE2008 were 0.55 and 0.58, respectively.

These correlations are still, however, too weak to replace or inform exercises like the RAE or REF. To demonstrate that, we decided to use the departmental h-index to predict outcomes, in terms of rankings, of RAE2014 in advance of their being announced by HEFCE. If simple citation-based metrics can be used as some sort of proxy for peer review, one would expect them to be able to predict at least some such aspects of the outcomes of such exercises. Even limited success might suggest that a citation metric could serve at least as a "navigator" – to help guide research institutes as they prepare for the expert exercises.

We placed our predictions for the rankings in Biology, Chemistry, Physics and Sociology on the arXiv in November 2014 (they were subsequently published as Ref.[16].) The REF results were announced in December 2014. We revisited our study in January 2015 and found that our predictions failed to anticipate REF outcomes. For example, the Pearson and Spearman correlation coefficients between the REF-measured s values and the departmental h-index in Physics were 0.55 and 0.50, respectively. One submission which was ranked 27th in a certain subject area according to the departmental h-index, actually came in in seventh place in the REF.

We also sought to predict whether institutions would move up or down in the rankings between RAE2008 and REF2014. For Physics, the correlation between our predictions and the actual results was 0.26. For Chemistry it was 0.05 and for Biology it was -0.15. As commented later, managers would find better estimates of movement in the league tables by tossing dice. Our results are published as Ref.[17].

Journal of Physics: Conference Series **681** (2016) 012004 doi:10.1088/1742-6596/681/1/012004

4. Conclusions

In the first part of this review, we reported upon the development of a precise definition of a minimum *critical mass* for academic research. (In our earlier publications on the topic, we referred to this quantity as "lower critical mass".) Contrary to previous intuition-based notions, this is not a threshold value marking a step change between low- and high-quality research. In fact it is not directly perceptible or measurable. Instead it is connected, via a scaling relation, to the Dunbar number for academic research. (We previously referred to this as the "upper critical mass".) This marks an average cognitive limit to the number of intra-group research relationships one can sustain; in a large department, one cannot cooperate with everybody all of the time. The Dunbar number is discipline dependent; e.g., it is less than or equal to four for Pure Mathematics and approximately twenty-five for Experimental Physics on average (with a standard deviation of five).

We showed that concentrating more staffing resources into already-large research groups does not lift the quality of those groups significantly, especially if the discipline has a high Dunbar number. This is because everyone already has sufficient links to maximise cognitive activity. Adding more resources to such a group only serves to increase the volume of high-quality research. If one wishes to increase quality more significantly, more research staff should be allocated to medium-sized research groups. The standard is raised by introducing new links, as relationships between existing staff and the new arrival are established over time.

We noticed in Fig.1, and similar figures for other disciplines, that the best performing groups are not always the biggest. Frequently groups with size just in excess of the Dunbar number perform exceptionally well - they are large groups behaving as if they were medium. For example, in REF2008, Lancaster University with 26 Physics staff submitted, was ranked highest in terms of quality but Cambridge, with 141.25 staff, was largest. We suggested that these are the groups within which communication flows easiest and we suggested that these should form a guide for managers at other institutions to seek to emulate.

We then compared measures of quality of research coming from peer review to a citation-based measures. The poor correlations mean that such citation indicators *cannot* be used as proxies for peer. Nor should such indicators be used to rank university departments or groups into league tables.

However, the NCI and h-index compare well to the outcomes of peer review when one is interested in measuring the strength or power, defined as volume of quality. Since this is the basis on which quality-related funding is distributed, one could argue that a suitable metric may be developed to replace invasive exercises or reduce their cost. Dorothy Bishop goes a step further; she suggests to abolish the REF altogether and revert to a block-grant system that existed before the RAE. She argues that besides not being cost-effective, the REF has numerous adverse effects on academia.

On the basis of evidence and rigorous analyses produced over the past five years, we are inclined to agree, but only for well-established universities with large research groups of solid and long-standing repute. Research evaluation exercises have, however, been useful for smaller and medium groups as they seek to establish themselves in newer universities which have less developed research cultures. Some such universities have learned – through the RAE and REF – that curiosity-driven, pure academic research has value. We suggest that peer review (not metric-based!) exercises for such groups can and should continue. But they should be extended to include measurements of the extent to which such institutes promote academic values such as academic freedom and bottom-up research. In other words, peer review should be more focused on institutions which can actually benefit from it. Those which are established, and who enjoy long-standing high reputations amongst their peers internationally, should be allowed to get on with research. Fig.1(a) shows that very little separates such groups in terms of quality; in ranking such groups, one is essentially ranking noise.

Journal of Physics: Conference Series **681** (2016) 012004 doi:10.1088/1742-6596/681/1/012004

At the risk of stating the obvious, citation-based metrics measure citation counts, not quality. These are very different things. We suggest that on no account should one use metrics to decide policy or inform management. In our opinions, these pose a serious and current threat to curiosity-driven research which is the very *raison d'être* for universities.

Finally we suggest that it is incumbent upon academics themselves, who have the training and expert knowledge, to turn their methodologies on the scientific process itself in order to produce evidence to influence the powers that be. The UK recently commissioned an independent review of the role of metrics in research assessment and management. The remit was "to investigate the current and potential future roles that quantitative indicators can play in the assessment and management of research." The report, "The Metric Tide", was published in July 2015 [18]. It concludes that peer review continues to command widespread support as the main basis for evaluating research and there is legitimate concern that indicators, including citation counts and rankings, can be misused or "gamed." The report also concluded that no current metric can provide a replacement for peer review.

A legacy of the Report is the establishment of a forum for ongoing discussion of issues related to metrics. This aims to celebrate good practices but also to name and shame bad ones. Inspired by the "Bad Sex in Fiction" award, a "Bad Metric" prize will be awarded to the most egregious example of inappropriate use of quantitative indicators in research management [19]. This demonstrates the power of turning the evaluation process on itself

References

[1] Kenna R and Berche B 2010 EPL **90** 58002.
[2] Kenna R and Berche B 2011 Scientometrics **86** 527 - 540.
[3] Kenna R and Berche B 2012 IMA Journal of Management Mathematics **23** 195-207.
[4] Bishop D 2014 http://deevybee.blogspot.co.at/2013/01/an-alternative-to-ref2014.html. Accessed 01 November 2015.
[5] Evidence Thomson Reuters http://www.evidence.co.uk. Accessed 01 November 2015.
[6] Evidence Thomson Reuters 2010 *The future of the UK University research base* http://www.universitiesuk.ac.uk/Publications/Documents/UUK-FutureOfResearch-LiteratureReview.pdf.
[7] Evidence Thomson Reuters 2011 *Funding research excellence: research group size, critical mass & performance.*
[8] Harrison M 2009 in The *Question of R&D Specialisation: Perspectives and Policy Implications,* edited by D Pontikakis, D Kriakou and R van Baval, JRC Technical and Scientific Reports (European Commission, Luxembourg), 57-59.
[9] Davide Castelvecchi 2015 Nature **256** 179-181.
[10] *REF 2014 rerun: who are the 'game players'?*, https://www.timeshighereducation.com/features/ref-2014-rerun-who-are-the-game-players/2017670.article. Accessed 01 november 2015.
[11] Dunbar R I M (1992) Journal of Human Evolution **22** 469493.
[12] Ringelmann M 1913 Annales de l'Institut National Agronomique **2** 2-39.
[13] Farla K and Simmonds P (2015) *REF Accountability Review: Costs, benefits and burden* Report by Technopolis to the four UK higher education funding bodies http://www.hefce.ac.uk/pubs/rereports/Year/2015/refreviewcosts/Title,104406,en.html. Accessed 31.10.2015.
[14] Mryglod O, Kenna R, Holovatch Yu and Berche B 2013 Scientometrics **95** 115-127.
[15] Mryglod O, Kenna R, Holovatch Yu and Berche B 2013 Scientometrics **97** 767-777.
[16] Mryglod O, Kenna R, Holovatch Yu and Berche B 2015 Scientometrics **102** 2165-2180.
[17] Mryglod O, Kenna R, Holovatch Yu and Berche B 2015 Scientometrics **104** 1013-1017.
[18] The Metric Tide: Report of the Independent Review of the Role of Metrics in Research Assessment and Management (July 2015) http://www.hefce.ac.uk/pubs/rereports/Year/2015/metrictide/Title,104463,en.html. Accessed 01 November 2015.
[19] Responsible Metrics www.ResponsibleMetrics.org. Accessed 01 November 2015.

International Conference on Computer Simulation in Physics and Beyond 2015 IOP Publishing
Journal of Physics: Conference Series **681** (2016) 012005 doi:10.1088/1742-6596/681/1/012005

Spanning Tree Calculations on D-Wave 2 Machines

M.A. Novotny[†], **L. Hobl**[‡], **J.S. Hall**[†], **and K. Michielsen**[‡]

[†] Department of Physics and Astronomy, and Center for Computational Sciences, Mississippi State University, Mississippi State, MS 39762-5167, USA

[‡] Institute for Advanced Simulation, Jülich Supercomputing Centre, Forschungszentrum Jülich, D-52425 Jülich, Germany, and RWTH Aachen University, D-52056 Aachen, Germany

E-mail: man40@msstate.edu

Abstract. Calculations on D-Wave machines are presented, both for the 500-qubit and the 1000-qubit machines. Results are presented for spanning trees on the available $K_{4,4}$ Chimera graphs of both machines. Comparing trees of approximately the same size, the frequency of finding the ground state for the 1000-qubit machine is significantly improved over the 500-qubit older generation machine. Spanning trees are difficult problems for solution by adiabatic quantum computers, so the enhanced frequency of finding the ground state for newer machine generations and larger machines is encouraging for this immature technology.

1. Introduction

The availability of an Adiabatic Quantum Computer (AQC) with thousands of qubits would be a disruptive technology. In complexity theory the class P stands for polynomial time problems and NP for non-deterministic polynomial time problems [1]. We assume, although no proof is known, that P$\neq$NP. The reason AQC would be a disruptive technology is that in theory an AQC would enable solutions to problems that belong to the complexity class NP-hard, whereas all classical computers allow solutions to problems of the complexity class P. In theory any NP-hard problem can be reduced in polynomial time to a chosen NP-hard problem. Therefore, a machine that can solve any particular NP-hard problem would be a disruptive technology.

The goal of an AQC is to solve one particular NP-hard problem, namely finding the ground state of an Ising spin glass. The AQC operates at a low temperature, but the functionality of an AQC is easiest understood by considering the zero-temperature dynamics of the qubit system that makes up the AQC. A bit can be represented as a vector of length two, with the binary variable 0 represented as the vector (0 1) and the binary variable 1 represented as the vector (1 0). A qubit can be represented as a vector of length two of the form $(a_1\ a_2)$ with the normalization $a_1^*a_1 + a_2^*a_2 = 1$ where the a_i are in general complex numbers. In other words, a qubit is a superposition of the two classical binary bit values 0 and 1. An AQC works by initializing the vector of all N qubits as a direct product of the ground state of a Hamiltonian $\mathcal{H}_x$ of the Pauli matrix σ^x. The goal is to find the ground state of the Ising spin glass Hamiltonian $\mathcal{H}_z$, a function of the Pauli matrix σ^z. The Hamiltonian $\mathcal{H}_\lambda=\lambda\mathcal{H}_x+(1-\lambda)\,\mathcal{H}_z$ is adiabatically adjusted by changing the variable λ with time. At time t=0 one starts with λ=1, and at the end of the annealing cycle one has λ=0. After the annealing cycle the vector is measured to provide a ground state of $\mathcal{H}_z$.

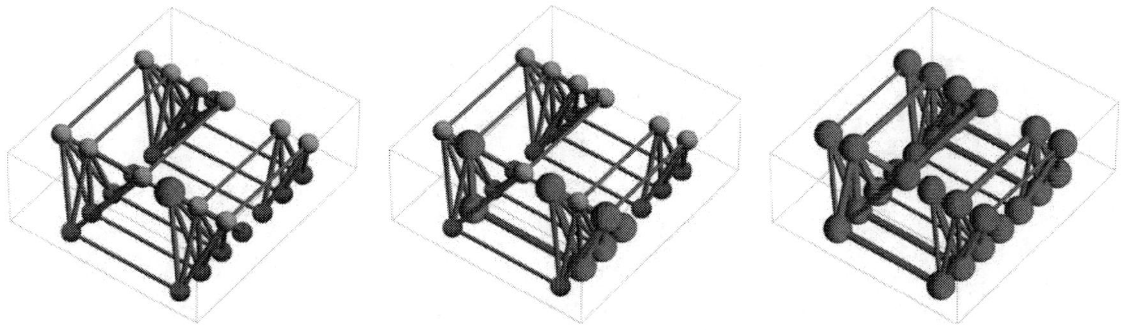

Figure 1. An $L = 2$ $K_{4,4}$ Chimera graph $\mathcal{G}$ is shown. The graph $\mathcal{G}$ is represented as a bi-layer system, with blue nodes in the bottom layer and green nodes in the top layer. Each complete $K_{4,4}$ unit cell has four green nodes and four blue notes, with every blue node connected to every green node. Also shown are three stages of the construction of a spanning tree on $\mathcal{G}$. The initial graph with the chosen tree root (red), the tree with seven of the of the 27 sites occupied by the growing tree, and the spanning tree with all 27 sites occupied.

Obtaining the ground state of an Ising spin glass is an NP-hard problem. Consider a connected undirected graph $\mathcal{G}$ with N vertices (nodes) with an Ising variable $s_j = \pm 1$ placed on each vertex. The ground state of the Ising spin glass is the state $\{s\}$ of Ising spins on $\mathcal{G}$ that minimizes the energy function [2]

$$E = -\sum_{i=1}^{N-1} \sum_{j=i+1}^{N} J_{i,j} s_i s_j - \sum_{j=1}^{N} h_j s_j \tag{1}$$

for fixed bias fields h_j and fixed bond weights $J_{i,j}$. If there is no edge in $\mathcal{G}$ connecting vertices i and j then $J_{i,j}$=0. The $J_{i,j}$ and h_j are the parameters in the Hamiltonian $\mathcal{H}_z$. If all bonds are ferromagnetic, $J_{i,j} \geq 0$, and all bias fields are zero, the ground state is degenerate, as both the state with all $s_j = +1$ and the state with all $s_j = -1$ have the same energy. An Ising spin glass requires that there is frustration in the model, in other words that it is not possible to place all spins on $\mathcal{G}$ in such a fashion that every term in the sums in Eq. (1) is positive.

If P$\neq$NP, then in the worst-case instance the solution of an NP-problem requires an exhaustive search of all possible states. For the Ising spin glass an exhaustive search requires calculating the energy in Eq. (1) 2^N times. This can be easily programmed in parallel for a Single-Instruction-Multiple-Data (SIMD) classical computer. The steps of the SIMD algorithm, assuming 2^N processors, are:

- Broadcast the same $J_{i,j}$ bond weights and h_j bias fields to each processor
- Scatter one configuration of the Ising variables s_j to each processor
- Each processor calculates the energy of Eq. (1) for its configuration of the Ising variables
- Gather the energies from each processor, and keep only the configuration(s) with the lowest energies.

The comparison (last) step can be performed with the order of the number of operations $\mathcal{O}(N)$. Therefore the number of computations required is given by the rate limiting step, the third step. The number of calculations which should be performed is $\mathcal{O}\left(N^2 2^N\right)$, since each energy calculation requires $\mathcal{O}(N^2)$ computations.

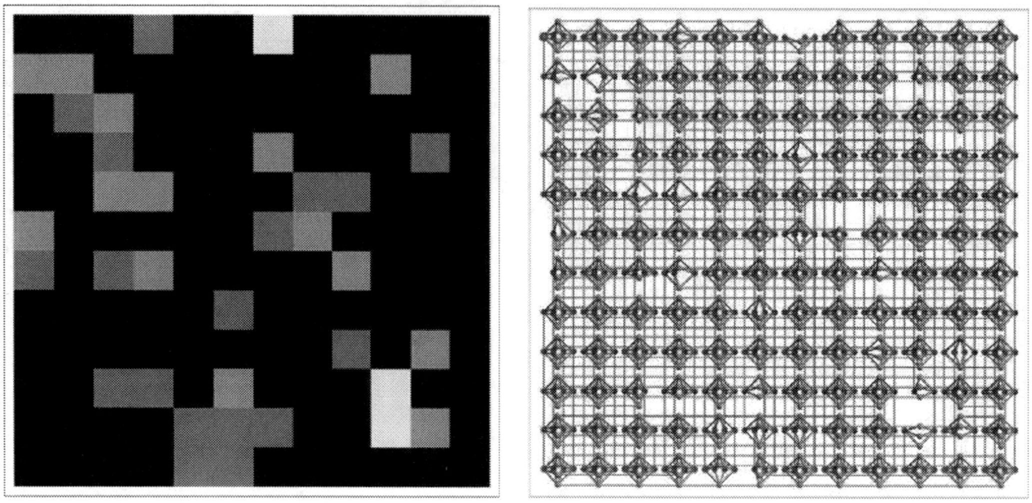

Figure 2. Configuration of the D-Wave 2X used to study spanning trees on 1000+ qubits. [Left] Each square is a $K_{4,4}$ unit cell of eight qubits. The chip is a 12×12 square array of unit cells (with free boundary conditions). The colors correspond to the number of functioning qubits in the unit cell: black=8, gray=7, pink=6, and yellow=5. [Right] The D-Wave 2X graph for the $K_{4,4}$ Chimera graph $\mathcal{G}$ used in this study, shown as a bilayer system. The blue qubits are in the bottom layer and the green qubits in the top layer. The chip has 1097 available qubits (1097 vertices in $\mathcal{G}$). The qubit labeled as zero on the D-Wave 2X is shown in red (top left).

The company D-Wave, Inc [3] recently announced a 1000-qubit machine [4]. The reason the SIMD algorithm above cannot be used in principle is that $2^{1000} \approx 10^{301}$, while the estimated number of particles in the visible universe is 'only' about 10^{80}. Clearly the SIMD algorithm is impracticle for 1000 vertices in $\mathcal{G}$. An ideal AQC can perform all 2^N calculations of the energy in parallel because of principles of quantum mechanics, in particular coherence, superposition, and tunneling [5]. Quantum mechanics is probabilistic, and hence any AQC is a probabilistic machine. An ideal AQC is possible only in the unphysical limits of the machine operating at zero degrees Kelvin (absolute zero) and taking an infinite time to perform the required adiabatic calculation. Therefore the question that needs to be addressed for any practical AQC is whether it provides groundstates with a high enough frequency of occurrence that it is useful to calculate NP-hard problems.

This conference proceedings article, based on a Sept. 9, 2015 plenary talk at the *International Conference on Computer Simulation in Physics and beyond* in Moscow, Russia, seeks to address the practical question of applicability of D-Wave machines. The results of that talk, as well as of this paper, include only preliminary results for the 1000 qubit D-Wave 2X machine due to the short amount of time the machine has been available. Nevertheless, even the preliminary results can shed light on the functionality of the D-Wave 2X.

The approach we have taken to test the D-Wave machines is to use spanning trees on the D-Wave graph $\mathcal{G}$. Section 2 details the graph $\mathcal{G}$ of the D-Wave machines. Section 3 presents results of placing ferromagnetic bonds on the 1000 qubit D-Wave 2X. Section 4 describes spanning trees on D-Wave machines. Results of solving spanning tree problems on D-Wave machines are presented in Section 5 for ferromagnetic bonds and in Section 6 for random ±1 bonds. Section 7 contains a discussion, outlook, and conclusions.

2. D-Wave Machines

The architecture of the D-Wave chip is an $L \times L$ $K_{4,4}$ Chimera lattice [3, 4]. The number of qubits on the chip is $N = 8L^2$. The Chimera lattice can be represented as a bilayer system, as shown in Fig. 1 for an L=2 lattice. Some of the qubits on a chip are not accessible to the programmer, so the $K_{4,4}$ Chimera lattice has disorder, just as depicted in Fig. 1.

In Fig. 1 the graph $\mathcal{G}$ is taken to be a $K_{4,4}$ Chimera lattice, composed of a 2×2 arrangement of unit cells. The graph is represented as a bilayer system. Each unit cell has four vertices in the top layer (green vertices) and four vertices in the bottom layer (blue vertices). In each unit cell the top four vertices are each connected (brown edges) to every one of the four vertices in the bottom layer. The vertices in the bottom layer are connected (blue edges) to corresponding bottom layer vertices in nearest neighbor unit cells. Similary, the vertices in the top layer are connected (gray edges) to corresponding top layer vertices in nearest neighbor unit cells. The graph $\mathcal{G}$ in Fig. 1 has disorder, as only 27 of the possible $2^2 8$=32 vertices are present. If a vertex from the full $K_{4,4}$ Chimera lattice is missing, all of its associated edges are also missing in $\mathcal{G}$. The D-Wave machines have the same $K_{4,4}$ structure as in Fig. 1, but are on an 8×8 arrangement of unit cells for the 500 qubit machine (the total number of qubits is 8^3=512) and on a 12×12 arrangement of unit cells for the 1000 qubit machine (the total number of qubits is $12^2 8$=1152).

Results from three different D-Wave chips will be reported. Two of the chips are last generation 512 qubit chips. The number of functioning qubits is 496 of one chip [7] and of the other is 476 qubits. Both chips were operated at the company headquarters in Burnaby, British Columbia, Canada, and were so-called in-house 'test' chips. The 1152 qubit D-Wave 2X chip actually had 1097 accessible qubits. This computer is located at the Ames Research Center located at Moffett Field in California, USA. The D-Wave machine is managed by a collaboration of NASA (National Aeronautics and Space Administration) [8], USRA (Universities Space Research Association) [9], and Google [10].

The configuration of the 1000 qubit D-Wave 2X is shown in Fig. 2. Of the $L^2 = 144$ $K_{4,4}$ unit cells, 109 have all eight qubits available. The details of the number of available qubits in the unit cells are in Table 1. Any qubit j which is not available results in all the $J_{i,j}$ connections not available. In addition, there is one unavailable $J_{i,j}$ connection between two available qubits. This makes that there are 3060 bonds available from the 3360 bonds in an undisordered $12^2 K_{4,4}$ Chimera lattice. The bilayer representation of the graph of the 1097 qubit machine is shown in Fig. 2. In Fig. 2(Right) the qubit in red (top left) marks the qubit that is assigned to be qubit number zero. The D-Wave 2X is an analog machine, and has a 3% analog control error on the h_j and the $J_{i,j}$.

Total qubits in a unit cell	8	7	6	5
144	109	18	14	3

Table 1. The number of available qubits in each unit cell of the 1000 qubit D-Wave 2X.

3. Ferromagnetic Computations on a 1000 Qubit D-Wave Machine

The first test we performed on the 1097 qubit D-Wave 2X was to place $J_{i,j}$=1 on all available bonds, resulting in overall ferromagnetic bonds and to put the same bias fields h_j on each qubit. One thousand anneals were performed for each value of h_j (set to be uniform). All anneals found the expected ground state, each having the energy E=−3060. For zero field the two ground states with all s_j=+1 and s_j=−1 should be found equally often. However, as shown in Table 2 only one of the two ground states was found. By gradually changing the uniform value of the bias field h_j on each qubit, for the bias field h_j=0.07 both ground states were found. For

Journal of Physics: Conference Series **681** (2016) 012005 doi:10.1088/1742-6596/681/1/012005

the D-Wave 2X there is a 3% analog control error, so quantitatively the results in Table 2 are reasonable.

h_j	0.00	0.06	0.07	0.08
s_j=−1	1000	1000	823	0
s_j=+1	0	0	177	1000

Table 2. The number of times each of the two ground states were found for different applied fields h_j for ferromagnetic $J_{i,j}$=1 bonds.

4. Spanning Trees

A tree is a graph with no loops. A spanning tree of $\mathcal{G}$ is a subgraph of $\mathcal{G}$ that has a tree structure and that includes all N vertices of $\mathcal{G}$. We have generated our spanning trees using an algorithm of A. Broder [6]. The algorithm has an expected run time of $\mathcal{O}\left(N\log N\right)$ for almost all graphs, and a run time per generated tree of $\mathcal{O}\left(N^3\right)$ for the worst graphs. Three steps during a run of the algorithm are sketched in Fig. 1.

The algorithm to find a spanning tree is

- Choose one vertex of $\mathcal{G}$ uniformly at random. In Fig. 1(left) the chosen vertex is red. Let this vertex be labeled by j.
- Repeat the steps below until all N vertices belong to the tree:
  - Pick an edge uniformly at random that leaves the vertex j
  - Walk along the chosen edge to a vertex labeled as k
  - If this is the first time vertex k has been visited, add vertex k and the edge j,k to the growing tree. In Fig. 1(center) this is shown by coloring the vertices and edges red.
  - Change the label j to the label k, whether or not the vertex k was visited for the first time.

In other words, a random walk starting at any vertex is generated, and the first time a vertex is visited it is added to the tree and the edge that led to this first visit of the vertex is added to the tree. The algorithm stops when all vertices belong to the tree [see Fig. 1(right)]. Any spanning tree has $N-1$ edges. The spanning trees generated by this algorithm, for example the spanning tree in Fig. 1, are those drawn uniformly from the distribution of all spanning trees of $\mathcal{G}$ [6].

One way to test an AQC machine is to give problems to the machine for which one knows the answer. This strategy has been followed by others in terms of chains of qubits [11] or frustrated optimization problems with tunable hardness [12]. The spanning tree planted graphs have the advantage that:

- All vertices are included in every spanning tree.
- The ensemble of spanning trees covers all edges.
- The ensemble of spanning trees is well defined.
- Spanning trees exist for any connected graph $\mathcal{G}$.
- The ground state for a spanning tree is known. For example, if all h_j=0 and all $J_{i,j}$=±1 in the tree, then the ground state energy is the negative of one less than the number of edges in the spanning tree, $E=-N+1$.
- The spin arrangement of the ground state of the spanning tree is easy to calculate.
- Obtaining a ground state for a spanning tree problem is a difficult problem for an AQC.

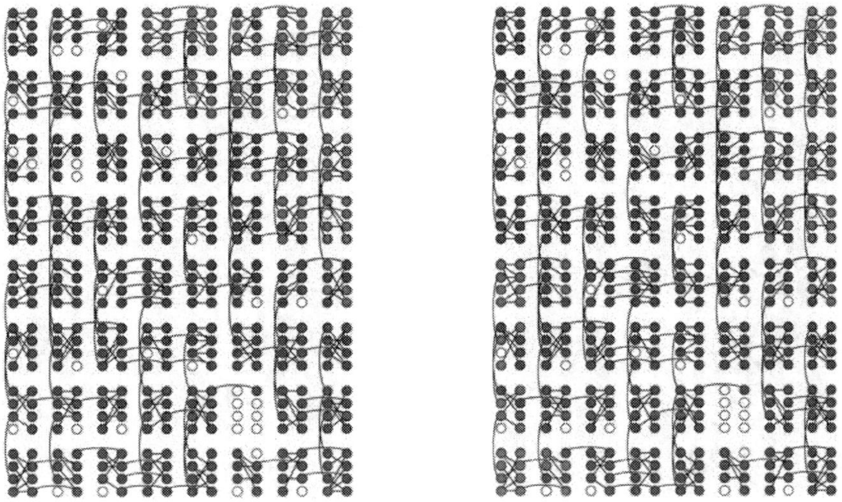

Figure 3. Two solutions of a ferromagnetic spanning tree on the 476 qubit D-Wave 2. The 8×8 arrangment of unit cells are represented with all qubits in a single plane, with blue qubits +1 and red qubits −1. All available bonds have non-zero weight $J_{i,j}$=1, and only these bonds are drawn (black lines and curves). The ground state should have all qubits in the same state (all qubits should be either blue or all should be red). In order to bias which ground state has the minimum energy, a non-zero bias field h_{452}=−1 was placed on qubit number 452 (at the bottom-left), while all other bias fields were zero. This bias should have forced all qubits to be s_j=−1 (red). This is not true for either solution of this spanning tree.

Some results for spanning trees on a 500 qubit D-Wave machine have recently been reported [7]. There are two disadvantages of using spanning trees to test AQC machines. The trees do not contain any loops, and hence there is no frustration in the model, in other words the problem is in P not NP. The NP-hard problems that an AQC should solve contain P problems like the spanning tree, but they are designed to provide answers to the NP-hard problem of finding the ground state of the Ising spin glass. Nevertheless, the spanning trees give a feeling for how much to trust the answers obtained from an AQC.

5. Ferromagnetic Spanning Trees on D-Wave Machines

On the 476 chip we placed one spanning tree. Figure 3 shows the spanning tree on the 476 qubit chip, here represented with all qubits in a single plane. The 8×8 arrangement of unit cells is shown, with only bonds in the spanning tree shown. Note that the spanning tree graph as represented here is not drawn in a planar fashion, even though any spanning tree can be drawn as a planar graph with non-crossing bonds. The $K_{4,4}$ Chimera graph is shown in a way the D-Wave 2 presents the qubits on the web interface. All bonds were ferromagnetic, $J_{i,j}$=+1.

We examined the number of times each energy E was found following the anneal cycle. All bias fields were set to zero. A total of 10^3 submissions, each with 10^3 anneals, was performed. The ground state, with energy E_{gs}=−475, was only found 237 times out of the 10^6 anneals. Fig. 4 shows the occurrence of each energy for this particular spanning tree.

Since all bias fields were set to zero, the ferromagnetic tree for the 476 qubit chip should find the two ground states the same number of times. In that case the magnetization averaged over all anneals should be $\langle m_j \rangle$=0. Alternatively if only one state should be found, as in the ferromagnetic case of Sec. 4, then for the ground state all qubits should have the same $\langle m_j \rangle$. For

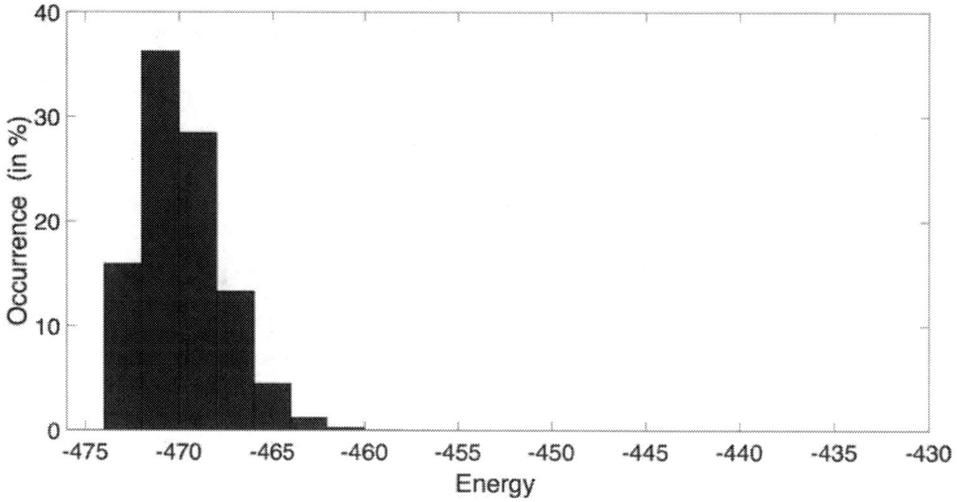

Figure 4. Occurrence in % of the energy found for a 476 qubit D-Wave 2 for a ferromagnetic spanning tree. The tree was submitted 10^3 times each with 10^3 anneals. The ground state has an energy $E_{\rm gs} = -475$ and has a relative occurrence frequency of 0.02%.

the 476 qubit machine, as seen in Fig. 5, neither of these assumptions are true. The behavior of $\langle m_j \rangle$ may be understood as follows. The analog nature of the D-Wave machine may leave a non-zero field h_j on each qubit even when the h_j value has been set to zero. This slightly different from zero field for each qubit is sufficient to decrease the number of times the ground state is found.

In order to have a non-degenerate ground state, we set all bias fields to $h_j = 0$, except for qubit number 452, which was set to $h_{452}{=}{-1}$. The bias field should have enabled the D-Wave 2 to find the ground state with all $s_j{=}{-1}$ (red in Fig. 3). For this chip and spanning tree, not even the qubit number 452 was forced to be in the $s_j{=}{-1}$ (red) state. Neither the ground state energy ($E_{\rm gs}{=}{-476}$) nor the first excited state was returned by the D-Wave machine. Fig. 3 shows two representative returned low-energy configurations, both of a second excited state (energy $E_2{=}{-472}$). Fig. 3 shows that this chip seems to be inhomogeneous, since $h_{452}{=}{-1}$ did not bias it to find the non-degenerate ground state. This chip may not have been fine tuned, because it was an in-house 'test' chip.

The analog nature of the D-Wave 2 presents certain issues in terms of programming the machine. One question one could ask is whether setting bonds to zero disconnects the two qubits. Because of the 3% analog setting, the degree of isolation should be tested. We have considered the 476 qubit chip to be four $L{=}4$ $K_{4,4}$ Chimera lattices. A missing qubit on any $L{=}4$ Chimera quadrant was replicated by setting the $J_{i,j}{=}0$, leading to four identical graphs $\mathcal{G}_{L=4}$ on each quadrant. The same ferromagnetic spanning tree was placed on each quadrant. As seen in Fig. 6 each quadrant was found to obtain the ground state a different fraction of time. The success frequency of each quadrant was labeled as P_1, P_2, P_3, and P_4. Assuming the success frequency of finding the ground state in each quadrant was independent of the ground state found in the other quadrants, the frequency of simultaneously finding the ground state in quadrants 1 and 4 would be

$$P_{1\,4} = P_1\,P_4\,. \tag{2}$$

Similarly for any two quadrants. Fig. 6 shows results for $P_{1\,2}$, $P_{1\,3}$, $P_{1\,4}$, $P_{2\,3}$, $P_{2\,4}$, and $P_{3\,4}$. The

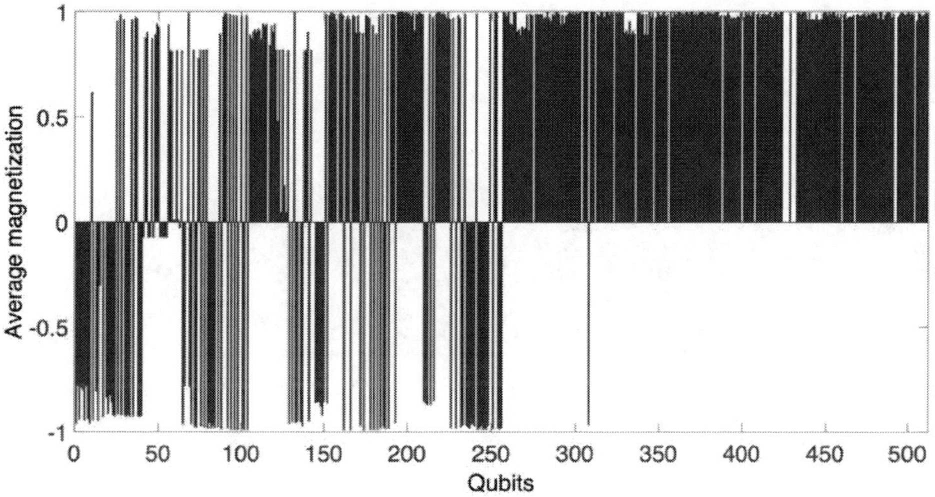

Figure 5. The average magnetization $\langle m_j \rangle$ for a 476 qubit D-Wave 2 for a ferromagnetic spanning tree. The tree was submitted 10^3 times each with 10^3 anneals. For a system with no bias, the average magnetization should be zero.

measured frequency of success (red bars in Fig. 6) correspond within the measurement errors to the predicted results. Similarly for simultaneously finding the ground state in three quadrants, labeled $P_{1\,2\,3}$, $P_{1\,2\,4}$, $P_{1\,3\,4}$, and $P_{2\,3\,4}$. The frequency of simultaneously finding the ground state in all four quadrants is $P_{1\,2\,3\,4}$. If the probability of finding a ground state in each quadrant is independent, $P_{1\,2\,3\,4} = P_1 P_2 P_3 P_4$. Fig. 6 compares the measured frequencies (red bars) to the expected probabilities obtained by multiplying the frequency of finding the ground state in each quadrant (right yellow bars). The conclusion is that within the measurement errors the ground state in each quadrant is found to be independent of what is occuring in the other quadrants.

6. Spanning Tree Computations with ±1 Bonds on D-Wave Machines

We performed spanning tree calculations on $L{\times}L$ $K_{4,4}$ Chimera subgraphs of two different chips, the one with 496 qubits and the one with 1097 qubits. The bond weights were uniformly randomly chosen to be $J_{i,j}{=}{\pm}1$. All magnetic fields h_j were set to zero. The $L{\times}L$ subgraphs were chosen to all include the $K_{4,4}$ unit cell in the upper-left corner of the chip. This choice is similar to the gauge transformations used by others [13] to remove local biases on the chip. Results for the 496 qubit machine were published previously [7]. We submitted a number of spanning trees for each L, and each submission requested 10^3 ground states (10^3 anneal cycles for the given tree and $J_{i,j}$ values). Each tree was submitted with different $J_{i,j}$ values up to 10^2 times. The probability the D-Wave 2 returned the ground state, $p_{\rm gs}$, was estimated by the frequency of occurrence of success from all anneals. Results are shown in Table 3 and Fig. 7.

The D-Wave machine can be modeled to a first approximation as an uncorrelated probabalistic machine. Let the random variaible be $X{=}1$ if the ground state solution is returned in a particular anneal and $X{=}0$ otherwise. The expectation value for X is

$$\mathcal{E}\,(X) \;=\; 1\cdot p_{\rm gs} + 0\cdot(1-p_{\rm gs}) \;=\; p_{\rm gs}\,. \tag{3}$$

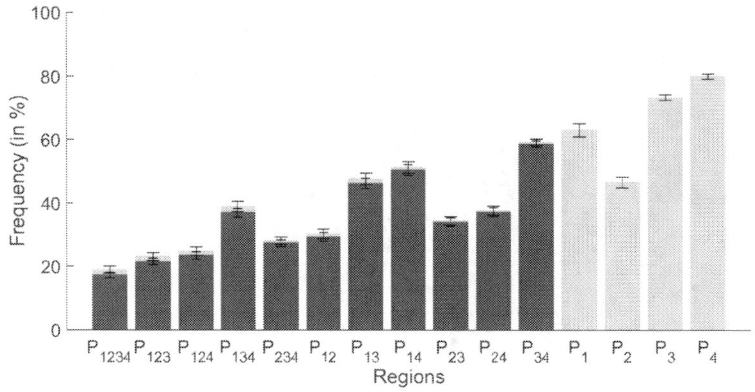

Figure 6. Results on the 476 qubit D-Wave 2. The success frequencies of finding the ground state for each $\mathcal{G}_{L=4}$ quadrant is labeled as P_1 through P_4 (yellow bars on right). The success frequency of simultaneously finding the ground state in two quadrants m and m' is labeled as $P_{m\,m'}$ (red middle bars) and should be equal to $P_{m\,m'}{=}P_m P_{m'}$ (the prediction of the superimposed yellow bars) if the solution frequencies are independent of each other. Similarly for simultaneously finding the ground state for three quadrants, and for all four quadrants (red left bar labeled P_{1234}).

Similarly, $\mathcal{E}\left(X^2\right) = p_{\rm gs}$. Therefore the standard deviation σ is

$$\sigma^2 = \mathcal{E}\left(X^2\right) - \left[\mathcal{E}\left(X\right)\right]^2 = p_{\rm gs}\left(1 - p_{\rm gs}\right) . \tag{4}$$

The standard error from the mean is thus estimated to be $\sigma/\sqrt{M-1}$ when there are M samples. The number of anneals are not uniform for the 496 qubit chip due to program or chip or network issues. There are two ways of deciding on the value of M. One way would be to use $M{=}M_{\rm anneal}$ to be all anneal requests for a given L. The other way would be to use $M{=}M_{\rm tree}$, the number of different spanning trees submitted for each L. The error estimate with $M{=}M_{\rm tree}$, with each tree submitted with up to 10^2 different $J_{i,j}{=}\pm1$ and each submitted for 10^3 anneal cycles, is shown in Fig. 7 and listed in Table 3.

Two striking results are present in Fig. 7 and Table 3. The first is that the $8{\times}8$ $K_{4,4}$ Chimera lattice results are very different for the 496 qubit D-Wave 2 and the 1097 qubit D-Wave 2X. A marked improvement in the success frequency of finding the groundstate, $p_{\rm gs}$, is obtained for the newer generation machine. Some of these differences may be due to the smaller errors in setting the values of the h_j and $J_{i,j}$ on the D-Wave 2X compared with the D-Wave 2.

The second striking result is the small frequency of success for the full graphs, particularly for the 1097 qubit D-Wave 2X. As described above, a spanning tree is a difficult problem for any AQC to solve. That the ground state is found at all for the full graph, even with this low average success frequency is encouraging.

Assume that the probabilities of success on any anneal cycle are independent random variables, and are the same for every tree. Consequently the probability of not finding the ground state is $1 - p_{\rm gs}$. For M measurements the probability of not finding the ground state is $\left(1 - p_{\rm gs}\right)^M$. The probability of finding the ground state after M measurements is

$$P_{\rm gs} = 1 - \left(1 - p_{\rm gs}\right)^M . \tag{5}$$

Journal of Physics: Conference Series **681** (2016) 012005 doi:10.1088/1742-6596/681/1/012005

# qubits	L	# trees	Million anneals	$p_{\rm gs}$	standard error
496	2	100	10.0	1.0000	0.0
496	3	85	8.4	0.9999	0.0011
496	4	38	3.8	0.9779	0.0242
496	5	100	10.0	0.8953	0.0308
496	6	100	7.9	0.7335	0.0444
496	7	80	10.0	0.5889	0.0554
496	8	100	10.0	0.3261	0.0471
1097	7	100	10.0	0.9944	0.0075
1097	8	100	10.0	0.8443	0.0364
1097	9	100	10.0	0.5800	0.0496
1097	10	100	10.0	0.3444	0.0478
1097	11	100	10.0	0.2017	0.0403
1097	12	100	10.0	0.0443	0.0207

Table 3. Details of the results plotted in Fig. 7. The number of spanning trees as well as the total number of annealing cycles is shown for each $K_{4,4}$ Chimera size, together with the frequency a ground state was found and the calculated standard deviation of the results. The spanning trees all had h_j=0 and randomly chosen bond weights $J_{i,j}=\pm 1$.

Thus the number M of measurements required to obtain a given frequency $P_{\rm gs}$ that the ground state was found is

$$M = \frac{\ln(1 - P_{\rm gs})}{\ln(1 - p_{\rm gs})} \,. \qquad (6)$$

Every submission to the D-Wave by default requests 10^3 anneal cycles. Even for an extremely small average frequency of success $p_{\rm gs}$=0.0005, the number of measurements required to find the ground state 99.99% of the time would require about 18 submissions. However, we find that the 1097 qubit D-Wave 2X probability of successfully finding the ground state is most probably not independent of the particular spanning tree. We have performed 10^6 anneal cycles (10^3 submissions with different $J_{i,j}=\pm 1$) on at least one spanning tree without the D-Wave returning the ground state.

7. Discussion, Outlook, and Conclusions

We have performed tests of D-Wave adiabatic quantum computers (AQCs) by detailing the results of placing spanning trees on the underlying $L \times L$ $K_{4,4}$ disordered Chimera graph of the D-Wave machines. Other researchers have put planted solutions on D-Wave machines [12]. A spanning tree test has a number of advantages over other planted solutions on AQCs. First, spanning trees are a difficult problem for any AQC machine. The other advantages of spanning trees are that they work on any graph, each tree includes all qubits, the ensemble of spanning trees include all couplings $J_{i,j}$, and even for large graphs it is easy to obtain a spanning tree drawn uniformly from the ensemble of all spanning trees. The disadvantage is that spanning trees have no loops, and hence no frustration is present.

The comparison of the two generations of D-Wave machines is encouraging. Spanning trees for 8×8 $K_{4,4}$ Chimera graphs were obtained both on the 512 qubit D-Wave 2 (with 496 functioning qubits) and the 1152 qubit D-Wave 2X (with 1097 functioning qubits). The frequency of finding the ground state for a tree with one anneal increased from $p_{\rm gs}$=0.33 for the D-Wave 2 to $p_{\rm gs}$=0.84 for the D-Wave 2X, both for the L=8 Chimera lattice. Some of this increase may be due to the

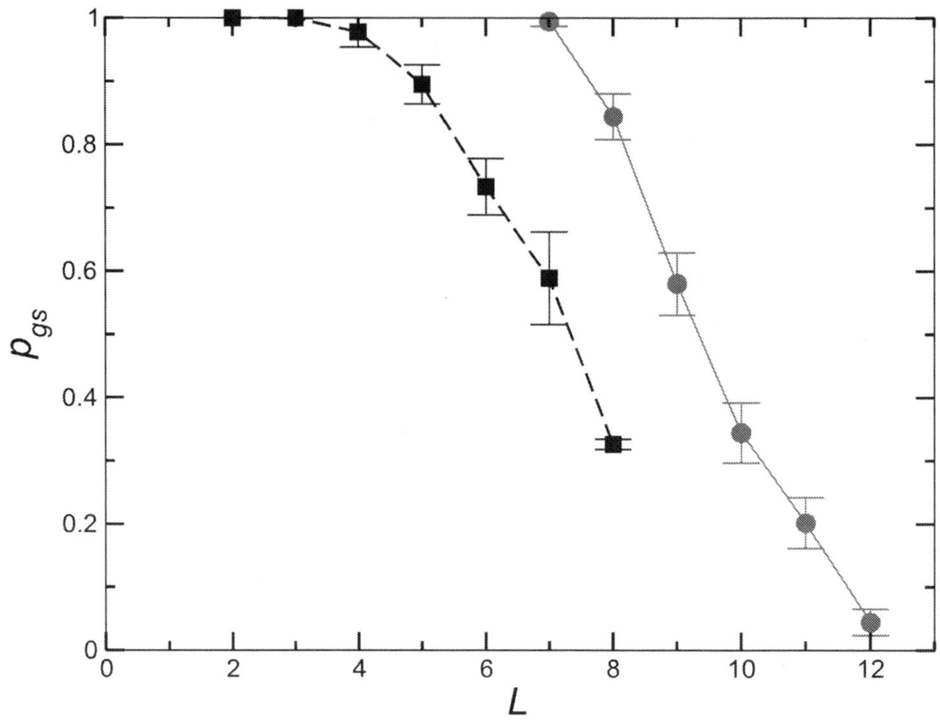

Figure 7. Success frequency of finding the ground state as a function of the size of the Chimera $K_{4,4}$ subgraph. The data are for the D-Wave 2 with 496 qubits (black squares) and the D-Wave 2X with 1097 qubits (red circles). The error bars for the success frequency are calculated as described in the text. The dashed and solid lines are guides for the eye. The results are tabulated in Table 3.

increased accuracy of setting the h_j and the $J_{i,j}$ parameters on the D-Wave 2X.

AQCs hold the promise that they can find the ground state of an Ising spin glass, and in the unphysical limits of operating at zero degrees Kelvin and taking an infinitely long time for each anneal you are gauranteed to obtain the ground state. Classical computers and advanced algorithms can obtain solutions that are close to the Ising spin glass ground state, and this property has been used to benchmark a D-Wave 2X machine [14]. The property of obtaining *the* global minimum is what, in the authors opinion, makes AQCs a disruptive computer technology. The ability of AQCs to find the solution of one particular NP-hard optimization problem, *the* ground state of an Ising spin glass, means that in principle other NP-hard problems can be solved by mapping them onto an Ising spin glass NP-hard problem. Any (non-ideal) AQC is a probabilistic machine, since it operates using quantum mechanical principles and quantum mechanics is a theory that only allows an observer to make predictions about probabilities of measureable quantities.

In this paper we only deal with spanning trees, so the number of edges between qubits is one, and there are no loops. This could also be called a tree with a tree width of one. Numerical transfer methods have been used to study Ising systems in two and three dimensions [15, 16] as well as in non-integer dimensions [17, 18], including studies of metastable phenomena [19, 20]. It would be possible to utilize numerical transfer matrix methods to study spanning trees with different widths. For example, one could place a tree of width four on the 12^2=144 square array of $K_{4,4}$ unit cells shown in Fig. 2. The size of the requried transfer matrix, since every unit cell would connect to its (at most two) neighboring unit cells with four $J_{i,j}$ connections can be obtained by multiplying matrices of size $2^4 \times 2^4$ (after utilizing generalized matrix products to exactly renormalize the tree branches [21, 22]). These types of numerical transfer matrix calculations would allow loops within the subgraphs, and hence can be used to study the model with frustration. Numerical transfer matrix methods have been used previously to study Ising glasses on regular lattices [23, 24, 25, 26] and on hierarchical lattices with non-integer fractal dimensions [26].

The question to be asked is whether the glass is half-full or half-empty in terms of usability of the current generation of D-Wave machines.

The glass-half-full outlook for AQCs is the promise given by Eq. (6). Even for a very low average success frequency, say $p_{\rm gs}$=0.0005, if you want to find the ground state 99.99% of the time, you only need to submit the tree to the D-Wave 2X eighteen times with the default 10^3 anneals per submission. Hence, AQCs are a disruptive technology as no known classical computer with advanced algorithms could make a similar claim for the $2^{1097} \approx 1.7 \times 10^{330}$ state space of the D-Wave 2X machine. The fact that this paper utilizes a D-Wave machine with more than 1000 qubits while published work on D-Wave machines with about 100 qubits appeared in print in 2014 [11, 13] shows the speed at which AQC is becoming a reality.

The glass half-empty outlook for AQCs is that for the 1097 qubit D-Wave 2X the ground state average success frequency for the spanning trees studied was only found on the 1097 qubit D-Wave 2X with an average frequency of $p_{\rm gs}$=0.0443. Furthermore, there was at least one spanning tree on the 1097 D-Wave 2X that did not find the ground state after 10^3 submissions (10^6 anneal cycles). One might expect that general spin-glass problems on the D-Wave 2X may have a similar difficulty, namely some ground states are much more difficult for the D-Wave 2X to find than are other ground states. There is still work to do to quantify whether the conclusions based on defining and attempting to detect quantum speedup in a D-Wave machine with 503 qubits [27] hold when applied to a 1000 qubit D-Wave 2X.

Acknowledgements

MAN is supported in part by NSF grant DMR-1206233. JSH is supported in part by Pacific Northwest Laboratories. The authors thank D-Wave Systems, Inc for access to the 500 qubit D-Wave 2 machines. Time on the D-Wave 2X was granted through USRA (University Space Research Association).

References

[1] C.H. Papadimitriou and K. Steiglitz, *Combinatorial Optimization: Algorithms and Complexity* (Prentis-Hall, New Jersey, 1982), Dover edition published 1998.
[2] D.L. Stein and C.M. Newman *Spin Glasses and Complexity* (Princeton University Press, Princeton, NJ, 2013).
[3] D-Wave Systems, Inc, http://www.dwavesys.com
[4] "Quantum chip doubles the qubits", Science, volume 349, page 906 August 28, 2015.
[5] E.G. Rieffel and W.H. Polak, *Quantum Computing: A Gentle Introduction* (MIT Press, Cambridge, MA, 2011).
[6] A. Broder "Generating random spanning trees", in FOCS, pages 442-447 (1989).
[7] J.S. Hall, M.A. Novotny, T. Neuhaus, and K. Michielsen, "A study of spanning trees on a D-Wave quantum computer", *Physics Procedia*, volume 68, pages 56-60 (2015).
[8] NASA Quantum Artificial Intelligence Laboratory (QuAIL), http://www.nas.nasa.gov/quantum/quantumcomp.html
[9] Quantum computing at USRA, http://www.usra.edu/quantum/
[10] Google quantum A.I. lab, https://plus.google.com/+QuantumAILab/posts
[11] K.L. Pudenz, T. Albash, and D.A. Lidar, "Error-corrected quantum annealing with hundreds of qubits", Nature Communications **5**, article 3243 [10 pages] (2014).
[12] I. Hen, J. Job, J. Job, T.F. Røoonw, M. Troyer, and D. Lidar, "Probing for quantum speedup in spin glas problems with planted solutions", arXiv:1502.01663v2.

[13] S. Boizo, T.F. Rönnow, S.V. Isakov, Z. Wang, D. Wecker, D.A. Lidar, J.M. Marinis, and M. Troyer, "Evidence for quantum annealing with more than one hundred qubits", Nature Physics **10**, pages 218-224 (2014).

[14] J. King, S. Yarkoni, M.M. Nevisi, J.P. Hilton, and C.C. McGeoch, "Benchmarking a quantum annealing processor with the time-to-target metric", arXiv:1508.05087.

[15] M.P. Nightingale, "Transfer matrices, phase transitions, and critical phenomena: numerical methods and applications", in *Finite Size Scaling and Numerical Simulations of Statistical Systems*, edited by V. Privman, pages 287-351 (World Scientific, Singapore, 1990).

[16] M.P. Nightingale and H.W.J. Blöte, "Transfer-matrix Monte Carlo estimates of critical points in the simple-cubic Ising, planar, and Heisenberg models", Physical Review B **54**, pages 1001-1008 (1996).

[17] M.A. Novotny, "Numerical transfer matrix study of the Ising model between one and two dimensions", Europhysics Letters **17**, pages 297-302 (1992); Erratum **17**, page 297 (1992).

[18] M.A. Novotny, "Critical exponents for the Ising model between one and two dimensions", Physical Review B **46**, pages 2939-2950 (1992).

[19] C.C.A. Günther, P.A. Rikvold, and M.A. Novotny, "Numerical transfer-matrix study of metastability in the d=2 Ising model", Physical Review Letters **71**, pages 3898-3901 (1993).

[20] T. Fiig, B.M. Gorman, P.A. Rikvold, and M.A. Novotny, "Numerical transfer-matrix study of a model with competing metastable states", Physical Review E **50**, pages 1930-1947 (1994)

[21] M.A. Novotny, "Matrix products with applications to classical statistical mechanics", Journal of Mathematical Physics **20**, pages 1146-1150 (1979).

[22] M.A. Novotny, unpublished.

[23] D.A. Huse and I. Morgenstern, "Finite-size scaling study of the two-dimensional Ising spin glass", Physical Review B **32**, pages 3032-3034 (1985).

[24] L. Saul and M. Kardar, "Exact integer algorithm for the two-dimensional $\pm J$ Ising spin glass", Physical Review E **48**, pages R3221-R3224 (1993).

[25] H. Rieger, L. Santen, U. Blasum, M. Diehl, M. Jünger, and G. Rinaldi, "Critical exponents of the two-dimensional Ising spin glass revisited: exact ground-state calculations and Monte Carlo simulations", Journal of Physics A: Mathematical and General **29**, pages 3939-3950 (1996).

[26] R.F.S. Andrade, E. Nogueira, Jr., and S. Coutinho, "Ising sping glass by the transfer matrix approach", Physical Review B **68**, article 104523 [9 pages] (2003).

[27] T.F. Rønnow, Z. Wang, J. Job, S. Boixo, S.V. Isakov, D. Wecker, J.M. Martinis, D.A. Lidar, and M. Troyer, "Defining and detecting quantum speedup", Science **345**, pages 420-424 (2014).

Minimal kinetic theory: a mathematical framework for non-equilibrium flowing matter

Sauro Succi
IAC-CNR, via dei Taurini, 19,00159, Rome, Italy
E-mail: succi@iac.cnr.it

Abstract. We discuss the intriguing ability of minimal kinetic theory to describe a broad variety of complex non-equilibrium flows across scales of motion. It is argued that, besides major computational progress, minimal kinetic theory also provides a new conceptual framework to investigate the complexities of flowing matter far from equilibrium.

1. Introduction

Statistical physics is the branch of theoretical physics in charge of unraveling the connections between the macroscopic world, as we perceive it, and its microscopic constituents (atoms, molecules and below). Kinetic theory, in turn, is the sector of statistical physics which deals with the transport phenomena by which macroscopic systems attain their thermodynamic equilibrium [1, 2]. The undisputed hero is Ludwig Boltzmann (1844-1906), who first laid down the mathematical bridge between the micro and macroworlds. The Boltzmann equation, the mathematical cornerstone of kinetic theory, is however very difficult to solve, on account of its high-dimensionality (six-dimensional phase-space plus time), non-linearity and non-locality in velocity space. Moreover, it was derived for dilute gases only, and the extension to dense fluids and liquids has proven pretty difficult beyond expectation [3].

In the last two decades, however, several forms of *minimal kinetic theory* (MKT), have been developed, which provide a very fertile background not only for the numerical simulation of complex non-equilibrium phenomena, but also for their conceptual study. In this paper, we wish to provide a brief survey of the leading themes behind these developments.

In the recent years, MKT, and most notably the Lattice Boltzmann method (LB), have made proof of an amazing and largely unanticipated capability of straddling across a broad range of fluid scales, from fully developed turbulence, all the way down to nanoscopic flows of biological interest, and lately even quantum-relativistic subnuclear fluids, spanning over ten spatial decades in the process [4].

This multi-scale capability often takes MKT into a theoretically 'unprotected' territory, where the conditions for a rigorous analysis of the continuum limit (Chapman-Enskog asymptotics) are no longer available [5]. For simplicity, we shall call this the *non-hydrodynamic* regime, which is typically but not exclusively, associated with the molecular mean free-path becoming comparable with the smallest macroscopic length-scale.

This non-hydrodynamic capability is still only partially understood and represents one the most interesting aspects of modern MKT. In this paper, starting from the by now largely

Journal of Physics: Conference Series **681** (2016) 012006 doi:10.1088/1742-6596/681/1/012006

documented versatility of LB, we shall argue that, besides its practical impact as a computational tool, the most far-reaching contribution of MKT is to provide a conceptually elegant and computationally powerful *mathematical framework* for the study of non-equilibrium phenomena in flowing matter across many scales of motion, including the quantum/classical border [6].

2. Boltzmann's kinetic theory

Let us begin by writing the Boltzmann equation (BE) in its standard form (vector notation relaxed for simplicity):

$$\frac{df}{dt} = \frac{1}{\tau}Q(f, f) \tag{1}$$

where $df/dt \equiv \partial_t f + v\nabla_x f + a\nabla_v f$ is the streaming term in phase-space and $C \equiv Q(f, f)/\tau$ is the collision term, where the characteristic collisional time scale τ is exposed for dimensional clarity.

The central lengthscale of Boltzmann's kinetic theory is the molecular mean-free-path, i.e. the averaged distance traveled by a representative molecule before hitting into another molecule. The mean-free path is given by

$$\lambda \sim v_{th}\tau$$

where $v_{th} = \sqrt{D\frac{k_B T}{m}}$ is the thermal speed in D spatial dimensions, m the mass of the molecule, T the temperature and k_B Boltzmann's constant. The collisional time τ, in turn, is given by

$$\tau = nv_{th}\sigma(v_{th})$$

where n is the number density of the gas and σ its cross section (length square), namely the effective "size" of the molecule as dynamically perceived by its partner molecules in a collisional event. For short-range potentials such as those pertaining to the Boltzmann scenario, $\sigma \sim r_0^2$, r_0 being the range of the interaction potential.

Regardless of the complexity of the interaction potential, the collision operator obeys mass-momentum-energy conservation laws, namely:

$$\int C(f, f)\{1, mv_a, \frac{mv^2}{2}\}d_3v = 0 \tag{2}$$

where the latin subscript $a = x, y, z$ labels the three spatial coordinates. In other words, mass, momentum and energy, are *microscopic invariant*, i.e. they are left unchanged by molecular collisions, no matter how complicated. This insensitivity to molecular details (Universality) stands at the very roots of macroscopic hydrodynamics. Another central property of the collision operator is the fact of vanishing at local equilibrium:

$$C(f^{eq}, f^{eq}) = 0 \tag{3}$$

This condition implies a dynamical balance between direct and inverse collisions, sometimes also called Loss and Gain terms (a direct collision takes a molecule away from the phase-space element $dxdv$, while an inverse one puts it back). Due to the existence of the microscopic invariants, the local equilibrium takes the universal Maxwell-Boltzmann distribution (for classical=non quantum and non-relativistic particles):

$$f^{eq}(x, v; t) = n\frac{e^{-c^2/2}}{(2\pi v_{th}^2)^{D/2}} \tag{4}$$

where $c \equiv \frac{v-u}{v_{th}}$, u being the local flow speed, is the so-called peculiar speed, in units of the thermal speed. Note that the five macroscopic quantities, (n, u_a, T) associated with the microscopic invariants can be arbitrary functions of space and time, as long as their scale of variation remains well above the mean-free path, (slow modes, at the molecular scale).

Finally, Boltzmann's collision operator obeys the famous H-theorem, namely:

$$\dot{s} \equiv -\int C(f,f) log f \; d_3 v \geq 0 \tag{5}$$

where $\dot{s}$ is the local rate of entropy production. The H-theorem stands as possibly the deepest outcome of Boltzmann's equation, offering the first mathematical bridge between the hitherthen disjoint fields of mechanics and thermodynamics. A bridge over "troubled waters" indeed, as history of science informs us, and yet a monumental one [7].

Conceptually, the BE is limited to dilute gases with binary short-range collisions, so that triple collisions of the form $Q(f, f, f)$ and higher can safely be neglected at the right-hand side. This formally rules out dense gases and liquids, or, in a modern language, strongly interacting fluids, in which the mean-free path is comparable with the range of molecular interactions.

Even so, the BE is computationally highly demanding, for it lives in $6+1$ dimensions and collisions come in the form of a quadratic integral operator in velocity space. Because of the above, the use of the Boltzmann equation has been traditionally confined to the study of weakly interacting fluids, i.e. dilute gases and rarefied states of matter in which kinetic energy dominates over interparticle interactions. The computational tool of the trade is Bird's Direct Simulation Monte Carlo [8], a stochastic particle technique whereby computational molecules are advanced along single particle trajectories and undergo binary collisions according to probabilistic rules controlled by the cross section and the resulting molecular mean free path.

3. Minimal kinetic theory

The idea of minimal kinetic theory is not new: model equations aiming at relinquishing the mathematical complexity of the BE without surrendering the main physics, have been available for over half a century. Perhaps, the most popular one is the celebrated Bhatnagar-Gross-Krook version, in which the quadratic integral $Q(f, f)$ is replaced by the departure of the distribution from its local equilibrium [9], namely:

$$C_{BGK} = (f^{eq} - f)/\tau$$

What is new though, is the realization of the breadth of complex non-equilibrium problems that can be handled by suitable generalizations of such minimal models. In this respect, the lattice has proven an invaluable ally in affording computational efficiency, even though the conceptual picture disclosed by MKT is not restricted to lattice realizations.

Lattice versions have shown that the hydrodynamic complexity of macroscopic equations, typically Navier-Stokes, can be captured by a very stylized representation of velocity space. This is a gift of collisional invariants, which secure lasting order out of molecular chaos (microscopic conservation laws). In the following we shall refer to the Lattice Boltzmann equation (LBE) for the sake of concreteness, but other forms of minimal kinetic theory, say Stochastic Rotation methods [10] and Dissipative Particle Dynamics [11], inscribe to the same conceptual framework.

The LBE, in BGK form for simplicity, reads as follows [12, 13, 14, 15]

$$f_i(x + c_i\Delta t, c_i; t + \Delta t) - f_i(x, c_i; t) = \omega(f_i^{eq} - f_i) + \gamma S_i \tag{6}$$

where the index i labels the set of discrete velocities. In the above $\omega = \Delta t/\tau$ is the ratio of free-streaming version collision timescales, S_i is a source term associated with external/internal forces and $\gamma = \Delta/\tau_S$ is a measure of the strength of the external versus collisional relaxation.

The lattice local equilibrium comes in the form of a truncated polynomial expansion of the local Maxwell-Boltzmann distribution in the local Mach number. This expansion is typically, but

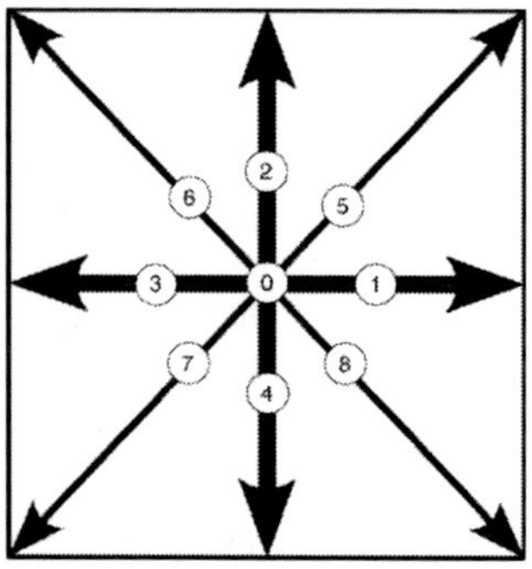

Figure 1. The D2Q9 lattice, i.e. nine discrete velocities in two spatial dimensions. The lattice affords fourth order isotropy and permits to recover the isothermal, quasi-incompressible two-dimensional Navier-Stokes equations.

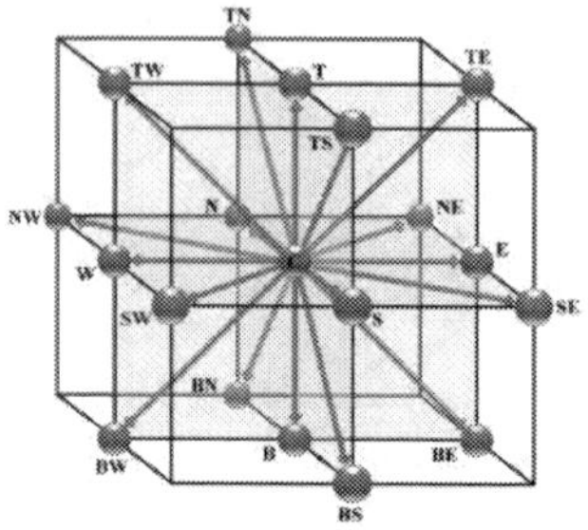

Figure 2. The $D3Q19$ lattice, i.e. nineteen discrete velocities in three spatial dimensions. The lattice provides fourth order isotropy and permits to recover the isothermal, quasi-incompressible three-dimensional Navier-Stokes equations.

not necessarily, truncated to second order since the Navier-Stokes equations are quadratically non linear. The lattice local equilibrium reads as follows:

$$f_i^{eq}(x;t) = \sum_{k=0}^{2} H_{ik} M_k^{eq}(x;t) \tag{7}$$

where $M_0^{eq} = \rho$ is the fluid mass density, $M_1^{eq} = J \equiv \rho u$ is the flow current density and $M_2^{eq} = \rho u^2 + P$ is the momentum flux, P being the fluid pressure. In D spatial dimensions, the current J_a is a D-dimensional vector and the momentum flux P_{ab} is a second-rank tensor. In the above H_{ik} is a set of lattice basis functions in velocity space, the analogue of Hermite polynomials $H_k(v)$, evaluated at $v = c_i$.

Once the set of discrete velocities c_i (typically of order 10 in two dimensions and 20 in three), provides enough symmetry to recover the isotropy of fourth order lattice tensors, and the lattice local equilibria ensure local mass and momentum conservation, the lattice BGK equation can be shown to reproduce the (weakly-compressible) Navier-Stokes equations of macroscopic hydrodynamics in the limit of small Knudsen numbers. This is usually defined in terms of the collisional relaxation time, i.e $\lambda = c_s \tau$, where c_s is the lattice sound speed, typically $O(1)$ in lattice units. In the strongly coupled (hydrodynamic) regime, the mean free path is typically much smaller than the lattice spacing, i.e. $\lambda << \Delta x$. As a result, the effective Knudsen number is most appropriately defined as $Kn_\Delta = \Delta x/L$, L being the smallest macroscale of the problem.

This means that near-grid scales $L \sim \Delta x$ are not guaranteed to obey macroscopic hydrodynamics, a point to which we shall return shortly. For now, let's stay with the bona-fide statement that for lengthscales such that $Kn_\Delta << 1$, the LBE provides a quantitative description of Navier-Stokes hydrodynamics in kinetic vests.

The LB has made proof of a series of remarkable properties, both conceptual and computational. Without delving into specialistic details, in the following we shall try to convey the essence of the story.

4. The benefits of extra-dimensions

By a mere counting of degrees of freedom, the fact that just ten or twenty discrete distributions f_i are sufficient to recover hydrodynamics is hardly a surprise. Indeed, it is well known from continuum kinetic theory that hydrodynamics is just the emergent tip of a kinetic iceberg involving a virtually infinite hierarchy of kinetic moments:

$$M_n(x;t) = \int f(x,v;t)H_n(v)dv \tag{8}$$

where $H_n(v)$ is a suitable basis in kinetic (Hilbert) space, typically Hermite's polynomials. Tensor notation is again relaxed for simplicity, but it should be borne in mind that subscript n labels n-th order tensors.

The point, however, is not the *number* of degrees of freedom, but rather their structural organization! Kinetic theory exhibits a very elegant organization, directly subtended by the Stream-Collide structure of the Boltzmann equation, which contrasts with the Advection-Diffusion form of macroscopic transport equations.

In a catchy motto, in the stream-collide kinetic paradigm: *Non-linearity is local and non-locality is linear*, a property which is inevitably lost by any hydrodynamic representation.

Differently restated, Stream-Collide in (minimal) phase space is more transparent (and efficient) than the emergent Advection-Diffusion picture of macroscopic transport.

Let us spell this important point out.

4.1. Streaming versus Advection

The kinetic moments obey recursive PDE's of the form:

$$\partial_t M_n + \nabla_x N_{n+1} = C_n \tag{9}$$

where $M_n(x;t) = \int f(x,v;t)H_n(v)dv$ is the kinetic moment of order n, $N_{n+1} = \int f(x,v;t)vH_n(v)dv$ is the corresponding current and C_n is the collisional contribution to the rate of change of M_n. By setting $v = u + c$, where u is the local fluid velocity (space-time dependence omitted for simplicity)

$$J_{n+1} = M_n u + J'_{n+1} \tag{10}$$

where

$$J'_{n+1} = \int f(x,u+c;t)cH_n(u+c)dc \tag{11}$$

is the non-equilibrium component of the generalized current. The key point is the following; leaving aside the non-equilibrium current, information travels along the material streamlines defined by the flow velocity $u(x;t)$. In a complex flow, say turbulence, or multiphase flows with complex interfaces, the material streamlines are overly complex, as they carry the complexity of macroscopic hydrodynamics, as expressed by the non-linear inertial acceleration:

$$A_a = u_b \nabla_b u_a$$

where repeated indices are summed upon. Now, contrast this with the peaceful form of the (force-free) kinetic streaming operator:

$$\hat{S}f \equiv v_a \nabla_a f$$

All the complexity is now embedded in $f(x,v;t)$ and the streaming is happily unaware of the macroscopic complexity in configuration spacetime (x,t): information always travels along the straight streamlines $dx_v = vdt$ as opposed to the material line $dx_u = u(x;t)dt$.

The price to pay are the extra degrees of freedom: there are always more discrete velocities than hydrodynamic moments, but the reward is major: streaming is linear and can be solved *exactly*. Thus instead of $u\nabla u$ which is non-linear and non-local at a time, in kinetic theory, non-linearity is local (collision operator) and non-locality is linear (streaming operator).

4.2. Relaxation versus Diffusion

It is well known that hydrodynamics results from kinetic theory in the limit where the momentum flux tensor $P_{ab} = m \int f v_a v_b d3v$ is enslaved to its hydrodynamic equilibrium form, $P^{eq}_{ab} = \rho u_a u_b + p\delta_{ab}$, where $p = p(\rho, T)$ is the pressure. As a result of this enslaving, dissipative forces take the form of a second-order spatial derivatives, coupled to the fluid viscosity:

$$F^D_a \sim \tau \partial_b \partial_c Q^{eq}_{abc} = \nu \partial_b \partial_c (J_a \delta_{bc})$$

In the above, $Q^{eq}_{abc} = \int f^{eq} v_a v_b v_c d3v$ is the third order energy-flux tensor, evaluated at local equilibrium. We have further taken τ and $\nu = c^\tau_s$ as constant.

Second order spatial derivatives in combination with first order time derivatives imply irreversibility, hence dissipation.

In kinetic theory, irreversibility and dissipation are built-in in the collision operator via the H-theorem, securing monotonic entropy growth to equilibrium and compliance with the Second Principle of thermodynamics. To be noted that the H-theorem does not impose any non-local structure, such as the Laplacian. This is another major advantage of the kinetic formulation: dissipation is local, as it results from local enslaving of the momentum flux tensor, with no need of second order spatial derivatives. This is computationally convenient, especially near solid boundaries, as it permits to keep the boundary procedure first-order in space. It is also appealing from the analytical viewpoint, as it permits to work with discrete propagators of the form $P_i(k, \omega) = [e^{j(kc_i - \omega)} - 1]^{-1}$, which are Galilean invariant at all orders. In the above, j stands for the imaginary unit, to avoid confusion with the discrete velocity index i.

5. Beyond hydrodynamics

The discussion so far has been confined to macroscopic hydrodynamics. While it is very pleasing to observe that MKT has shown major competiteveness on a subject, computational fluid dynamics, which was considered a "no fly zone" for the Boltzmann equation, one may still argue that this is "just" a computational achievement.

This author would hardly subscribe to the "just" above, but this is another paper. So, let us proceed to show how MKT can eventually be taken into the theoretically unprotected non-hydrodynamic regime.

Hydrodynamics emerges from Boltzmann's kinetic theory in the limit of vanishing Knudsen number, i.e. weak non-equilibrium, weak inhomogeneities. The theoretical backup is known as Chapman-Enskog asymptotics. Given that LB was originally targeted to hydrodynamics only, some authors have concluded that LB cannot and actually -should- not be used outside this specific realm. Yet, for more than a decade, a wide body of evidence has been accumulated showing that suitable LB extensions *do carry* valuable non-equilibrium information beyond the hydrodynamic level.

These extensions involve larger sets of discrete velocities [16], ensuring isotropy to six or eighth order, in combination with boundary conditions catering for momentum exchange between the fluid molecules and solid walls [17]. This is again not so surprising after all, since higher-order lattices contain many more discrete distributions than the thirteen moments of the standard Grad's generalized hydrodynamics [18].

However, as discussed earlier on, the mere counting of the degrees of freedom alone does not say much: the right structural organization matters way more. Indeed the discrete-velocity organization gives rise to realizable boundary conditions, a property which is not shared by generalized hydrodynamics.

This is basically the Streaming versus Advection item discussed earlier on.

One further point regards the so called Regularization procedure [19, 20].

This consists in filtering out non-hydrodynamic moments after the streaming step, the one fueling non-equilibrium.

Journal of Physics: Conference Series **681** (2016) 012006 doi:10.1088/1742-6596/681/1/012006

In equations:

(i) *Stream*: $f_i(x,t) = f_i'(x - c_i\Delta t, t - \Delta t)$

(ii) *Regularize*: $h_i(x,t) = \mathcal{P} f_i(x;t)$

(iii) *Collide*: $f_i'(x,t) = (1-\omega)h_i(x;t) + \omega h_i^{eq}(x;t)$

In the above, $\mathcal{P}$, denotes a projector in kinetic space, filtering out the non-hydrodynamic (ghost) modes, i.e. $\mathcal{P}f = h$ and $(1-\mathcal{P})f = g$, $f = h+g$. By hydrodynamic modes, we imply here density, momentum and momentum-flux, while all remaining moments classify under the ghost rubric (the kinetic moments lacking a direct macroscopic interpretation). In this respect, the hydro-ghost representation follows in the steps of Grad's 13-moment generalized hydrodynamics [18], with two crucial twists: i) it contains a larger set of kinetic moments, ii) these moments are organized by the symmetry of the discrete lattice and not by the Hermite expansion in Hilbert space.

While still under exploration, the Regularized LB shows encouraging promise to reproduce at least the global features of finite-Knudsen flows, such as the transition from collective hydrodynamics to single-particle molecular regimes in finite Knudsen flows between parallel plates.

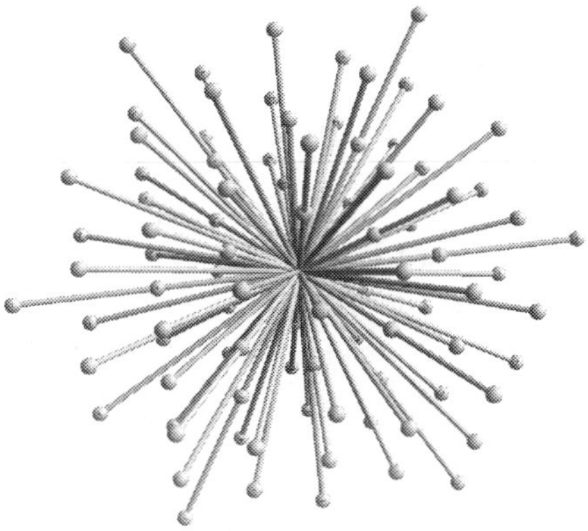

Figure 3. A high-order lattice featuring 93 discrete velocities ensuring 8th order isotropy. Courtesy of A. Montessori.

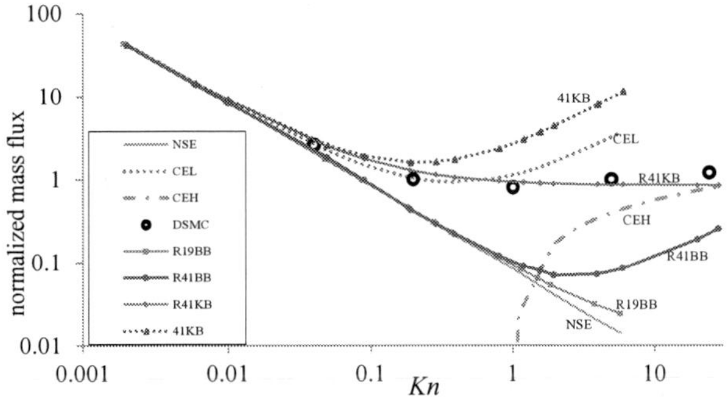

Figure 4. The Knudsen paradox: Flow across a plate channel as a function of the Knudsen number for various LB Models. DnQm stands for LB in d-spatial dimensions with m discrete velocities. BB and KBC stand for bounce (no-slip) and Kinetic boundary conditions, while R stands for Regularized.

Some Regularized LB's prove capable of reproducing the Knudsen paradox, i.e. an increasing flow rate beyond a given Knudsen number, in close agreement with the Direct Simulation Monte Carlo (DSMC) solution of the Boltzmann equation. CEL and CEH refer to Cercignani's asymptotic solutions in the low (high) Knudsen regimes, respectively. Finally, NSE denotes the Navier-Stokes solution. Courtesy of A. Montessori.

Does this mean that LB can reproduce the BE?
The answer is a most certain NO!

Journal of Physics: Conference Series **681** (2016) 012006 doi:10.1088/1742-6596/681/1/012006

What it means though, is that non-hydro kinetic models can be formulated which provide useful quantitative information, without resorting to the full Boltzmann equation.

This is a recurrent theme in MKT: filling the (many) gaps where hydrodynamics is no longer adequate, and yet full molecular details are not needed either. In this case, solving the full BE would be a computational overkill.

MKT has sometimes been heralded under the catchy motto *"simple models for complex flows"*. This is nice, but only works as long as Universality stands by our side; out of equilibrium Universality recedes, by definition. The good news is that it often does it in graceful way, i.e. Universality is only weakly broken: *weakly broken universality* is possibly the richest hunting ground for MKT. In this sense, MKT really seems to offer a new computational and *conceptual* path in non-equilibrium statistical mechanics.

By "new path" we imply discrete features which can *not* be reconduced to direct discretizations of the Boltzmann equation.

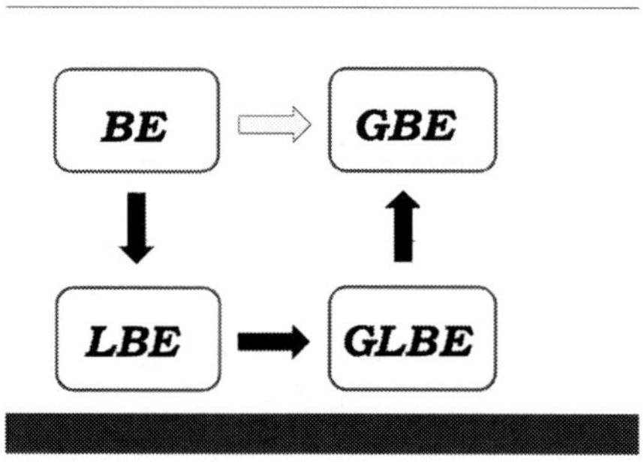

Figure 5. Sketch of the conceptual scenario by which minimal kinetic theory would generate new kinetic equations in the continuum. Take a LBE descending directly from BE by discretization (not all LBE's do!), generalize it in the minimal (lattice in this figure) framework (GLBE), and export it back to the continuum (GBE). This gives rise to new kinetic equations, which may or may not be derivable as an extension of the BE in the continuum (empty horizontal arrow). Some GLBE's may not admit a well-defined continuum limit, in which case they should be regarded as discrete dynamical models of complex moving matter, in their own right.

A very elegant example in this direction is offered by recent developments of the entropic version of LB, see [22]. Here, the authors show that the LBE is a direct implication of kinetic theory when relaxation terms are retained to all orders, a step which is usually *not* taken in the context of standard continuum kinetic theory.

The Regularization proceure provides another example of discrete benefits which are not (at least, not easily) returned to the continuum. Indeed regularized versions of Grad's 13-moment generalized hydrodynamic have been developed in the last decades [18, 21]. Even though such R13 regularizations do indeed mend Grad's original weaknesses, i.e. bulk instability and hardly realizable boundary conditions, they still face with a number of difficulties. Arguably, much of these difficulties might stem from the fact of sticking to the advection-diffusion paradigm, rather than the stream-collide kinetic formalism. For exact LB solutions beyond Navier-Stokes, see [23].

6. Minimal kinetic theory for dense fluids

A major limitation of the Boltzmann's equation is its (alleged) lack of applicability beyond the context of dilute gases. Indeed, extensions of Boltzmann equations to dense fluids and liquids have proven exceedingly difficult, to the point that, historically, the research community in the mid 50's turned to Molecular Dynamics simulation.

To a large extent, this picture is still with us: use Navier-Stokes as much as you can, typically by resort to generalized boundary conditions, and stick to the macroscopic PDE's as much as possible.

The point is that dense fluids exhibit *non-perturbative* effects on both equilibrium (equation of state) and non-equilibrium (transport coefficients) properties. Capturing these corrections would, in principle, require the solution of the full BBGKY hierarchy, or at least the two-body Liouville equation. Helas, this is a 12+1 dimensional problem...

Some authors have argued that, being in fact a cartoon/caricature of Boltzmann kinetic theory, MKT inherits all of its limitations, at best. This intuitive view proves however overly simplistic and misses a great deal of interesting physics.

Indeed, somehow paradoxically, suitably enriched "caricatures" of the Boltzmann equation prove more general than the original Boltzmann equation itself!

This should by no means sound as a detrimental statement towards the BE: quite the opposite, *it speaks even more assertively for the greatness of the Boltzmann equation!* Indeed, it uncovers a whole world of mathematical richness which allows the BE equation to spread its wings across an amazingly broad spectrum of applications, far beyond the original framework of dilute gases!

So, how does one incorporate many (two)-body physics in MKT?

Here, one turns to another chapter of kinetic theory where Boltzmann is flanked by a towering figure of modern kinetic theory, Anatoly Vlasov (1908-1975).

The Boltzmann-Vlasov equation is characterized by the appearance of a mean-field force in the streaming operator:

$$\partial_t f + v\nabla_x f + a\nabla_v f = C$$

where $a = F/m$ is the acceleration experienced by a representative molecule. In the simplest instance, this is just an electric or gravitational field, but most interesting physics is opened up by generalizations in the spirit of density functional theory.

Perhaps, the most popular one is the so-called pseudo-potential approach, which takes the form [24]:

$$a(x) = G\psi(x)\sum_i w_i\psi(x + r_i) \tag{12}$$

In the above, $\psi(x) = \psi[\rho(x)]$ is a local functional of the fluid density, sometimes also known as generalized density, and G is the coupling strength associated with the pseudo-potential. This expression, with suitable choices of the generalized density, has proven immensely valuable for the study of a broad variety complex multiphase and multicomponent flows. Once again, a subject where one would have not even dreamt of using Boltzmann-like equations until two decades ago!

As an example, out of a huge class of applications, we mention here recent study of soft glassy materials, such as foams and emulsions [26], a rich and fascinating subject where the complexity of interfacial dynamics stands in the way of most analytical and numerical methods.

Unlike Regularized LB, here one can apply a "reverse MKT" argument, i.e return the lattice benefits back to the continuum, but in the form of *new* continuum equations.

More precisely, we argue that a Vlasov term of the form:

$$a(x) = \psi(x)\int G(x,y)\psi(y)dy \tag{13}$$

may prove extremely useful for the study of complex flows with dynamic phase-transitions, no need to solve it on the lattice! To the best of our knowledge, the study of such kinetic equations in the continuum is largely (maybe totally) unexplored, both analytically and numerically.

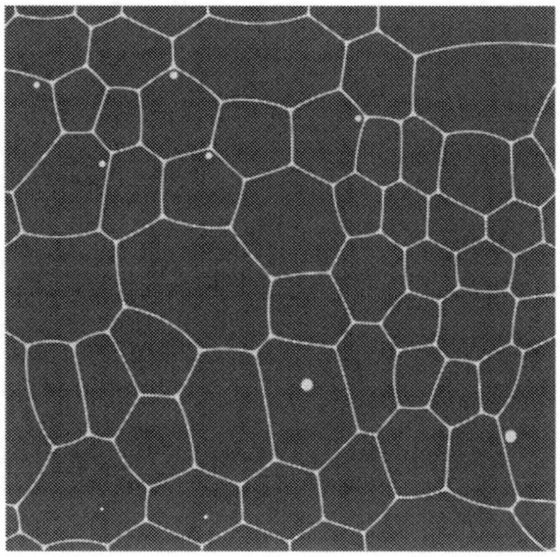

Figure 6. LB simulation of a dry foam. The blue region represents the gas phase, with the liquid flowing along the sides of the polygon (Plateau borders). The simulation was performed via a two-component Shan-Chen LB with competing short-range attraction and mid-range repulsion [25]. Courtesy of M. Sbragaglia.

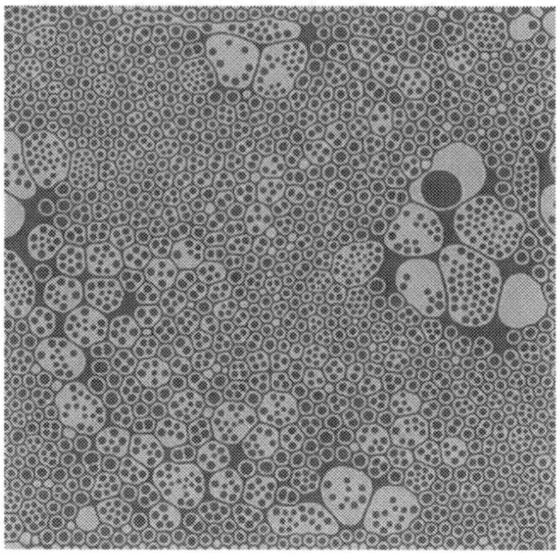

Figure 7. LB simulation of a wet foam. Here the fraction of liquid is 0.5, with the result that the Plateau borders invade the bulk fluid and take the shape of liquid droplets within gas droplets. Courtesy of M. Sbragaglia.

Of course, the approach remains heuristic, in that the "Green function" $G(x, y)$ is not derived ab-initio from the two-body Liouville equation, and most notably from the two-body radial correlation function. Nevertheless, a top-down approach, whereby one would experiment with a whole class of plausible model Green functions, might deliver a wealth of new knowledge in non-equilibrium thermodynamics.

And again, the progress is conceptual in the first place, no compelling need to solve the equations on the lattice.

7. Down to the atomistic scale

Here, we start by stating the obvious: MKT is not Molecular Dynamics!

For one, it does not contain molecular fluctuations. However, it is liable to generalizations whereby fluctuations can be inserted back, essentially through a stochastic source term. Again nothing new under the sky, fluctuating kinetic theory was proposed half a century ago by a number of authors, particularly Fox and Uhlenbeck [27]. Such fluctuating kinetic equations surely played a major role in advancing our basic knowledge of fluctuating nano-hydrodynamics. No less goes for fluctuating LB, as pioneered by A.J.C Ladd [28] and subsequently improved by several other authors, which have opened entirely new perspectives in the study of colloidal suspensions and nanofluids in general.

8. Summary

It is sometimes said that the scientific import of an equation is proportional to the conceptual distance between its left and right hand side. Very true for Boltzmann: non-equilibrium versus equilibrium, respectively, the "Evershifting battle" in Boltzmann's own words. But it is also true that the actual physics is not contained in the equations themselves, but in their solutions instead. And solutions for truly complex problems are preciously rare. Thus, it appears plausible to argue that the scientific import of an equation rests largely with its amenability to workable approximations. Possibly, one of the most profound merits of minimal kinetic theory is to have unveiled the greatness of the Boltzmann equation in this respect too. Given how appreciative Boltzmann himself was of practical aspects of science, I cherish the idea that he would be delighted at the unsuspected advances spawned by suitable "caricatures" of his own equation in the last two decades.

9. Acknowledgements

The author wishes to acknowledge illuminating discussions over the years with R. Benzi, B. Boghosian, H. Chen, I. Karlin, A. Montessori, M. Sbragaglia, V. Yakhot and J. Wettlaufer.

References

[1] L. Boltzmann, Lectures on Gas Theory, (English translation by S. Brush), Univ. of California Press, 1964.
[2] C. Cercignani, Theory and applications of the Boltzmann equation, Scottish Academic Press, 1975
[3] J.R. Dorfman, E.G.D. Cohen, J. Math. Phys., 8, 282 (1967)
[4] S. Succi, Eur. Phys. J. B: 64, 471, (2008)
[5] S. Chapman and T.G. Cowling, The mathematical Theory of Non-Uniform Gases, Cambridge Univ. Press, (1952)
[6] S. Succi, EPL 109, 50001, (2015)
[7] J. Lebowitz, Phys. Today, 46, 32 (1993)
[8] G. Bird, G. Bird, Molecular Gas Dynamics and the Direct Simulation of Gas Flows, Claredon, Oxford (1994)
[9] P. Bhatnagar, E. Gross and M. Krook, Phys Rev A, 94, 511, (1954)
[10] R. Kapral, A. Malevanets, J. Chem. Phys. 110, 8605, (1999)
[11] P. Hoogerbruke and J. Koelman, Europhys. Lett., 19, 150, (1992)
[12] R. Benzi, S. Succi, M. Vergassola, Phys. Rep., 222, 145, (1992)
[13] S. Chen, G. Doolen, Ann. Rev. Fluid Mech., 30, 329 (1998)
[14] G.Mc Namara, G. Zanetti, Phys. Rev. Lett., 61, 2332 (1988)
[15] F. Higuera and S. Succi, Europhys. Lett. 8, 517, (1989)
[16] X. Shan, XF Juan, H. Chen, J. Fluid Mech., 550, 413 (2006)
[17] S. Ansumali, I.V. Karlin, Phys. Rev. E 66 , 026311 (2002)
[18] H. Grad, Commun. Pure App. Math., 2, 331 (1949)
[19] J. Latt and B. Chopard, Math. Comput. Simul., 72, 165 (2006)
[20] A. Montessori et al, Phys Rev E, 89, 053317 (2014)
[21] H. Struchtrup and M.Torrilhon, Phys. Fluids, 15, 2668, (2003)
[22] F. Bosch and I.V. Karlin, Phys. Rev. Lett., 111, 090601, (2013)
[23] S. Ansumali, I.V. Karlin et al, Phys. Rev. Lett., 98, 124502, (2007)
[24] H. Chen, X. Shan, Phys. Rev. E, 47, 1815, (1993)
[25] G. Falcucci et al, Comm. in Comp. Phys., 2, 1071 (2007)
[26] R. Benzi et al, J. Chem. Phys., 131, 104903, (2009)
[27] R. Fox and G. Uhlenbeck, Phys. of Fluids, 13, 1893, (1970)
[28] A. Ladd, J. Fluid Mech., 271, 285, (1994)

International Conference on Computer Simulation in Physics and Beyond 2015 IOP Publishing
Journal of Physics: Conference Series **681** (2016) 012007 doi:10.1088/1742-6596/681/1/012007

Assessment of the dynamics of Asian and European option on the hybrid system

A V Bogdanov[1], E A Stepanov[2] and D S Khmel[3]

[1]Nauki pr., 12-4-29, St. Petersburg, 195257, RUSSIA
E-mail: bogdanov@csa.ru
[2]Gakkelevskaya st., 20-1-51, St. Petersburg, 197372, RUSSIA
E-mail: e.an.stepanov@gmail.com
[3]Halturina st., 9-12, Peterhof, St. Petersburg, 198504, RUSSIA
E-mail: dskhmel@cc.spbu.ru

Abstract

In this article the problem of performance optimization for estimation of European and Asian options pricing is discussed. The main goal is to substantially improve the performance in solving the problems on the hybrid system. The authors optimized the algorithms of the Monte Carlo method for solving stochastic differential equations and path integral derived from Black-Scholes model for pricing options.

1. Introduction

One of the typical spheres of "complex applications", i.e. computational tasks that deal with large amounts of significantly irregular information when the rate of input data subject to processing in a reasonably limited time varies by several orders is financial mathematics. Finally, the end user to make a decision should have information in a comprehensible form. Such applications are characterized by the following factors.

- Tremendous number of end users (brokers)
- Large variety of heterogeneous sources of information
- Unpredictable moments of sudden data volume "explosion"
- Necessity to keep in mind as long prehistory as possible to make the prognoses more precise
- Limited time to make decisions, in practice in a real time manner

The current state and dynamics of changes in global financial system based on the financial markets play nowadays a significant role in the life of the world economic community. The financial markets are as active as never before. In modern electronic markets, stock prices may change several times within a few milliseconds. Participating traders (that can also be computers) have to evaluate the prices and react very quickly in order to get the highest profit, which requires a lot of computational effort. Information of huge volume received from a large variety of heterogeneous sources is then to be processed using properly adequate mathematical tools.

International Conference on Computer Simulation in Physics and Beyond 2015 IOP Publishing
Journal of Physics: Conference Series **681** (2016) 012007 doi:10.1088/1742-6596/681/1/012007

Over the years, increasingly sophisticated mathematical models and derivative pricing strategies have been developed, but their credibility was violated by the financial crisis of 2007–2010. Contemporary practice of mathematical finance has been subjected to criticism from such a notable figures within the field as Nassim Nicholas Taleb in his book "The Black Swan" [1]. The book focuses on the extreme impact of certain kinds of rare and unpredictable events and humans' tendency to find simplistic explanations for these events retrospectively. Taleb claims that the prices of financial assets cannot be characterized by the simple models currently in use, rendering much of current practice at best irrelevant, and, at worst, dangerously misleading. Many mathematicians and applied fields scientists are now attempting to establish more effective theories and methods.

Generally speaking, the fundamental computational problem for adequate providing activity of the army of brokers consists in huge amount and large variety of input heterogeneous sources of information still drastically enlarged by archives' stored data, limited time to make decisions, in practice in a real time manner, and in unpredictable moments of sudden data volume "explosion".

2. Formulation of the problem

In general, running such a computations on servers or clusters with standard CPUs is not feasible due to either long run times or high energy consumption. One of the typical in technical facilities approaches to solve the mentioned problems is to implement hardware accelerators, which is rather novel for financial mathematics issues. Using general purpose graphics processing units as accelerators helps to increase the speed.

A big challenge for hardware accelerators is the complexity of many models used to estimate the future price behavior of financial products. In many cases no mathematical closed-form solution exists so that approximation methods like Monte Carlo simulations or the finite difference method must be employed

We will discuss the procedures of estimation for prices of European and Asian option in two ways, through the stochastic differential equation and path integral.

Stochastic differential equations. For a European option Blake-Scholes pricing model is used. In accordance with the stochastic differential equation the dynamics of the underlying asset *S(t)* is represented as follows:

$$dS(t) = rS(t)dt + \sigma S(t)dB(t), \tag{1}$$

where *S (t)* being the price of the underlying asset at time *t*, *r* being risk-free interest rate, σ being volatility, *r*, σ being constants, and *B (t)* is a standard Wiener process, $B(t) \sim N(0, 1)$.

The result of solution of (1) is famous formula:

$$S\ (T) = S_0\ exp\left\{\left(r - \frac{\sigma^2}{2}\right)T + \sigma\sqrt{T}\epsilon\right\} \tag{2}$$

where ϵ is random variable with standard normal distribution.

Once the value has been generated price of the underlying asset is established it is necessary to calculate the payment of the cash flow of the option [2],

$$\mathrm{C}(S_0, T) = \max\left(S(T) - K, 0\right), \tag{3}$$

with *S(T)* being the generated random price of the underlying asset at the date of expiration of the option, and *K* is the strike price.

*** Formatting Issue - Best Available Paper/Graphic ***

International Conference on Computer Simulation in Physics and Beyond 2015 IOP Publishing
Journal of Physics: Conference Series **681** (2016) 012007 doi:10.1088/1742-6596/681/1/012007

According to the Monte Carlo method for pricing of the options we need to generate as exact, as possible the price of the underlying asset value by (2) and calculate the cash flow of the option (3). It then gives the discounted average price of the option by the formula

$$C = e^{-rT} S_{mid} \, , \tag{4}$$

where S_{mid} is the arithmetic mean of payments of the option.

For the Asian option algorithm is more complicated. The difference is that if the usual option was enough to generate a random value of the price of the underlying asset at the date of expiration, in the case of the Asian option you need to generate a random time series containing the price of the underlying asset to all observation dates.

Path integral. The partial differential equation for calculating the price of the European option is as follows:

$$\frac{\sigma^2}{2} S^2 \frac{\partial^2 C}{\partial S^2} + rS \frac{\partial C}{\partial S} - rC = -\frac{\partial C}{\partial t} \tag{5}$$

After the change of variables $x = ln\, S$ it is reduced to the form:

$$\frac{\sigma^2}{2} \frac{\partial^2 C}{\partial x^2} + \mu \frac{\partial C}{\partial x} - rC = -\frac{\partial C}{\partial t} \tag{6}$$

$$\mu = r - \frac{\sigma^2}{2}$$

$$C(e^{X_T}, T) = F(S_T) = \max\{e^{x_T} - K, 0\}$$

The solution of (6) can be represented using the Feynman-Kac formula:

$$C(S,t) = e^{-rT} E[F(S_T)], \tag{7}$$

where E [·] is expectation value [3]

$$E[F(S_T)] = \int_{-\infty}^{+\infty} \dots \int_{-\infty}^{+\infty} dx_1 \dots dx_n \, dx_{n+1} max\{e^{x_{n+1}} - K, 0\} \cdot$$

$$\cdot \frac{1}{\sqrt{(2\pi\sigma^2\Delta t)^{n+1}}} exp\left\{-\frac{1}{2\sigma^2\Delta t} \sum_{k=1}^{n+1} [x_k - (x_{k-1} + \mu\Delta t)]^2\right\}, \tag{8}$$

Integral (8) can be computed by Monte Carlo procedure as follows. First we need to generate the set of random numbers $(x_0, x_1, \dots, x_n, x_{n+1})$, so that the distribution density x_i is dependent on x_{i-1} via the formula

$$f(x_i) = \frac{1}{\sqrt{(2\pi\sigma^2\Delta t)}} exp\left\{-\frac{1}{2\sigma^2\Delta t} [x_i - (x_{i-1} + \mu\Delta t)]^2\right\}. \tag{9}$$

Using the obtained set the integrand $S(T) = max\, \{e^{x_{n+1}} - K, 0\}$ is calculated. For more accurate results you must perform these actions numerous times and find the arithmetic mean. Then, the option price will be equal to:

*** Formatting Issue - Best Available Paper/Graphic ***

$$C(S,t) = e^{-rT}\frac{1}{n}\sum_{i=1}^{N} S_i \tag{10}$$

In the case of an Asian option the procedure is similar. The only difference is that the integrand is more complicated:

$$S(T) = max\left(\frac{\Delta t \sum_{n=1}^{N} \int_0^1 d\tau\omega(\tau)e^{(x_n - x_{n-1})\tau + x_{n-1}}\left(1 + \sigma^2\Delta t(1-\tau)\frac{\tau}{2}\right)}{T} - K, 0\right) \tag{11}$$

3. Experiments

To improve performance, the solution of the problem by Monte Carlo method could be split into a large number of independent operations. To get optimal acceleration available on the GPU-system we implemented serial and parallel algorithm for this problem with C ++ and CUDA. Performance results are compared with the parallel implementation of coherent systems based on the CPU. All experiments were realized on GPU NVidia Tesla M2050.

In this study was used four programs on Asian and European pricing options prepared for calculations on the GPU. Empirically derived optimal set of random numbers was estimated at 50 million. On the picture one can see the final option price convergence (fig 1).

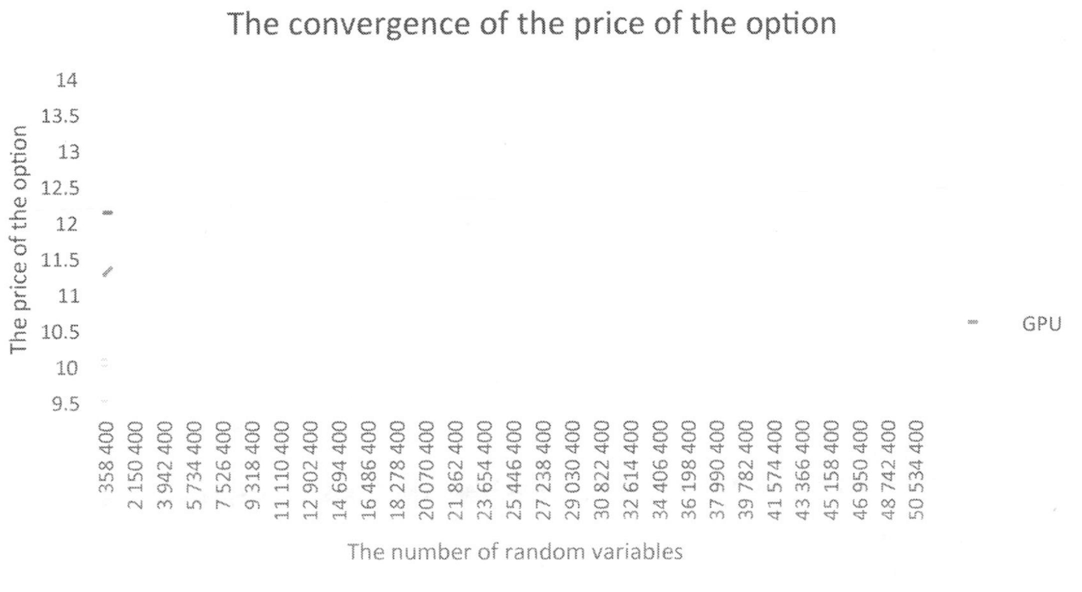

Figure 1 – The convergence of the price of an option on different sets of random numbers.

Immediately it should be noted that not all implementations of GPU showed impressive acceleration. Table 1 shows the results of all programs.

Table 1 – Acceleration obtained on single GPU versus CPU.

	European option	**Asian options**
Stochastic differential equations	106x	1.9x
Path integral	72x	77x

From the table it can be noted that in general the GPU can seriously speed up computations in problems of options pricing. But it must be borne in mind that even though the GPU has more than a hundred times the number of CPU cores, the entire use of the GPU does not occur. This is due to the fact that the GPU core are the "lungs" i. e., they can not effectively carry out a large number of instructions as the CPU core. Therefore GPU show advantage only on the algorithms without large parallel portions. Thus, in Asian option pricing by stochastic differential equation could not get rid of data transfer between individual cores. The result was a slight acceleration in 1.9 times. Figure 2 is a graph of the computation time in this experiment.

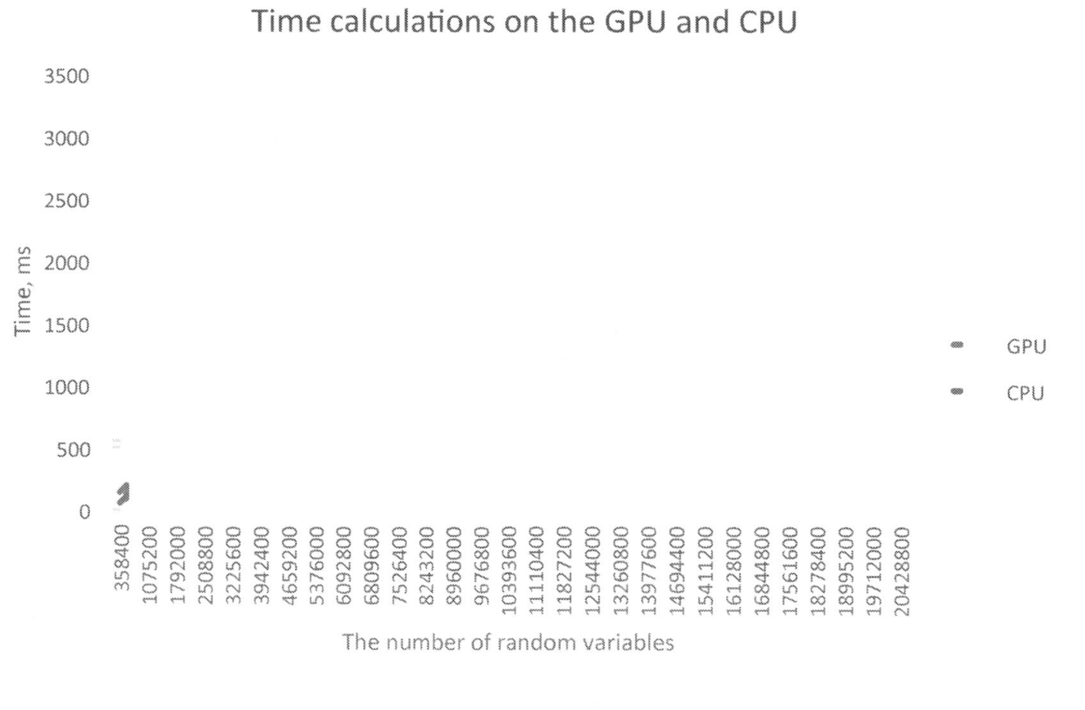

Figure 2 – Time for the calculations of the Asian option pricing by stochastic differential equation.

In other experiments, the problem of data dependence between operations was solved. For example, a European option via the path integral shows almost linear GPU acceleration with respect to one CPU (Fig. 3).

Also this problem has been scaled for three GPU's (Fig 4). As a result the acceleration of 180 times was obtained.

International Conference on Computer Simulation in Physics and Beyond 2015 IOP Publishing
Journal of Physics: Conference Series **681** (2016) 012007 doi:10.1088/1742-6596/681/1/012007

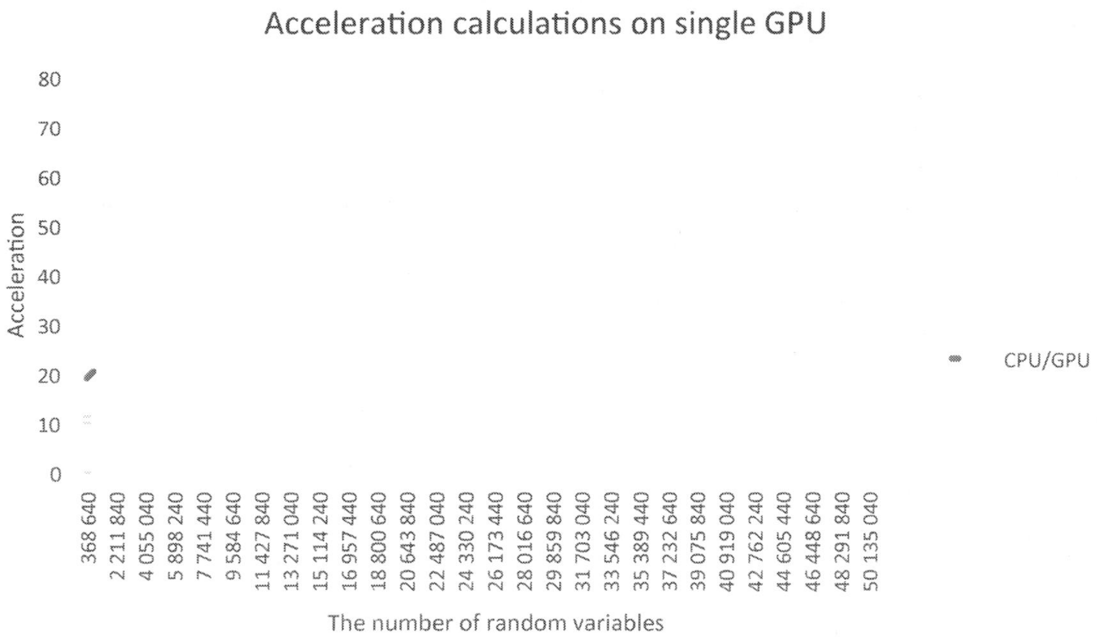

Figure 3 – Acceleration, obtained on single GPU

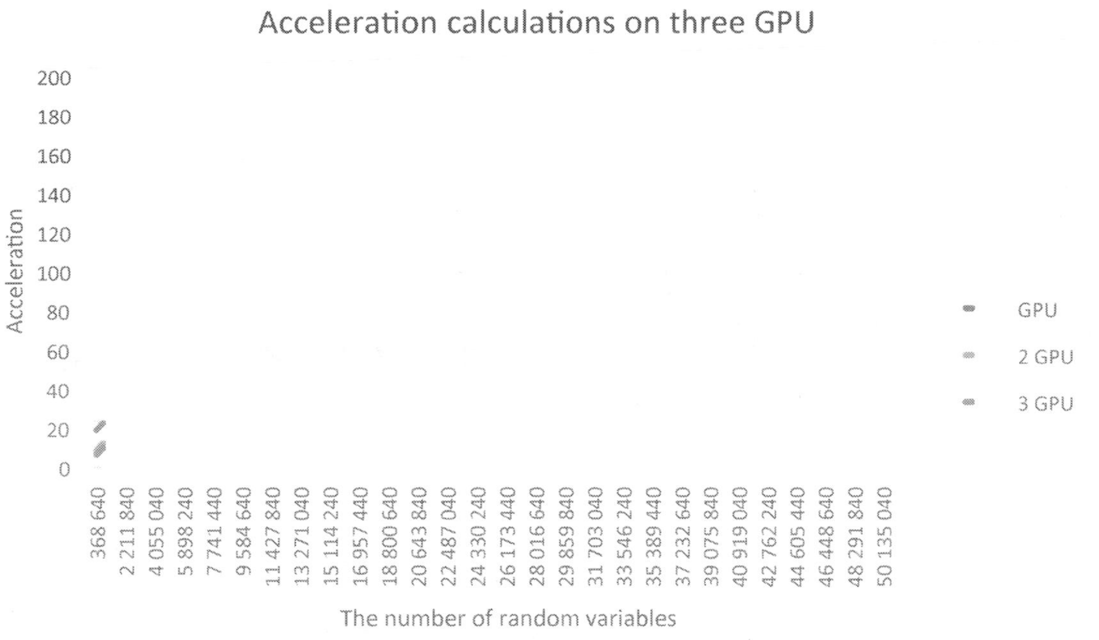

Figure 4 – Acceleration, resulting from the scaling of GPU's.

International Conference on Computer Simulation in Physics and Beyond 2015 IOP Publishing
Journal of Physics: Conference Series **681** (2016) 012007 doi:10.1088/1742-6596/681/1/012007

Both graphs show a clear logarithmic dependence. Thus, given this fact, it is possible to choose the optimal number of the GPU to a specific task.

4. Conclusion

In this paper, the authors applied Monte Carlo method to solve the problem of options pricing. Thus, the search of option price is reduced to the implementation of a large number of independent operations. For such problems solution general purpose graphics processor seems to be very efficient. Experiments have shown that a single GPU can give an increase by 100 times. But not all problems can be realized with such acceleration. In every parallel algorithm, it is necessary to get rid of the dependence between operations performed on a single core. It was also found, that the problem of options pricing easily scale to multiple GPU's. As a result, the performance obtained can be used to improve the accuracy, while increasing the number of random numbers, or to increase the computing speed.

References

1. Taleb, Nassim Nicholas. The Black Swan: The Impact of the Highly Improbable. Random House, 2007.
2. Gluhov M. Option Pricing Monte Carlo. – Futures & Options № 4 April 2009. – 50 page.
3. Montagna G., Nicrosini O. A Path Integral Way to Option Pricing 2008. – 20 page.

Virtual network as excitable medium

Taisiya S. Shinyaeva, Yuri Yu. Tarasevich
Astrakhan State University, Astrakhan, Russia
E-mail: danilova.taisiya@gmail.com, tarasevich@asu.edu.ru

Abstract. We simulated the spread of an activity in a virtual group using the model of excitable medium. We assumed that the structure of the virtual group corresponds to a scale-free network. In our simulation, the network consists of 100 nodes, the average degree of the nodes is 1.98. We considered the propagation of excitation both in a homogeneous and an inhomogeneous excitable medium. The simulation showed that the initial conditions have a little effect on the behaviour of the model. In inhomogeneous medium, fraction of the excited nodes increases, when permanent excited elements ('active' centres) appear in the network. The fraction of the excited nodes increases, when we increase the number of the permanent excited elements. Locations of the active centres do not affect at the level of excitation. External source of activator increases the fraction of the excited nodes in the scale-free network with distribution of parameters.

1. Introduction

Dissemination of ideas and opinions in virtual networks, e.g., in social networks, the Internet, academic networks, the blogosphere, etc. is of special interest [1]. A virtual network can be considered as the active medium [2, 3]. Model by Wiener and Rosenblueth is one of the possible approaches for describing such systems [4]. It is a cellular automaton model that allows explaining the main features of the processes in excitable media, such as electrical excitation's wave propagation through the heart. Elements of the systems have three possible states, of rest, excitation, and refractoriness, and implement a series of successive transitions from one state to another. The elements of a cellular automaton may be connected with different number of the neighbours, so the structure of networks can be not only a regular lattice, but also arbitrarily complex network.

Recently, the spread of certain ideas in a professional virtual group was simulated using generalized Wiener-Rosenblueth model [3]. A particular case has been investigated when a structure of the virtual group corresponded to a complete graph and each member (site) had an individual set of parameters. The simulation showed that interest in the idea can vanish or fluctuate depending on the set of parameters in the virtual group.

We perform simulations of activity in a virtual group using the model of excitable medium. In contrast with [3], we suppose that the structure of the virtual group corresponds to a scale-free network. We considered the propagation of excitation in a homogeneous and an inhomogeneous excitable medium. We studied also impact of initial conditions, the effect of 'active' centre (permanent excited element) and external influences on the behaviour of the model.

2. Model

2.1. Structure of the networks

We considered the discrete systems, i.e. the networks. A scale-free network is a network whose degree distribution follows a power law [5]. There are some nodes in a scale-free network which have much more connections than others; such nodes are called 'hubs'. Many networks have been reported to be scale-free [5]. We suppose that the structure of the virtual group corresponds to a scale-free network. In our model, scale-free network was generated using the function *igraph barabasi game* from the library for analysis of social networks *igraph C library*[1].

2.2. Model of excitable medium

We used generalized Wiener-Rosenblueth model of excitable medium [4, 6][2]. Each element of the network have three possible states, i.e. rest, excitation, refractoriness. Initially, elements are in the rest state. The i-th element becomes excited under the influence of an external excitation whose intensity is not lower than the threshold value, h_i. The i-th element states in the excited state during the time τ_i^e, then it goes into the refractory state in which it states during the time τ_i^r, then it comes back to the state of the rest. A state of any element is specified by the integer phase, Φ_i^n, and the activator concentration, u_i^n, where integer n indicates the discrete time step. Activator decays with time, g_i is the rate of decay.

Transitions between the states obey the following set of rules

$$\Phi_i^{n+1} = \begin{cases} \Phi_i^n + 1, & \text{when } 0 < \Phi_i^n < \tau_i^e + \tau_i^r, \\ 0, & \text{when } \Phi_i^n = \tau_i^e + \tau_i^r, \\ 0, & \text{when } \Phi_i^n = 0 \text{ and } u_i^{n+1} < h_i, \\ 1, & \text{when } \Phi_i^n = 0 \text{ and } u_i^{n+1} \geqslant h_i. \end{cases}$$

A site gets activator from its associated active neighbours of the network.

$$u_i^{n+1} = g_i u_i^n + \sum_j I_j^n,$$

where j is the number of the neighbour site i,

$$I_j^n = \begin{cases} 1, & \text{when } 0 < \Phi_j^n \leqslant \tau_j^e, \\ 0, & \text{when } \tau_j^e < \Phi_j^n \leqslant \tau_j^e + \tau_j^r \text{ or } \Phi_j^n = 0. \end{cases}$$

3. Results

3.1. Homogeneous excitable medium

In homogeneous excitable medium, all elements (sites) have the same parameters thereby the subscripts in the notations of all quantities may be omitted.

In our simulation, we observed the following modes.

- The excitation does not propagate, i.e. an initially excited site returns to the state of rest without passing excitation to any of its neighbours.
- Initial excitation propagates to a small area of the network and then vanishes.
- A solitary wave of excitation propagates through the entire network only once (Fig. 1 and Supplementary multimedia file Wave.gif).

[1] http://igraph.org/c

[2] Original description of the generalized Wiener–Rosenblueth model [6] is a bibliographic rarity. Nevertheless, it is described in several books, see, e.g. [7]

Journal of Physics: Conference Series **681** (2016) 012008 doi:10.1088/1742-6596/681/1/012008

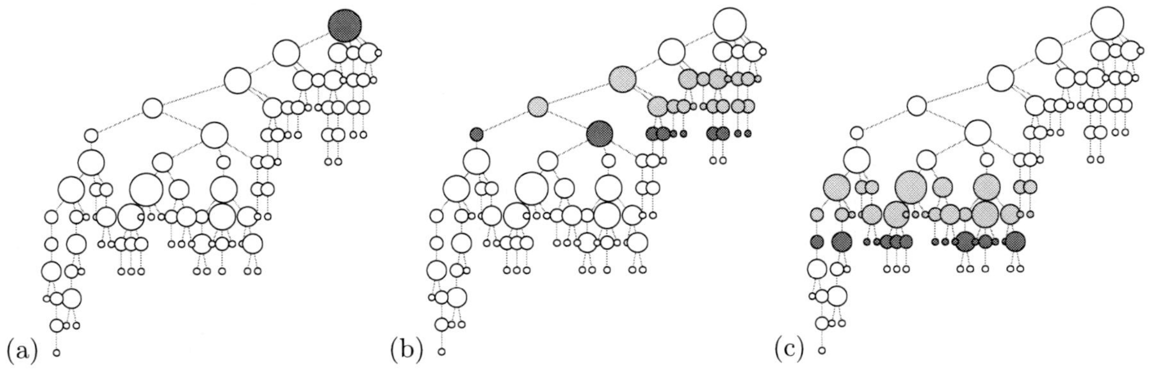

Figure 1. Excitation wave propagation in a homogeneous scale-free network: (a) step 0, (b) step 4, (c) step 9. Parameters of the model are $h = 0.75$, $g = 0.12$, $\tau_e = 1$, $\tau_r = 2$. Initial conditions are one node (hub) is in the excited state.

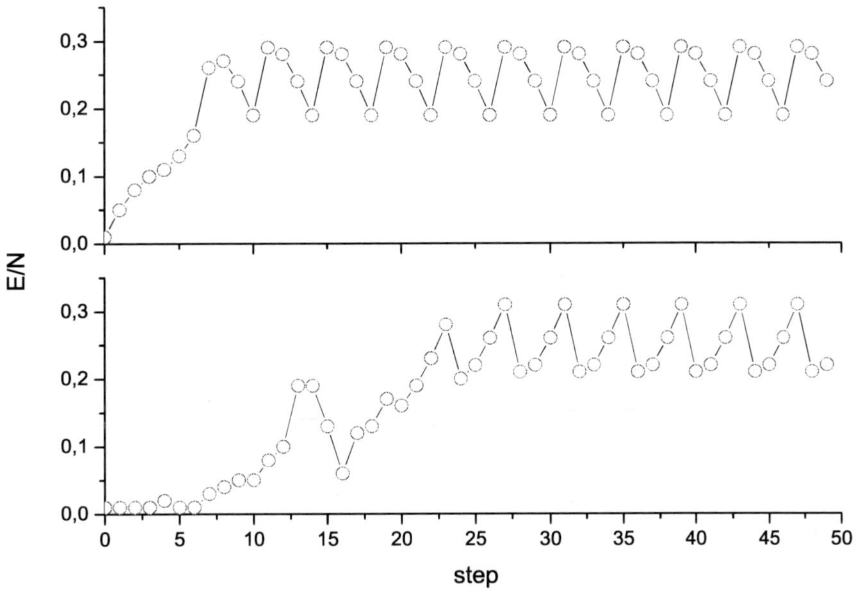

Figure 2. Travelling waves in a homogeneous scale-free network. Initial conditions are (a) one node (hub) is in the excited state, (b) one node with one connection is in the excited state. E is the number of excited sites, N is the total number of sites. Parameters of the model are $h = 1$, $g = 0.5$, $\tau_e = 1$, $\tau_r = 2$.

- Travelling waves propagate through the entire network one by one (Fig. 2 and Supplementary multimedia files Waves.gif, Waves2.gif).

Fig. 2 shows, that travelling waves appear for a given set of parameters regardless of the initial conditions, i.e. when the initial excitation is localized in a hub or in a node with a small number of bonds. Nevertheless, the excitation propagates with the higher spead, when the initial excitation is localized in a hub.

Fig. 3(a) shows how the fraction of the excited nodes changes with time in a scale-free network. The excitation starts with a single node (hub) and propagates throughout the network only once. When the node states always in an excited state then the travelling wave will propagate in the

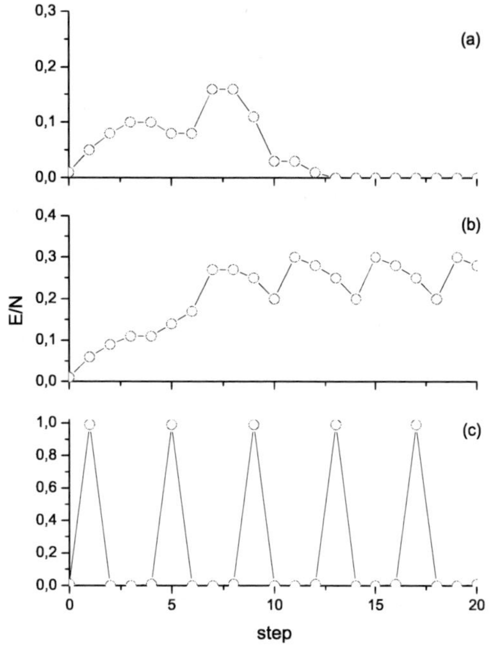

Figure 3. Variation of the fraction of the excited nodes with time (a) in a scale-free network, (b) in a scale-free network with active centre (single node), (c) in a scale-free network with the external source of the activator. Parameters of the model are $h = 0.75$, $g = 0.12$, $\tau_e = 1$, $\tau_r = 2$.

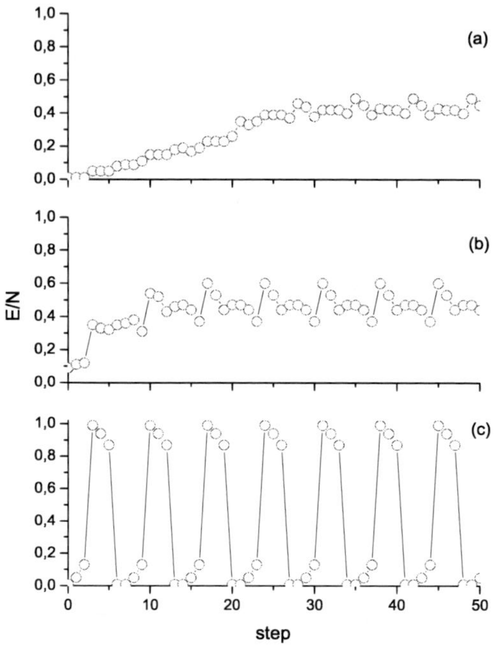

Figure 4. Variation of the fraction of the excited nodes with time (a) in a scale-free network, (b) in a scale-free network with active centre (nodes with 4 connections), (c) in a scale-free network with the external source of activator. Parameters of the model are $h = 2$, $g = 0.75$, $\tau_e = 3$, $\tau_r = 3$.

network (Fig. 3(b)). External source of the activator leads to oscillations, whose amplitude depends on the initial fraction of the excited nodes (Fig. 3(c)).

The values of model parameters have a drastic influence on the spread of excitation over the network. For example, the excitation does not spread across the network when the parameters of the model are $h = 5$, $g = 0.5$, $\tau_e = 3$, $\tau_r = 4$ and a randomly selected node is initially excited. Activator produced only by one node is not enough to excite the neighbouring nodes (high threshold of excitation, h). In this case, either presence of an active centre or receiving the activator from the external permanent source at each step may not affect the propagation of the excitation. Nevertheless, the excitation will propagate when the parameter g is greater, or when each node receives the amount of activator equals to 3 at each step. However, it is possible to increase the fraction of the nodes in the excited state with a help of the 'active' nodes or external influence (Fig. 4).

3.2. Inhomogeneous excitable medium

In inhomogeneous excitable medium, each member (site) has an individual set of parameters, i.e. each person (site) has their own temper and manner. Parameter h is a real positive number. We suppose that the values of the parameter h among the group have a lognormal distribution. σ_h is the standard deviation of the distribution. Parameter g is also a real positive number, $0 \leqslant g \leqslant 1$. In our simulation, we use normal distribution for this parameter. σ_g is the standard deviation of the distribution. Parameters τ_e, τ_r are positive integers. We used the Poisson

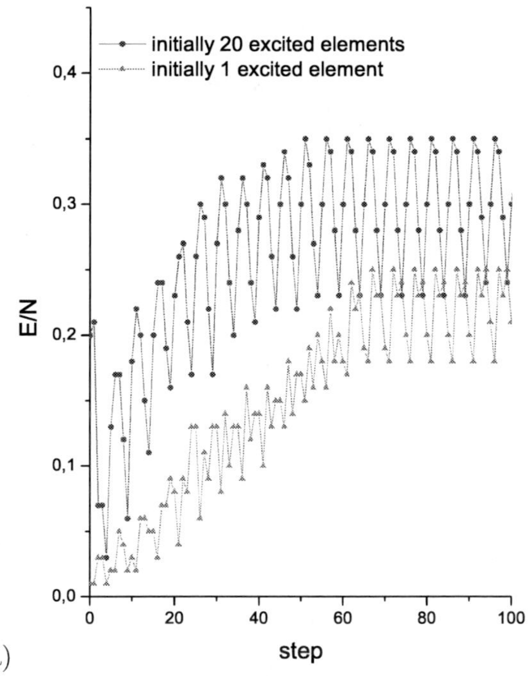

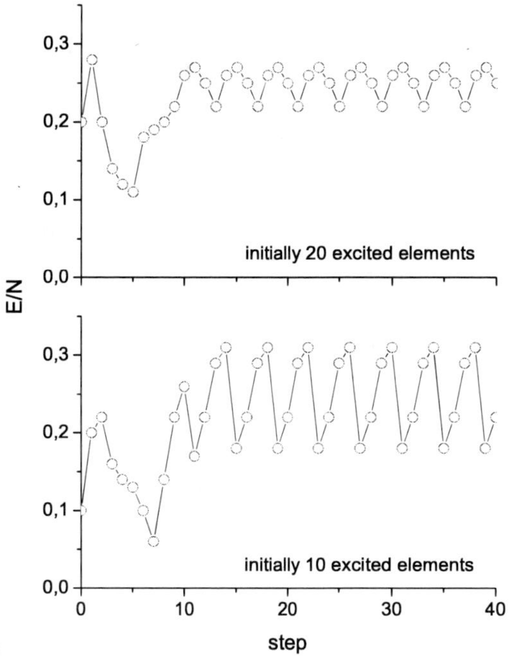

Figure 5. Variation of the fraction of the excited nodes with time in scale-free network. (a) Lognormal distribution of the parameter h. Parameters of the model are $h = 1, \sigma_h = 0.5, g = 0.9, \tau_e = 2, \tau_r = 2$. (b) Normal distribution of the parameter g. Parameters of the model are $h = 1, g = 0.5, \sigma_g = 0.1, \tau_e = 1, \tau_r = 2$.

distribution for these parameters. λ_{τ_e}, λ_{τ_r} are parameters for the Poisson distribution. We shift the values, obtained by using the Poisson distribution, by one unit to avoid zero values for parameters τ_e, τ_r.

We examined the effect of distribution of each parameter individually, i.e. only values of the one parameter have a distribution. The biggest impact has distribution of parameter h, the smallest impact has the distribution of the parameter g. The fraction of the excited nodes decreases as the number of initially excited nodes decreases, when we use lognormal distribution of the parameter h (Fig. 5(a)). Herewith the time of excitation propagation increases. In this case, there appears a mode in which only a fraction of the nodes is involved in the transmission of excitation through the network. Initial conditions can affect the amplitude of oscillation when we use normal distribution of the parameter g (Fig. 5(b)). The initial conditions have a little effect on the behaviour of the model with the distribution of parameter τ_e. Initial conditions have a little impact when we use distribution of all parameters.

We examined the effect of 'active' centre and external source of activator in scale-free network with a distribution of all parameters (Fig. 6). External source of the activator increases the fraction of the excited nodes (Fig. 6(b)). Increasing the amount of the permanent excited elements increases the fraction of the excited nodes. Location of the active centres does not affect at the level of excitation (Fig. 6(c)).

4. Conclusion

For the homogeneous scale-free network, we found that

- the initial conditions have a little effect on the behaviour of the model;

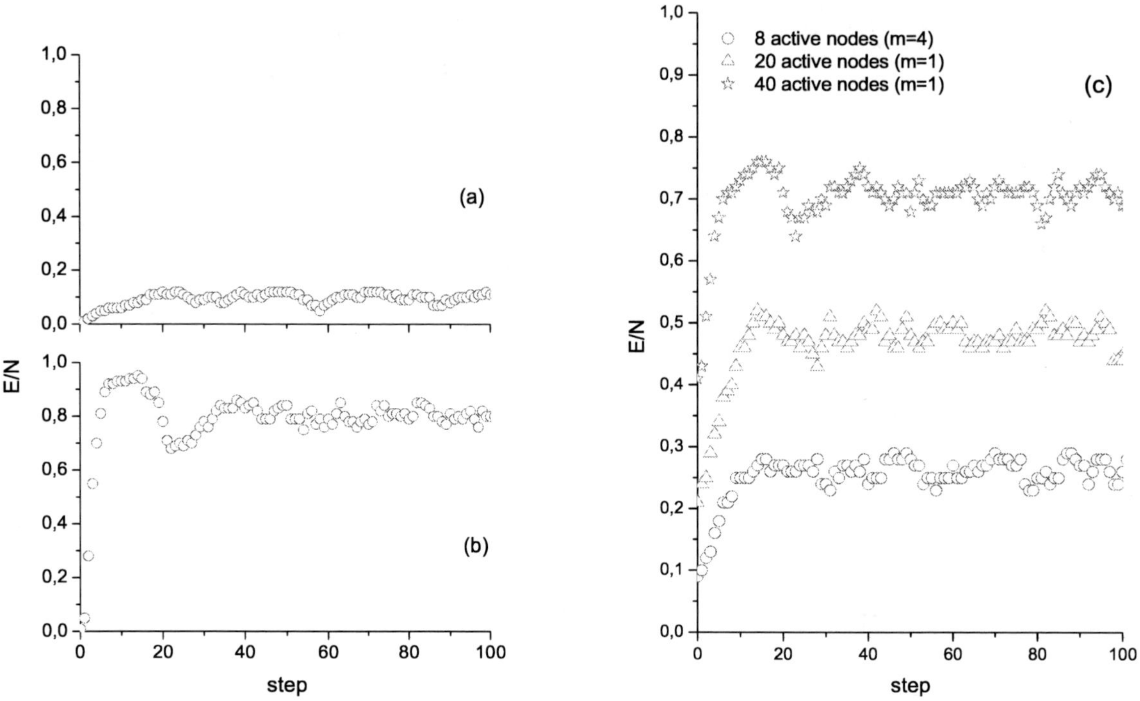

Figure 6. Inhomogeneous scale-free network. The effect of 'active' centre and external source of activator: (a) there is no external influence, (b) external source of the activator, (c) active centre (variable m means a number of connections). Initial conditions are single node (hub) is in the excited state. Parameters of the model are $h = 1, \sigma_h = 0.5, g = 0.6, \sigma_g = 0.06, \lambda_{\tau_e} = 20, \lambda_{\tau_r} = 1$.

- behaviour of the model is more dependent on the parameters;
- in the cases, when the excitation does not propagate, active centre cannot change the behaviour of the model;
- external source of the activator leads to oscillations, whose amplitude depends on the initial fraction of the excited nodes.

For the inhomogeneous scale-free network, we found that

- the initial conditions have a little effect on the behaviour of the model;
- 'active' nodes can increase the fraction of the excited nodes by increasing the number of 'active' nodes, rather than by their location on the network;
- external source of the activator increases the fraction of the excited nodes.

References

[1] Castellano C, Fortunato S and Loreto V 2009 *Rev. Mod. Phys.* **81**(2) 591–646
[2] Klimek P, Bayer W and Thurner S 2011 *Physica A* **390** 3870–3875
[3] Tarasevich Y Y and Zelepukhina V A 2015 *Comp. Res. Model.* **7** 177–183 in Russian
[4] Wiener N and Rosenblueth A 1946 *Arch. Inst. Cardiol. Mex.* **16** 205–265
[5] Albert R and Barabási A L 2002 *Rev. Mod. Phys.* **74**(1) 47–97
[6] Zykov V S and Mikhailov A S 1986 *Sov. Phys. Doklady* **31**(1) 51–55
[7] Mikhailov A S 1990 *Foundations of Synergetics I: Distributed active systems* Springer series in synergetics (Springer) ISBN 978-3-642-78558-0

International Conference on Computer Simulation in Physics and Beyond 2015 IOP Publishing
Journal of Physics: Conference Series **681** (2016) 012009 doi:10.1088/1742-6596/681/1/012009

Optimization problems for WSNs: trade-off between synchronization errors and energy consumption

Larisa Manita
Higher School of Economics (HSE) Moscow Institute of Electronics and Mathematics, Russia
E-mail: lmanita@hse.ru

Abstract. We discuss a class of optimization problems related to stochastic models of wireless sensor networks (WSNs). We consider a sensor network that consists of a single server node and m groups of identical client nodes. The goal is to minimize the cost functional which accumulates synchronization errors and energy consumption over a given time interval. The control function $u(t) = (u_1(t), \dots, u_m(t))$ corresponds to the power of the server node transmitting synchronization signals to the groups of clients. We find the structure of extremal trajectories. We show that optimal solutions for such models can contain singular arcs.

1. Model

Power consumption, sychronization and network lifetime are very important objectives in analysis of wireless sensor networks [1]–[6]. The majority of studies on WSNs focus on communication protocols, network topology and technical designs. Some papers are concerned with optimization problems for WSNs; for example, in [4, 5] the network lifetime maximization problem is studied. In this paper we consider a mathematical model related to networks whose nodes are equipped with noisy non-perfect clocks. The task of optimal clock sychronization in such networks is reduced to the classical control problem. Its functional is based on the trade-off between energy consumption and mean-square synchronization error.

We will consider the model of clock synchronization in large wireless sensor network proposed in [2]. The sensor network consists of a single server node (denoted by 0) and m groups

$$\Gamma_i = \{(i,1), (i,2), \dots, (i,N_i)\}, \qquad i = \overline{1,m},$$

of identical client nodes labeled by (i,j). Let $x_j^{(i)} \in \mathbb{R}^1$ be a state of the node (i,j) having the meaning of a local clock value at this node. The network evolves in time $t \in \mathbb{R}_+$ as follows.

1) The node 0 is a time server with the perfect clock: $\frac{d}{dt} x^{(0)}(t) = v > 0$.

2) The client nodes (i,j) are equipped with non-perfect clocks with random Gaussian noise

$$\frac{d\,x_j^{(i)}(t)}{dt} = v + \sigma_i\, dB_j^{(i)}(t) + \text{synchronizing jumps},$$

where $B_j^{(i)}(t)$, $i = 1, \dots, m$, $j = 1, \dots, N_i$, are independent standard Wiener processes, $\sigma_i > 0$ corresponds to the strength of the noise and "synchronizing jumps" are explained below.

3) At random time moments the server node 0 sends messages to client nodes. Let u_i be the intensity of the message flow issued from the server to the group Γ_i. A message directed to the group Γ_i choses a destination node $(i,j) \in \Gamma_i$ at random (with probability $1/N_i$). The client (i,j) that receives at time τ the message from the node 0 immediately ajusts its clock to the current value of $x^{(0)}$: $x_j^{(i)}(\tau+0) = x^{(0)}(\tau)$, $x_l^{(k)}(\tau+0) = x_l^{(k)}(\tau)$, $(k,l) \neq (i,j)$. Hence the client clocks $x_j^{(i)}(t)$, $t \geq 0$, are stochastic processes which interact with the time server. The function $R_i(t) = \mathsf{E}\frac{1}{N_i}\sum_{j=1}^{N_i}\left(x_j^{(i)}(t) - x^{(0)}(t)\right)^2$ is a cumulative measure of desynchronization between the group of clients Γ_i and the server node. Here E stands for the expectation. It was proved in [2] that the function $R_i(t)$ satisfies the differential equation $\dot{R}_i = -u_i R_i + N_i \sigma_i^2$.

In this paper we consider the problem of minimizing of the functional which accumulates clock synchronization errors in the clients nodes and the energy consumption of the server over some time interval $[0,T]$. The control function $u(t) = (u_1(t), \ldots, u_m(t))$ corresponds to the power of the server node transmitting synchronization signals to the groups of clients. We prove that optimal solutions in the control problem for the wireless sensor network with a single time server node and m groups of identical client nodes exist. We use the Pontryagin Maximum Principle to define the structure of the optimal control. We show that optimal solutions contain singular arcs [7, 8], i.e., trajectories on which the control is not uniquely determined from the maximum condition. For the case $m = 1$ (the network with a single time server node and one group of client nodes) we find the structure of the optimal trajectories for all possible parameter values. We show that for sufficiently large u_0 the solutions contain singular arcs. We study the the behaviour of the optimal solutions as $T \to \infty$. Condition, under which the optimal solutions necessarily contain the singular arc, is given. For the case $m = 2$ (the network with a single time server node and two groups of client nodes) we study a control problem with two-dimensional control. Using the structure of the optimal control we obtain numerical solutions.

2. Optimal control problem

Consider the following optimal control problem

$$\int_0^T \left(\sum_{i=1}^m \alpha_i R_i(t) + \beta \sum_{i=1}^m u_i(t)\right) dt \to \inf \tag{1}$$

$$\dot{R}_i(t) = -u_i(t)R_i(t) + g_i, \quad R_i(0) = R_{i0}, \quad i = 1, \ldots, m \tag{2}$$

$$\sum_{i=1}^m u_i(t) \leq u_0, \quad u_i(t) \geq 0, \quad i = 1, \ldots, m \tag{3}$$

where $\alpha_1, \ldots, \alpha_m$, β are some positive constants.

Denote $R(t) = (R_1(t), \ldots, R_m(t))$, $R_0 = R(0)$, $\alpha = (\alpha_1, \ldots, \alpha_m)$, $g = (g_1, \ldots, g_m)$.

Proposition 1.

(i) For any R_0 and any parameter values T, u_0, β, α, g there exists a solution $(\hat{R}(t), \hat{u}(t))$ to the problem (1)-(3).

(ii) For any optimal solution $\widehat{R}(t)$ there exists a time moment τ, $0 \leq \tau < T$, such that $\hat{u}(t) = 0$, $t \in [\tau, T]$, i.e., the sending messages at times close to T is not optimal.

The proof of this proposition is similar to the proof in the case $m = 1$ (see [9]).

To construct an optimal control we apply the Pontryagin Maximum Principle [11] to the problem (1)-(3). Let $\left(\widehat{R}(t), \widehat{u}(t)\right)$ be an optimal solution. Then there exists a continuous function $\psi(t) = (\psi_1(t), \ldots, \psi_m(t))$ such that for all $t \in (0,T)$ the following ***maximum condition*** holds

$$H\left(\widehat{R}(t), \psi(t), \widehat{u}(t)\right) = \max_{u \in U} H\left(\widehat{R}(t), \psi(t), u\right) \tag{4}$$

Here H is the Hamiltonian function and U is the control set

$$H(R, \psi, u) = -\Big(\sum_{i=1}^{m} \alpha_i R_i(t) + \beta \sum_{i=1}^{m} u_i(t) + \sum_{i=1}^{m} \psi_i(-u_i R_i + g_i)\Big)$$

$$U = \Big\{u(t) = (u_1(t), \ldots, u_m(t)) : u_i(t) \geq 0,\ i = 1, \ldots, m,\ \sum_{i=1}^{m} u_i(t) \leq u_0\Big\}$$

The function $\psi(t)$ satisfies the adjoint equation $\dot{\psi}(t) = -\frac{\partial H(\widehat{R}(t), \psi(t), \widehat{u}(t))}{\partial R}$ (*a.e.*) and the transversality condition $\psi(T) = 0$.

The solutions $(R(t), \psi(t))$ of the Hamiltonian system

$$\dot{\psi} = -\frac{\partial}{\partial R} H(R, \psi, \widehat{u}), \qquad \dot{R} = \frac{\partial}{\partial \psi} H(R, \psi, \widehat{u}) \tag{5}$$

where $\widehat{u}(t)$ satisfies the maximum condition (4) are called *extremals*.

Denote $H_0(R, \psi) = -\sum_{i=1}^{m} (\alpha_i R_i + g_i \psi_i)$, $H_i(R, \psi) = \beta + R_i \psi_i$ $(i = 1, \ldots, m)$. Then $H = H_0 - \sum_{i=1}^{m} u_i H_i$. The maximum condition (4) leads to

$$\widehat{u}_i(t) = \begin{cases} 0, & H_i(t) > 0 \\ u_0, & H_i(t) < 0 \text{ and } H_i(t) < \min_{j \neq i} H_j(t) \end{cases} \tag{6}$$

The optimal control $\widehat{u}$ is undetermined if there exists an interval (t_1, t_2) and indexes k, i, $k \neq i$, such that $H_i(t) = H_k(t) < 0$ or $H_i(t) = 0 \leq \min_{j \neq i} H_j(t)$ for all $t \in (t_1, t_2)$. The corresponding extremal $(R(t), \psi(t))$ is called a *singular* one on the interval (t_1, t_2).

2.1. Case $m = 1$

Let the sensor network contain one group of client nodes. Then the optimal control problem (1)-(3) has the form

$$\int_0^T (\alpha_1 R_1(t) + \beta u_1(t))\, dt \to \inf \tag{7}$$

$$\dot{R}_1(t) = -u_1(t) R_1(t) + N_1 \sigma_1^2,\ R_1(0) = R_{10} \tag{8}$$

$$0 \leq u_1(t) \leq u_0 \tag{9}$$

The maximum condition (6) yields: $\hat{u}_1(t) = 0$ if $H_1(t) > 0$ and $\hat{u}_1(t) = u_0$ if $H_1(t) < 0$ where $H_1(R, \psi) = -\beta - R_1 \psi_1$. Suppose that $H_1(t)$ vanishes over an interval (t_1, t_2). Then the maximum condition gives no information about the value of the optimal control $\hat{u}_1(t)$. To compute a singular control we differentiate the identity $H_1(R(t), \psi(t)) \equiv 0$. It is known that a non-zero coefficient of control u_1 can arise for the first time in the derivative of H_1 even order $2q$. The number q is called the *order* [10] of the singular extremal.

It was proved that for suffiently large u_0 a singular control is realised in the problem (7)-(9).

Proposition 2 ([9]). *Let* $\sqrt{\frac{\alpha_1 g_1}{\beta}} \leq u_0$.

(i) In the problem (7)-(9) there exists a singular solution of order 1

$$\hat{R_{1s}}(t) \equiv \sqrt{\frac{g_1\beta}{\alpha_1}}, \quad \psi_{1s}(t) \equiv -\sqrt{\frac{\alpha_1\beta}{g_1}} \tag{10}$$

and the corresponding singular control is $u_{1s} = \sqrt{\frac{\alpha_1 g_1}{\beta}}$.

(ii) The optimal control $\hat{u}_1(t)$ *has one of the following forms*

$$(a) \quad \hat{u}_1(t) = 0,\ t \in (0,T);$$

$$(b) \quad \hat{u}_1(t) = \begin{cases} u_0, & t \in (0,t_1) \\ 0, & t \in (t_1,T) \end{cases}; \quad (c) \quad \hat{u}_1(t) = \begin{cases} u_s, & t \in (0,t_1) \\ 0, & t \in (t_1,T) \end{cases}$$

$$(d) \quad \hat{u}_1(t) = \begin{cases} 0, & t \in (0,t_1) \\ u_s, & t \in (t_1,t_2) \\ 0, & t \in (t_2,T) \end{cases}; \quad (e) \quad \hat{u}_1(t) = \begin{cases} u_0, & t \in (0,t_1) \\ u_s, & t \in (t_1,t_2) \\ 0, & t \in (t_2,T) \end{cases}$$

i.e., the number of control switchings does not exceed 2 and the optimal solutions may contain the singular solution (*cases* $(c)-(e)$).

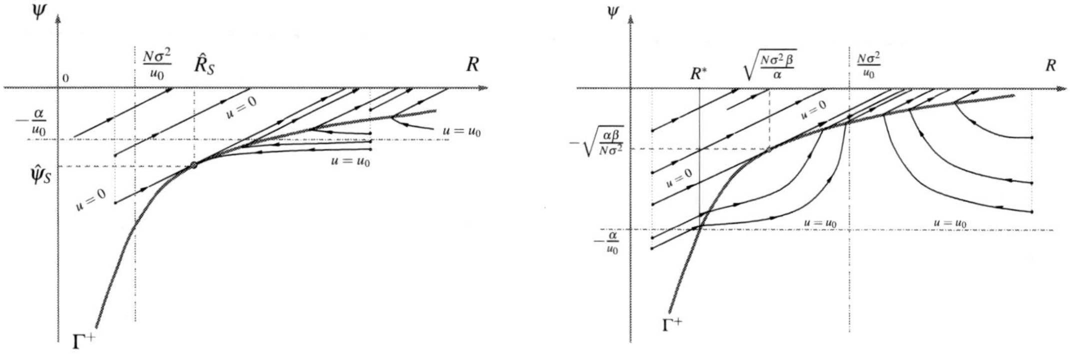

Figure 1: The plots show the optimal extremals and optimal controls in singular case (left) and nonsingular case (right). Here Γ^+ is the switching curve, i.e., a set of points such that $H_1 = 0$.

Proposition 3 ([9]). *Let* $\sqrt{\frac{\alpha N\sigma^2}{\beta}} > u_0$. *Then the optimal solutions are nonsingular. The optimal control* $\hat{u}(t)$ *has one of the following forms*

$$\hat{u}(t) = \begin{cases} 0, & t \in (0,t_1) \\ u_0, & t \in (t_1,t_2) \\ 0, & t \in (t_2,T) \end{cases} \quad \hat{u}(t) = \begin{cases} u_0, & t \in (0,t_1) \\ 0, & t \in (t_1,T) \end{cases}; \quad \hat{u}(t) = 0,\ t \in (0,T);$$

i.e., the optimal control switches between $u = 0$ *and* $u = u_0$ *and the number of switchings does not exceed 2.*

The next statement describes the behaviour of the optimal solutions as $T \to \infty$. Let $J(T)$ be an optimal cost functional value.

Proposition 4. *For any* $R(0)$ *we have*

1. $R(T) \longrightarrow R_T$ *and* $\frac{J(T)}{T} \longrightarrow \alpha_1 R_T$ *where* $R_T = \begin{cases} \frac{g_1}{u_0} + \frac{\beta u_0}{\alpha_1}, & \text{if } \sqrt{\frac{\alpha_1 g_1}{\beta}} > u_0 \\ 2\sqrt{\frac{\beta g_1}{\alpha_1}}, & \text{if } \sqrt{\frac{\alpha_1 g_1}{\beta}} \le u_0 \end{cases}$
2. *If* $\sqrt{\frac{\alpha_1 g_1}{\beta}} \le u_0$ *then optimal solutions contain the singular solution (10).*

2.2. Case $m = 2$. Numerical examples

Let the sensor network contain two groups of client nodes. Then the optimal control problem (1)-(3) has the form

$$\int_0^T (\alpha_1 R_1(t) + \alpha_2 R_2(t) + \beta (u_1(t) + u_2(t)))\, dt \to \inf$$

$$\dot{R}_i(t) = -u_i(t) R_i(t) + g_i, \quad R_i(0) = R_{i0} \quad (i = 1, 2)$$

$$u_1(t) + u_2(t) \le u_0, \quad u_1(t) \ge 0, \quad u_2(t) \ge 0$$

Even for $m = 2$ the behaviour of the optimal solutions can be very complicated [7, 8]. So we solve the problem numerically. To do this we write the Hamiltonian system (5)

$$\dot{\psi}_i = \alpha_i + \widehat{u}_i(t)\, \psi_i, \qquad \dot{R}_i = -\widehat{u}_i(t) R_i(t) + g_i, \qquad i = 1, 2 \tag{11}$$

where $\hat{u}(t) = (\hat{u}_1(t), \hat{u}_2(t))$ satisfies the maximum condition (6), i.e.,

$$\hat{u}(t) = \begin{cases} (u_0, 0), & (H_1(t) < 0)\ \&\ (H_1(t) < H_2(t)) \\ (0, u_0), & (H_2(t) < 0)\ \&\ (H_2(t) < H_1(t)) \\ (0, 0), & H_1(t) > 0, \quad H_2(t) > 0 \end{cases} \tag{12}$$

Here $H_i(t) = \beta + R_i(t)\psi_i(t)$ $(i = 1, 2)$. If $H_1(t) = 0$, $H_2(t) > 0$ over an interval (t_1, t_2), then the corresponding extremal is called *singular on the interval* (t_1, t_2) *with respect to component* (1) or (1)-*singular.* If $H_2(t) = 0$, $H_1(t) > 0$, $t \in (t_1, t_2)$, then the corresponding extremal is called (2)-*singular.* If $H_1(t) = H_2(t) < 0$, $t \in (t_1, t_2)$, then the corresponding extremal is called (1, 2)-*singular.*

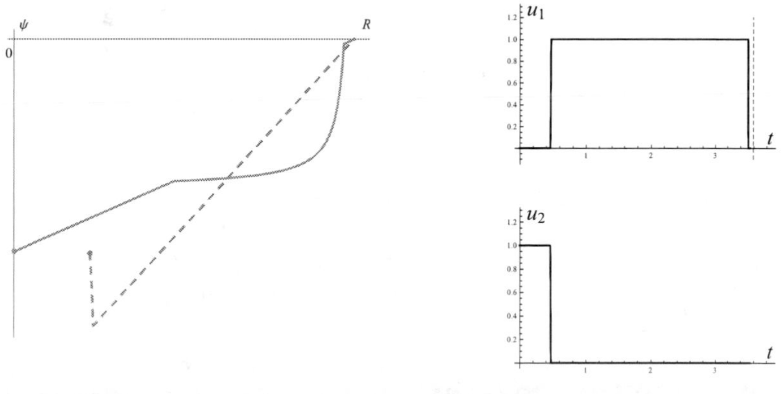

Figure 2: For this plots $R_1(0) = 0$, $R_2(0) = 4.49$, $g_1 = 20$, $g_2 = 5$, $\alpha_1 = 1.6$, $\alpha_2 = 1$, $T = 3.55$, $u_0 = 1$, $\beta = 1$

We conclude this section by some numerical results. We solve (11)-(12) with the boundary conditions $R_i(0) = R_{i0}$, $\psi_i(T) = 0$ $(i = 1, 2)$. Fig. 2 and Fig. 3 represent optimal extremals and optimal controls. (R_1, ψ_1) is the red solid line and (R_2, ψ_2) is the brown dashed line.

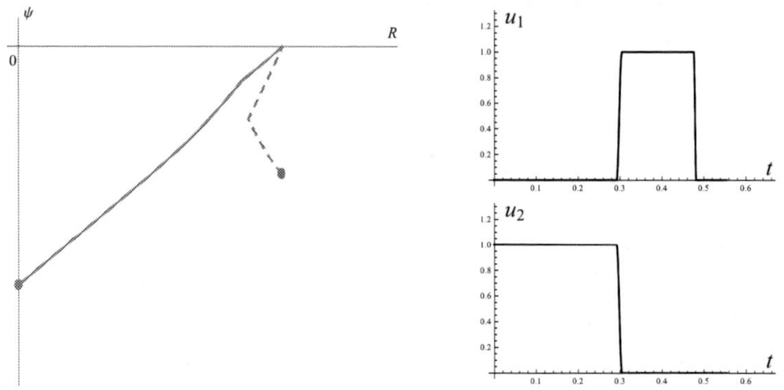

Figure 3: For this plots $R_1(0) = 0$, $R_2(0) = 9.75$, $g_1 = 20$, $g_2 = 5$, $\alpha_1 = 1.6$, $\alpha_2 = 1$, $T = 0.55$, $u_0 = 1$, $\beta = 1$

3. Conclusions

We considered the control problem for wireless sensor networks with a single time server node and m groups of identical client nodes. The cost functional of this control problem accumulates clock synchronization errors in the clients nodes and the energy consumption of the server over some time interval $[0, T]$. We proved that the sending messages at times close to T is not optimal, i.e., for any optimal solution $\widehat{R}(t)$ there exists a time moment τ, $0 \leq \tau < T$, such that $\hat{u}(t) = 0$, $t \in [\tau, T]$. We showed that the optimal solutions can contain singular arcs.

Acknowledgement

This research was supported by the Russian Foundation for Basic Research (projects No. 14-01-00784)

References

[1] Sundararaman B, Buy U, Kshemkalyani A D 2005 Clock synchronization for wireless sensor networks: a survey Ad Hoc Networks 3 pp 281–323
[2] Manita A 2014 Clock synchronization in symmetric stochastic networks Queueing Systems V 76 2 pp 149-180
[3] Feistel A, Wiczanowski M, Stanczak S 2007 Optimization of Energy Consumption in WSNs, Proc. ITG/IEEE Int. Workshop on Smart Antennas (WSA, Wien)
[4] Pourazarm S, Cassandras C G and Wang T 2015 Optimal routing and charging of energy-limited vehicles in traffic networks Int. J. Robust. Nonlinear Control doi: 10.1002/rnc.3409
[5] Pourazarm S, Cassandras C G 2015 Lifetime Maximization of Wireless Sensor Networks with a Mobile Source Node arXiv:1508.07343
[6] Wang L, Xiao Y 2005 Energy Saving Mechanisms in Sensor Networks BroadNets 2005 V 1 pp 724 - 732
[7] Zelikin M I, Borisov V F 2003 Optimal chattering feedback control J. Math. Sci. (N. Y.) 114:3 pp 1227–1344
[8] Lokutsievskiy L V 2014 Singular regimes in controlled systems with multidimensional control in a polyhedron Izvestiya: Mathematics 78(5):1006
[9] Manita L 2014 Controlling of clock synchronization in WSNs: structure of optimal solutions arXiv:1408.1434 p 14
[10] Powers W F 1980 On the Order of Singular Optimal Control Problems J. of Opt. Theory and Applications V 32 4 pp 479-489
[11] Pontryagin L S, Boltyanskii V G, Gamkrelidze R V, Mishchenko E F 1962 The Mathematical Theory of Optimal Processes (John Wiley)

Journal of Physics: Conference Series **681** (2016) 012010 doi:10.1088/1742-6596/681/1/012010

Open Marketplace for Simulation Software on the Basis of a Web Platform

A P Kryukov[1] and A P Demichev[2]

Skobeltsyn Institute of Nuclear Physics, Lomonosov Moscow State University, Leninskie gory, 1, Moscow, 119991, Russia

E-mail: [1] kryukov@theory.sinp.msu.ru, [2] demichev@theory.sinp.msu.ru

Abstract. The focus in development of a new generation of middleware shifts from the global grid systems to building convenient and efficient web platforms for remote access to individual computing resources. Further line of their development, suggested in this work, is related not only with the quantitative increase in their number and with the expansion of scientific, engineering, and manufacturing areas in which they are used, but also with improved technology for remote deployment of application software on the resources interacting with the web platforms. Currently, the services for providers of application software in the context of scientific-oriented web platforms is not developed enough. The proposed in this work new web platforms of application software market should have all the features of the existing web platforms for submissions of jobs to remote resources plus the provision of specific web services for interaction on market principles between the providers and consumers of application packages. The suggested approach will be approved on the example of simulation applications in the field of nonlinear optics.

1. Introduction

Currently the development of scientific web applications is booming around the world. In particular, today scientific services in the Internet provide the access to various data (for example, http://www.nature.com/sdata), scientific experimental facilities (for example, High Speed Networking with Subaru Telescope in Hawaii, http://www.naoj.org/), and educational materials (Virtual Learning Environment [1]). In many cases researchers need to run a large number of similar computing tasks to solve some problem in a particular application area. Quite often, these tasks are performed by using already existing specialized application packages, the tasks being identical in form and differing only in the values of the input data (parameters). In this case, pre-installation and configuration of commonly used application packages on a computing resource and the availability of specialized web services as well as appropriate user web interfaces allow owners of computing resources to increase efficiency of their use (Software as a Service model; SaaS). The set of specialized interconnected web services and web application interfaces is called a web platform for remote access to computing resources (see, e.g., a short survey in [2]). Development of web platforms for scientific research in different areas are widely underway around the world. Examples of such developments: nanoHUB (http://www.nanohub.org); Yabi (https://ccg.murdoch.edu.au/yabi); e-Science Central (http://www.esciencecentral.co.uk); other examples and discussions see in [2].

Each individual pre-arranged task (a launch of an application package, an access to a data storage, and so on) is called "a tool". With the access to the tools provided by a platform, a user only has to specify particular values of the input parameters or input files, and the rest of the task description is generated automatically. Thus the user can carry out the generating of the task, its submission, monitoring, and getting the results via a standard web browser. This approach proves to be particularly effective and convenient when the full research requires a set of application packages rather than just one of them. Moreover, often the output of some package is the basis for the generating of the task for the next application package in a series of stages of the research. Such sets of tasks are called workflows. Besides the launch of computing packages, workflow may include other operations such as a query to relational database, visualization of the results or obtaining online data from an operating scientific facility. In addition, providers can offer prepared in advance debugged and ready to execution workflows, constructed of an application software from third-party providers. This is in direct compliance with actively developing concept of Business Process as a Service (BPaaS) [3].

The main functional requirements for the platforms are the following: management of user credentials granting the right to use the available resources; remote administration of the web platform via a web browser; job execution management; data files transfer management; tools (services) management. These requirements correspond to the web platforms with minimal functionality, namely the ones intended solely for remote job execution. In the case of web platforms with more general functionality additional requirements arise.

Further line of development of the web toolkit, suggested in this work, may be related with the improvement of the technology of remote deployment of new application software on resources interacting with the web platforms. This approach will help to overcome an important problem associated with the use of the SaaS model in scientific areas, namely, limited set of application packages offered by SaaS providers. Often, these providers focus on mass servicing of single-type customers and scientific activity is beyond the scope of their interests. Currently, the provision of services for providers of application software in the context of scientific-oriented web platforms is not developed enough. Although some implementations (for example, the nanoHUB and e-Science Central) have services for remote application software deployment, they are still insufficient to ensure the creation of a web platform capable of performing the whole range of tasks characteristic for a free open market of application softwares.

General principles of constructing of the web platforms providing open marketplace for scientific application sofware in the framework of the SaaS model are discussed in the next section. Section 3 briefly describes possible operation algorithms of such a web platform.

2. Open Marketplace Paradigm

In this paper we propose basic features of the design of the web platform through which consumers of web services and their providers can interact with each other, and the platform needs to perform a range of functionalities, which are characteristic for a free open market. These tasks and the corresponding functional modules of the platform under development are presented in the Table 1.

We shall use the following nomenclature: consumer is a platform user requesting software as a service via the platform, provider is a user providing a software to consumers by means of the platform. There are two ways for offering an application software by means of the web platform.

In the first case, the software is deployed on provider's own resources. Thus the provider offers not only a software, but also the computing resources. In this case the provider can use different types of resources, e.g., supercomputers, clusters, standalone servers, cloud systems. Of course, these computing facilities are not necessarily in property of the provider but can be leased resources (e.g., in a public cloud). This way is especially suits those providers who want to exclude completely access both users and platform administrators to the software (both

Table 1. Tasks which are characteristic for a free open market and the corresponding functional modules of the platform under development.

Open market tasks	Content of the task	Modules (functionality) of the
Information	Marketplace gives participants the information about quantity and quality of producs	service catalog with search engine;
Matchmaking	Matchmaking of users and SW providers	search engine of the service catalog; module for blog posting and for assigning rating marks
Price setting	The price is formed on the market on the basis of supply and demand given the competition	price list in the catalog; billing module;
Regulation	Market balances the supply and demand	service catalog with search engine; billing module; module of verifying compliance with the terms of service; module of e-wallets;
Stimulating innovation products	Market stimulates the development of innovation products, the reduction of costs and increase the quality	billing module; module for blog posting and for assigning rating marks
Coordination	Market encourages service providers to create the necessary public services with the lowest cost	module for blog posting and for assigning rating marks
Optimization of market supply	Identification of inefficient products	logging module; billing module;

to source code and binary executables). It is worth noting that in the case of provider's own resources a problem of reliability of delivery of services may appear.

In the second case, the software is deployed on platform resources. In this case provider offers only a software as a service; a fee for the use of computing resources goes to the owners of the platform resources. The web platform have to provide a service for remote installation of the application software by users (providers) themselves. In this case the type of the resources is practically limited to the cloud systems. Indeed, an independent remote installation of application software on supercomputers by users is highly problematic both in terms of security and from a technical point of view: for a deployment of a software the user must deeply know the features of the architecture and software of the specific supercomputer. A natural solution to these problems is the use of virtual machines [5] that provide both safety (isolation of installed software) and feasibility of using the operating system required for the application software to be installed. Thus, since cloud infrastructure provides tools for managing virtual machines, the web platforms allowing software installation must provide access to the cloud systems in the framework "Platform as a Service" (PaaS) model.

3. Web Platform Architecture and Algorithms

Basically, all the web platforms have the typical three-layer architecture (see, e.g., [4]). The first layer is the frontend that provides the user web interface; the second layer is the platform engine responsible for job submitting, obtaining results, execution of workfows, etc. as well as the administration module that is responsible for job management, tool configuration, audit trails and user management; the third layer is a resource manager that exposes data and compute resources to the preceding layer. Also all web platforms have more or less sophisticated security infrastructure.

We will not discuss in details here the standard modules of a web platform responsible for job submission to remote resources (see, e.g., [2] and refs. therein). Instead we concentrate in this work on the specific modules needed for the web platform to operate as a marketplace. Part of such modules were indicated in Table 1. However, in addition the marketplace-like platform have to include a number of other auxiliary services for fully-featured operation. The list of the modules providing marketplace features includes:

- module of software registration;
- service catalog with search engine;
- logging module;
- billing module;
- integrator of conditions for workflows;
- module of verifying complinece with the terms of service;
- software deployment module;
- module of a tool setup;
- module of e-wallets;
- repository of the software codes;
- module for review and blog posting and for assigning rating marks.

Algorithm of the platform operation when a provider is offering his SaaS for consumers depends on the type of resources where the provider deploys the software. In the case of provider's own resources the algorithms reads as follows (see figure 1; the numbers near the arrows corresponds to the items in the list below). The provider

1. registers himself at the platform and acquires a restricted access to the administrative interface;
2. A. in the case of provider's own resources, deploys the software on this resources and provide an access to the software via the SSH protocol;
   (a) if the resource is not a standalone server but an HPC system with a batch manager, the latter should be in the list of managers supported by the platform;
2. B. deploys his software into the platform cloud by means of appropriate platform module (this implies the provider gets an access to a virtual mashine with an appropriate operating system, etc.; PaaS model)
3. sets up his software and converts it in the web tool;
4. registers his web tool in the platform catalog togeter with
   (a) its description in a machine-readable form (e.g., via a completion of an appropriate web form),
   (b) terms of the service (including the price);
5. on regular basis checks his e-wallet and transfers the digital cash obtained to his bank account; this action should be preceded by consumption and logging/billing of the tools offered by the given provider (the corresponding arrows in the figure are marked with (5)).

The steps 3 and 4 may include setting up and registering debugged and ready to execution workflows, constructed of provider's own application software or/and software from third-party providers in the spirit of the BPaaS model [3].

We suppose that each user of the platform (both a customer and a provider) has the electronic (virtual) wallet (e-wallet) containing the conditional means (digital cash) within a platform. Users can transfer their funds to the e-wallets of other users. In reality users also shall have opportunity to transfer funds from the e-wallets to real payment systems (like PayPal) or to bank accounts, but in this work we do not consider this process.

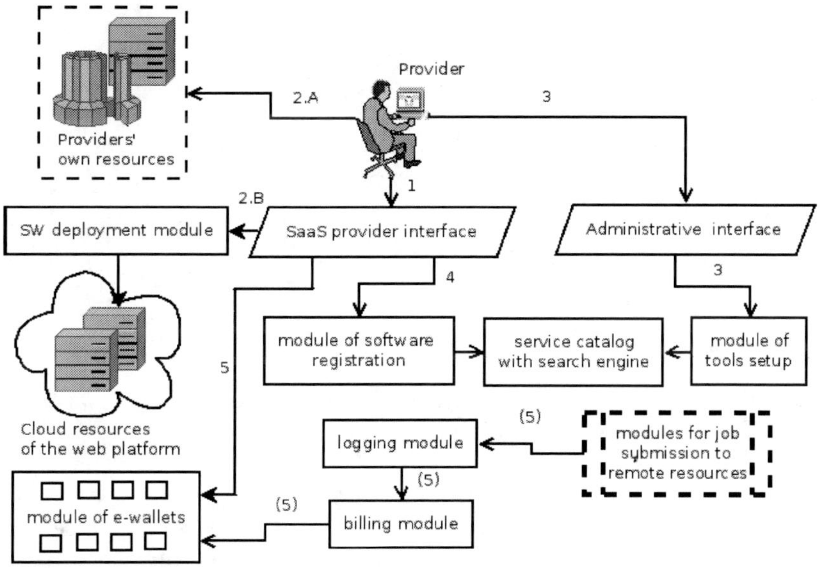

Figure 1. The algorithm of a provider and market modules operation.

For a consumer the both cases (that is separate resources or platform cloud) are equivalent (the type of resources is tansparent for consumers). To use the platform tools, a consumer (see figure 2; the numbers near the arrows corresponds to the items in the list below):

1. chooses from the catalog appropriate tools;
2. designs out of these tools a workflow or select predefine business process (BPaaS);
3. gets terms of services and price for their use;
4. accepts or deny the terms of conditions;
5. if accepts, transfer digital money from his e-wallet to the provider's e-wallet
   (a) the module of verifying compliance with the terms of services checks that the conditions are accepted and ample means have been transferred to the providers' e-wallets;
6. starts the workflow and obtains a result.

An administrator of the web platform provides user registration, moderation of software installation, user support and Quality of Service (QoS) obligations from the side of providers offering their own computing resources.

4. Conclusion

A prototype of such a platform, entitled eSciMart, is currently being developed at SINP MSU. The basis of the prototype platform is web service technologies and, in particular, the service-oriented architecture (SOA) [6], as well as the REST architectural style [7]. The approach

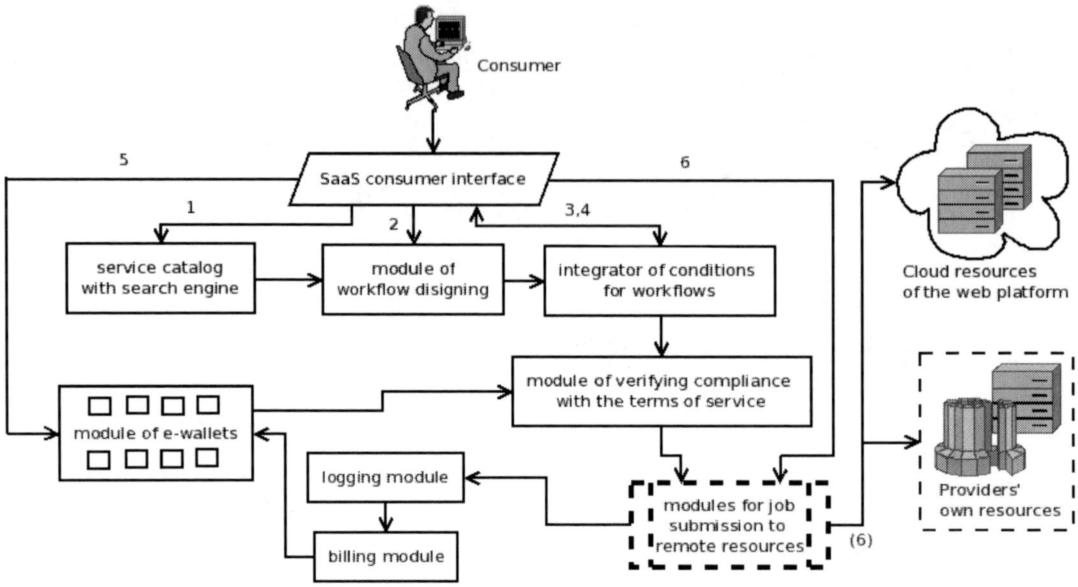

Figure 2. The algorithm of a consumer and market modules operation.

to creation of this web platforms market of application software is based both on the original solutions and on the synthesis and adaptation of the solutions used in research hubs (e.g., nanoHUB; nanohub.org), cloud and grid systems, as well as in on-line application stores. However, unlike the on-line app stores, the platform will not only provide information services for searching the tools needed by users, but also provide the feasibility of direct using of the necessary tools. Thus, the future eSciMart web platform will provide a single entry point both for web service providers and for their customers.

Acknowledgments
This work is supported by the RFBR grant No.15-07-09309.

References
[1] Weller M 2007 *Virtual learning environments: using, choosing and developing your VLE* (London: Routledge)
[2] Polyakov S P, Demichev A P, and Kryukov A P 2015 *Proceedings of the 4th International Young Scientists Conference on Computational Science* (Amsterdam: Elsevier) (to be published)
[3] Barton T and Seel C 2014 *Proceedings of the Conference on Enterprise Modelling and Information Systems Architectures* (Available at: http://subs.emis.de/LNI/Proceedings/Proceedings234/145.pdf)
[4] Hunter A A, Macgregor A B, Szabo T O, Wellington C A and Bellgard M I 2012 *High Performance and Cloud computing. Source Code for Biology and Medicine* **7** 22
[5] James E S and Ravi N 2005 *Virtual Machines: Versatile Platforms For Systems And Processes* (San Francisco: Morgan Kaufmann)
[6] Erl T 2009 *SOA Design Patterns* (New York: Prentice Hall)
[7] Richardson L and Ruby S 2007 *RESTful Web Services* (Cambridge: O'Reilly Media)

The complex of neural networks and probabilistic methods for mathematical modeling of the syntactic structure of a sentence of natural language

A Sboev, R Rybka, I Moloshnikov, D Gudovskikh
NRC "Kurchatov Institute" (Academic Kurachatov sq. 1, Moscow), Department of mathematical modeling and information technologies, Russia

E-mail: sag111@mail.ru, RybkaRB@gmail.com

Abstract. The formalized model to construct the syntactic structure of sentences of a natural language is presented. On base of this model the complex algorithm with use of neural networks founded on data of Russian National language Corpus and set of parameters extracted from this data was developed. The resulted accuracy along with possible accuracy which theoretically could be received with these parameters is presented.

1. Introduction

The growing intensity of information exchange in the virtual environment, in particular in social networks, makes it necessary to use mathematical modeling techniques and complex software for automated analysis of data from the virtual environment, and for displaying social phenomena in them. In this area, typical application tasks are: automated annotation of texts, content analysis of business information, sentiment analysis, an analysis of emotive text, identifying threats in social networks. To solve these problems, the computerization of the main stages of the analysis of unstructured data is crucial. In particular, it refers to a problem of determining the syntactic structure of the text, written in natural language. It is a key addition to the arsenal of BigData technologies for texts analysis. The classic approach to automating analysis of objects and phenomena is their mathematical modeling.

Syntactic Parsing is performed by linguists on the basis of the natural language rules, for which it is not possible to fully build a rigorous mathematical theory. Existing solutions for obtaining the syntactic structure of a sentence are based on two main approaches: the first includes the software implementation of linguistic parsing rules, and the second includes mathematical algorithms with pre-training on a set of examples with syntactic markup. The first are not universal and poorly tolerated in other languages, and time-consuming in implementation (such as the ETAP-3 for the Russian language has been being created more than 20 years). Syntactic analysis methods based on algorithms with pre training include: systems based on a combination of parsing algorithms such as CKY or EARLY, as well as their modifications, with probabilistic grammars (Link Grammar, PCFG); neural network method based on a combination of RAAM and SRN; algorithms based on combination of transition systems and pre-trained classifier. Application of the above techniques in conjunction with the parametric description of the words of sentence has its peculiarities. In particular, the methods of formal grammars are mainly used for languages without direct order of words and with projective connections. Parsers based on recurrent neural networks [1] lose their analysis accuracy with increasing the number of words in a sentence. On the whole, the approach based on neural network models has some advantages because neural networks have some generalizing properties. The authors of [2] [3] have shown that application of convolution neural networks to extract features of sentence in conjunction with methods of deep learning [4] contributes to increase in quality of the shallow parsing and determination of parts of speech when analyzing sentences of the English language. The specialists at Stanford University proposed a method based on the replacement of words and parts of speech by their probability vectors formed based on the Word2Vec model [5] that is trained on a corpus of texts. They created parsers for some languages (English, Chinese, German, Arabic) [6], which are based on this idea. Parsers based on models of transitions [7] are widely used at present, but a full-fledged research of classification models in its composition is not submitted.

This paper proposes a combined approach to the creation of a mathematical model of the syntactic structure of the sentence, which involves:

- mathematical formalization of linguistic procedure of syntactic analysis;
- development of methodology of parameterization of words of sentence for choosing the parameters allowing to establish syntactic relations with minimal ambiguity;
- assessment of the achievable accuracy of syntactic parsing on the basis of used corpus dataset;
- selection of methods for determining syntactic relations and formation of syntactic structure of sentence.

In this study the base "SinTagRus" from the Russian National Corpus [8] is used for training neural network models in the composition of a model of syntactic structure of sentence. It contains sentences with unique morphological and syntactic markup. Examples from the RNC can be regarded as a space of pairs of words that include syntactic relations between words with markings of the relations, the word in its composition, and supplementary information of sentence. In this space, when using neural network methods, actually a measure is created that can be used to assess the syntactic proximity. This leads to application of probabilistic methods to eliminate ambiguities arising after parsing.

2. Material and methods

2.1. Mathematical formalization of linguistic procedure of syntactic parsing

The syntactic structure of natural language includes structures of individual sentences. Syntactic structure of the sentence $F = \{f_1 \dots f_n\}$ (where f_i is word in this sentence) can be formalized in a tree with oriented edges (arcs) $T(W, D_T)$. Here W is set of vertices $w_i = \{f_i, l_i, \overrightarrow{m_i}\}$. The linguistic characteristics of w_i are:

- l_i – the normal form of the word (lemma);
- $\overrightarrow{m_i}$– Morphological characteristics: the part of speech, gender, number, case, etc;
- D_T – a plurality of oriented arcs in the parse tree. In the tree, the number of arcs $N_D = n - 1$, where n is the number of words in the considered sentence.

Thus, linguistic parsing procedure of sentence can be formalized as a sequence of graphs G_i, which should end with the construction of the tree: $G_0 \Rightarrow G_1 \Rightarrow \cdots \Rightarrow G_i \dots \Rightarrow T$. According to the presented formalization, we described an automatic procedure for determining of syntactic relations and for building of syntactic structure of sentences (Syntactic parsing tree) in general forms:

1) subgraph g that consists of 2 vertices and their environment is selected from the graph G_i;
2) subgraph g analyzed by formal neural networks for establishing syntactic relation between 2 selected vertices;
3) if neural network analysis of subgraph g has shown that syntactic link is established between the two selected vertices, then the new arc is added to the graph G_i, and a new graph G_{i+1} is created. If the establishment of syntactic relation does not happen, then the other subgraph g is selected from the graph G_i.

In the event that all vertices that have no relations from examined vertices of graph, and all possible subgraphs g in the graph G_i have been enumerated, but the addition of a new arc does not happen, then following options are available depending on the algorithm of busting subgraphs g from G_i:

1) G_i is a tree. Then parsing of sentence is completed;
2) G_i can be represented in the form of several trees $T_k(W, D_T)$. Then one of them, which is accepted as the final result of the parsing of sentence, is selected using probabilistic methods;
3) G_i cannot be decomposed into a set of several trees. In this case, the syntactic structure of sentences is not determined.

2.2. Parameterization methodology

The purpose of this technique is the choice of parameterization of words of sentence to form an input vector for a neural network. Criterion for selection of a particular parameterization is based on the number of clear determined syntactic relations by the statistical analysis of the texts of RNC. The parameterization was held on four groups of parameters [9] (table 1). They contain common parameters (morphological characteristics of words; punctuation signs and indicator of capital letter) and parameters supplemented by us. The latter includes parameters of potential syntactic relationships (p_sinto) that are established on the basis of morphological characters of two words. P_sinto indicates the possible syntactic connections within a sentence. Thus, efficacy was evaluated by the number of ambiguous relationships: the higher is the number of ambiguous relations, the worse is the efficacy of a set of parameters.

Table 1. Description of sets of parameters

Parameters \ set №	1	2	3	4
Morphological characters	+	+	+	+
Additional		+	+	+
The displacement of the main word to the dependent word of the pair in sentence			+	+
Potential syntactic relations (p_sinto) between main and dependent words (pair of words)			+	+
Potential syntactic relations from the words of the pair to other words in expression				+

2.3. Methods of syntactic parsing

2.3.1. Approaches to building a syntactic parse tree and to formation of training examples to determine syntactic relations

Two approaches were investigated to construct a syntactic parse tree: the first one is based on exhaustive enumerating all possible options for establishing Sinto between words in a sentence; the second one is based on the Covington scheme [7] of incremental parsing. In the first case, training set consists of pairs of words from all RNC sentences divided into two classes: 1) form Sinto; 2) do not form a Sinto. The number of examples in the second class is much larger than in the first. Therefore we build a method of filtering the pairs of words that form sinto. After that we determine syntactic relations in the filtered set of examples.

The essence of the second approach lies in building a model of transitions in three lists of words: the Right list consisting of all unparsed words, the Left list comprising the word for finding relationship between the heads of Right and Left lists, and the intermediate list. If relation between head of Right and Left lists has not been found, then the intermediate list will be replenished with the word from the head of the Left list. Four classes of actions (No-Arc, Shift, Right-arc, Left-arc) are used for relocation of words between the lists and the definition of syntactic context. In this case, the training set consists of examples corresponding to these actions according to the type of sinto.

2.3.2. Determining syntactic relations

The following solutions are used for classification of Sinto and action of transitions:

1) Creating a sequence of classification neural network models to determine a Sinto or action independently (binary classification). The sequence is formed based on the number of examples for each class: from biggest to smallest;

2) Combining Sinto into several groups based on the number of examples for them, and on the accuracy of models for an independent classification of syntactic relations (or actions). Neural network models are created for each group of Sinto;
3) Creating a single model for multiclass classification for all Sinto (or actions).

In the first and second cases, each next classifier is trained on base of a training set which is free from the examples used for learning previous models. For constructing classification model, several methods where investigated. The main goal of them is to determine the syntactic relations of sentence.

We explored a large variety of methods for improving the accuracy of the classification, including AdaBoost, Boosting by filtering, for reducing the input space of attributes, including PCA, ICA, RBFsampler. Methods MLP, SGD [10], SVM strategy one-vs-all [11], PNN, GNT [12], ensembles of decision trees (RFC) [13] in combination with the methods of reducing the dimension of the input space (Nystroem) [14] were also investigated and demonstrated high scores (see section Experiments). For estimates of the models we used indicators of Positive and negative predictive values (PPV and NPV).

3. Experiments.

3.1. Parameterization of words of sentence

The results of Table 2 show that the fourth set of parameters has the best effectiveness. Therefore, the fourth set of parameters will be used for subsequent parameterization of words of the sentence in the input vector for the classifier based on neural network. After selecting the parameterization of words, the RNC data is analyzed to assess the achievable accuracy of syntactic parsing. Syntactic trees in RNC have several properties:

- the parse tree has only one vertex;
- there is only one input connection for all words in a sentence except for vertex;
- syntactic parse tree includes all the words in the sentence.

Table 2. Comparing sets of parameters

№ set of parameters	Average number of ambiguous syntactic relations for word of sentences	Percentage of clear syntactic relations determined
1	102,29	58,48
2	56,28	78,9
3	8,84	85,72
4	1,43	98,91

From this perspective, we have formed the criteria for the effectiveness of a selected set of parameters to syntactic parsing:

- the number of sentences with unambiguous parsing;
- the number of sentences with ambiguous parsing;
- the average number of syntactic trees for ambiguous-parsed sentences.

The evaluation results show that the proportion of uniquely-parsed sentences is 79.9% of the total number of proposals (42.9 thousand). The average number of parses of ambiguous-parsed sentences (20.1%) is 21.3. Classifications based on neural network PNN, MLP and SVM are designed to establish parameter of p_sinto. Average of PPV, NPV of setting p_sinto are very high (PPV=99.9%, NPV=99.8%)

3.2. Building procedure of syntactic parsing

For determine syntactic relation we investigate several approach and classification methods (see chapter 2.3). The best results demonstrate approach based on incremental parsing scheme (Table 3). In this case the training sample is about 1.35 million examples.

Table 3. PPV and NPV for defining actions in incremental parsing scheme. Multiclass is model based on ensembles of decision trees or SVM with strategies such as One-vs-all. Binary is the decision on the basis of a binary classification action when a classifier is constructed for each action.

Variant of descent	Using methods	PPV	NPV
multiclass	SVM (linear kernel)	90.1	91.2
multiclass	Nystroem + RFC	83.8	84.1
binary	SVM+SGD	87.3	88.1
binary	Nystroem + RFC and SVM+SGD	89.2	90.1

Thus, the approach based on combination of the incremental parsing scheme and the classifier based on SVM with linear kernel was selected for the implementation of the methods of modelling syntactic structure of Russian language sentence.

3.3. Testing

The results of testing the implemented model and comparison with other systems are presented in Table 4.

Table 4. Estimation of syntactic parsing and comparison with estimation from literature sources "Estimation 1" obtained by using method presented on previous section. For obtaining "Estimation 2" we add word forms in set of parameters. Estimation from literature sources 1 is based on the use of ETAP-3 [15]. Estimation from literature sources 2 is presented in [16]. UAS – unlabeled attachment score; LAS – labeled attachment score; TRD – is the ratio of number sentences with correct root determination to number of sentences; TSSP –is the ratio of number sentences with right structure of syntactic parse tree to the total number of sentences; TSPT – is the ratio of number sentences with right syntactic parsing tree to the total number of sentences.

Task description	UAS (%)	TRD (%)	LAS (%)	TSPT (%)	TSSP (%)
Our estimations 1	85.81	82.23	79.33	14.05	29.47
Our estimations 2	91.73	88.84	89.39	35.91	52.38
Estimation from literature sources 1	94.3	---	92.3	29.7	37.4
Estimation from literature sources 2	89.4	---	83.4	21.8	33.3

Testing of models to determine syntactic relations on the corpus sentences unused during training showed that the accuracy of determining the type of syntactic relations is equal to 79.33%, and the accuracy of the forming syntactic tree is equal to 14.05%. Adding word forms to the selected set of parameters increases the accuracy of the determination of syntactic relations by 10% and the one of forming syntactic tree by 20.7%.

4. Conclusion

A combined approach based on proposed formalization of linguistic procedure to modeling of syntactic structure of a sentence in Russian language was presented in this paper. Set of combinations of methods of constructing a syntactic structure, of establishment of syntactic relations, and parameterization of words were investigated. The best accuracy of syntactic parsing was obtained by using a combination of: selected parameterization of the words, including the potential syntactic relations; the neural network methods for their establishment; the classifier based on support vector machines with linear kernel and incremental parsing scheme. In general, the results of testing of the syntactic parsing algorithm on the basis of the developed model are higher than ones from the literature sources (see [17], [16]). Estimates of the achievable accuracy of parsing have shown that the quality of the parsing can be significantly improved by using a more accurate algorithm for determining the syntactic relations. In further, our work will focus on improving methods of establishing syntactic relations in a sentence.

5. Acknowledgements

Numerical simulations were performed at supercomputing resources in NRC "Kurchatov Institute", which are supported as the centre for collective usage (project RFMEFI62114X0006, funded by Ministry of Science and Education of Russia).

6. References

[1] Wong Chun Kit, "Recursive Auto-Associative Memory as Connectionist Language Processing Model-Training Improvements via Hybrid Neural-Genetic Schemata," City University of Hong Kong, Hong Kong, 2004.

[2] R. Collobert. , "AISTATS," in *Deep Learning for Efficient Discriminative Parsing*, 2011.

[3] Bengio Y. et al., "A neural probabilistic language model," *The Journal of Machine Learning Research,* vol. 3, pp. 1137-1155, 2003.

[4] LeCun Y., Bengio Y.,Hinton G., "Deep learning," *Nature 521,* p. 436–444, 2015.

[5] "Tool for computing continuous distributed representations of words.," Google inc., [Online]. Available: https://code.google.com/p/word2vec/.

[6] Chen D., Manning C.D., "A fast and accurate dependency parser using neural networks," *Proc. EMNLP,* p. 740–750, 2014.

[7] Nivre J., "Incrementality in Deterministic Dependency Parsing," School of Mathematics and Systems Engineering, Vaxjo, Sweden, 2004.

[8] "Russian National Corpus," [Online]. Available: http://www.ruscorpora.ru/instruction-syntax.html.

[9] Rybka R. et al., "Statistically selected set of parameters for definitions of syntactic relations in Russian language sentences," *System analysis and information technologies.Journal of the Voronezh State University.,* no. 2, pp. 117-124, 2014.

[10] Zhang T., "Solving Large Scale Linear Prediction Problems Using Stochastic Gradient Descent Algorithms," in *ICML 2004*, 2004.

[11] Rocha A., Goldenstein S., "Multiclass from Binary: Expanding One-vs-All, One-vs-One and ECOC-based Approaches," in *IEEE Transactions on Neural Networks and Learning Systems*, 2013.

[12] Sboev A., et al, "Neuronetwork package Neurotree adapted for the segment of the Russian grid network," *Informatization and Communication,* 2012.

[13] L. Breiman, "Random forest," University of California, Berkeley, 2001.

[14] Kumar S., Mohri M., "Ensemble Nystrom Method," Courant Institute of Mathematical Sciences, New York, 2009.

[15] Jurij Apresjan et al., "ETAP-3 Linguistic Processor: a Full-Fledged NLP Implementation of the Meaning <>Text Theory," Laboratory of Computational Linguistics Institute for Information Transmission Problems, Russian Academy of Sciences, Moscow.

[16] Sharoff S., Nivre J., "The proper place of men and machines in language technology Processing Russian without any linguistic knowledge," in *Proc. Dialogue 2011, Russian Conference on Computational Linguistics.*, Moscow, 2011.

[17] Kazennikov A., "A comparative analysis of machine learning," *Papers from the Annual International Conference "Dialogue",* no. 9 (16), pp. 157-163, 2010.

[18] Kübler S. et al., Dependency Parsing Synthesis Lectures on Human Language Technologies, 2009.

Journal of Physics: Conference Series **681** (2016) 012012 doi:10.1088/1742-6596/681/1/012012

A probabilistic-entropy approach of finding thematically similar documents with creating context-semantic graph for investigating evolution of society opinion

Moloshnikov I.A., Sboev A.G., Rybka R.B., Gydovskikh D.V.

NRC "Kurchatov Institute" (Academic Kurachatov sq. 1, Moscow), Department of mathematical modelling and information technologies, Russia

E-mail: `ivan-rus@yandex.ru`, `sag111@mail.ru`, `rybkarb@gmail.com`, `dmitrygagus@gmail.com`

Abstract. The composite algorithm integrating, on one hand, the algorithm of finding documents on a given topic, and, on the other hand, the method of emotiveness evaluation of topical texts is presented. This method is convenient for analysis of people opinions expressed in social media and, as a result, for automated analysis of event evolutions in social media. Some examples of such analysing are demonstrated and discussed.

1. Introduction

The analysis of people opinions expressed in social media is the very important tool to investigate the level of social tension and indicate emerging critical situations. This analysis is complex one, it may be divided into the following four steps:

(1) selecting topical texts,
(2) selecting subtopics,
(3) emotivness analysis of the subtopics,
(4) analysis of the temporary evolution of the subtopics and the whole theme.

Currently it is difficult to point out some systems to solve this problem in whole. Concerning points 1 and 2, the search engines, based on Apache Lucene, with implementations of the "bag of words" method may be mentioned. But despite such their advantages, as fast performance and universality, it can be difficult to use them for analysis of key word evolution in time and, as a result, for analysis of temporally changing topics (point 4). As for the use of statistical LDA, PLSA [1] and Doc2vec [3] based on neural network, their shortcomings are the following: the great training corpus must be used, not high precision of results and difficulty to determine necessary level of proximity. In this work, the method based on a combination of several probabilitic and entropic algorithms for extracting keywords and collocations with including additional information sources, such as the Russian National Corpus, is used for topic analyzes.

The emotiveness evaluation of topics which has been implemented in this work, is based on the set of special markers and the context-independent approach, proposed in [8] and briefly

Journal of Physics: Conference Series **681** (2016) 012012 doi:10.1088/1742-6596/681/1/012012

described in chapter 5. It allows to determine the emotional background of topics according to point 4 and demonstrate the emotional strengths of subtopic connections of documents in collection and also connections of keywords in document collection.

2. Search thematically similar documents

To find thematically similar documents the full-text search in Solar system, with setting weights of keywords and phrases in the boost query were used [7]. For this purpose, 15 words and 15 bigrams from the reference collection, with the highest weights calculated by the ranking function, were selected. This approach had shown the best results in comparing of different approaches on the corpus SCTM-ru [2], which is a set of labelled news topics from the free news source Wikinews.

2.1. Theme modelling

To simulate the theme for searching we compute relevant weights for keywords and key phrases (combinations of two or three words in one sentence), using the following indicators for terms:

- Kullback-Leibler divergence for the comparing of the term distribution in documents with theoretical distribution;
- Information entropy represents uniformity distribution in documents;
- The Bernoulli Model of Randomness distribution of terms;
- The Ginzburg semantic algorithm to determine the thematic proximity of the words.

The word "term" means a "word" in case of using the algorithm to extract keywords, and in case of using the algorithm to extract key phrases the word "term" means a "phrase".

We combine this indicators to one term weight using ranking functions. The algorithm of ranking function:

Algorithm 1 Algorithm of ranking function

for each parameters **do**
  Number the unique value of the parameter in the sort order, from 0 to the number of unique values;
  Normalize the obtained number in the range from 0 to 1;
end for
for each term **do**
  term weight = summary of normalized parameters;
end for

This approach allows one to combine multiple parameters, smoothing out differences in the scales. The word weight shows how the word is important in the theme. And for bigrams this weight shows the strength of relationship between two words in a theme.

3. Extracting of subtopics

A subtopic is an automatically allocated set of weighted words and bigrams defining subtopics in the main theme of analysed documents collection. Furthermore, this topic must be significantly represented in the collection of analysed documents. Also, our proposed method allows to compute the strength of the relationship between nested topics.

The algorithm of extracting of subtopics is based on the construction of a graph of the relationship between the keywords of theme (bigrams). It consists of several steps:

(1) allocation of 100 words with the highest weight of relevance to topic,

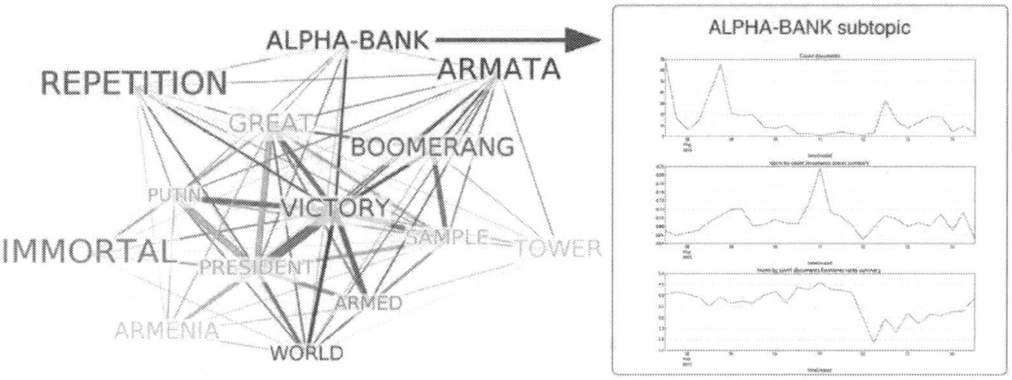

a b

Figure 1. The part of the revision graph

(2) formation bigrams based on previously selected keywords,
(3) filtering bigrams with weight below the arithmetic mean
(4) generating the affinity matrix, based on bigrams weights,
(5) applying the Affinity Propagation [5] clustering method for extracting subtopics

This algorithm does not require to set a number of clusters or centres of clusters. This is a big advantage in the analysis of the various thematic collections of documents when the user does not know the number of subtopics. The centres of cluster are the words which are most specific for subtopics, i.e. strongly associated with the other words and clusters through a common context.

To compute relationships between the clusters, the weights of the edges for the nodes of different clusters are combined into one bond that reflects the relationship between subtopics.

4. Visualizing

We construct a context-semantic graph for visual annotations of large thematic collection of documents. The nodes of the graph are keywords, and the edges are key bigramms obtained during the analysis of search results using presented earlier methods. The size of the nodes, the distance and the thickness of the lines reflect how words and phrases characterize the collection of documents.

5. Methods of emotiveness evaluation

The approach presented in this work is based on the psycholinguistic diagnosis of stylistic features of the text with application of natural language processing methods. Initially we chose the following types of psycholinguistic markers from the article "A quantitative method of text emotiveness evaluation on base of the psycholinguistic markers founded on morphological features" [8]. This paper shows that the emotiveness value of a text reflects the degree of emotional exhilaration of the author of the text at the time of writing. Values are calculated based on psycholinguistic markers of text that are allocated using morphological characters of words. There are indicators that reflect the psychological state of the author at the time of writing the text. In this article we use:

- the ratio of verbs to adjectives,
- the ratio of verbs to nouns,

Journal of Physics: Conference Series **681** (2016) 012012 doi:10.1088/1742-6596/681/1/012012

- the ratio of the number of verbs and their forms (participles and gerunds) to the number of all words,
- the ratio of prepositions to the total number of words.

We combine this markers to calculate emotive weight of document using ranking function similar to 1.

6. An analysis of the dynamics of topics

We propose a method for visualising of the dynamics of topics. It is based on sequential analysis of the flow of messages on a sliding time window algorithm 2.

Algorithm 2 Algorithm of dynamic graph

```
DynGraph = dynamic graph
list_of_time_windows = overlapping time windows with size of 6 hours, and step of 1 hour
for time window in list_of_time_windows do
  wgraph = graph for one window;
  extract 10 keywords from time window documents;
  add keywords with weights as nodes into wgraph;
  extract key bigrams on base of 10 keywords;
  add bigrams with weights as edges into wgraph;
  for each node in wgraph do
    node visual size = node weight * sum(of adjoining edges weight)/count(of adjoining
    edges);
    if node in list of nodes of DynGraph then
      append to list of node visual size with time of start and end for window;
    else
      add new node to DynGraph with visual size and time of start and end for window;
    end if
  end for
  for each edge in wgraph do
    if edge in list of edges of DynGraph then
      append to list of edges new weight with time of start and end for window;
    else
      add new node to DynGraph with weight and time of start and end for window;
    end if
  end for
end for
```

The graph is generated in a GEXF format(http://gexf.net/format/), which allows one to specify different values of weights for nodes and edges in the corresponding periods of time. For visualization we use the program Gephi [4] and online laying algorithm Force Atlas 2 [6]. An example of such a graph is shown in figure 2. It is seen as a term of "Greenland" is growing. This is associated with a gradual increase in number of messages containing a discussion of the ice cover in the Arctic and Greenland.

7. The emotiveness evaluation of subtopics

To find documents of the subtopic we form a boost query to Solr. It consists of the key words and phrases of the cluster with given weights. We analyze only the documents which contain at least one phrase from a subtopic. Solr calculates a score for each document from the search results. This reflects the weight of whether the document belongs to the cluster.

Journal of Physics: Conference Series **681** (2016) 012012 doi:10.1088/1742-6596/681/1/012012

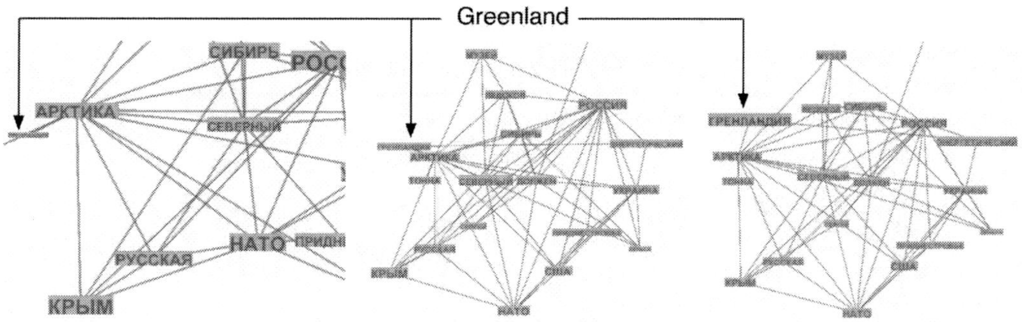

Figure 2. Dynamic graph example for theme "Arctic"

7.1. Emotiveness of subtopics

We combined the proposed approach for searching documents by cluster with an approach for analysis of emotiveness of texts. For each document of cluster overall rank of emotiveness is calculated, based on the selected psycholinguistic markers and ranking algorithm, similar to 1. Total emotive weight of the subtopic is calculated by the formula 1.

$$E(c) = \frac{\sum_{d \in D(c)} S(d) \cdot M(d)}{N(c)} \tag{1}$$

Where $E(c)$ is the total weight of the cluster c. And d is the document of subtopic, $D(c)$ is a collection of documents belonging to the cluster, $M(d)$ is the total weight of emotiveness of document, calculated on the basis of psycholinguistic markers, $N(c)$ is the number of documents in the cluster

We analysed the corpus of news and blogs (about 7,000 documents) on the topic "Armata tank" for days near the Victory Parade date. Graph of subtopics is shown in figure 1(a). The thickness of the lines shows the context strength of the connection between embedded themes. The size of the nodes show the summary of emotiveness rank for subtopics, that is presented in table 1

"Immortal Regiment" is a public event, held in Russia on Victory Day, during which the participants carry banners with photographic portraits of their relatives who participated in the Great Patriotic War. Subtopics "Repetition" and "Armata" refer to the description of Victory Parade rehearsal and acsendent with the breakdown of the new Armata tank. As you can see, the themes related to the parade and the new tanks have a greater emotiveness value, therefore it is a hotly debated topic with a large number of excited users reviews.

7.2. An example of dynamic of subtopics

To reflect the dynamics of the embedded threads, we use 3 characteristics with a window at 6 hours:

- the number of documents in the window
- the relative score of documents in the cluster for window
- relative emotiveness weight $M(d)$ of documents for windows

For the subtopic of "Alfa-Bank" these graphs are shown in figure 1(b). This topic relates to the discussion of financial claims of Alfa Bank to the manufacturer of the new Armata tank. From these graphs it can be concluded that the main peak of messages came in the dates from

Table 1. Emotiveness of subtopics

Subtopics	Emotiveness	Subtopics	Emotiveness
IMMORTAL	0.60	TOWER	0.27
REPETITION	0.58	ARMENIA	0.24
ARMATA	0.47	SAMPLE	0.16
ALPHA-BANK	0.33	PRESIDENT	0.15
VICTORY	0.31	WORLD	0.11
GREAT	0.31	ARMED	0.09
BOOMERANG	0.30	PUTIN	0.06

8 to 9, but basically, the topic was discussed in passing, it is seen by the average values in the second chart. On the 3rd graph of relative emotiveness, there is a clearly visible decline of emotional stress in the subject with the beginning of the working week, which means rather small interest in the topic.

8. Conclusion

The system which is suitable for analysis of people opinions expressed in social media based on selection of thematic documents of given topic and emotiveness evaluations has been designed and tested. The presented algorithms have been applied for emotiveness evaluation of subtopics in frame of whole topic for some thematic examples. The results demonstrate that hotly debated subtopics with large number of reviews of excited users have higher values of emotiveness. Thus, the proposed composite approach is good to reflect a given topic in the form of a weighted set of keywords and phrases along with evaluation of the emotiveness of subtopics. It is convenient for the analysis of themes developing in time and social processes. The further development of this algorithm will be directed to adding a visualization in form of thematic-emotion graph with brief annotation for subtopics.

Acknowledgment

This study was partially supported by the Russian Foundation for Basic Research (project 15-29-01173 ofi_m).

References

[1] Blei, David M and Ng, Andrew Y and Jordan, Michael I, 2003, "Latent dirichlet allocation", *the Journal of machine Learning research*, 993–1022

[2] Karpovich S.N., 2015, "The Russian language text corpus for testing algorithms of topic model", *SPIIRAS Proceedings*, 123–142

[3] Le, Quoc and Mikolov, Tomas, 2014, "Distributed Representations of Sentences and Documents", *Proceedings of The 31st International Conference on Machine Learning*, 1188–96

[4] Mathieu Bastian and Sebastien Heymann and Mathieu Jacomy, 2009, "Gephi: An Open Source Software for Exploring and Manipulating Networks", *International AAAI Conference on Weblogs and Social Media*

[5] Frey B. J., Dueck D., 2007, "Clustering by passing messages between data points" *science*, **315.5814**, 972-976.

[6] Jacomy M. et al., 2011, "Forceatlas2, a continuous graph layout algorithm for handy network visualization", *Medialab center of research.*

[7] I. Moloshnikov, A. Sboev, D. Gudovskikh, R. Rybka, 2015, "An algorithm of finding thematically similar documents with creating context-semantic graph based on probabilistic-entropy approach", *Procedia Computer Science*, in press

[8] A. Sboev, D. Gudovskikh, R. Rybka, and I. Moloshnikov, 2015, "A quantitative method of text emotiveness evaluation on base of the psycholinguistic markers founded on morphological features", *Procedia Computer Science*, in press.

A comparison of learning abilities of spiking networks with different spike timing-dependent plasticity forms

Alexander Sboev[1,2] Danila Vlasov[1], Alexey Serenko[2], Roman Rybka[2], Ivan Moloshnikov[2]

[1]MEPhI National Research Nuclear University, Moscow, Russia
[2]NRC "Kurchatov Institute", Moscow, Russia

E-mail: vfked0d@gmail.com, serenko@phystech.edu, rybkarb@gmail.com, sag111@mail.ru, ivan-rus@yandex.ru

Abstract. A study of possibility to model the learning process on base of different forms of timing-dependent plasticity (STDP) was performed. It is shown that the learning ability depends on the choice of spike pairing scheme and the type of input signal used for learning. The comparison of performance of several STDP rules along with several neuron models (leaky integrate-and-fire, static, Izhikevich and Hodgkin-Huxley) was carried out using the NEST simulator. The combinations of input signal and STDP spike pairing scheme, which demonstrate the best learning abilities, were extracted.

1. Introduction

Despite the great amount of devoted research, the problem of development for spiking network a learning mechanism, which is biologically founded and practically effective, is far from resolution. There is a number of approaches [3, 4] to make spiking neural networks learn to respond with the desired output to a given input with the help of different variations of Hebbian-inspired learning. This task involves two problems:

- firstly, what input-output transformations a network can in principle implement with some stable values of its synaptic weights;
- secondly, whether the network can learn the desired transformation with the chosen learning protocol.

Obviously, a neuron cannot produce any output spike train in response to arbitrary input spike trains. A less obvious problem, discussed in the section 4.5, is that the desired input-output transformation may be possible but require synaptic weights that are unstable with the chosen synaptic plasticity form.

Provided that the desired input-output transformation can be performed by the neuron with some stable synaptic weights, in [1] a supervised learning protocol is suggested. It is based on clamping the neuron output to desired, and the ability of any transformation, meeting the requirements stated above, to be learnt by this protocol was proven for Leaky Integrate-and-Fire neuron model for the case of all synapses receiving Poisson input trains with same mean frequency.

In this study we show that in case of different neuron models and more complex inputs learning is also possible (Section 4.2). However, the learnability is heavily affected by the choice of the scheme of pairing spikes in the STDP weight change rule (Section 4.1).

The reason in some cases is that the protocol of forcing the neuron to fire in desired moments by stimulating it with current impulses may, firstly, sometimes fail to cause a desired spike, and, secondly, not prevent the neuron from undesired spikes, caused by incoming synaptic currents instead of teacher stimulation. To prove this a learning protocol without a neuron, in which the weight change rule is just applied to the input trains and the desired output, is considered in Section 4.3, and it provides more successful learning.

2. Neuron and synapse models

In the Leaky Integrate-and-Fire neuron model the membrane potential V changes with time as following:

$$\frac{dV}{dt} = \frac{-\left(V(t) - V_{resting}\right)}{\tau_m} + \frac{I_{syn}(t)}{C_m} + \frac{I_{ext}}{C_m}, \tag{1}$$

When the potential reaches the threshold value V_{th}, the neuron fires a spike, then its potential is instantaneously dropped to the value V_{reset}, and during the refractory period τ_{ref} the synaptic current has no effect on the potential. $V_{rest} = 0$, V_{th} was chosen to be 15.3 mV in all sections unless other stated, the other LIF neuron parameters were taken as in [1].

We also used the Hodgkin-Huxley [5] and Izhikevich [6] neuron models, taking the constants same as in the original works cited here, and static adder as a neuron, which simply accumulates weights of synapses from which it receives spikes, and when the accumulated value reaches the threshold V_{th}, the sum is dropped to zero and the neuron fires a spike.

As the synapse model we used postsynaptic current of exponentially decaying form along with the Maass-Markram short-term plasticity [8], taking all parameters as in [1].

2.1. Spike-timing-dependent plasticity

STDP is a biologically inspired long-term plasticity model [7], in which each synapse is given a weight $0 \le w \le w_{max}$, characterizing its strength, and its change depends on the exact moments t_{pre} of presynaptic spikes and t_{post} of postsynaptic spikes:

$$\Delta w = \begin{cases} -W_- \cdot \dfrac{w}{w_{max}}^{\mu_-} \cdot \exp\left(-\dfrac{t_{pre} - t_{post}}{\tau_-}\right), \text{if } t_{pre} - t_{post} > 0, \\ W_+ \cdot (1 - \dfrac{w}{w_{max}})^{\mu_+} \cdot \exp\left(-\dfrac{t_{post} - t_{pre}}{\tau_+}\right), \text{if } t_{pre} - t_{post} < 0; \end{cases} \tag{2}$$

where $W_+ = 0.3$ and $W_- = 1.035 \cdot W_+$. The rule with $\mu_+ = \mu_- = 0$ is called additive STDP, with $\mu = \mu_- = 1$ — multiplicative, the one with $\mu_+ = 0$ and $\mu_- = 1$ is the van Rossum rule; intermediate values $0 < \mu < 1$ are also possible.

In case of additive STDP the auxiliary clause is added to prevent the weight from falling below zero or exceeding the maximum value w_{max}:

$$\text{if } w + \Delta w > w_{max} \text{ or } w + \Delta w < 0, \text{ then } \Delta w = 0.$$

2.2. Spike pairing schemes in the STDP rule

An important part of STDP rule is the scheme of pairing pre-and postsynaptic spikes when evaluating weight change according to the rule (2). Three nearest-neighbour schemes [7] are shown on fig. 1: symmetric A, presynaptic-centered B and reduced symmetric C. Black tics denote moments of spikes, and gray lines mean taking into account that pair of spikes in the STDP weight update rule.

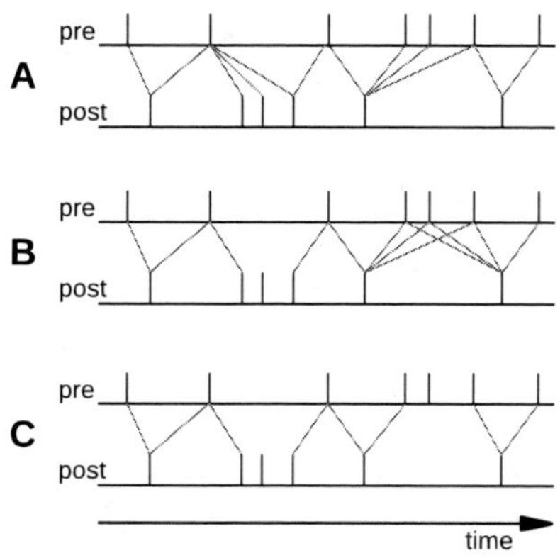

Figure 1. Spike pairing schemes: A – symmetric, B – presynaptic-centered, C – reduced symmetric. Redrawn from [7]

3. Experiment technique

3.1. The learning protocol

The following protocol suggested in [1] is to force the synaptic weights of a neuron to converge to the target weights:

(i) *Obtaining the teacher signal*
The neuron's weights are set equal to the target, STDP is disabled, and the neuron's response to the input trains is recorded as the desired output.

(ii) *Learning*
The neuron's weights are set random (but about 4 times smaller than the target ones), STDP is turned on, and the same input trains are given to the neuron's incoming synapses. During this the neuron is stimulated by the teacher signal, obtained from the desired output train by replacing spikes with 0.2-ms-duration current impulses of 2 mA (which is thousands of orders more than typical magnitude of synaptic current).

Another protocol, further called Learning without neuron, is not to use any neuron at all on the Learning stage, just to calculate weights change by the STDP rule using the input trains and the desired output train.

3.2. The experiment configuration

The one neuron had 90 excitatory and 10 inhibitory synapses. The maximum synaptic weights in the STDP rule (2) were chosen from N(54, 10.8), values less than 21.6 and more than 86.4 being replaced by 21.6 and 86.4 correspondingly. The target weights $\overrightarrow{W}_{target}$ were chosen to obey bimodal distribution: half of them were equal to zero, and the other half – to their maximum values; because such weights distribution is known [7] to be a steady state settled by the additive STDP rule.

For the implementation of neuron and synapse models we used the NEST simulator [9], modified by the authors of [2] to implement different spike pairing schemes.

4. Results

4.1. Impact of the spike pairing scheme on the learning performance

Legeistein et al. in [1] used not a nearest-neighbour, but the all-to-all spike pairing scheme, and used 20-Hz Poisson spike trains as input. In [2] the same experiment was conducted with three nearest-neighbour schemes shown on Fig. 1, and the ability of weights to converge to the target was shown to depend on the choice of spike pairing scheme.

As a measure for learning performance the deviation β between current and target weights was used:

$$\beta(t) = \frac{\sum_{i=1}^{90} |W^i(t) - W^i_{target}|}{\sum_{i=1}^{90} W^i_{target}}.$$

Table 1. β after learning with different spike pairing schemes in STDP rule and different input types

STDP scheme	uniform	normal	Poisson
all-to-all		0.11	0.11
A	1	1	0.8
B	0.10	0.10	0.08
C	0.08	0.08	0.05

So, the closer β is to 0, the more successful the learning is.

In Table 1 β after 3,000 seconds of training LIF neuron having additive STDP with different spike pairing schemes and receiving input of 20-Hz-mean trains to all synapses is shown, averaged over 5 independent simulations. The neuron threshold V_{th} was chosen about 15.6 mV, so that the mean output frequency was about 25 Hz. Normal input means that the interspike intervals in the input trains were normally distributed, and uniform input means that on every simulation step the neuron had a fixed probability of receiving a spike from that input.

The scheme C demonstrates the best learning performance, so it is the one used for the rest of the paper.

4.2. Possibility of learning with different types of input signals and different neuron models

As can be seen from Table 2, different neuron models can learn target weights not only in case all synapses of the neuron receive trains of one constant mean frequency, but also if frequency of each input train changes every 100 s. The amplitude of current impulses the neuron was stimulated with on the Learning stage was 2 mA. The static neuron threshold was 15300, and, since such type of a neuron cannot be stimulated with current, the Learning stage in this case was performed according to the Without a neuron protocol.

Table 2. β after training different neuron models with different input types

	10 Hz		10, then 60 Hz		10, 30, 50, 100 Hz	
Neuron model	normal	poisson	normal	poisson	normal	poisson
LIF neuron	0.09	0.07	0.04	0.04	0.04	0.05
Hodgkin-Huxley neuron	0.10	0.07	0.03	0.04	0.06	0.11
Static neuron	0.03	0.03	0.05	0.08	0.19	0.35

4.3. Learning without a neuron, based only on input and output signals and STDP rule

Slightly less successful learning is observed if different synapses receive inputs of different mean frequencies. On Fig. 2 is an example of learning with input that causes rather poor weights convergence: half of synapses received Poisson trains with mean frequency of 30 Hz, and the other half — 10 Hz. It can be seen that increasing the neuron threshold on the Learning stage by 1 mV compared to the Obtaining teacher signal stage improved the weights convergence. Further increase in threshold does not provide more improvement. So the reason was supposed to be in avoiding unforced spikes, i.e. caused by the incoming synaptic currents instead of teacher impulses, and therefore differing from the desired output.

Another fact we found is that choosing the magnitude of the teacher current impulses to be 1 mA instead of 2 mA impairs learning, but increasing it above 2 mA does not improve the learning more. The reason was supposed to be that high teacher current avoids cases when a teacher impulse does not force the neuron to fire.

To confirm the explanations of these two facts we introduced the learning protocol without neuron, using just the inputs and the desired output. In the context of the two facts being discussed in this section, such a protocol provides the idealised way of learning, with no difference between the postsynaptic train and the desired output. Fig. 2 shows that learning without a

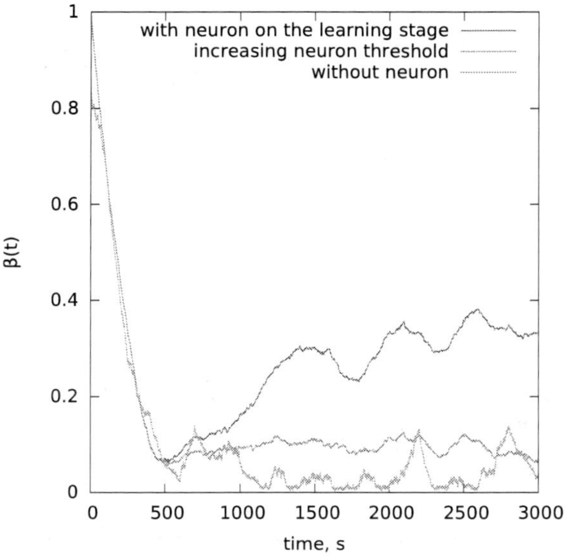

Figure 2. $\beta(t)$ during training LIF neuron with additive STDP. In this case the neuron had 100 excitatory and 90 inhibitory synapses; half of excitatory and all inhibitory synapses received Poisson trains with mean frequency of 30 Hz, and another half of excitatory synapses – of 10 Hz. The LIF neuron threshold was 15.3 mV on both stages (violet), then increased to 16.3 on the Learning stage (green), then learning without a neuron was performed (blue)

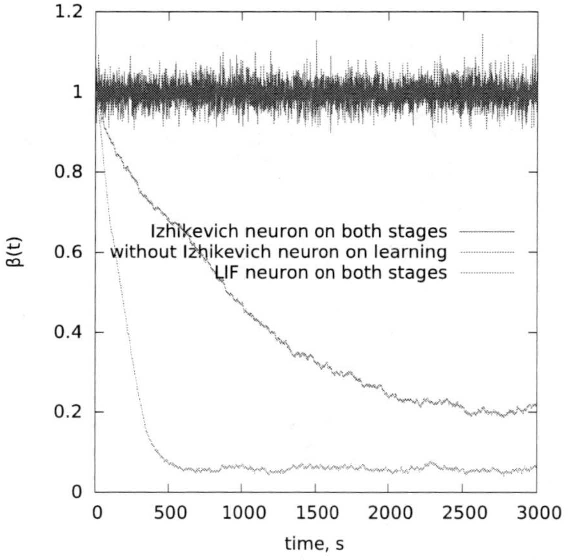

Figure 3. $\beta(t)$ during learning when Izhikevich neuron was used on both stages; when Izhikevich neuron was used on the Getting teacher signal stage, and learning was performed without neuron; and, for comparison, when LIF neuron was used on both stages. Both neurons were producing output with roughly same mean frequency

neuron is even more successful than the one with neuron and with increasing its threshold on the learning stage.

As a result, we conclude that differences in the learning performance of the protocols with and without the neuron are caused by the disadvantages of the protocol of presenting the teacher signal to the neuron in the form of current impulses, while, if the learning with some input and output trains was unsuccessful even without neuron, the reason would be that such input-output transformation cannot be learnt ever.

4.4. Learning with the Izhikevich neuron

No learning is possible using the Izhikevich neuron model on both Getting reinforcement signal and Learning stages, but in the case of using the protocol without a neuron with teacher signal obtained with Izhikevich neuron learning does take place (Fig. 3). So, the protocol of stimulating the neuron with current impulses does not work with the Izhikevich neuron, though output trains it produces can be used as desired during learning.

In case of additive STDP used in [1] and in this study (except the current section), only weight values of zero and maximum are equilibrium points, so target weights must obey the bimodal distribution in order to be stable. It must restrict the variety of possible input-output transformations.

Journal of Physics: Conference Series **681** (2016) 012013 doi:10.1088/1742-6596/681/1/012013

4.5. Impossibility of learning with non-additive STDP forms

If learning with multiplicative STDP, i. e. with $\mu_+ = \mu_- = 1$ in weight change rule (2), was possible, intermediate values of weights could be reached during learning. However, in case of multiplicative STDP weights do not converge to the target (see Fig. 4). Increasing the dependency of weight change Δw on w makes learning less successful. Choosing $\mu_+ = 0.12$ and $\mu_- = 0.1236$ (as in [1]) leads to rather poor weights convergence, despite this values are still below the critical value of $\mu = 0.023$ exceeding which the steady state of weights from bimodal becomes [7] unimodal. Therefore, making intermediate values able to be reached as a result of learning still remains a question.

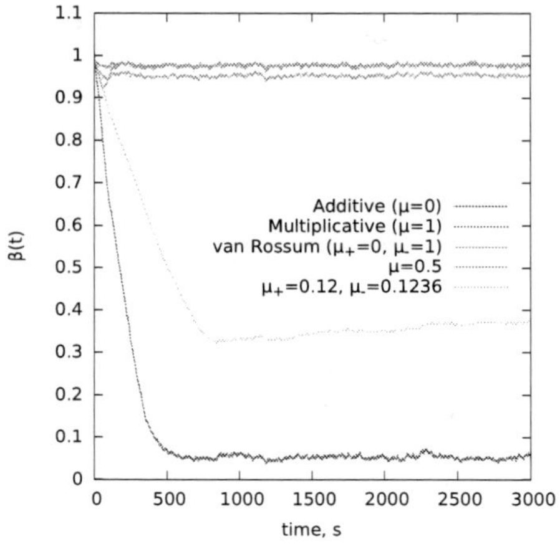

Figure 4. $\beta(t)$ during training without neuron with 20 Hz Poisson trains as inputs with different μ_+ and μ_- in the STDP weight update rule: additive $\mu_+ = \mu_- = 0$, multiplicative $\mu_+ = \mu_- = 1$, van Rossum $\mu_+ = 0$ and $\mu_- = 1$, The desired output was obtained with LIF neuron having threshold of 15.3 mV.

5. Conclusion

An output train can be used as desired during learning with some input only if it was obtained in response to that input, but regardless of the neuron model it was obtained with. In other words, output must be somehow correlated with the input. Nevertheless, learning may be impossible even with desired output correlated with input, if such input-output transformation can only be realised by a neuron with non-bimodal weights.

During learning by stimulating the neuron to fire at desired moments, deviations of the neuron's spike times from the desired output severely impairs learning performance compared to the case when output during learning exactly satisfies the desired one, as in the protocol without a neuron. The reason is that in STDP, like in any Hebbian learning rule, a postsynaptic spike can only lead to increasing synaptic weights due to "pre-before-post" part of the rule, regardless of whether it is desired or not. To provide the mechanism of penalizing synapses for causing undesired spikes, in most works [3, 4] some anti-Hebbian clauses are introduced, while their biological plausibility has not yet been proved. Development of both efficient and biologically plausible learning protocol still remains an unsolved problem.

References

[1] R. Legenstein, C. Naeger, W. Maass, "What Can a Neuron Learn with Spike-Timing-Dependent Plasticity?", *Neural Computation*, vol. 17, 2005, pp. 2337–2382.

[2] K. Kukin, A. Sboev, "Comparison of Learning Methods for Spiking Neural Networks", *Optical Memory and Neural Networks (Information Optics)*, vol. 24, 2015, No. 2, pp. 123–129 (2015)

[3] R. Gütig, H. Sompolinsky, "The tempotron: a neuron that learns spike timing-based decisions" *Nat. Neurosci.* 9 (3), 2006, pp. 420–428

[4] S. Mitra, S. Fusi, G. Indiveri, "Real-Time classification of complex patterns using spike-based learning in neuromorphic VLSI", *IEEE Transactions on biomedical circuits and systems*, vol. 3, 2009, No. 1

[5] A. L. Hodgkin, A. F. Huxley, "A quantitative description of membrane current and its application to conduction and excitation in nerve", *J. Physiol.*, vol. 117, 1952, pp. 500–544

[6] Eugene M. Izhikevich, "Simple model of spiking neurons", *IEEE Transactions on Neural Networks*, vol. 14, No. 6, Nov. 2003

[7] A. Morrison, M. Diesmann, W. Gerstner, "Phenomenological models of synaptic plasticity based on spike timing", *Biol. Cybern.*, vol. 98, 2008, pp. 459–478.

[8] W. Maass, H. Markram, "Synapses as dynamic memory buffers", *Neural Networks*, vol. 15, 2002, pp. 155-161

[9] Gewaltig M.-O., Diesmann M., "NEST (Neural Simulation Tool)", *Scholarpedia* 2(4), 2007, p. 1430.

Journal of Physics: Conference Series **681** (2016) 012014 doi:10.1088/1742-6596/681/1/012014

Connectivity properties of the random-cluster model

Martin Weigel[1], Eren Metin Elçi[1,2], and Nikolaos G. Fytas[1]

[1]Applied Mathematics Research Centre, Coventry University, Coventry CV1 5FB, United Kingdom

[2]School of Mathematical Sciences, Monash University, Clayton, Victoria 3800, Australia

E-mail: Martin.Weigel@coventry.ac.uk

Abstract.
We investigate the connectivity properties of the random-cluster model mediated by bridge bonds that, if removed, lead to the generation of new connected components. We study numerically the density of bridges and the fragmentation kernel, i.e., the relative sizes of the generated fragments, and find that these quantities follow a scaling description. The corresponding scaling exponents are related to well known equilibrium critical exponents of the model. Using the Russo-Margulis formalism, we derive an exact relation between the expected density of bridges and the number of active edges. The same approach allows us to study the fluctuations in the numbers of bridges, thereby uncovering a new singularity in the random-cluster model as $q < 4\cos^2(\pi/\sqrt{3})$ in two dimensions. For numerical simulations of the model directly in the language of individual bonds, known as Sweeny's algorithm, the prevalence of bridges and the scaling of the sizes of clusters connected by bridges and candidate-bridges play a pivotal role. We discuss several different implementations of the necessary connectivity algorithms and assess their relative performance.

1. Introduction

The random-cluster model was introduced by Fortuin and Kasteleyn as a generalization of the much studied Ising model of ferromagnetism and the percolation model introduced in the 1950s [1]. Through its relation to the Potts model and, ultimately, as a subject of study in its own right it developed into an important generalized model particularly for the study of continuous and first-order phase transitions [2, 3]. While a significant number of exact results are available in two dimensions, and in particular for the special cases $q \to 0$, $q \to 1$ and $q = 2$, the main approach for studying the random-cluster model away from such islands of exact solutions are Markov chain Monte Carlo simulations [4]. While these are quite efficient in general, studies of systems close to points of phase transitions are affected by a dynamical loss of ergodicity: for second-order transitions this effect is known as critical slowing down. It is caused by the diverging length scale at criticality that, in turn, leads to a divergence of time scales for the decorrelation of the system in the Monte Carlo dynamics. For this case, *cluster updates* turn out to be extremely efficient at speeding up the simulation. The algorithms of Swendsen and Wang [5] (and its generalization by Chayes and Machta [6]) and the single-cluster variant by Wolff [7] are rather well-known and frequently applied, but they are not applicable for $q < 1$. On the other hand, an algorithm due to Sweeny [8] to simulate the system directly in the random-cluster language is significantly less well-known, although a potentially superior performance has been demonstrated previously [8, 9] and it is applicable for any $q > 0$. This is due to the algorithmic complication that for this single-bond update one requires information about the large-scale connectivity structure of the current cluster configuration at each time step. Efficient implementations of such algorithms, as the ones developed here, have only recently become available.

The performance of such connectivity algorithms crucially depends on the cluster structure at criticality where such algorithms are typically applied. How strongly connected are such clusters? What happens if a random bond is removed? These questions are relevant for the understanding of the fragmentation properties of the random-cluster model. While fragmentation has been studied for bond percolation, the additional variability in the connectivity structure implied by the tunable cluster weight q allows for the modeling of a wider variety of different random structures. Using the highly efficient simulation algorithms developed in the first part, we will present a detailed study of the fragmentation properties of critical RC clusters below. The fragmentation scaling exponents λ and ϕ are determined and derived from the set of equilibrium critical exponents. Bridges are bonds that upon removal lead to a fragmentation event. Using a probabilistic approach known as Russo-Margulis formalism [3], we find that the expected density of such bonds can be rigorously related to the mean number of active (or open) edges, revealing that the number of bridge bonds is essentially an energetic quantity. The same approach also allows us study the fluctuations in the number of bridges, thereby uncovering a new singularity in the random-cluster model occurring (in two dimensions) for $q < \tilde{q} = 4\cos^2(\pi/\sqrt{3}) = 0.2315891\cdots$.

The rest of the paper is organized as follows. In Sec. 2 we introduce the random-cluster model and discuss its relation to other models of statistical mechanics. In Sec. 3 we discuss Sweeny's algorithm for simulations of the random-cluster model and compare a number of algorithmic approaches for the required connectivity checks. Section 4 is devoted to the behavior of clusters of the random-cluster under the random removal of bonds, i.e., a fragmentation process. In Sec. 5 we outline some of the analytical results on the role of bridge bonds. Finally, Sec. 6 contains our conclusions.

2. The random-cluster model

The random-cluster (RC) model [3], introduced by Fortuin and Kasteleyn in a series of seminal papers [1, 10, 11], serves as a generalization of percolation and the Ising model to a continuum of models parameterized by the cluster weight $q \geq 0$. Its partition function is

$$Z_{\mathrm{RC}} = \sum_{A \subseteq G} v^{|A|} q^{K(A)}, \tag{1}$$

where $A \subseteq E$ denotes the set activated edges on a graph $G = (V, E)$ with vertex set V and edge set E, resulting in $K(A)$ connected components, and the bond weight $v = p/(1-p)$. Hence the RC model assigns weights to (spanning) sub-graph configurations A, i.e., subsets of activated edges and the complete set of vertices, of the underlying graph G. A special limit of Eq. (1) is $q \to 1$ in which case the model reduces to uncorrelated percolation.

As shown by Fortuin and Kasteleyn [1] for integer values of q the model is equivalent to the q-state Potts model with Hamiltonian

$$\mathcal{H} = -J \sum_{e \in E} \delta_{\sigma_i, \sigma_j}, \tag{2}$$

where $e = (i, j)$ is an edge in the graph G, and $\sigma_i \in \{1, \ldots, q\}$. This, in turn, up to a constant reduces to the Ising model for the case of only two possible states, i.e., $q = 2$. For the purposes of the present discussion, we restrict ourselves to systems on finite patches of the square lattice with dimensions $L \times L$, applying periodic boundary conditions. For this case, the ordering transition of the Potts model occurs at the coupling $J/k_B T = \ln(1 + \sqrt{q})$, corresponding to the

Journal of Physics: Conference Series **681** (2016) 012014 doi:10.1088/1742-6596/681/1/012014

critical bond weight $v_c = \sqrt{q}$ in (1). This transition is continuous for $q \leq 4$ and first-order for $q > 4$ [2].

3. Simulation using Sweeny's algorithm

While the construction of Swendsen and Wang [5], published in 1987, of a rejection-free update of spin clusters in connection with bond variables implied by the random-cluster representation is well known and widely used to simulate Ising, Potts and (through the cluster embedding trick) also continuous spin models [7], a simpler and more direct approach was proposed by Sweeny [8] in 1983 already. His approach is rooted in the results of Fortuin and Kasteleyn [1] in directly attempting to sample bond configurations of the RC model according to the weight

$$w_{\mathrm{RC}}(A) = q^{K(A)} v^{|A|}, \tag{3}$$

implied by the partition sum (1). For a given sub-graph or bond configuration A, the basic update operation is then given by the deletion of an occupied edge or the insertion of an unoccupied edge. As is seen from Eq. (3), the resulting transition probabilities are determined by the change $\Delta|A| = \pm 1$ in open edges and the change ΔK in connected components. While the value of $\Delta|A|$ is trivially clear depending on whether one is attempting to open or to close a given edge, the change in cluster number depends on whether the chosen edge is *pivotal* [12], i.e., whether it is a *bridge* bond that increases the number of connected components on its closing or a *candidate bridge* that amalgamates two components upon its opening. For such pivotal edges, one has $\Delta K = \pm 1$, while for *non-pivotal* edges the cluster number remains unchanged, $\Delta K = 0$. The construction and implementation of data structures supporting the efficient calculation of ΔK constitutes the intricacy of Sweeny's algorithm.

3.1. Connectivity algorithms

We implemented three main types of connectivity algorithms to facilitate the calculation of ΔK for edge moves: breadth-first searches (BFS), a union-and-find algorithm (UF), and a dynamic connectivity algorithm (DC).

3.1.1. Breadth-first search In the simplest case, one does not keep any state information relating to the connectivity of points in the graph, but checks for the pivotality of a given edge $e = (i, j)$ each time an opening or closing of e is proposed. This entails a traversal of the components connected to i and j, respectively, to decide whether these are connected through paths not crossing e. We implement such traversals in a breadth-first (BFS) manner [13], but depth-first variants lead to essentially identical runtime scaling. Regarding the typical runtime for such searches for the RC model at criticality, arguments relating to the scaling of the shortest path ℓ between points i and j and the so-called *spreading dimension* $\hat{d}$ [14] imply that the average number of sites touched by the BFS for a non-pivotal edge is [15]

$$\langle \ell^{\hat{d}} \rangle \sim L^{d_F - x_2}. \tag{4}$$

Here, d_F is the cluster fractal dimension [16], while x_2 corresponds to the two-arm exponent [17, 18]. For pivotal edges, on the other hand, a complete traversal of the encountered clusters is required to terminate, and hence the expected number of operations scales as the typical cluster size $M_2 \sim L^{\gamma/\nu}$. As $\gamma/\nu > d_F - x_2$ everywhere for the RC model, the operations on pivotal edges dominate the asymptotic runtime, see the scaling results compiled in Table 1.

An improvement of the BFS approach can be achieved through the *interleaving* of searches in the two cluster arms joined by $e = (i, j)$ [19, 20]. In this case cluster traversal for a pivotal edge can stop once the smaller cluster has been exhausted. The size of this smaller cluster is

Table 1. Asymptotic runtime scaling at criticality of the elementary operations of opening or closing of pivotal and non-pivotal edges, respectively, using sequential breadth-first search (SBFS), interleaved BFS (IBFS), union-and-find (UF) or the fully dynamic connectivity algorithm (DC) as a function of the linear system size L.

move	SBFS	IBFS	UF	DC
non-bridge insertion	$L^{d_F-x_2}$	$L^{d_F-x_2}$	const.	$\log^2 L$
bridge insertion	$L^{\gamma/\nu}$	$L^{d_F-x_2}$	const.	$\log^2 L$
non-bridge deletion	$L^{d_F-x_2}$	$L^{d_F-x_2}$	$L^{d_F-x_2}$	$\log^2 L$
bridge deletion	$L^{\gamma/\nu}$	$L^{d_F-x_2}$	$L^{\gamma/\nu}$	$\log^2 L$
dominant scaling	$L^{\gamma/\nu}$	$L^{d_F-x_2}$	$L^{\gamma/\nu}$	$\log^2 L$

argued in Ref. [20] to scale $\sim L^{d_F-x_2}$, thus leading to a uniform $L^{d_F-x_2}$ scaling of all operations for this variant of the algorithm. This behavior is indeed found in the numerical simulations [20, 15, 21].

3.2. Union-and-find

The use of suitably designed data structures can reduce the runtime of at least some of the required operations by storing and updating connectivity information continuously. If one restricts the move set to the successive opening of edges, this leads to a continuous amalgamation of clusters if pivotal edges are selected. A suitable data structure are balanced trees using path compression that have previously been used for highly efficient simulations of the uncorrelated bond percolation problem [22]. Using these, opening moves can be performed in a runtime that is practically independent of the size of the system. The closing of edges, however, requires more expensive updates to the data structure [15]. In the case of closing a non-pivotal edge, we can use interleaved BFS to traverse both clusters, resulting in a runtime scaling proportional to $L^{d_F-x_2}$. For the case of a pivotal edge, leading to fragmentation of the original cluster, a complete re-labeling of both new clusters is required, however, resulting in a total scaling of $\sim L^{\gamma/\nu}$ for this step. As Table 1 illustrates, the asymptotically dominant runtime scaling for this approach is hence $L^{\gamma/\nu}$, provided the different edge types appear all with (asymptotically) constant probability (which is indeed the case as we show below). We note that a further improvement of this scaling can be achieved by augmenting the union-and-find data structure by cluster identifier labels, thus reducing the runtime for the closing of a bridge bond to the form $L^{d_F-x_2}$ as well. Details will be presented elsewhere [23].

3.2.1. Dynamic connectivity algorithm The goal of maintaining connectivity information for a dynamically changing graph in constant or at most (poly-)logarithmic time per operation is known as *dynamic connectivity* problem in computer science [24, 25]. We use the approach suggested in Ref. [25] which is deterministic and features amortized runtimes of $\mathcal{O}(\log N)$ for connectivity queries and $\mathcal{O}(\log^2 N)$ for deletions and insertions on graphs of N nodes. The time complexity for connectivity queries depends on the underlying binary search tree used to encode the graphs. In our case we used splay trees [26] which result in the amortized bound. An identical worst-case bound holds for balanced binary search trees [25]. The algorithm represents the edge set in form of Euler tours, stored in balanced trees, thus allowing for the efficient implementation of the necessary tree manipulations. To facilitate the search for replacement edges for the case of the closing of edges, a level hierarchy of a depth that is logarithmic in the size of the graph is introduced. The edge hierarchy is dynamically adapted as edges are opened and closed to ensure that replacement edges can always be found efficiently. A more in-depth discussion of the algorithm can be found in Ref. [21]. For simulations of the RC model, we can therefore perform each operation in a runtime asymptotically proportional to $\log^2 L$, hence clearly outperforming the other approaches at criticality, cf. Tab. 1.

3.3. Simulation results

The runtime scaling of the connectivity algorithms discussed above essentially determine the asymptotic performance of the corresponding implementations of Sweeny's algorithm[1]. The actual efficiency of a simulation algorithm, however, also depends on its abilities to efficiently decorrelate the configuration of the system. This latter property is encoded in the autocorrelation times or, more generally, the mixing time of the algorithm [27]. We considered the *integrated autocorrelation times* [28],

$$\tau_{\mathrm{int},\mathcal{O}} = \frac{1}{2} + \sum_{t=1}^{M} \rho_{\mathcal{O}}(t), \tag{5}$$

where M is the length of the time series and $\rho_{\mathcal{O}}(t)$ is the normalized autocorrelation function for the observable $\mathcal{O}$. $\tau_{\mathrm{int},\mathcal{O}}$ controls the efficiency of an algorithm in determining the number of steps it takes to create a statistically independent sample. Close to a critical point, we expect scaling of the form $\tau_{\mathrm{int},\mathcal{O}} \sim L^{z_{\mathrm{int},\mathcal{O}}}$, thus defining the dynamical critical exponent $z_{\mathrm{int},\mathcal{O}}$. The ultimate efficiency of our implementation is then gauged by studying the quantity

$$T_{\mathcal{O}} \equiv \tau_{\mathrm{int},\mathcal{O}}\bar{t}, \tag{6}$$

where $\bar{t}$ is the runtime per operation and hence $T_{\mathcal{O}}$ measures the effective time, per edge, it takes to generate a statistically independent sample. While this is clearly hardware-dependent, we expect the relative performance of different implementations to be largely unaffected by these details [15, 21].

The behavior of the autocorrelation times for the Sweeny update alone has a number of very interesting features, most prominently that of *critical speeding up*, i.e., a decrease of autocorrelation times with system size (in natural *sweep* time units), for certain magnetic quantities, such that $z_{\mathrm{int},\mathcal{O}} < 0$ there. These effects are discussed in detail in Refs. [9, 15, 21, 27]. Regarding the performance of the slowest mode which is given by the number of open edges (the energy) [27], we find slightly smaller values for the dynamical critical exponent than for the Swendsen-Wang-Chayes-Machta algorithm [5, 6] where it is applicable. The question is whether taking into account the *algorithmic* slowing down induced by the expense of frequent connectivity queries destroys this asymptotic advantage, which will be the case if $\bar{t}$ strongly increases with the linear system size L. As a look at Table 1 indicates, for our four different implementations of connectivity algorithms — sequential breadth-first search (SBFS), interleaved breadth-first search (IBFS), union-and-find (UF) and the dynamic connectivity algorithm based on Euler tours and an edge level hierarchy (DC) — the advantage is always preserved asymptotically for the DC approach with poly-logarithmic scaling, whereas for the other three approaches the answer depends on the values of $z_{\mathrm{int},\mathcal{O}}$ and γ/ν resp. $d_F - x_2$ for the specific value of q under consideration. In two dimensions, we find these values to be such that the advantage of the single-bond dynamics is destroyed for the SBFS, IBFS and UF approaches apart from, maybe, in the vicinity of the tricritical point $q = 4$.

For $q < 1$, on the other hand, Sweeny's algorithm is the only available method. In Fig. 1 we compare the runtime $\bar{t}$ of operations per edge for systems with very small $q = 0.005$ between

[1] Additional optimizations concern cases where connectivity queries are not necessary and move proposals can be accepted unconditionally, but these do not affect the asymptotic scaling [15].

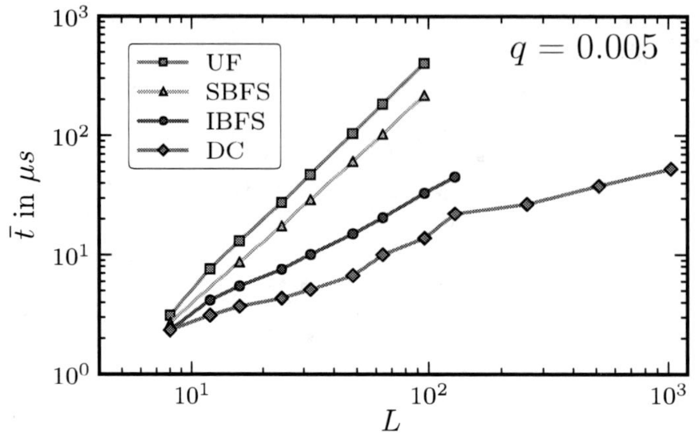

Figure 1. (Color online) Run-time per edge operation of simulations of the $q = 0.005$ square-lattice RCM and the bond algorithm employing the SBFS, IBFS, UF and DC connectivity implementations, respectively.

the four different implementations of connectivity algorithms. While the first three approaches clearly show power-law scaling (see also Table 1), we expect poly-logarithmic behavior for DC. Within the range of studied system sizes, however, this is not clearly visible. Still, the new implementation outperforms the other approaches. Further optimizations for the DC approach, in particular a reduction of the level hierarchy to only one level, further increase this advantage [27, 23].

4. Fragmentation properties

The distinction between pivotal and non-pivotal edges was found to be crucial for the asymptotic runtime scaling of most of the connectivity algorithms discussed above. The question of whether the closing (or removal) of a certain edge from the subgraph (or bond configuration) leads to the creation of a new connected component also is pivotal for the understanding of the fragmentation properties of clusters in the RC model. If one removes a single bond at the critical point $p = p_c$ this might or might not lead to a fragmentation event. The number of such *fragmenting edges* or *bridges* might depend on the cluster size and we assume the scaling form [29]

$$a(s) \sim s^{\lambda}.$$

To determine λ for the critical RC model, consider the total number of bridges,

$$\frac{\sum_s s\, n_c(s) a(s)}{\sum_s s\, n_c(s)} \sim \int s^{-\tau+1+\lambda} e^{-cs}\, \mathrm{d}s \sim L^{(\tau-\lambda)/(\sigma\nu)}, \tag{7}$$

where we have used the scaling form of the critical cluster size distribution, $n_c(s) \sim s^{-\tau} e^{-cs}$ as well as the relations $c \sim |p - p_c|^{\frac{1}{\sigma}}$ and $|p - p_c| \sim L^{-\frac{1}{\nu}}$ [16], where L is the linear dimension. Using simulations with the Sweeny and Swendsen-Wang-Chayes-Machta algorithms, we studied the density of bridges of the RC model on the square lattice numerically [30]. From the results, summarized in Fig. 2, it is clear that the critical density of bridges is asymptotically non-vanishing. Hence the average number of bridges in (7) must grow as L^d, implying $d = (\tau - \lambda)/\sigma\nu$.

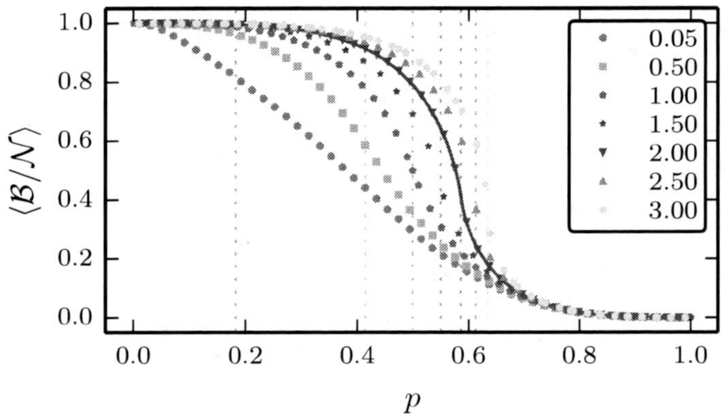

Figure 2. Proportion of bridges among active edges in the random-cluster model for different values of q. Simulation data are for systems sizes $L = 64$ ($q \neq 1$) and $L = 2048$ ($q = 1$), respectively. The solid line denotes the exact result for $q = 2$ and $L \to \infty$. The vertical dashed lines specify the location of the critical point.

With the exponent identities $\sigma\nu = 1/d_F$ and $\tau = 1 + d/d_F$, where d_F is the critical cluster fractal dimension, this shows that

$$\lambda = 1, \quad (8)$$

independent of q. This generalizes numerical results previously found for the percolation model in two and higher dimensions [31, 32].

If a fragmentation event occurs, it is natural to ask for the relative sizes of the two fragments created. This is encoded in the probability $b_{s',s}$ (the *fragmentation kernel*) of creating a fragment of size s' on fragmenting a cluster of size s. The scale-free nature of the critical RC model suggests a large-s scaling form

$$b_{s',s} \sim s^{-\phi}\mathcal{G}\left(\frac{s'}{s}, \frac{s}{L^{d_F}}\right), \quad (9)$$

which is compatible with exact results for percolation in 1D and on the Bethe lattice [33]. To connect ϕ to previously established critical exponents, we multiply Eq. (9) by s' and then integrate to find that $\mu_s \sim s^{2-\phi}\mathcal{H}(s/L^{d_F})$. From using a finite-size scaling form of the overall cluster-size distribution [20] we conclude that the scaling of the ensemble average daughter cluster size is $\langle s' \rangle \sim L^{d_F(3-d/d_F-\phi)}$. On the other hand, one can show [27] that this is proportional to the average of $C_{\min,2}$, the size of the smaller of the two clusters attached to two neighboring disconnected vertices [20, 15]. In Ref. [20] it was shown that $\langle C_{\min,2} \rangle \sim L^{d_F - x_2}$, which implies

$$\phi = 2 + (x_2 - d)/d_F = 2 - d_R/d_F, \quad (10)$$

where $d_R = d - x_2$ is the red-bond fractal dimension. This is consistent with previous results for bond percolation [31, 34, 32]. However, the scaling form suggested there, $\phi = 2 - 1/(\nu d_F)$, is only correct for percolation itself as $d_R = 1/\nu$ in this case [35]. For general values of q, on the other hand, only the form (10) gives the correct result. We also studied the fragmentation kernel using numerical simulations of the square-lattice RC model [30]. These data are summarized in Fig. 3 (left panel), demonstrating the validity of the scaling form of Eq. (9), showing an excellent collapse of data for different cluster and system sizes onto scaling functions parametrized by q. As is illustrated in Fig. 3 (right panel), our estimates for the fragmentation exponent ϕ are in

Journal of Physics: Conference Series **681** (2016) 012014 doi:10.1088/1742-6596/681/1/012014

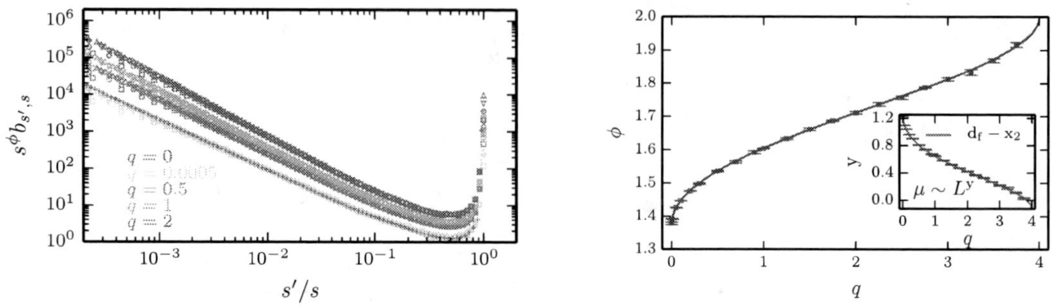

Figure 3. Left: Re-scaled conditional fragmentation probability $b_{s',s}$ according to Eq. (9) for different values of the cluster coupling q. Right: Scaling exponent ϕ of daughter clusters in the fragmentation of the square-lattice RC model as compared to the exact result (10). The inset shows the scaling exponent of the ensemble average daughter cluster size $\langle s' \rangle$.

excellent agreement with the form (10) proposed above if we substitute the exact values known from Coulomb gas arguments for d_R and d_F.

5. The role of bridges

Our arguments for the value $\lambda = 1$ of the exponent governing the density of fragmenting bonds were based on the numerical observation of a value $\langle \mathcal{B}/\mathcal{N} \rangle$ of the expected relative density of bridge bonds among all open edges that is strictly greater than zero and less than one. It turns out that this property follows as a corollary from a much more general property of bridge bonds in the random-cluster model [12].

This property can be derived starting from a well-known identity in percolation theory known as Russo-Margulis formula [36],

$$\frac{\mathrm{d}}{\mathrm{d}p} \mathbb{E}_{p,G}[X] = \sum_{e \in E} \mathbb{E}_{p,G}[\delta_e X], \tag{11}$$

where $X(A) : \Omega \to \mathbb{R}$ is an arbitrary observable and the quantity $\delta_e X$ is called the *influence* of e on X and is given by

$$(\delta_e X)(A) \equiv X(A^e) - X(A_e),$$

where A is a given bond configuration and A^e refers to A modified by opening e while A_e correspond to A with edge e closed. Here, $\mathbb{E}_{p,G}[X]$ refers to an expectation with respect to the probability measure of uncorrelated bond percolation on a graph G. Applying this formalism to the cluster-number observable K, for which $K(A) - K(A_e) = -1$ if e is a bridge, $e \in B(A)$, and 0 otherwise, it is possible to show that [12]

$$\mathbb{E}_{p,q,G}[\mathcal{B}] = \frac{\mathbb{E}_{p,q,G}[\mathcal{N}] - p}{(1-p)(1-q)}, \tag{12}$$

which we refer to as *bridge-edge formula*. Here, $\mathbb{E}_{p,q,G}[X]$ refers to the expectation of X with respect to the random-cluster weight (3). This is a general relation, valid for any graph and any value of $q \neq 1$ (the case of $q \to 1$ is discussed in Ref. [12]). Since we know that for $0 < p < 1$ the edge density $\mathbb{E}_{p,q,G}[\mathcal{N}]$ satisfies $0 < \mathbb{E}_{p,q,G}[\mathcal{N}] < 1$, it follows that $\lambda = 1$ exactly for all graphs and for critical as well as off-critical conditions.

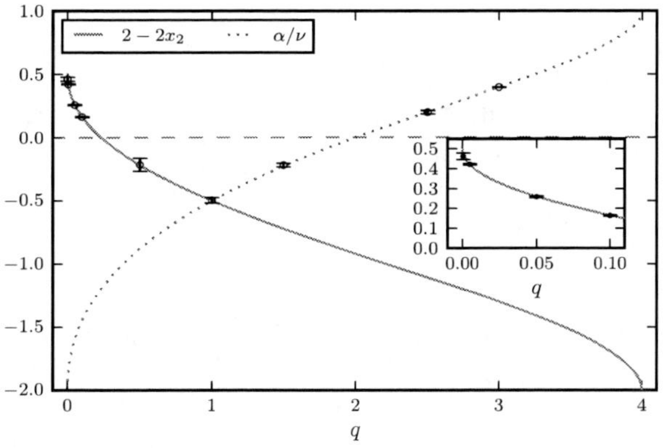

Figure 4. The exponents α/ν and $2-2x_2$ appearing in the system size scaling of $\mathrm{Var}_q[|B|]/|E|$ at criticality. The solid and dotted lines show the exact value of $2-2x_2$ and α/ν, respectively, following from the Coulomb gas mapping. The symbols denote our numerical estimates from fitting to the variance of the bridge density, cf. Ref. [12].

As the number of open edges in the RC model is an energy-like quantity, through Eq. (12) the same holds true for the density of bridges. Exact results for the energy and hence the density of open edges at the self-dual, critical point [2] allow us to exactly find the critical density of bridges,

$$\mathbb{E}_{p_{\mathrm{sd}}(q),q,\mathbb{Z}^2}[\mathcal{B}] = \frac{1}{2(1+\sqrt{q})}. \tag{13}$$

From the same connection to the internal energy of the Potts model, one may also deduce the finite-size corrections to the asymptotic result (13), and one finds that the dominant term is proportional to $L^{1/\nu-d}$ [12]. Similarly, it is possible to derive analogous relations for other types of edges, i.e., non-bridges as well as candidate bridges and candidate non-bridges. Details can be found in Ref. [12].

It turns out that the Russo-Margulis formalism allows for even more advanced investigations. Following along similar lines as for the derivation of the bridge-edge formula (12) one can investigate the *fluctuations* in the number of bridge bonds. The variance $\mathrm{Var}_{p,q,G}[|B|]/|E|$ is found to be exactly given by

$$\frac{\mathbb{E}_{p,q,G}[\mathcal{N}]\,(2p-1)-p^2+\mathrm{Var}_{p,q,G}[N]/|E|}{(1-q)^2(1-p)^2} + \mathbb{E}_{p,q,G}[\mathcal{B}] + \frac{1}{1-q}\frac{1}{|E|}\sum_{e\neq f\in E}\mathbb{P}_{p,q,G}[e \Leftrightarrow f], \tag{14}$$

where we write $e \overset{A}{\Leftrightarrow} f$ for the event that both e and f are non-bridges in the same cycle such that removing e will cast f into a bridge. The first term in Eq. (14) is essentially the fluctuation in the number of open edges which, in turn, corresponds to the internal energy. Hence we expect asymptotic critical scaling proportional to $L^{\alpha/\nu}$. The last term, on the other hand, can be shown to lead to scaling proportional to L^{d-2x_2} [37, 12]. The values of these exponents are illustrated for the square-lattice model in Fig.4. In two dimensions we hence find that the term proportional to $L^{\alpha/\nu}$ dominates for $q>1$ and, as is well known, becomes positive for $q>2$. For $q<1$, on the other hand, the new term proportional to L^{2-2x_2} dominates and, in particular, leads to a new divergence for $2-2x_2>0$, which occurs for $q<\tilde{q}=4\cos^2(\pi/\sqrt{3})=0.2315891\cdots$.

6. Conclusions

We have presented a general implementation of Sweeny's algorithm for the random cluster model that works independent of the structure and dimensionality of the lattice and features, at least asymptotically, an improved efficiency in decorrelating the system as compared to the better known Swendsen-Wang-Chayes-Machta algorithm. In addition, it can be applied to the regime $q < 1$ which is inaccessible to all other known simulation approaches. It would be interesting to study the dynamic behavior of the algorithm in more detail, in particular for the random-cluster model in three dimensions.

In the course of developing and comparing different implementations of the required connectivity checks we discovered the crucial importance of bridge bonds or singly connected bonds for the dynamic as well as the static behavior of the RC model. Fragmentation processes can be understood from studying the bridge bonds and the properties of the clusters connected by bridges. Studying fragmentation we found that the density of bridges is asymptotically independent of cluster size, thus rendering the fragmentation probability upon random bond removal directly proportional to the size of clusters. We investigated the fragmentation kernel, i.e., the conditional probability of a fragmentation event on a cluster of size s to generate a daughter cluster of size s'. This is found to be highly asymmetric, i.e., an fragmentation event is very unlikely to yield fragments of comparable size, in contrast to recent claims in Ref. [38]. This property also explains our recent observation of the fact that, under suitable initial conditions, the equilibrium fragmentation exponent ϕ also governs the final fragment size distribution in a non-equilibrium process [30, 27].

Using the Russo-Margulis formalism of percolation theory, we were able to relate the expected density of bridges to the density of open edges. Some of the manifold consequences of this observation are discussed in Ref. [12]. Additionally, the study of fluctuations in the number of bridges allowed us to uncover a previously unknown singularity in the random-cluster model occurring, in two dimensions, for $q < 4\cos^2(\pi/\sqrt{3})$. Removing *all* bridges from critical RC model configurations leads to a new ensemble of structures and the sizes of the resulting fragments are found to scale according to the backbone critical exponent of the model [12]. This promises an efficient new approach for determining this otherwise hardly accessible critical exponent, especially for the case of the random-cluster model in three dimensions.

Acknowledgments

The authors acknowledge funding from the EC FP7 Programme (PIRSES-GA-2013-612707).

References

[1] Fortuin C M and Kasteleyn P W 1972 *Physica* **57** 536–564
[2] Wu F Y 1982 *Rev. Mod. Phys.* **54** 235–268
[3] Grimmett G 2006 *The random-cluster model* (Berlin: Springer)
[4] Binder K and Landau D P 2009 *A Guide to Monte Carlo Simulations in Statistical Physics* 3rd ed (Cambridge: Cambridge University Press)
[5] Swendsen R H and Wang J S 1987 *Phys. Rev. Lett.* **58** 86–88
[6] Chayes L and Machta J 1998 *Physica A* **254** 477
[7] Wolff U 1989 *Phys. Rev. Lett.* **62** 361–364
[8] Sweeny M 1983 *Phys. Rev. B* **27** 4445
[9] Deng Y, Garoni T M and Sokal A D 2007 *Phys. Rev. Lett.* **98** 230602
[10] Fortuin C M 1972 *Physica* **58** 393–418
[11] Fortuin C M 1972 *Physica* **59** 545–570
[12] Elçi E M, Weigel M and Fytas N G 2015 *Preprint arXiv:1509.00668*, to appear in Nuclear Physics B
[13] Cormen T H, Leiserson C E, Rivest R L and Stein C 2009 *Introduction to Algorithms* 3rd ed (Cambridge, MA: MIT Press)
[14] Grassberger P 1992 *J. Phys. A* **25** 5475
[15] Elçi E M and Weigel M 2013 *Phys. Rev. E* **88** 033303
[16] Stauffer D and Aharony A 1994 *Introduction to Percolation Theory* 2nd ed (London: Taylor & Francis)

[17] Cardy J 1998 *J. Phys. A* **31** L105
[18] Smirnov S and Werner W 2001 *Math. Res. Lett.* **8** 729–744
[19] Weigel M 2010 *Physics Procedia* **3** 1499–1513
[20] Deng Y, Zhang W, Garoni T M, Sokal A D and Sportiello A 2010 *Phys. Rev. E* **81** 020102
[21] Elçi E M and Weigel M 2014 *J. Phys.: Conf. Ser.* **510** 012013
[22] Newman M E J and Ziff R M 2001 *Phys. Rev. E* **64** 016706
[23] Elçi E M and Weigel M In preparation.
[24] Henzinger M R and King V 1999 *J. ACM* **46** 502–516
[25] Holm J, de Lichtenberg K and Thorup M 2001 *J. ACM* **48** 723–760
[26] Sleator D D and Tarjan R E 1985 *J. ACM* **32** 652–686
[27] Elçi E M 2015 *Algorithmic and geometric aspects of the random-cluster model* Ph.D. thesis Coventry University
[28] Sokal A D 1997 Monte carlo methods in statistical mechanics: Foundations and new algorithms *Functional Integration: Basics and Applications* Proceedings of the 1996 NATO Advanced Study Institute in Cargèse ed DeWitt-Morette C, Cartier P and Folacci A (New York: Plenum Press) pp 131–192
[29] Redner S 1990 Fragmentation *Statistical Models for the Fracture of Disordered Media* ed Herrmann H J and Roux S (Amsterdam: Elsevier) pp 321–348
[30] Elçi E M, Weigel M and Fytas N G 2015 *Phys. Rev. Lett.* **114** 115701
[31] Edwards B F, Gyure M F and Ferer M 1992 *Phys. Rev. A* **46** 6252
[32] Cheon M, Heo M, Chang I and Stauffer D 1999 *Phys. Rev. E* **59** R4733
[33] Gyure M F and Edwards B F 1992 *Phys. Rev. Lett.* **68** 2692–2695
[34] Debierre J M 1997 *Phys. Rev. Lett.* **78** 3145
[35] Coniglio A 1989 *Phys. Rev. Lett.* **62** 3054
[36] Grimmett G 2010 *Probability on Graphs – Stochastic Processes on Graphs and Lattices* (Cambridge: Cambridge University Press)
[37] Vasseur R, Jacobsen J L and Saleur H 2012 *J. Stat. Mech.: Theory Exp.* **2012** L07001
[38] Schröder M, Rahbari S E and Nagler J 2013 *Nat. Commun.* **4** 2222

Journal of Physics: Conference Series **681** (2016) 012015 doi:10.1088/1742-6596/681/1/012015

Non-equilibrium critical vortex dynamics of disordered 2D XY-model

Ivan S. Popov[†], **Pavel V. Prudnikov**[‡], **Vladimir V. Prudnikov**[§]

Department of Theoretical Physics, Omsk State University, Omsk 644077, Russia

E-mail: [†]diphosgen@mail.ru, [‡]prudnikp@univer.omsk.su, [§]prudnikv@univer.omsk.su

Abstract. Vortex dynamics and clustering in non-equilibrium critical relaxation of disordered 2D XY-model are investigated for different initial states. Time dependencies of vortex concentration and clusters sizes are studied for different spin concentrations. The anomalous slow down of clustering in disordered system are explained by pinning of vortices on defects. The calculated temperature dependence of transverse stiffness allows to estimate critical temperature T_{BKT} and applicability of spin-wave approximation for disordered system.

1. Introduction

Investigations of low-dimensional systems with continuous symmetry are represents a considerable scientific interest. These systems are characterized by specific topological phase transitions with realization of special phases without long-range order [1]. Non-equilibrium behavior of these systems demonstrates slow dynamics with such characteristic effects as an aging [2–4] and the memory about the initial states and about any intermediate state in the course of the relaxation process [5, 6]. The critical coarsening is an important non-equilibrium phenomenon in such systems, characterized by the emergence of larger scale structures from smaller ones and accompanied by the decrease of total number of these structures [4, 7]. Systems with the coarsening process are called "coarsening dynamical systems" [8–10]. As opposed to "coarsening" can be introduced the concept of "fragmentation" as a process with fragmentation large structures to smaller ones with an overall increase in the number of objects.

The 2D XY-model takes an important position in the set of an low-dimensional systems [1]. This model describes the behavior of a wide class of physical systems, such as the planar magnetics [11, 12] and ultrathin magnetic films [13], in particular the films of Co and Ni on Cu-substrate and Fe on Au-substrate [13]. Also, the 2D XY-model is used for description of various properties of the superconducting films [14] and the superfluid helium films [15], Josephson-junction arrays [16], especially tailored nematic liquid crystals [17], toy models for two-dimensional turbulence [18], and many others [1, 19].

The 2D XY-model demonstrates the Beresinskii-Kosterlitz-Thouless topological phase transition [20, 21] which is associated with the dissociation of coupled vortex-antivortex pairs at the transition point T_{BKT}. An important feature of the Beresinskii-Kosterlitz-Thouless phase transition in the 2D XY model is that each temperature in the low-temperature phase becomes critical; i.e., a continuous cascade of second-order phase transitions is observed [20, 21]. This allows a certain analogy between non-equilibrium effects in the 2D XY model and the features of spin glass [22].

The equilibrium behavior and statical properties of the 2D XY-model are relatively complete explored scientific area, but investigations of non-equilibrium properties and influence of structural disorder on these properties are of significant interest. According to the Harris criterion [23], the presence of structural defects can be significant if the critical exponent of the heat capacity α is positive. It is predicted that the presence of structural defects does not affect to the 2D XY-model behavior near the critical temperature $T_{\rm BKT}$. However, in low-temperature phase for $T < T_{\rm BKT}$, as shown by analytical and numerical study of the equilibrium and non-equilibrium properties of the model [24–26], the presence of defects leads to changing of the equilibrium characteristics for the correlation functions and their concentration dependence.

2. Description of the model and methods

In this work, we take the Hamiltonian of the disordered system in the form

$$H = -J \sum_{<i,j>} p_i p_j \mathbf{S}_i \mathbf{S}_j, \tag{1}$$

where J is a exchange integral, $\mathbf{S}_i$ is the classical planar spin fixed in ith site of square lattice with linear size L, p_i are the occupation numbers: $p_i = 1$ and 0 if the ith site is occupied by spin and defect, respectively. Summation occurs on a nearest neighbors of sites $\langle i, j \rangle$. The distribution of defects are taken as uncorrelated and the concentration of defects c_{imp} is determined by spin concentration p: $c_{imp} = 1 - p$.

For simulation of non-equilibrium relaxation it was used the Metropolis algorithm of flipping individual spins which appropriately realizes dynamics of 2D XY-model [27] in low-temperature phase for $T < T_{\rm BKT}$. The dynamics of the Metropolis algorithm corresponds to dissipative processes which described by dynamic model A in Halperin-Hohenberg classification [28].

3. The transverse stiffness

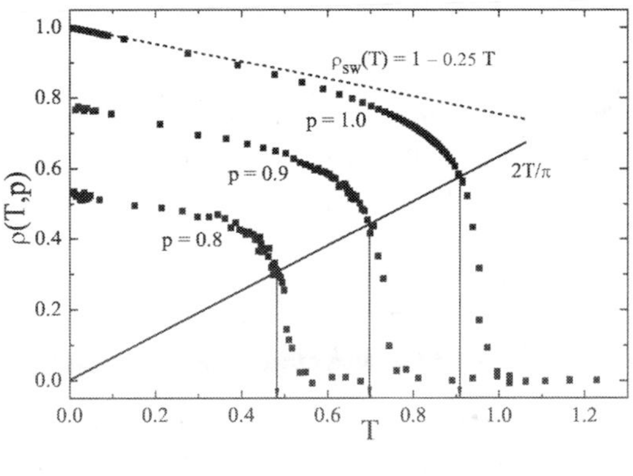

Figure 1. Temperature dependence of the stiffness for different spin concentrations p. $\rho_{SW}(T)$ shows the results of the spin-wave approximation.

Low-temperature phase $T < T_{\rm BKT}$ is characterized by non zero transverse stiffness $\rho(T,p)$ [20, 27, 29]. For calculation of stiffness it was used numerical method from [30]. The simulation were carryed out for linear size of lattice $L = 256$, full random initial conditions, spin concentration $p = 1.0$, 0.9, 0.8. It was used $3 \cdot 10^7$ Monte Carlo steps per spin (MCS/s) for equilibration and $5 \cdot 10^7$ MCS/s for statistical averaging. Results of simulation are presented on Fig. 1. These curves have been obtained by averaging over 100 different initial configurations for pure system and by averaging over 50 defects configurations for disordered system with 5 statistical runs for each sample.

The relation [29] $\rho(T_{\rm BKT}) = (2/\pi)T_{\rm BKT}$ allows to calculate the value of $T_{\rm BKT}$. Values $T_{\rm BKT}(p)$ for different spin concentrations are marked by arrows on Fig. 1. Obtained values are in well agreement with the results from [25] which $T_{\rm BKT}(p)$ were determined by method of correlation functions.

In [29] it was introduced relation for pure system $\rho(T) = e^{-T/4\rho(T)}$ in spin-wave approximation (i.e. without vortex component). It let us conclude that $\rho(T = 0) = 1.0$. Spin-wave approximation is works well in low-temperature limit $T \ll T_{\rm BKT}$. If we consider the case $T/4\rho(T) \ll 1$, then $e^{-T/4\rho(T)} \simeq 1 - T/4\rho(T)$ with assuming the relation $\rho_{SW}(T) = 1 + \alpha_{SW}(T)$,

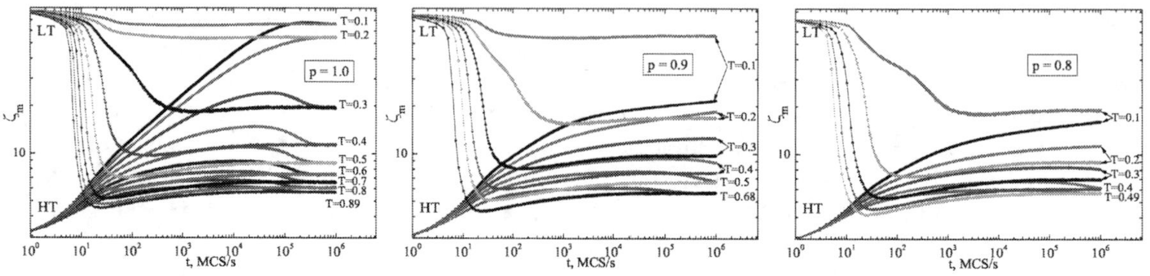

Figure 2. Time dependence of large clusters size $\zeta_{\rm m}$ in system for different temperature $T < T_{BKT}(p)$, different spin concentration p and different initial state: low-temperature initial state (LT) and high-temperature initial state (HT).

where $|\alpha_{SW}(T)| \ll 1$ it was obtained the following relation $1 + \alpha_{SW}(T) \simeq 1 - \frac{T}{4[1+\alpha_{SW}(T)]} \simeq 1 - T/4[1-\alpha_{SW}(T)]$ and the final solution $\alpha_{SW}(T) = -\frac{T/4}{1-T/4} \simeq -T/4$. Thus, the spin-wave form of ρ is $\rho_{SW}(T) \simeq 1 - T/4$ in low-temperature limit $T \ll T_{\rm BKT}$. The temperature dependence of $\rho_{SW}(T)$ is presented in Fig. 1 by dash line.

The spin-wave form $\rho_{SW}(T) \simeq 1 - T/4$ doesn't work for disordered system with $p < 1$ by reason of $\rho(T = 0, p < 1.0) \neq 1$.

4. Non-equilibrium vortex dynamics and clustering

In the 2D XY-model where are two different types of critical coarsening: vortex coarsening [7, 31] and quasi-long-range order (quasi-LR order) coarsening [24, 32]. Applied to lattice models it can using concept of "clustering", where coarsening is the emergence of larger scale structures. The "fragmentation" is the fragmentation large structures on smaller.

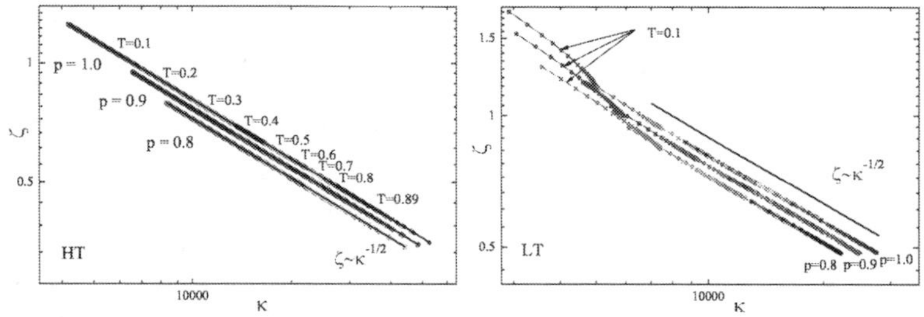

Figure 3. Parametric dependence ζ on κ in system for different temperature $T < T_{BKT}(p)$, different spin concentration p and different initial state: high-temperature initial state (left) and low-temperature initial state (right).

Numerical simulation are performed for 2D XY-model with $L = 256$ and spin concentration $p = 1.0$ (pure system) and 0.9 and 0.8 (disordered system). Values $T_{\rm BKT}(p)$ for this spin concentrations p are used from [25]. Simulation of critical relaxation are performed from two different non-equilibrium initial states: low-temperature initial state ($T_0 = 0$) and high-temperature initial state ($T_0 \gg T_{BKT}(p)$). For relaxation from high-temperature initial state (HT) excessive amount of vortex excitations are present in system originally. As opposed to relaxation from low-temperature initial state (LT) vortex excitation in system are not presented originally and are generated in system in the process of relaxation by energy of temperature fluctuations.

Non-equilibrium relaxation are investigated until the time 10^6 MCS/s. For pure system ($p = 1.0$) it was used 20000 different initial configuration, 10000 different defects configuration for $p = 0.9$, and 15000 different defects configuration for $p = 0.8$ and 5 different initial configuration for low-temperature phase $T \leq T_{BKT}(p)$.

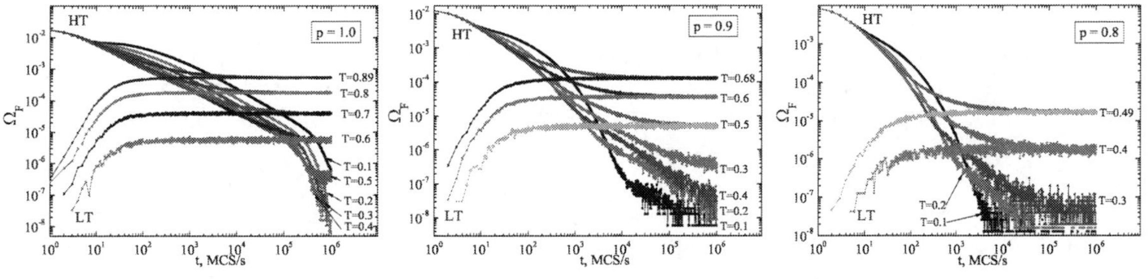

Figure 4. Time dependence of free vortex concentration Ω_F in system for different temperature $T < T_{BKT}(p)$, different spin concentration p and different initial state.

For investigation of quasi-LR order clustering it was created special algorithm. Continuous phase ϕ_i are discretized on discrete set $\{\phi_k\}_i$, where k is enumerates a set of phases. It was realized algorithm of search clusters in system with using this discrete set of phases on 2D-lattice. The algorithm calculate geometric characteristics of clusters in system. As the defining parameters it have been selected large clusters size ζ_m – linear size of largest cluster in system, average cluster size ζ and quantity of clusters κ in system. Linear size of cluster was selected as square root of square of cluster.

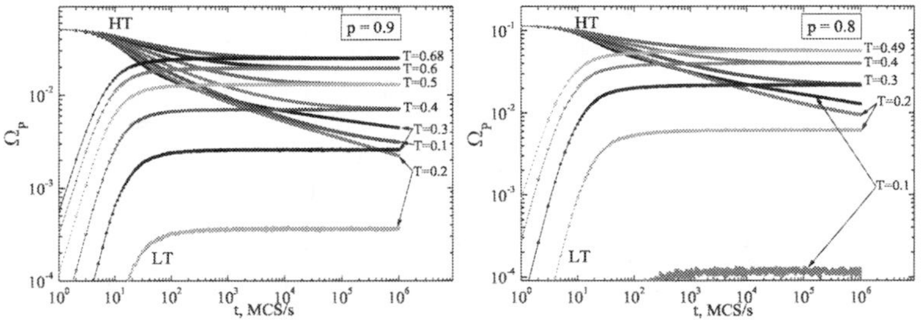

Figure 5. Time dependence of pinning vortex concentration Ω_P in system for different parameters of simulation.

On Fig. 2 it were presented time dependencies of large clusters size ζ_m in system for different temperature $T < T_{BKT}(p)$, different spin concentration p and different initial state. As seen from this dependencies the dynamics of relaxation of system is significantly slow because curves of time dependencies for different initial states are intersected and tend to one line for large times for pure system $\sim 10^6\ MCS/s$, but for disordered system and for low-temperature T these times $\gg 10^6\ MCS/s$ and takes values $\sim 10^{12} - 10^{18}\ MCS/s$. When defects are introduced to system, the size of clusters is decrease, but this effect can be explained by violation of the translational symmetry in disordered system i.e. areas of quasi-LR order become smaller. The violation of the translational symmetry can not explained by slow down effects of cluster relaxation.

On Fig. 3 it were presented parametric dependencies average clusters size ζ on quantity of clusters κ for all studied temperatures $T < T_{BKT}(p)$ all spin concentration p and different initial state. As seen that this dependencies well described dependencies by the relation of the form $\zeta \sim \kappa^{-1/2}$. It should pay attention that ζ is determined as $\sqrt{S}$, where S is the square of cluster and only after the averaging of different clusters and not vice versa. If averaging over different clusters designated as $\|\ldots\|$ then dependencies on the type $\zeta \sim \kappa^{-1/2}$ are indicate the relation $\|\sqrt{S}\| \sim \sqrt{\|S\|}$, because $\sqrt{\|S\|} \sim \kappa^{-1/2}$ obvious from the geometric considerations about clustering of square lattice. Relation $\|\sqrt{S}\| \sim \sqrt{\|S\|}$ is demonstrated that clustering in lattice of 2D XY-model in non-equilibrium relaxation can be represented as everywhere dense in the lattice of mapping without self-intersections clustered curve. So problem of quasi-LR order clustering in non-equilibrium critical relaxation of 2D XY-model can be reduced to two simpler problems: problem of clustering curve and problem of everywhere dense mapping curve without self-intersections in the plane. However as seen from Fig. 3 this arguments does not apply to $T = 0.1$ for all spin concentration and for relaxation of system from low-temperature initial state. It can be concluded that in low-temperature limit $T \ll T_{BKT}(p)$ for low-temperature started dynamics of system the problem of clustering more complex and can not be reduced to two different independent problems. For low-temperature dynamics of system with relaxation from high-temperature initial state this effect is not arising possibly because of existence subsystem of vortex excitations originally in system.

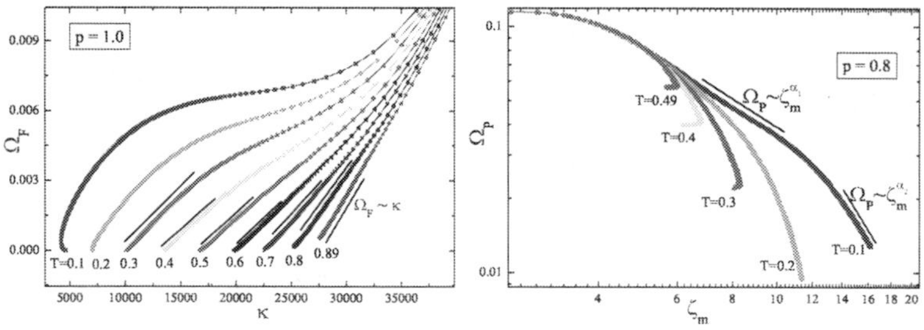

Figure 6. Parametric dependence of free vortex concentration $\Omega_{\rm F}$ on quantity of clusters κ for pure system (left) and parametric dependence of pinning vortex concentration $\Omega_{\rm P}$ on large clusters size $\zeta_{\rm m}$ for disordered system with $p = 1.0$ (right). Relaxation from high-temperature initial state.

For investigation of vortex dynamics it was created special algorithm, when vortex is identified as contour with phase increment $\Delta\varphi$ which multiple 2π. In structure disordered system it were searched all minimal contours around defects clusters and vortex are identified by the phase increment. This algorithm can calculated quantity of free and pinning vortex and antivortex in pure and disordered system. With using this algorithm it were calculated dynamical dependencies of free vortex excitation concentration (quantity of vortex excitations divided on quantity of spins $N = pL^2$) $\Omega_{\rm F}$, pinning vortex excitation concentration $\Omega_{\rm P}$.

On Fig. 4 and Fig. 5 are presented time dependencies of free vortex in system for different temperature $T < T_{BKT}(p)$, different spin concentration p and different initial state. Of these dependencies can draw conclusions that concentration of free and pinning vortex have a multimode time dependencies for relaxation from high-temperature initial state. It is connected with complex process of non-equilibrium annihilation vortex and antivortex and non-equilibrium pinning of vortex excitations on defects of structure. In disordered system significantly slowing are observed, i.e. relaxation times in order of magnitude equal to relaxation times of quasi-LR order clustering.

This slowing can well explained by non-equilibrium pinning process. The vortex excitations are "pinned" on defects and their mobility is reduced considerably. It can be explaine the slowing process of quasi-LR order clustering in disordered system. On Fig. 6 it were presented parametric dependence of free vortex concentration $\Omega_{\rm F}$ on quantity of clusters κ for pure system and parametric dependence of pinning vortex concentration $\Omega_{\rm P}$ on large clusters size $\zeta_{\rm m}$ for disordered system with $p = 1.0$. From these dependencies it can conclude that vortex excitations constrained growth of quasi-LR order areas – quantity of clusters κ is increase and large clusters size $\zeta_{\rm m}$ is decrease with increase vortex concentration $\Omega_{\rm F}$. Thus, slowing of vortex dynamics in disordered system by non-equilibrium pinning is slowing quasi-LR order clustering by boundaring effect. As seen on Fig. 6 the power parametric dependencies $\Omega_{\rm F} \sim \kappa$ for pure system and multimode regimes $\Omega_{\rm P} \sim \zeta_{\rm m}^{\alpha_1}$ and $\Omega_{\rm P} \sim \zeta_{\rm m}^{\alpha_2}$ on different time intervals for disordered system are appeared.

5. Conclusion

Vortex dynamics and clustering in non-equilibrium critical relaxation of disordered 2D XY-model are investigated for low-temperature and high-temperature initial states. It was showed for the first time that quasi-LR order clustering in non-equilibrium critical relaxation of 2D XY-model can be reduced to problem of clustering curve and problem of everywhere dense mapping curve without self-intersections on the plane. The slow down effect of quasi-LR order clustering and slow dynamics of vortex subsystem were demonstrated for disordered system. It was found boundaring effect of vortex for clustering process. The calculated temperature dependence of transverse stiffness allows to estimate values of critical temperature $T_{\rm BKT}(p)$ for disordered system.

Acknowledgments

Work is supported by the Ministry of Education and Science of the Russian Federation, project No. 1627 and grant of Omsk State University for young scientists, project MU-4/2015. Numerical studies were performed with the support by supercomputer center of Moscow State University, Moscow and St.Petersburg Joint Supercomputer Center.

References

[1] Korshunov S E 2006 *Phys. Usp.* **49** 225.
[2] Ciliberto S, Gomez-Solano R and Petrosyan A 2013 *Annu. Rev. Condens. Matter Phys.* **4** 11.
[3] Berthier L and Kurchan J 2013 *Nature Phys.* **9** 310.
[4] Afzal N and Pleimling M 2013 *Phys. Rev. B* **87** 012114.
[5] Khan N, Mandal P and Prabhakaran D 2014 *Phys. Rev. B* **90** 024421.
[6] Berthier L 2005 *Phys. Rev. B* **71** 214429.
[7] Jelic A and Cugliandolo L F 2011 *J. Stat. Mech.* **02** 02032.
[8] Gratton M B and Witelski T P 2009 *Phys. D: Nonlinear Phenomena* **238** 2380.
[9] Lifshitz I M and Slyozov V V 1961 J. Phys. Chem. Solids **19** 50.
[10] Watson S J, Otto F, Rubinstein B Y and Davis S H 2003 Physica D: Nonlinear Phenomena **178** 148.
[11] Kawabat C and Bishop A R 1986 *Solid State Commun.* **60** 167.
[12] Elmers H-J, Hauschild J, Liu G H and Gradmann U 1996 *J. Appl. Phys.* **79** 4984.
[13] Vaz C A F, Bland J A C and Lauhoff G 2008 *Rep. Progr. Phys.* **71** 056501.
[14] Bishop D J and Reppy J D 1980 *Phys. Rev. B* **22** 5171.
[15] Bishop D J and Reppy J D 1978 *Phys. Rev. Lett.* **40** 1727.
[16] Hebard A F and Fiory A T 1980 *Phys. Rev. Lett.* **44** 291.
[17] Pargellis A N, Green S and Yurke B 1994 *Phys. Rev. E* **49** 4250.
[18] Tabeling P 2002 *Phys. Rep.* **362** 1.
[19] Taroni A, Bramwell S T and Holdsworth P C W 2008 *J. Phys.: Condens. Matter* **20** 275233.
[20] Berezinskii V L 1971 *Sov. Phys. JETP* **32** 493.
[21] Kosterlitz J M and Thouless D J 1973 *J. Phys. C: Solid State Phys.* **6** 1181.
[22] Franz S, Mezard M, Parisi G and Peliti L 1998 *Phys. Rev. Lett.* **81** 1758.
[23] Harris A B 1974 *J. Phys. C.* **7** 1671.
[24] Berche B, Farinas-Sanchez A I, Holovatch Yu and Paredes R 2003 *Eur. Phys. J. B.* **36** 91.
[25] Prudnikov P V and Popov I S 2014 *J. Phys.: Conf. Ser.* **510** 1742.
[26] Prudnikov P V, Prudnikov V V and Popov I S 2015 *JETP Lett.* **101** 596.
[27] Prudnikov V V, Prudnikov P V, Alekseev S V and Popov I S 2014 *Phys. Metal. Metallogr.* **115** 1186.
[28] Hohenberg P C and Halperin B I 1977 *Rev. Modern Phys.* **49** 435.
[29] Pokrovsky V L and Uimin G V 1973 *Phys. Lett. A* **45** 467.
[30] Teitel S and Jayaprakash C 1983 *Phys. Rev. B* **27** 598.
[31] Yurke B, Pargellis A N, Kovacs T and Huse D A 1993 *Phys. Rev. E* **47** 1525.
[32] Prudnikov P V and Popov I S 2015 *Solid State Phenomena* **233-234** 8.

Journal of Physics: Conference Series **681** (2016) 012016 doi:10.1088/1742-6596/681/1/012016

Monte Carlo simulation of magnetic multilayered structures with giant magnetoresistance effects

V V Prudnikov†, P V Prudnikov‡, and D E Romanovskiy§
Omsk State University, Pr. Mira 55A, Omsk 644077, Russia
E-mail: †prudnikv@univer.omsk.su, ‡prudnikp@univer.omsk.su, §romfizz898@gmail.com

Abstract. Description of giant magnetoresistance effects in magnetic multilayered structures with the use of the anisotropic Heisenberg model for determination of magnetic properties of thin ferromagnetic films forming these structures is given. Monte Carlo simulations of magnetic properties for structures, which are constructed from two ferromagnetic films divided by nonmagnetic film, are carried out. The temperature and magnetic field dependencies are considered for ferromagnetic and antiferromagnetic configurations of these structures. The calculation of the magnetoresistance is carried out for different thicknesses of the ferromagnetic films. It was shown, that the obtained temperature dependence for the magnetoresistance is agreed very well with experimental results, measured for the magnetic multilayered structures similar to structures, which are considered in our investigations.

The physics of ultrathin magnetic films with thickness from one and two to several tens of monolayers appears as a direction of intensive research over the past two decades (for a review see [1]). This heightened interest of scientists is caused by quite a number of unique properties of films different from properties of bulk materials, that defines an importance of these objects for the new facilities of fundamental physics of magnetism, surface physics and their practical applications [2, 3]. Thus, the ultrathin films of magnetic metals and alloys are used as the building blocks for magnetic multilayer structures. These artificial created magnetic superlattices has become of great interest in wide range of applications based on the phenomena of the giant magnetoresistance (GMR) [4–6] and the tunneling magnetoresistance (TMR) [7–9].

The multilayer structures with GMR effects consist of the ferromagnetic layers separated by nonmagnetic layers. The thickness of these nonmagnetic metal layers are selected in such a way that the long-range and oscillating RKKY exchange interaction between the spins of the ferromagnetic layers is effective antiferromagnetic. Through this interaction, the magnetizations of the adjacent ferromagnetic layers are oriented opposite to each other. When this structure is placed in an external magnetic field, the magnetization of layers begin to orient in parallel, that leads to a significant change in electrical resistance. Devices based on the GMR effect are widely used as read heads of hard disks, memory devices, sensors, etc [2, 10, 11].

Achievements in the development of technology give possibility now to receive a high-quality ultrathin films and multilayer coatings on the basis of the magnetic transition metals Fe, Co, and Ni [1]. Investigation of the nature of magnetism in ultrathin films and multilayer structures has a large fundamental interest because the dimensional dependence of the magnetic characteristics which demostrate the transition from the specific bulk values for films with thickness of several tens monolayers ($d \geq 10$ nm) to the two-dimensional values in films with thickness less than 4-6 monolayers ($d \leq 1 - 2$ nm) [12, 13]. The magnetic properties of ultrathin films and superstructures sensitive to the effects of anisotropy generated by the crystal field of the substrate or by the nonmagnetic layers. Therefore, the physical properties of ultrathin films on the basis of Fe, Co and Ni can be described by the anisotropic Heisenberg model.

At present time, the statistical Monte Carlo methods have proved as successful methods for simulation behavior and describing the physical properties of various magnetic systems with different dimensionality with identifying their features during phase transitions [15]. The Monte Carlo study of dimensional crossover effects in the critical properties of multilayer Heisenberg films have been performed in works [16, 17]. For films with different thicknesses it was taken into account the effect of anisotropy generated by the crystal field of the substrate. The values of calculated critical exponents demonstrated a dimensional crossover from two-dimensional to three-dimensional properties of the films with increase in number of monolayers.

In this paper we apply the methodology and results of the multilayer Heisenberg films studies obtained in [16, 17] to Monte-Carlo simulations and calculation of magnetic properties in magnetic multilayer and spin valve sandwich structures. At first, data of the microscopic spin configurations of ferromagnetic films were used for the calculation of the magnetoresistance temperature dependence for these structures with different thicknesses of the ferromagnetic films.

For the statistical Monte Carlo description of the Heisenberg ferromagnetic films we introduce the Hamiltonian of the spin system in the form:

$$H = -J \sum_{<i,j>} [(1 - \Delta(N))(S_i^x S_j^x + S_i^y S_j^y) + S_i^z S_j^z] - \mathbf{h} \sum_i \mathbf{S}_i, \tag{1}$$

where $J > 0$ characterizes the short-range exchange interaction between the spins $\mathbf{S}_i$ fixed in the sites of lattice; $\mathbf{S}_i = (S_i^x, S_i^y, S_i^z)$ is introduced as a three-dimensional unit vector, $\mathbf{h}$ is the external magnetic field , Δ is an anisotropy parameter. The dependence of the anisotropy parameter $\Delta(N)$ from the film thickness N is selected in accordance with the results in articles [16, 17].

Figure 1. The model of the multilayer structure (a) which consists of two ferromagnetic films separated by nonmagnetic metal film; L and N are linear sizes of films, J_1, J_2 are the exchange integrals. Model of the spin valve structure (b) which consists of two ferromagnetic films separated by non-magnetic metal film, and a layer of antiferromagnetic adjacented to one of the ferromagnetic films; L, N and N_{AF} are linear sizes of films, J_1, J_2, J_3, J_4 are the exchange integrals.

At first stage, we considered in this paper the multilayer structure which consists of two ferromagnetic films separated by a nonmagnetic metal film (Fig. 1). The simulations were performed for the films with linear sizes $L \times L \times N$ with applied periodic boundary conditions in the film plane. $L \times L$ gives a number of spins in each layer, and N is the number of monolayers in the thin film. The value of the exchange integral that determines the interaction of neighboring spins in the ferromagnetic film set as $J_1 = 1$, the interaction between the films $J_2 = -0.1$. The temperature T in this system are measured in units of exchange integral J_1. The spin configurations of the Heisenberg ferromagnetic films in the magnetic structure (Fig. 1) are updated using the Swendsen-Wang cluster algorithm [18].

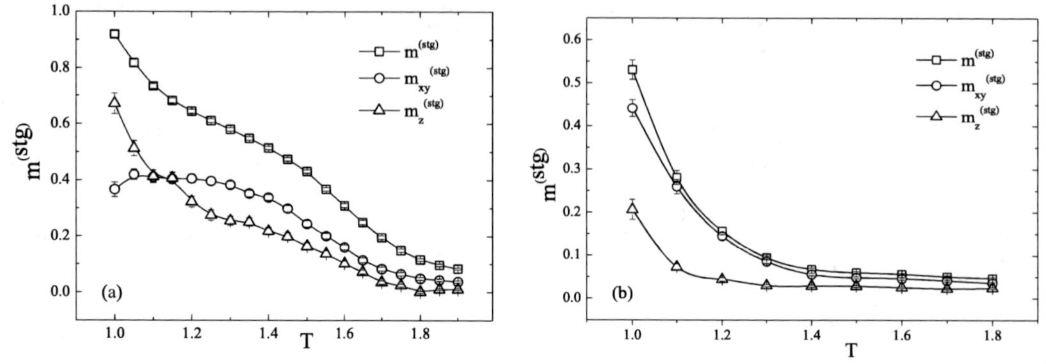

Figure 2. The temperature dependence of the staggered magnetization $m^{(stg)}$ and its projections $m_z^{(stg)}$ and $m_{\rm xy}^{(stg)}$ for threelayer structure with linear sizes of films $N = 3$, $L = 32$ for values of the external magnetic field $h = 0$ (a), $h_z = 0.4$ (b).

We measured the magnetization of films $m_{1,2}$ and their components m_z and m_{xy}, which are determined by the relations:

$$m = \left\langle \frac{1}{N_s} \left[\sum_{\alpha \in \{x,y,z\}} \left(\sum_i^{N_s} S_i^\alpha \right)^2 \right]^{1/2} \right\rangle, \qquad (2)$$

$$m_z = \left\langle \frac{1}{N_s} \sum_i^{N_s} S_i^{\rm z} \right\rangle, \qquad m_{\rm xy} = \left\langle \frac{1}{N_s} \left[\left(\sum_i^{N_s} S_i^{\rm x} \right)^2 + \left(\sum_i^{N_s} S_i^{\rm y} \right)^2 \right]^{1/2} \right\rangle,$$

where $N_s = NL^2$ is a total number of spins in film, angle brackets denote the statistical averaging.

In the case of absence of external magnetic field h, the antiferromagnetic configuration is realized in threelayer structure, therefore for description of the magnetic ordering in the system we measured the staggered magnetization $m^{(stg)} =\mid \mathbf{m}_1 - \mathbf{m}_2 \mid$ and projections of the staggered magnetization m^z_{stg} and m^{xy}_{stg}.

We have calculated the temperature dependence of the staggered magnetization for threelayer structures with different thicknesses N of the ferromagnetic films in the range of the $N = 1 \div 30$ monolayers. As an example, we give in Fig. 2a the resulting temperature dependence of the staggered magnetization and its projections for the films with the sizes of $N = 3, L = 32$ in the absence of a magnetic field. Also, we have carried out the study of the influence of an external magnetic field h_z on the magnetic characteristics of the same structure. The graphs presented in Fig. 2b show that under the influence of an external magnetic field the z-component of staggered magnetization is weakened, but the the xy-component of staggered magnetization in plane of the film gains strength on the contrary.

To reduce the saturation field in threelayer magnetic structure, usually an one of the ferromagnetic films is coupled with an antiferromagnetic metal, which constrains the direction of magnetization in the film by exchange coupling. Such structures are called as spin-valves. The spin-valve structures were first introduced in 1991 [19] and are now widely used, for example, in the construction of read heads for hard disks. In the spin-valve structures, the exchange coupling between the ferromagnetic layers is selected ferromagnetic and are weakened in comparison with threelayer structures through the use of a nonmagnetic conductive film from some noble metals (Cu, Ag or Au) with thicknesses of 1.5 - 5 nm. The ferromagnetic layers are produced from a soft ferromagnetic material such as $Ni_{80}Fe_{20}$ (permalloy) with 1.5-5 nm thick, at that one of these layers is coupled with a layer of antiferromagnetic alloy.

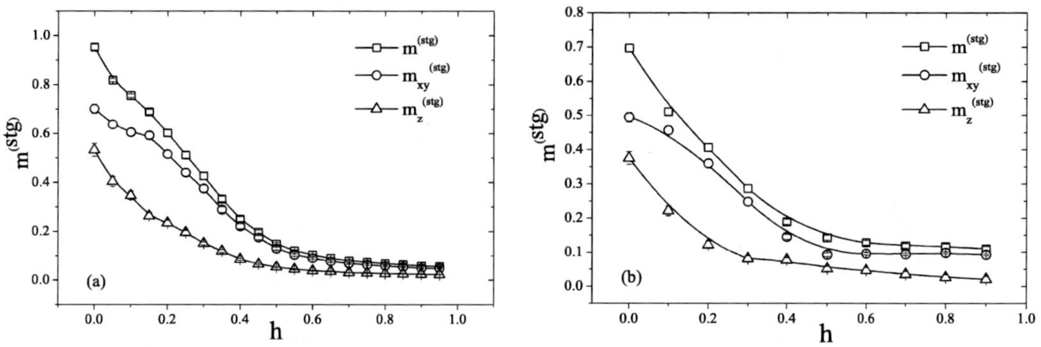

Figure 3. The dependence of the staggered magnetization on the external magnetic field h_z for threelayer structure (a) and spin-valve structure (b) at temperature $T = 1.2$ and for linear sizes of the ferromagnetic films $N = 3$, $L = 32$.

The materials which are used for the antiferromagnetic film that are alloys FeMn, NiO, TbCo, IrMn. As an example, the spin-valve sandwich $Ni_{80}Fe_{20}/Cu/Ni_{80}Fe_{20}/FeMn$ has the property that the magnetization of insulated copper layer $Ni_{80}Fe_{20}$ is free to rotate under the influence of an external magnetic field with respect to the magnetization $Ni_{80}Fe_{20}$ layer, magnetization of which is constrained by exchange coupling to the antiferromagnet FeMn.

We calculated the magnetization in spin-valve, for which the structure is shown in Fig. 1 b and is characterized by addition of the antiferromagnetic film with a constant of exchange interaction J_3 = -2.0, fixing the orientation of the magnetization in a coupled ferromagnetic film due to the exchange interaction with the constant $J_4 = 0.1$. From comparison of the graphs in Figs. 3a and 3b, it is seen that with the increase of the external magnetic field the z-component of the staggered magnetization for spin-valve decreases faster than for threelayer structure.

Henceforward, we calculate the magnetoresistance for an multilayer structures which is introduced by relation

$$\delta_h = \frac{R_{AP} - R_P}{R_P}, \tag{3}$$

where R_{AP} is the resistance of structure when the magnetization of adjacent ferromagnetic layers are aligned antiparallel, and R_P is the resistance of structure for parallel orientation of the magnetization of ferromagnetic layers. For the threelayer structure with antiferromagnetic exchange coupling of adjacent ferromagnetic layers, the R_{AP} characterizes the resistance of structure without an external magnetic field, and the R_P is the resistance in field h above the ferromagnetic layer saturation field h_s. The situation for the spin-valve structure is reversed, and, namely, the parallel orientation of the films magnetization with resistance R_P is realized in the absence of a magnetic field, and the antiparallel ordering with R_{AP} is in the magnetic field h, but with $h \ll h_s$.

The multilayer structure can connect in current line by two ways for measurements of resistance. For measurements of the current in plane (CIP), the current conduction is realized along layers and the electrodes are situated on a single side of structure. For measurements of the current perpendicular to plane (CPP), the current conduction is realized perpendicular layers of superstructure and the electrodes are situated on both sides of structure. It was shown [20, 21] that CPP magnetoresistance is characterized by larger values than the CIP magnetoresistance approximately to twice as much and becomes more competitive as device size shrinks. Regardless of the fact that the the CPP way of measurements is complicated for technical realization, now this method has a great practical interest because of the CPP magnetoresistive sensors

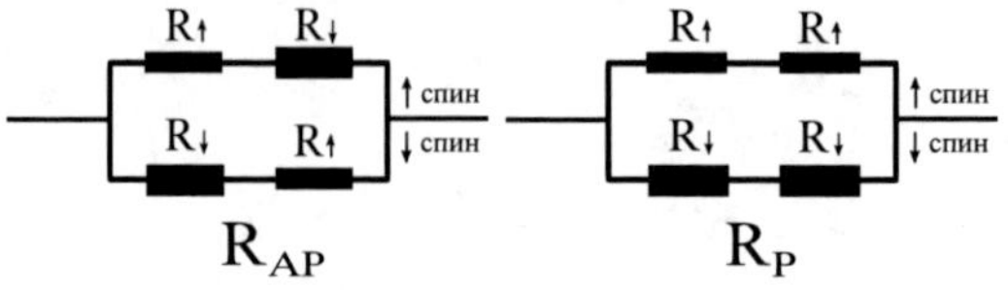

Figure 4. Equivalent scheme for determination of the resistance for antiparallel R_{AP} and parallel R_P configurations of the layers magnetizaion in threelayer magnetic structure

demonstrate more sensitivity than CIP-MR sensors [2]. At present paper we present the Monte Carlo realization of calculation of the CPP magnetoresistance.

Let us denote the resistance of an ferromagnetic film for two groups of electrons with spins up and down as $R_\uparrow$ and $R_\downarrow$, respectively, and use for calculation of the magnetoresistance δ_h of threelayer structure the simple two-current Mott model [22] for description of the resistance of different conductance channels [23]. This model suggests the conservation of orientation of electron spin moments during penetration of structure and better corresponds to measurement of the CPP magnetoresistance. The model was developed in papers [24–26] for description of the CPP-GMR in multilayer magnetic structures. Within the bounds of two-current model we neglect by scattering of current carriers from interfaces and by resistance of the nonmagnetic metal in comparison with resistance of the ferromagnetic metal. The total resistance of threelayer structure for the antiparallel configuration which is realized in the absence of a magnetic field is determined in concordance with equivalent resistor scheme in Fig. 4 by relation $R_{AP} = (R_\uparrow + R_\downarrow)/2$. The parallel configuration of the threelayer structure for the magnetic field $h \geq h_s$ is characterized by the resistance in the form $R_P = (2R_\uparrow R_\downarrow)/(R_\uparrow + R_\downarrow)$. Consequently, the magnetoresistance of threelayer structure is determined by relation

$$\delta_h = \frac{(R_\uparrow - R_\downarrow)^2}{4R_\uparrow R_\downarrow} = \frac{(\mathbf{J}_\uparrow - \mathbf{J}_\downarrow)^2}{4\mathbf{J}_\uparrow \mathbf{J}_\downarrow}, \tag{4}$$

where $\mathbf{J}_{\uparrow,\downarrow} = en_{\uparrow,\downarrow} \langle \mathbf{V}_{\uparrow,\downarrow} \rangle$ is the current density. Hear, $n_{\uparrow,\downarrow}$ is the density of electrons with z-components of spin moment equal - $+1/2$ and $-1/2$, $n = n_\uparrow + n_\downarrow$ is the total electron density, $\langle \mathbf{V}_{\uparrow,\downarrow} \rangle$ are the averaged velocity of electrons with corresponding spin projections. The electron densities with spin up and down can be expressed through the magnetization of film $n_{\uparrow,\downarrow}/n = (1 \pm m)/2$ determined in process of the Monte Carlo simulation. The averaged electron velocity $\langle \mathbf{V}_{\uparrow,\downarrow} \rangle$ can be expressed through an electron mobility and the external electric field intensity $\mathbf{E}$, and after that through a probability of electron jump in unit time from i-cell to a neighbouring cell in the direction of electric field [27] with averaging over all film cells:

$$\langle \mathbf{V}_{\uparrow,\downarrow} \rangle = \mu_{\uparrow,\downarrow} \mathbf{E} = \frac{e}{T} \mathbf{E} \Big\langle \exp\Big(-\frac{\Delta E_{i,\uparrow,\downarrow}}{T} \Big) \Big\rangle, \tag{5}$$

where μ is the electron mobility, ΔE_i characterizes the change of system energy connected with electron jump from i-cell to a neighbouring cell. $E_{i,\uparrow,\downarrow}$ is determined by relation

$$E_{i,\uparrow,\downarrow} = \mp J_1 \Big[\sum_{j \neq i} S_j^z (n_{j,\uparrow} - n_{j,\downarrow}) + S_i^z (n_{i,\uparrow} - n_{i,\downarrow}) \Big], \tag{6}$$

where the summation is realized of cells j neighbouring to i cell. In relation (6) for $E_{i,\uparrow,\downarrow}$ it was realized the fundamental idea of Nevill Mott [22], according to which an electrons of different spin bands realize a current only if their spin projection coincides with direction of the local magnetization of system, characterized at this case by spin S_j^z of the cell j.

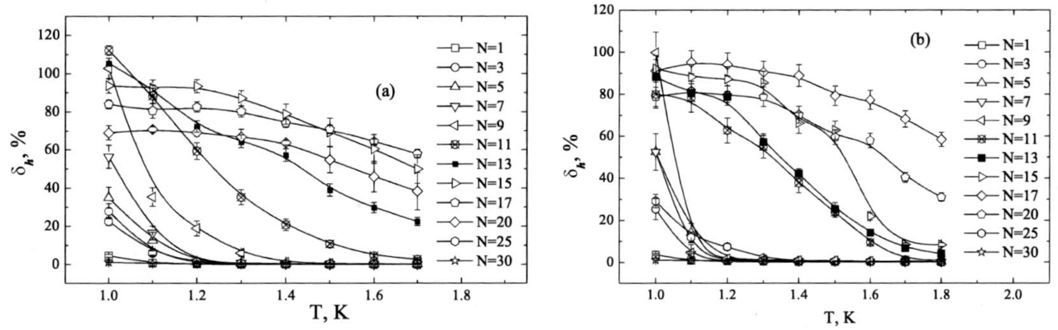

Figure 5. Temperature dependence of the magnetoresistance δ_h for threelayer structure (a) and spin-valve (b) with different thicknesses N of the ferromagnetic films

On basis of above presented relations we calculated the temperature dependence of magnetoresistance for the threelayer structure with different thicknesses N of ferromagnetic films. The procedure of calculation consists of the following steps: first step is connected with Monte Carlo simulations of an magnetic structure in equilibrium state at temperature T with determination of the magnetization of ferromagnetic films m_1 and m_2, which gives possibility to calculate the electron densities $n_{\uparrow,\downarrow}$ for film cells; at second step the average electron velocities $\langle \mathbf{V}_{\uparrow,\downarrow} \rangle$ and the current densities $\mathbf{J}_{\uparrow,\downarrow}$ are calculated under relations (5)-(6) subject to spin configuration, realized at given time of simulation, and averaged over Monte Carlo steps at times of equilibrium state simulation; at last step the calculation of the magnetoresistance is carried out under relation (4).

The graphs of calculated temperature dependence of the magnetoresistance are presented in Fig. 5a for the threelayer structure with different thicknesses N of the magnetic films. The obtained results demonstrate that for the low temperature with $T = 1$ the structure with film thicknesses in interval $N = 1 \div 11$ is characterized by sequential increase of the magnetoresistance from value $\delta_h = 4.5\%$ for $N = 1$ before value $\delta_h = 112\%$ for $N = 11$ and by fast falling down of $\delta_h(T)$ when temperature is increased. For films with thicknesses in interval $N = 13 \div 20$, the magnetoresistance for temperature $T = 1$ begin to be decreased with increasing thickness N, but the temperature falling down of magnetoresistance becomes more flat. However, for structures with $N \leq 11$ and $N \geq 25$ the magnetoresistance is characterized by fast falling down practically before null in the same interval of temperatures. Note that the observable changes of the magnetoresistance with increasing thickness of ferromagnetic films N correspond to dimensional changes of the critical temperature of ferromagnetic phase transition in Heisenberg films revealed in papers [16, 17] and agree with results of the experimental investigations of dimensional phenomena in the ultrathin films Fe, Co, and Ni on substrates from nonmagnetic metals [1]. In accordance with papers [16,17], the films with thicknesses in interval $N = 13 \div 20$ in which the magnetoresistance takes on a maximal values with very slow temperature falling down correspond to Heisenberg films demonstrating the critical behavior of the 3D Ising model.

The graph of calculated temperature dependence of the magnetoresistance δ_h for spin-valve structure is presented in Fig. 5b with the same thicknesses N of the ferromagnetic films. Comparison of data in Fig. 5a and b shows that the values of δ_h for spin-valve structure with the same thickness N rather less but generally comparable with values δ_h for multilayer structure and agree with their temperature dependence trend, described above.

The graph of calculated magnetoresistance is given in Fig. 6a for the multilayer structure from the iron films with thickness $N = 10$ with determination of temperature scale through the

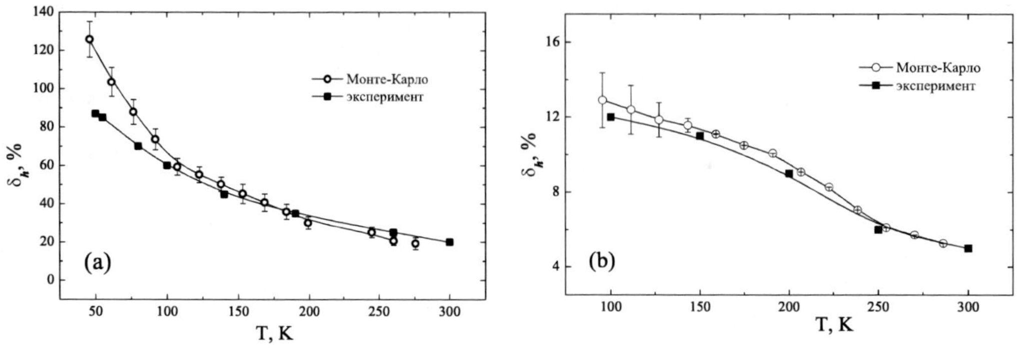

Figure 6. Comparison of the temperature dependensies of the CPP-magnetoresistance for multilayer structure Fe/Cr with thickness equal 3 nm for the iron films (a), and for spin-valve structure CFAS/Ag/CFAS/IrMn, obtained by experimentally in [21] (a), in [30] (b), and calculated by the Monte Carlo methods.

value of exchange integral $J_1 \simeq 2 \cdot 10^{-14}$ erg corresponding to the iron. In the same Fig. 6a the results of experimental measurements of the CPP-MR are presented for the multilayer structure Fe(001)/Cr(001) with thickness of the iron films equal 3 nm corresponding to $N = 10$ [21]. One can see that the calculated temperature dependence of the magnetoresistance agrees very well with experimental data.

Recently, a highly promising material for the use in spintronic devices have been allocated the Co_2 based Heusler alloys $Co_2FeAl_xSi_{1-x}$, characterized by the highest Curie temperatures (T_c = 1170 K, Co_2FeAl and Tc = 1100 K, Co_2FeSi) and by very high values of electron spin polarization at the Fermi surface (close to 100 % at $x \approx 0.5$), because it have been identified their half-metallic properties in ferromagnetic state [28]. In paper [29], it were presented the results that a magnetic tunnel junction (MTJ) showing 832 % tunneling magnetoresistance (TMR) at 2 K and 386 % at room temperature has been realized using the Heusler compound $Co_2FeAl_{0.5}Si_{0.5}$. These alloys are also recognized as very promising for production of read heads of ultrahigh density magnetic recording systems and magnetic sensors on basis of the CPP-GMR spin-valve structures because of the intrinsically low resistance of CPP-GMR devices with application of these alloys [30]. This property is suitable for achieving high-speed readout.

We used in this study the developed methodology for the calculation of the magnetoresistance for spin-valve structure with ferromagnetic layers from Heusler alloy $Co_2FeAl_{0.5}Si_0$.5 (CFAS). The graph of calculated temperature dependence of the magnetoresistance δ_h for spin-valve structure is presented in Fig. 6b for ferromagnetic layers with thickness $N = 23$ monolayers with determination of temperature scale through the value of exchange integral $J_1 \simeq 2.2 \cdot 10^{-14}$ erg, corresponding to the Curie temperature of the CFAS alloy $T_c \simeq 1140$ K. The same figure shows the results of experimental measurements of the temperature dependence of the CPP magnetoresistance for the spin-valve structure CFAS (20 nm) / Ag (5 nm) / CFAS (5 nm) / IrMn (10 nm) with an antiferromagnetic layer of an alloy $Ir_{22}Mn_{78}$, carried out in [30]. For this asymmetrical configuration of the spin-valve with a free CFAS layer thickness 20 nm and a layer CFAS with a fixed orientation of the magnetization with thickness 5 nm, we realized the calculation of magnetoresistance for symmetric spin-valve with a thickness of the ferromagnetic layers equal to the average value 12.5 nm corresponding to $N \approx 23$ CFAS alloy. The graphs in Fig. 6b show that calculated temperature dependence of the magnetoresistance by Monte Carlo method for this spin-valve structure is in good agreement with the experimental results [30].

Note in conclusion that in the present paper the Monte Carlo study of sandwich and spin-valve magnetic structures with GMR effects is carried out with applying an anisotropic Heisenberg model to describe the magnetic properties of the thin ferromagnetic films forming these structures. For ferromagnetic and antiferromagnetic configuration given structures, we have received the dependence of the magnetic characteristics on the temperature and the external magnetic field. With the use of two-current model to describe the resistance of different conduction channels, we have obtained an expression for the magnetoresistance of multilayer magnetic structures, determined by the probabilities of the electron jumps along the crystal cells subject to their spin projections. For the first time, it was developed a methodology of determination of the magnetoresistance with the use of the Monte Carlo method and was calculated the magnetoresistance temperature dependence for the threelayer and spin-valve structures with different thicknesses of ferromagnetic films. It is shown that the calculated temperature dependencies of the magnetoresistance are in good agreement with the experimental results obtained for multilayer structure Fe (001)/Cr (001) [21] and the spin-valve structure CFAS/Ag/CFAS/IrMn [30] with the Co_2 based half-metallic Heusler alloy $Co_2FeAl_{0.5}Si_0.5$.

Acknowledgments

The reported study was supported by the Russian Science Foundation through project No. 14-12-00562. The simulations were supported by the Supercomputing Center of Lomonosov Moscow State University and Joint Supercomputer Center of RAS.

References

[1] Vaz C A F, Bland J A C, Lauhoff G 2008 Rep. Prog. Phys. **71** 056501
[2] Bland J A C, Heinrich B (Eds) *Ultrathin Magnetic Structures*: (Berlin: Springer) 1994 vol I p 350; 1994 vol II p 350; 2005 vol III p 318; 2005 vol IV p 257
[3] Ustinov V V, Milyaev M A, Naumova L I 2014 SPIN **04** 1440001
[4] Baibich M N, Broto J M, Fert A, Van Dau F N, Petroff F, Etienne P, Creuzet G, Friederich A, Chazelas J 1988 Phys. Rev. Lett. **61** 2472
[5] Binash G, Grunberg P, Saurenbach F, Zinn W 1989 Phys. Rev. B **39** 4828
[6] Barthelemy A, Fert A 1991 Phys. Rev. B **43** 13124
[7] Julliere M 1975 Phys. Lett. A **54** 225
[8] Miyazaki T, Tezuka N 1995 JMMM **139** L231
[9] Sousa R C, Sun J J, Soares V, Freitas P P, Kling A, Silva M F, Soares J C 1998 Appl. Phys. Lett. **73** 3288
[10] Prinz G A 1999 JMMM **200** 57
[11] Chappert C, Fert A, Van Dau F N 2007 Nature Mater. **6** 813
[12] Li Y, Baberschke K 1992 Phys. Rev. Lett. **68** 1208
[13] Huang F, Kief M T, Mankey G J, Willis R F 1994 Phys. Rev. B **49** 3962
[14] Binder K, Landau D P 1976 Phys. Rev. B **13** 1140
[15] Prudnikov V V, Vakilov A N, Prudnikov P V 2009 *Phase transitions and methods their computer simulations* (Moscow: Fizmatlit) p 224
[16] Prudnikov P V, Prudnikov V V, Medvedeva M A 2014 JETP Lett. **100** 446
[17] Prudnikov P V, Prudnikov V V, Menshikova M A, Piskunova N I 2015 JMMM **387** 77
[18] Wang J-S, Swendsen R H 1990 Physica A **167** 565
[19] Dieny B, Speriosu V S, Parkin S S P, Gurney B A, Wilhoit D R, Mauri D 1991 Phys. Rev. B **43** 1297
[20] Gijs M A M, Bauer G E W 1997 Adv. Phys. **46** 285
[21] Bass J, Pratt W P 1999 JMMM **200** 274
[22] Mott N F 1936 Proc. Roy. Soc. (London) ser A **153** 699
[23] Mathon J 1991 Contemp. Phys. **32** 143
[24] Zhang S, Levy P M 1991 J. Appl. Phys. **69** 4786
[25] Bauer G E W 1992 Phys. Rev. Lett. **69** 1676
[26] Valet T, Fert A 1993 Phys. Rev. B **48** 7099
[27] Prudnikov V V, Prudnikov P V, Mamonova M V 2014 *Quantum-statistical theory of solid state* (Omsk: OmSU press) p 492
[28] Kandpal H C, Fecher G H, Felser C 2007 J. Phys. D **40** 1507
[29] Tezuka N, Ikeda N, Mitsuhashi F, Sugimoto S 2009 Appl. Phys. Lett. **94** 162504
[30] Furubayashi T, Kodama K, Sukegawa H, Takahashi Y, Inomata K, Hono K 2008 Appl. Phys. Lett. **93** 122507

Journal of Physics: Conference Series **681** (2016) 012017 doi:10.1088/1742-6596/681/1/012017

Monte-Carlo investigation of competition between uniaxial anisotropy, exchange and dipolar interactions in critical behavior of ultrathin magnetic films

Anna P. Soldusova[†], **Pavel V. Prudnikov**[‡], **Vladimir V. Prudnikov**[§]

Department of Theoretical Physics, Omsk State University, Omsk 644077, Russia

E-mail: [†]anka_omsk@mail.ru, [‡]prudnikp@univer.omsk.su, [§]prudnikv@univer.omsk.su,

Abstract. The investigation includes the observation of ferromagnetic-paramagnetic phase transition for a weak dipolar interaction. The system with strong dipolar interaction reveals the striped domain structure. Bilayer model with dipolar interlayer interaction demonstrates its effective antiferromagnetic character. Memory effects in ultrathin magnetic films are studied on basis of analysis of non-equilibrium relaxation of the autocorrelation function and magnetization.

1. Introduction

Ultrathin magnetic films are used in different applications such as storage and high-density record devices [1], high-sensitivity magnetic fields sensors [2], optical research of spin transmission [3], etc. The theoretical and experimental investigations of their properties become of great importance.

Competition between short-range exchange interaction, uniaxial anisotropy and long-range dipolar interaction results in appearance of different domain structures and in occurrence of the phase transitions. The strong enough dipolar interaction can provide the appearing of striped domain structures. Stripe phases are experimentally observed in high-temperature superconductors [4], and some computer studies are devoted to simulate these phenomena [5, 6, 7, 8].

This study includes the Monte-Carlo simulation of the critical behavior of ultrathin magnetic films. Memory effects can be observed in the non-equilibrium critical behavior of these systems.

2. Model and simulation

Magnetic film is constructed as a spin lattice with linear size L and thickness of the film N. The model is described by the Hamiltonian in the following form:

$$H = -J \sum_{<i,j>} \mathbf{S_i} \cdot \mathbf{S_j} - A \sum_i S_{i,z}^2 - \mathbf{h} \sum_i \mathbf{S_i} + D \sum_{i \neq j} \left(\frac{\mathbf{S_i} \cdot \mathbf{S_j}}{r_{ij}^3} - 3 \frac{(\mathbf{S_i} \cdot \mathbf{r_{ij}})(\mathbf{S_j} \cdot \mathbf{r_{ij}})}{r_{ij}^5} \right), \quad (1)$$

where $\mathbf{S_i}$ is a three-dimensional spin in the lattice cite i, J is exchange constant, A characterizes uniaxial anisotropy in out-of-plane direction, $\mathbf{h}$ is applied magnetic field, D characterizes the strength of dipolar interaction and $r_{ij} = r_i - r_j$ is the distance between cites i and j.

Magnetization of the system and its projections is determined by

$$m = \left\langle \frac{1}{N_s} \sum_i \sqrt{S_{i,x}^2 + S_{i,y}^2 + S_{i,z}^2} \right\rangle, \quad m_z = \left\langle \frac{1}{N_s} \sum_i S_{i,z} \right\rangle, \quad m_{xy} = \left\langle \frac{1}{N_s} \sum_i \sqrt{S_{i,x}^2 + S_{i,y}^2} \right\rangle, \tag{2}$$

where $N_s = NL^2$ represents the number of spins in the film and $\langle \ldots \rangle$ denotes the statistical averaging.

For critical temperatures evaluation it was used Binder's cumulant method determined by

$$U_\alpha = \frac{1}{2}\left(3 - \frac{\langle m_\alpha^4 \rangle}{\langle m_\alpha^2 \rangle^2}\right). \tag{3}$$

Orientational order parameter which characterizes striped domain structure with spins oriented in the direction α is determined by Eq. 4, where $n_h^\alpha = \frac{1}{2}\sum_{\mathbf{r}}\{1 - \text{sgn}[S_\alpha(r_x, r_y) \cdot S_\alpha(r_x + 1, r_y)]\}$ and $n_v^\alpha = \frac{1}{2}\sum_{\mathbf{r}}\{1 - \text{sgn}[S_\alpha(r_x, r_y) \cdot S_\alpha(r_x, r_y + 1)]\}$ represent the number of nearest spins with antiparallel alignment in the horizontal and vertical directions respectively.

$$O_\alpha = \left\langle \left| \frac{n_h^\alpha - n_v^\alpha}{n_h^\alpha + n_v^\alpha} \right| \right\rangle. \tag{4}$$

Susceptibilities of magnetization projections χ_m and that of the order parameter χ_O are calculated with following equations

$$\chi_m = \frac{\langle m_\alpha^2 \rangle - \langle m_\alpha \rangle^2}{N_s T}; \quad \chi_O = \frac{\langle O_\alpha^2 \rangle - \langle O_\alpha \rangle^2}{T}. \tag{5}$$

3. Results of modeling

Simulation of the system with $A = 0.1J$, $D = 0.01J$ results in observing phase transition from the paramagnetic state to the ferromagnetic state with all spin aligned in the out-of-plane direction. Temperature dependencies of magnetization projections m_z and m_{xy} and their susceptibilities are demonstrated in Fig. 1 for the system with linear size $L = 32$.

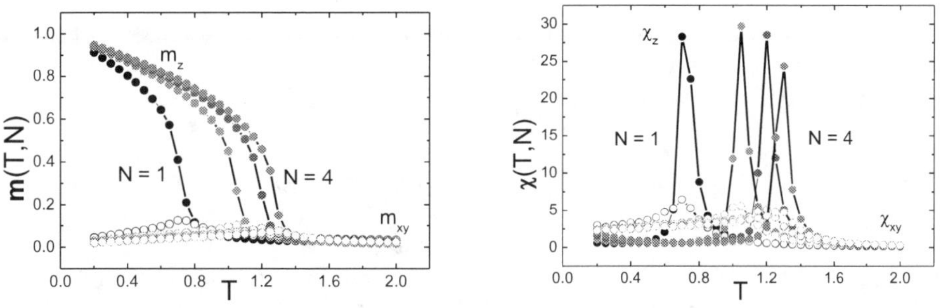

Figure 1. Temperature dependencies of magnetization projections m_z, m_{xy} and their susceptibilities χ_z, χ_{xy} for ultrathin film with $A = 0.1J$, $D = 0.01J$, and $N = 1 \div 4$

Temperature dependence of the orientational order parameter O_z and its susceptibility χ_O is shown in Fig. 2. There is no striped domain structure in this case. Dipolar interaction is too weak to form such a structure.

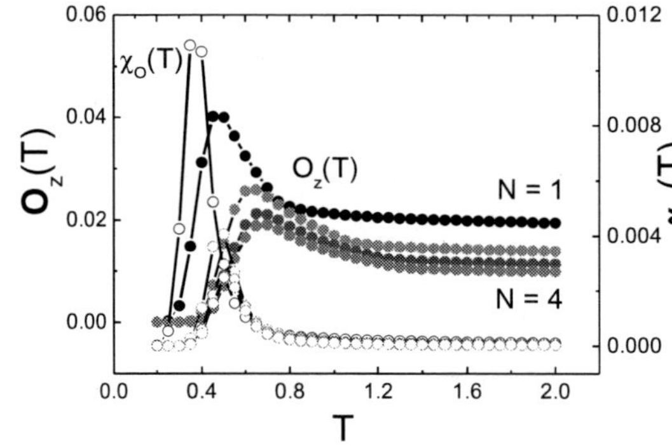

Figure 2. Temperature dependencies of orientational order parameter O_z and its susceptibility χ_O for ultrathin film with $A = 0.1J$, $D = 0.01J$, and the number of layers $N = 1 \div 4$

Binder's cumulants U provide calculating critical temperatures of ultrathin films with $A = 0.1J$, $D = 0.01J$: $T_c = 0.713(11)$ for $N = 1$; $T_c = 1.034(13)$ for $N = 2$; $T_c = 1.175(25)$ for $N = 3$; $T_c = 1.272(12)$ for $N = 4$.

3.1. Striped spin configurations for ultrathin film with strong dipolar interaction

The simulation was conducted for the system with $N = 2$ and strong dipolar interaction $D = 1.0J$, $10.0J$, $100.0J$. The system with $D = 1.0J$ demonstrates no striped domain structure. Instead, ferromagnetic order was observed between spins in the same layer and antiferromagnetic order between spins in neighboring layers at a low temperature. An example of such a configuration for $T = 0.7$ is shown in Fig. 3. In this figure colors show spin projections in the out-of-plane direction for both layers of the film. Spins has antiparallel alignment in neighboring stripes and in neighboring layers. At a high temperature paramagnetic state is observed.

For the system with $D = 10.0J$ and $D = 100.0J$ striped domain structure is appeared. Typical spin configurations are shown in Fig. 4 for $T = 0.7$. The increase of temperature leads to gradual destruction of stripes and transition to paramagnetic state. When the dipolar constant D becomes larger it leads to increase of the critical temperature.

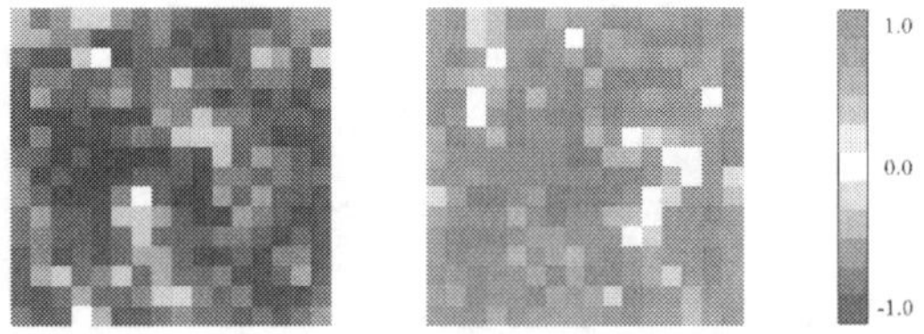

Figure 3. Striped spin configuration for ultrathin film with $N = 2$, $A = 0.1J$, $D = 1.0J$, $T = 0.7$

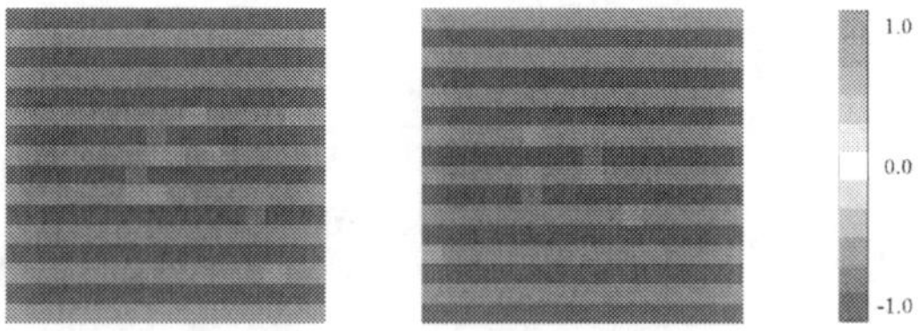

Figure 4. Striped spin configuration for ultrathin film with $N = 2$, $A = 0.1J$, $D = 10.0J$, $T = 0.7$

3.2. Bilayer structure with dipolar interlayer interaction

The study of bilayer structure was carried out. The scheme of such structure is presented in Fig. 5. The system consists of two films with thickness N and dipolar interaction between them.

At low temperature antiferromagnetic order is observed between two layers. Temperature dependence of out-of-plane layer magnetization m_z is demonstrated in Fig. 6. This order can be changed to ferromagnetic by applying of external magnetic field in the out-of-plane direction. Temperature dependence of out-of-plane layer magnetization m_z is shown in Fig. 7 for the case of $h_z = 1.0$.

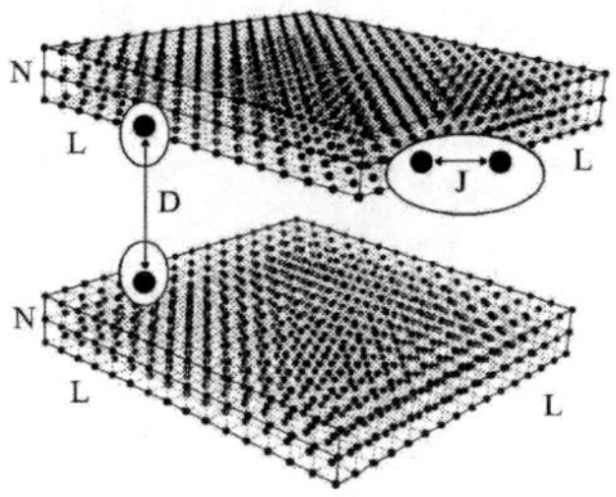

Figure 5. Scheme of bilayer ultrathin film with dipolar interlayer interaction

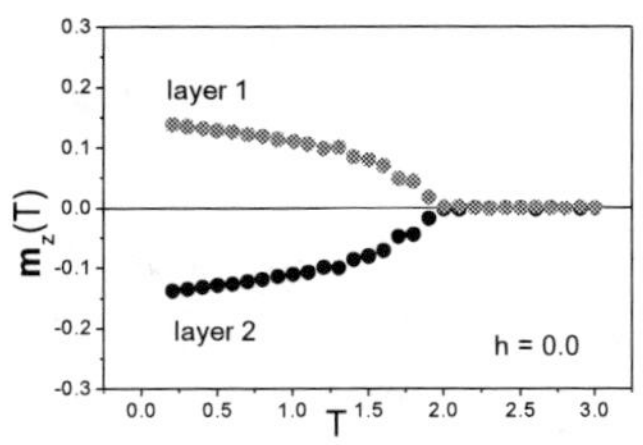

Figure 6. Temperature dependencies of out-of-plane layer magnetization m_z for $h = 0$

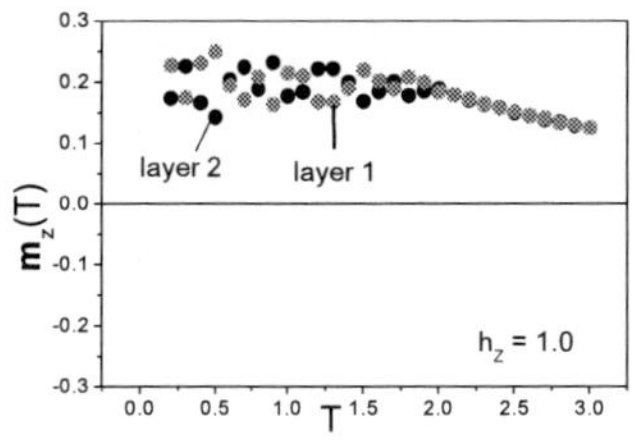

Figure 7. Temperature dependencies of out-of-plane layer magnetization m_z for $h_z = 1.0$

4. Memory effects

Memory effects were revealed by analysis of non-equilibrium critical relaxation of the autocorrelation function $C(t, t_w) = \left\langle \frac{1}{N_s} \sum_i \mathbf{S_i}(t)\mathbf{S_i}(t_w) \right\rangle$ and the magnetization from low- and high-temperature initial states. Time dependencies of these functions for low-temperature initial state ($m_0 = 1$) and waiting times $t_w = 20$, 40 MCs/s are shown in Fig. 8.

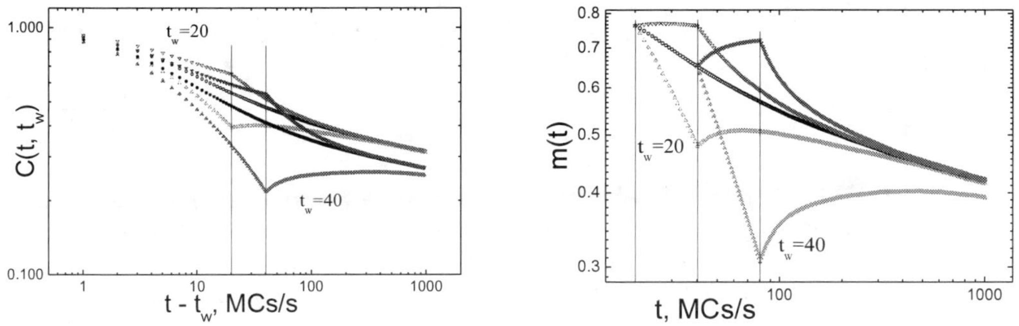

Figure 8. Non-equilibrium critical relaxation of autocorrelation function C and magnetization m for waiting times $t_w = 20$, 40 MCs/s. At the moment $t_{start} = t_w$ it were produced cooling (blue color) or heating (red color) cycling procedure for $\Delta T = 0.2$ during $\Delta t = t_w$ MCs/s. Relaxation starts from low-temperature initial state with $m_0 = 1.0$

Time dependencies of autocorrelation function and magnetization for high-temperature initial state ($m_0 \ll 1$) and waiting times $t_w = 20$, 100 MCs/s are shown in Fig. 9. It were concluded

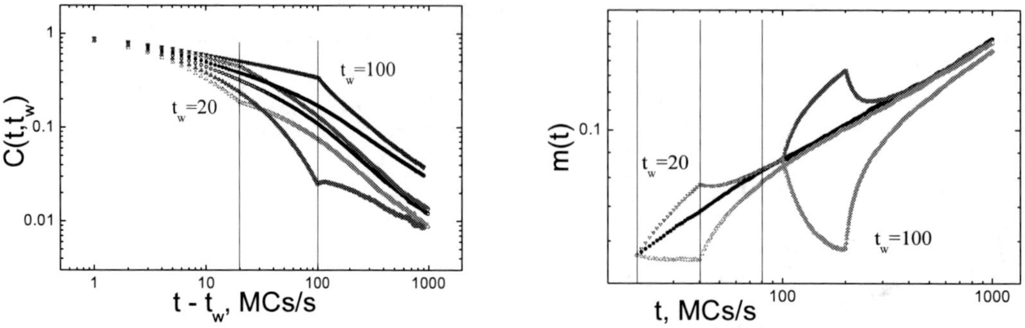

Figure 9. Non-equilibrium critical relaxation of autocorrelation function C and magnetization m for waiting times $t_w = 20$, 100 MCs/s. At the moment $t_{start} = t_w$ it were produced cooling (blue color) or heating (red color) cycling procedure for $\Delta T = 0.2$ during $\Delta t = t_w$ MCs/s. Evolution starts with high-temperature initial state with $m_0 = 0.01$

that memory effects brightly appear in cooling cycling procedure with relaxation from the low-temperature initial state.

5. Conclusions

The competition between exchange interaction, uniaxial anisotropy and dipolar interaction leads to appearance of wide range of phenomena. The phase transition from ferromagnetic to paramagnetic state is observed in systems with weak dipolar interaction with out-of-plane spin alignment in ferromagnetic phase. The increase of dipolar interaction strength leads to antiferromagnetic ordering between spins in adjacent layers with conserve of the ferromagnetic order in the same layer at a low temperature. Further increasing of dipolar interaction strength leads to striped domain structure phase.

The modeling of the bilayer film with dipolar interlayer interaction demonstrates the effective antiferromagnetic character of this interaction. Dipolar interaction is resulted in antiparallel orientations of adjacent ferromagnetic layers. This spin configuration may be changed by application of the external magnetic field.

The study of memory effects in thin magnetic films shows that these effects brightly appear in non-equilibrium relaxation from low-temperature initial state with cycling procedure of cooling.

Acknowledgments

This work was supported by Russian Science Foundation (project no. 14-12-00562). The simulations were supported by the Supercomputing Center of Lomonosov Moscow State University, Moscow and St. Petersburg Joint Supercomputer Center of the Russian Academy of Sciences.

References

[1] Cannas S A, Stariolo D A and Tamarit F A 2004 *Phys. Rev.* B **69** 092409
[2] Bauer U, Emori S and Beach S D 2012 *Appl. Phys. Lett.* **101** 172403
[3] Heinrich B, Bland J A C 2005 *Ultrathin Magnetic Structures* vol 4 (Berlin: Springer)
[4] Emery V J, Kivelson S A and Tranquada J M 1999 *Proc. Natl. Acad. Sci. USA* **96** 8814
[5] Seibold G and Lorenzana J 2005 *Phys. Rev. Lett.* **94** 107006
[6] Vojta M, Vojta T and Kaul R K 2006 *Phys. Rev. Lett* **97** 097001
[7] Carubelli M, Billoni O V, Pighin S A, *et al* 2008 *Phys. Rev.* B **77** 134417
[8] Belemuk A M, Udalov O G, Chtchelkatchev N M, and Beloborodov I S 2015 *Preprint* cond-mat:1505.05605

Journal of Physics: Conference Series **681** (2016) 012018 doi:10.1088/1742-6596/681/1/012018

Quantum fractional resonances in superconducting circuits with an embedded Josephson junction

M.V. Denisenko, V.O. Munyayev, A.M. Satanin

Lobachevsky State University of Nizhny Novgorod, Nizhny Novgorod, Russia

E-mail: sarkady@mail.ru

Abstract. We present a quantum electrodynamic treatment of the generation of fractional resonances in a planar waveguide with an embedded superconducting Josephson oscillator. We analyze the dynamics of the Josephson oscillator coupled with the electromagnetic pulse which is propagating along the waveguide. The calculations are carried out entirely in the Heisenberg picture. It is shown that the quantum Josephson oscillator excited by coherent pulse field at the pump frequency, can realize frequency down-conversion and emitting sub-harmonic multiples of the fundamental (fractional harmonics). The influence of dissipation on the phenomenon of resonance capture is discussed.

1. Introduction

Superconducting quantum circuits are a leading candidate for the engineering of quantum states in quantum logic devices [1–5]. Recently, there has been an increased activity in investigation of superconducting electrodynamics of waveguides and cavities with embedded artificial atoms - Josephson junction (Josephson oscillators) [6]. A Josephson oscillator is a macroscopic superconducting loop, interrupted by a weak link and may be threaded by a magnetic flux of electromagnetic wave propagating in a superconducting coplanar waveguide. Such an oscillator is characterized by energy $\hbar\omega_J$ that is much larger than the thermal energy at cryogenic temperatures and the system must be described by quantum mechanics. As it turned out, in the microwave range of frequencies can be achieved the strong coupling of the field with the artificial atoms, which opens up new possibilities for the control of quantum states. The other side of the interest due to the possibility of generating and analysis of Fock states up to a few photons [7].

In the present work we consider a coplanar waveguide (or resonator) with embedded Josephson junctions. The waveguides of chosen architecture may be used both as electromagnetic pumping and control of an inserted Josephson junction. The Josephson junction is supposed to be coupled directly with electromagnetic waves in the line with arbitrarily large strength. For a weak driving pulse a single Josephson junction operates as a linear oscillator and demonstrated the linear response for an excitation. At the same time, a strong driving pulses can cause the transition of Josephson oscillator in nonlinear regime of excitation. In this case in the first approximation only the Kerr type nonlinearity can be taken in account. It is shown that even in this case in the spectrum of excitation along with main resonance (1:1) also the fractional resonance (1:3) may appear. It means that the Josephson oscillator is trapped into nonlinear fractional resonance for long time and it can reemit microwave photons caused by a pulse of

Journal of Physics: Conference Series **681** (2016) 012018 doi:10.1088/1742-6596/681/1/012018

coherent photons. In conclusion we will discuss the application of correlated microwave photons for distribution of entanglement in quantum communication networks.

2. The model of the system and the basic equations

In this paper we will be interested in the coplanar waveguide, which consists of a superconducting wire evaporated on an insulating substrate and having superconducting ground planes adjacent to it on the same surface as shown in Fig. 1a. The planar electrical circuits can be constructed by using modern lithographic techniques and the problem of spatial mode is largely eliminated. It is well known that the current can flow through the junction: $I_s = I_0 \sin\varphi$ - where φ is the phase difference on the junction (the first Josephson relation) [6]. Since the wavelength of the microwave field is much larger than the size of the weak link, its effect is equivalent to the local potential jump (see. Fig. 1b), due to the Josephson effect [6]: $V_J(t) = \frac{\hbar}{2e}\frac{\partial\varphi(t)}{\partial t}$ (the second Josephson relation). It means that the embedded junction may be considered as a lumped element. We can thus assume the open boundary conditions for which the current produced by a current generator on the end of the waveguide.

We introduce a charge distributed over the waveguide line from a current point x up to ∞: $Q(x,t)$. The changing in the time and the space charge produces the current $I(x,t)$ and potential drop $V(x,t)$, which may be defined by equations: $I(x,t) = \frac{\partial Q(x,t)}{\partial t}$ and $V(x,t) = -\frac{\partial Q(x,t)}{C_l \partial t}$, where C_l is the capacitance per unit length. It means that the coordinate of the charge waves is taken to be $Q(x,t)$ and the conjugate momentum is $\Phi(x,t) = L_l \frac{\partial Q(x,t)}{\partial t}$, where $L_l = \frac{L_1}{c^2}$ and L_1 is the inductance per unit length. The classical Hamiltonian, describing the electromagnetic field Josephson junction and interaction between them, has form:

$$H = \int_{-\infty}^{\infty} dx \left\{ \frac{\Phi(x,t)^2}{2L_l} + \frac{1}{2C_l}\left(\frac{\partial Q(x,t)}{\partial x}\right)^2 \right\} + \frac{1}{2J}\left(p + \frac{\hbar}{2e}Q(0,t)\right)^2 + J\omega_J^2(1-\cos\varphi), \quad (1)$$

where p is the conjugate momentum to the coordinate of the oscillator φ; $J = C\left(\frac{\hbar}{2e}\right)^2$, C is the junction capacitance.

Note that in Eq. (1) the first two terms can be interpreted as the kinetic and potential energy of the field, and the third term is the Hamiltonian of Josephson junction including of

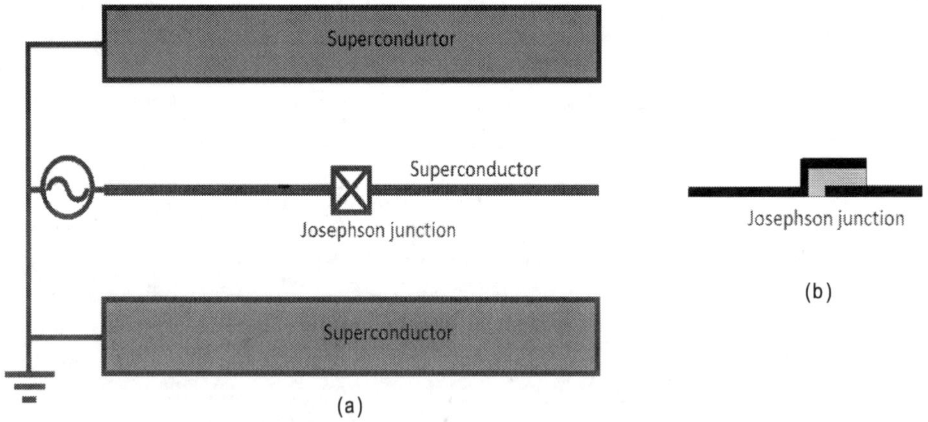

Figure 1. The coplanar waveguide is depicted schematically: a central superconducting wire in the origin of planar waveguide is interrupted (a) and a thin insulator (a weak link is imbedded) (b).

the interaction term between the field and the oscillator. It can be seen from Eq. (1) that the system under consideration is equivalent to a nonlinear pendulum interacting with a string.

To treat field and oscillator quantum mechanically we will use Heisenberg approach. The field coordinate $Q(x,t)$ field and the conjugate momentum $\Phi(x,t)$ are the conjugate variables, which should be replaced by operators with commutation relations: $\left[\hat{Q}(x,t),\hat{\Phi}(x',t)\right] = i\hbar\delta(x-x')$. The same way we introduce for the oscillator the coordinate $\varphi(t)$ and conjugate momentum $p(t)$: $[\hat{\varphi}(t)\hat{p}(t)] = i\hbar$. Before proceed, we shall simplify the Hamiltonian. The leading terms that we want to included is only quadratic term: $J\omega_J^2(1-\cos\varphi) \approx J\left(\omega_J^2\frac{\varphi^2}{2} - \frac{\beta}{4}\varphi(t)^4\right)$, $\beta = \omega_J^2/6$.

Finally, we have the quantized Hamiltonian to study in the form:

$$\hat{H}(t) = \int_{-\infty}^{\infty} dx \left\{ \frac{\hat{\Phi}(x,t)^2}{2L_l} + \frac{1}{2C_l}\left(\frac{\partial \hat{Q}(x,t)}{\partial x}\right)^2 \right\}$$
$$+\frac{1}{2J}\left(\hat{p}(t) + \frac{\hbar}{2e}\hat{Q}(0,t)\right)^2 + J\left(\omega_J^2\frac{\hat{\varphi}(t)^2}{2} - \frac{\beta}{4}\hat{\varphi}(t)^4\right). \tag{2}$$

The field operators may be expanded as:

$$\hat{Q}(x,t) = \sum_k c_k\left(e^{ikx}\hat{b}_k(t) + e^{-ikx}\hat{b}_k^\dagger(t)\right),$$
$$\hat{\Phi}(x,t) = -i\sum_k \omega(k)L_l c_k\left(e^{ikx}\hat{b}_k(t) - e^{-ikx}\hat{b}_k^\dagger(t)\right), \tag{3}$$

where L - the length of the wave guide, $c_k = \sqrt{\frac{\hbar}{2\omega(k)L_l L}}$, k - wavevector (as usually we will use the periodic boundary conditions for states classifications); $\omega(k) = v\,|k|$, $\hat{b}_k^\dagger(t)$ and $\hat{b}_k(t)$ are creation and annihilation bosonic operators: $\left[\hat{b}_k(t),\hat{b}_{k'}^\dagger(t)\right] = \delta_{k,k'}$ (for modes k and k'). Notice, that the waveguide modes form a one-dimensional continuum.

The same way for the Josephson junction oscillator the phase and moment can be written in terms of creation and annihilation with

$$\hat{\varphi}(t) = \sqrt{\frac{\hbar}{2J\omega_J}}\left(\hat{a}(t) + \hat{a}^\dagger(t)\right), \hat{p}(t) = -i\sqrt{\frac{\hbar J\omega_J}{2}}\left(\hat{a}(t) - \hat{a}^\dagger(t)\right), \tag{4}$$

where $\hat{a}^\dagger$ and $\hat{a}$ are the bosonic creation and annihilation operators, which are defined in Fock space: $|n\rangle$ $(\hat{a}^\dagger\hat{a}|n\rangle = n|n\rangle)$.

Expressing the quantum Hamiltonian in Eq. (2) in terms of these operators, we have

$$\hat{H} = \sum_k \hbar\omega(k)\hat{b}_k^\dagger\hat{b}_k + \hbar\omega_J\hat{a}^\dagger\hat{a} - \frac{1}{4}\hbar\mu\left(\hat{a}^\dagger + \hat{a}\right)^4 + i\hbar g\left(\hat{a}^\dagger - \hat{a}\right)\hat{Q}(0,t) + \frac{1}{2C}\hat{Q}(0,t)^2, \tag{5}$$

where: $g = \frac{1}{2e}\sqrt{\frac{\hbar\omega_J}{2J}}$, $\mu = \frac{\hbar\beta}{4J\omega_J^2}$;

$$\hat{Q}(0,t) = \sum_k c_k\left(\hat{b}(t) + \hat{b}^\dagger(t)\right). \tag{6}$$

3. The main resonance

We consider here the propagating of a pulse generated by the current-generator at the left end of the waveguide. Let us the carried frequency ω of the pulse is chosen close to the frequency of the Josephson junction ω_J. We will consider the case of wave packet initially in a coherent state.

They have the properties that they are eigenstates of the annihilation operator. The coherent state of a mode k defined by: $b_k|\beta_k\rangle = \beta_k|\beta_k\rangle$, and for a wave packet we will have: $|\beta\rangle = \prod_k |\beta_k\rangle$. Notice, that in coherent basis the propagating field looks like a classical wave.

The field of source is described by the charge operator $\hat{Q}_{ext}(t - x/v)$. We consider here only the case when at the initial time the field and the oscillator does not interact, and the system wave function has the form: $|\psi_{in}\rangle = |0\rangle \otimes |\{\beta\}\rangle$, where $|\{\beta\}\rangle$ is the coherent state of the field, $|0\rangle$ is the ground state of the oscillator.

The Heisenberg equations of motion for the operators in the problem are:

$$i\frac{d\hat{b}_k}{dt} = \omega(k)\hat{b}_k + \frac{c_k}{2eJ}\left(\hat{p}(t) + \frac{\hbar}{2e}\hat{Q}(0,t)\right), \tag{7}$$

$$i\frac{d\hat{a}}{dt} = \omega_J\hat{a} - \mu\left(\hat{a}^\dagger + \hat{a}\right)^3 + ig\hat{Q}(0,t). \tag{8}$$

The first equation is simply the bosonic oscillators evolution of the charge field coupled with the Josephson junction, while the second represents the nonlinear oscillator which is excited by the field propagating inside the waveguide.

We therefore begin an appropriate solution of these operators by formally integrating Eq. (7). The solution of Eq. (7)) is

$$\hat{b}_k(t) = \hat{b}_k(0)e^{-i\omega(k)t} - i\frac{c_k}{2e}e^{-i\omega(k)t}\int_0^t dt' e^{i\omega(k)t'}\frac{d\hat{\varphi}(t')}{dt'}, \tag{9}$$

where we chose to begin the problem at $t = 0$ and the Heisenberg equation for the phase is used:

$$\frac{d\hat{\varphi}(t)}{dt} = \frac{1}{J}\left(\hat{p}(t) + \frac{\hbar}{2e}\hat{Q}(0,t)\right). \tag{10}$$

The solution for $\hat{b}_k(t)$ is written as a sum of two contribution, one from free evolution of the field operator and the other from radiation of the oscillator. Substituting Eq. (9) into the Eq. (6) yields

$$\hat{Q}(0,t) = \hat{Q}_{ext}(t) - \frac{v}{2}C_l\frac{\hbar}{2e}\left(\hat{\varphi}(t) - \hat{\varphi}(0)\right). \tag{11}$$

Now, inserting the charge operator into the right side of Eq. (8), we obtain the equation of motion for the operator $\hat{a}$ in the form:

$$i\frac{d\hat{a}(t)}{dt} = \omega_J\hat{a}(t) - i\gamma\left(\hat{a}(t) + \hat{a}^\dagger(t)\right) - \mu\left(\hat{a}(t) + \hat{a}^\dagger(t)\right)^3 + ig\left(\hat{Q}_{ext}(0,t) + \frac{v}{2}C_l\frac{\hbar}{2e}\hat{\varphi}(0)\right), \tag{12}$$

where $\gamma = \frac{1}{4Z_lJ}\left(\frac{\hbar}{2e}\right)^2$, $Z_l = \sqrt{\frac{L_l}{C_l}}$ is the impedance per unit length. Note, that the interaction of the oscillator with an semi-infinity transmission line act like a interaction with a dissipative bath even though every one of its electrical elements is non-dissipative. By combining the solution of the Eq. (12) and the equation Eq. (11) corresponding reemitting field may be obtained.

The Eq. (12) shows that the linear oscillator free evolution operator $\hat{a}(t)$ is given by $\hat{a}(t) = \hat{a}(0)e^{-i\omega_Jt}$. For the chosen parameters it is supposed that term containing the frequency ω_J is dominant in right side of the Eq. (12). This means that we can neglect the rapidly oscillating terms in equations and keep only the slowly varying dynamical variables that are free from the rapid oscillations at resonant frequencies (rotating-wave approximation). In order to do this we decompose the field operator $\hat{Q}(0,t)$ into positive and negative frequency parts writing $\hat{Q}(0,t) = \hat{Q}^{(+)}(0,t) + \hat{Q}^{(-)}(0,t)$, where

$$\hat{Q}^{(-)}(0,t) = \sum_k c_k \hat{b}_k^\dagger(t), \qquad \hat{Q}^{(+)}(0,t) = \sum_k c_k \hat{b}_k(t). \tag{13}$$

At this point it is convenient to introduce slowly varying variables $\hat{a}_s(t) = \hat{a}(t)e^{i\omega t}$. In the terms of these new variables Eq. (12) become

$$i\frac{d\hat{a}_s(t)}{dt} = (\omega_J - 3\mu - \omega - i\gamma)\hat{a}_s(t) - 3\mu\left(\hat{a}_s^\dagger(t)\hat{a}_s(t)\right)\hat{a}_s(t) + ig\left(e^{i\omega t}\hat{Q}_{ext}^{(+)}(0,t)\right)_{av}, \tag{14}$$

where $(\cdots)_{av}$ means average per period in right side Eq. (14). Note that here and below all the non-resonant terms in the equations will be omitted.

We calculate now the backward current. As it is followed from Eq. (11) for $x \to -\infty$ we have

$$\hat{I}_R(x,t) = -\frac{v}{2}C_l \frac{\hbar}{2e}\frac{d\hat{\varphi}(t + x/v)}{dt}. \tag{15}$$

The relevant quantity reemission characteristic of the oscillator that may be calculated is the ratio of the averaged reflected power to the equivalent power, that is produced by the exited current:

$$R = \frac{\left(\langle\psi_{in}|I_R^2(t)|\psi_{in}\rangle\right)_{av}}{\left(\langle\psi_{in}|I_{ext}^2(t)|\psi_{in}\rangle\right)_{av}}. \tag{16}$$

Using Eqs. (4) and (14), the Eq. (15) may be rewritten as

$$R = 2R_0\frac{\omega_J^2}{g^2Q_m^2}\langle 0|\left(\hat{a}(t)\hat{a}^\dagger(t) + \hat{a}^\dagger(t)\hat{a}(t)\right)|0\rangle \qquad R_0 = \left(\frac{1}{2Z_lC\omega_J}\right)^2. \tag{17}$$

Here we are going to investigate a simple case when a coherent single mode in the state $|\beta_{k_\omega}\rangle$ with $k_\omega = \omega/v$ propagating from the current source. Then we have $\langle\beta_{k_\omega}|Q_{ext}(x - vt)|\beta_{k_\omega}\rangle = c_{k_\omega}\left(\beta_{k_\omega}e^{i(k_\omega(x-vt))} + \beta_{k_\omega}^* e^{-i(k_\omega(x-vt))}\right) = Q_m\cos(k_\omega(x - vt) + \theta)$, $\beta_{k_\omega} = |\beta_{k_\omega}|e^{i\theta}$, $Q_{k_\omega} = 2|\beta_{k_\omega}|c_{k_\omega}$ is the wave amplitude. Notice that the phase θ may be removed from wave equation by shift of time origin.

We calculate the correlation function Eq. (12) and time evoluation of the oscillator in coherent states of a single oscillator $\hat{a}|\alpha\rangle = \alpha|\alpha\rangle$. The equation of motion Eq. (14) for the nonlinear oscillator may be written as:

$$i\frac{d\alpha}{dt} = (\omega_J - 3\mu - \omega - i\gamma)\alpha(t) - 3\mu|\alpha(t)|^2\alpha(t) + i\frac{g}{2}Q_m, \tag{18}$$

We can rewrite Eq. (17) for the correlation function by using the expression $\langle 0|\left(\hat{a}(t)\hat{a}^\dagger(t) + \hat{a}^\dagger(t)\hat{a}(t)\right)|0\rangle = 2|\alpha(t)|^2 + 1$, ($|\alpha(t)|^2 \gg 1$) Subsisting steady state solutions Eq. (18) into Eq. (17) become

$$R = R_0\frac{\omega_J^2}{(\bar{\omega}_J - \omega)^2 + \gamma^2}, \tag{19}$$

$\bar{\omega}_J = \omega_J - 3\mu - 3\mu|\alpha|^2$, and $|\alpha|^2$ is a solution of equation $\left((\bar{\omega}_J - \omega)^2 + \gamma^2\right)|\alpha|^2 = (\frac{g}{2}Q_m)^2$. Thus, in this case the nonlinear oscillator is captured into a nonlinear resonance and emits at a frequency close to the ω_J. Notice, that when nonlinear term is negligibly small, the shape of line is Lorentzian (see Fig. 2).

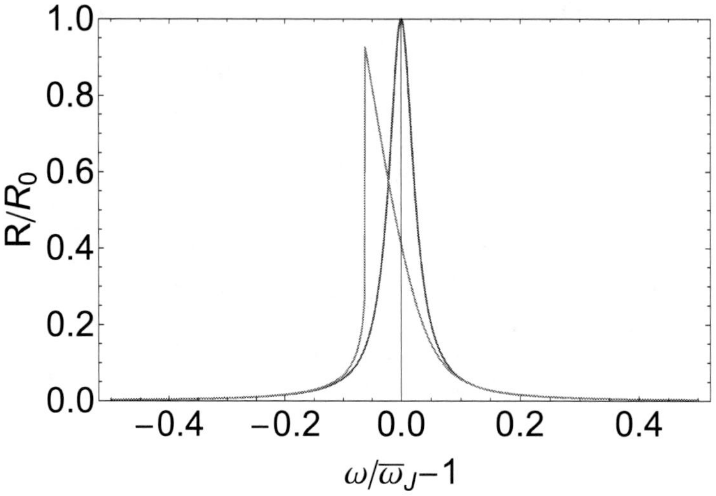

Figure 2. The curve of resonance reflection: blue line is linear ($\mu = 0$, $\gamma = 0.025\omega_J$) and the red line non-linear regime ($\mu = 0.0005$, $\frac{gQ_m}{2\omega_J} = 1$)

4. Fractional resonance

We now turn to the problem of down-conversion of the wave frequency in the wave guide with an embedded lumped nonlinear element. Let us consider the case when the characteristic frequency of the coherent incoming field $\hat{Q}_{ext}(t-x/v)$ (drawing field) is near 3ω and the resonant condition $\omega \sim \omega_J$ may be performed. To investigate the excitation of the nonlinear oscillator on the fractional resonance (1:3) we made a canonical transformation of the field and the oscillator variables. In order to extract the fractional resonance we introduce auxiliary variables oscillating on the frequency ω: $\hat{\phi}(x,t)$ is the conjugate momentum to the coordinate of the field $\hat{q}(x,t)$ (and the similar for the oscillator $\hat{P}$, $\hat{\varphi} = \hat{\varphi}_0 + \hat{\psi}$). The canonical transformation reads as: $\hat{Q}(x,t) = \hat{Q}_0(x,t) + \hat{q}(x,t)$, $\hat{\Phi}(x,t) = \hat{\Phi}_0(x,t) + \hat{\phi}(x,t)$, $\hat{p} = \hat{p}_0 + \hat{P}$, $\hat{\varphi} = \hat{\varphi}_0 + \hat{\psi}$. The transformed Hamiltonian takes form:

$$\hat{H}(t) = \int_{-\infty}^{\infty} dx \left\{ \frac{\hat{\phi}(x,t)^2}{2L_l} + \frac{1}{2C_l}\left(\frac{\partial \hat{q}(x,t)}{\partial x}\right)^2 \right\} + \frac{1}{2J}\left(\hat{P}(t) + \frac{\hbar}{2e}\hat{q}(x,t)\right)^2$$
$$+J\left(\omega_J^2 \frac{\hat{\psi}(t)^2}{2} - \frac{\beta}{4}\left(\hat{\psi}(t) + \hat{\varphi}_0(t)\right)^4\right), \tag{20}$$

where the auxiliary field $\hat{\varphi}_0$ is defined from equation:

$$\frac{\partial^2 \hat{\varphi}_0}{\partial t^2} + 2\gamma \frac{\partial \hat{\varphi}_0}{\partial t} + \omega_J^2 \hat{\varphi}_0 = \frac{\hbar}{2eJ} \hat{I}_{ext}(0,t), \tag{21}$$

In the modes representation the Hamiltonian is

$$\hat{H} = \sum_k \hbar\omega(k)\hat{b}_k^\dagger \hat{b}_k + \hbar\omega_J \hat{a}^\dagger \hat{a} - \frac{1}{4}\hbar\mu\left(\hat{a}^\dagger + \hat{a} + \xi\hat{\varphi}_0(t)\right)^4 + i\hbar g(\hat{a}^\dagger - \hat{a})\hat{q}(0,t) + \frac{1}{2C}\hat{q}(0,t)^2, \tag{22}$$

where $\xi = \sqrt{\frac{2J\omega_J}{\hbar}}$.

Comparison of Eqs. (5) and (20) shows that the main difference between the Hamiltonians of the main and fractional resonances is the presence of an operator $\hat{\varphi}_0$, that modulates the nonlinear term. Doing the same way as in Sec. 3 we can find the Heisenberg equation for the operator $\hat{a}(t)$:

$$i\frac{d\hat{a}(t)}{dt} = \omega_J\hat{a} - i\gamma\left(\hat{a}^\dagger + \hat{a}\right) - \mu\left(\hat{a}^\dagger + \hat{a}\right)^3 - \mu\left(\hat{a}^\dagger + \hat{a} + \xi\hat{\varphi}_0(t)\right)^3 + ig\hat{q}_{ext}(0,t). \tag{23}$$

The Eq. (23) should be solved simultaneously with the Eq. (21). Now we are ready to find sub-harmonics in the quantum system. Again restrict ourselves to the case when the drawing field is prepared in the coherent state and has the carrier frequency 3ω ($\omega \sim \omega_J$), i.e. $I_{ext}(0,t) = I_m\cos(3\omega t)$. Keeping resonant harmonics $\omega(k) \sim 3\omega$, i.e. $k_\omega \sim 3\omega/v$, we write the solution Eq. (21) in the form

$$\varphi_0(t) = A\cos(3\omega t) + B\sin(3\omega t), \tag{24}$$

where $A = \frac{\hbar}{2eJ}I_m\frac{D}{D^2+(6\omega\gamma)^2}$, $B = \frac{\hbar}{2eJ}I_m\frac{6\omega\gamma}{D^2+(6\omega\gamma)^2}$, $D = \omega_J^2 - (3\omega)^2$. Substituting Eq. (24) into the Eq. (23) and using slowly varying operators, we obtain:

$$\begin{aligned} i\tfrac{d\hat{a}_s(t)}{dt} = (\omega_J - \omega - 3\mu - i\gamma)\hat{a}_s(t) - \tfrac{3\mu}{2}\Big(\left(\hat{a}_s^\dagger(t)\hat{a}_s(t) + 2\xi^2(A^2+B^2)\right)\hat{a}_s(t) \\ +2\xi(A+iB)\left(\hat{a}_s^\dagger(t)\right)^2\Big) + ig\left(e^{i\omega t}\hat{q}_{ext}(0,t)\right)_{avg}. \end{aligned} \tag{25}$$

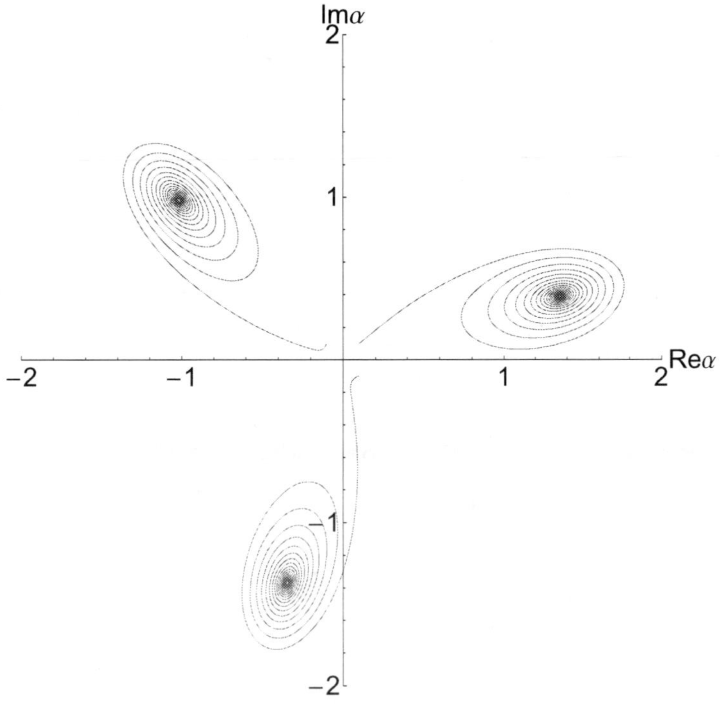

Figure 3. The phase plane Eq. (25), showing the presence of three stationary solutions (attractors) ($\mu = 0.0005$, $\gamma = 0.025\omega_J$, $\frac{gQ_m}{2\omega_J} = 1$). For instance, one of them has coordinates: $a_\infty^1 = (1.347, 0.391)$.

The Eq. (25) may be written in the coherent basis as

$$i\frac{d\alpha(t)}{dt} = (\omega_J - \omega - 3\mu - i\gamma)\alpha(t)$$
$$-\frac{3\mu}{2}\left(\left(|\alpha(t)|^2 + 2\xi^2(A^2+B^2)\right)\alpha(t) + 2\xi(A+iB)\left(\alpha^*(t)\right)^2\right) + i\frac{g}{2}\hat{q}_m. \quad (26)$$

The investigation of the phase plane $(\mathrm{Re}(\alpha(t)), \mathrm{Im}(\alpha(t))$ the nonlinear equation shows that depending on the parameters of the system in Eq. (26) possible existence of four fixed points. One point (0,0) corresponds to the trivial solution, which means that no generation of fractional frequency. When the parameter are changed in the equation may appear the three stable (under certain parameters of the system) points (see. Fig. 3), which correspond to the generation of the fractional resonance. The stationary Eq. (26), determines the reflection coefficient according to

$$R = 4R_0\frac{\omega_J^2}{g^2q_m^2}|\alpha|^2.$$

5. Conclusion

We have examined the interaction of a Josephson junction with the quantized electromagnetic field in a coplanar superconducting waveguide. We have used the Heisenberg picture to study system dynamics and rotating-wave approximation for the calculation of the field re-emittance by excited oscillator. We fond that it is possible to capture oscillator into main and fractional resonance. It is shown that when the Josephson oscillator is trapped to nonlinear fractional resonance it can reemit microwave photons with the fractional frequencies. This process may be used for the generation of entangled photons when the external source emit photons in the Fock states. In this case the Josephson oscillator can radiate three microwave photons for each incoming absorbed photon. Such kind of strong correlated (entangled) photons may useful for modern quantum communication networks [8].

Acknowledgments This work was supported by the Russian Science Foundation grant No. 14-07-00582.

References

[1] Blais A, Huang R-S, Wallraff A, Girvin S M and Schoelkopf R J 2004 *Physical Review A* **69**: 062320-1-12
[2] Wallraff A, Schuster D I, Blais A, Frunzio L, Huang R-S, Majer J, Kumer S, Girvin S M and Schoelkopf R J 2004 *Nature* **431**: 162-167
[3] Astafiev O, Zagoskin A M, Abdumalikov A A, Pashkin Y A, Yamamoto T, Inomata K, Nakamura Y, and Tsai J S 2010 *Science* **327**: 840-843
[4] Zagoskin A M 2011 *Quantum Engineering - Theory Design of Quantum Coherent Structures* (Cabridge: Cambridge University Press)
[5] Astafiev O V, Ioffe L B, Kafanov S, Pashkin Yu A, Arutyunov K Yu, Shahar D, Cohen O and Tsai J S 2012 *Nature* **484**:355-358
[6] Likharev K 1985 *Dynamics of Josephson Junctions and Circuits* (Gordon and Breach, New York)
[7] Pechal M, Huthmacher L, Eichler C, Zeytinoglu S, Abdumalikov A A Jr, Berger S, Wallraff A, and Filipp S 2014 *Physical Review X* **4**: 041010-1-8
[8] Edamatsu K 2007 *Japanese Journal of Applied Physics* **46**: 7175-7187

Journal of Physics: Conference Series **681** (2016) 012019 doi:10.1088/1742-6596/681/1/012019

On the balanced quantum hashing

F Ablayev, M Ablayev and A Vasiliev

Kazan Federal University, Kazan, Russian Federation

E-mail: `Alexander.KSU@gmail.com`

Abstract. In the paper we define a notion of a resistant quantum hash function which combines a notion of pre-image (one-way) resistance and the notion of collision resistance. In the quantum setting one-way resistance property and collision resistance property are correlated: the "more" a quantum function is one-way resistant the "less" it is collision resistant and vice versa. We present an explicit quantum hash function which is "balanced" one-way resistant and collision resistant and demonstrate how to build a large family of balanced quantum hash functions.

1. Introduction

Peter Shor's quantum factoring algorithm and quantum algorithm for finding discrete logarithm are results that use quantum mechanical effects to break cryptographic systems. The response from a cryptographic community is a "Post-quantum cryptography", which refers to research on problems (usually public-key cryptosystems) that are not efficiently solvable by quantum computers. Currently post-quantum cryptography includes different approaches, in particular, hash-based digital signature schemes such as Lamport signature and Merkle signature scheme.

Quantum key distribution and quantum digital signature are also part of Post-quantum cryptography. Gottesman and Chuang introduced a notion of a quantum one-way function and proposed a quantum digital signature protocol [1], which is based on such a function. This is also the case for other protocols (see for example [2] and [3]).

Recall that in the classical setting a cryptographic hash function h should have at least the following three properties (see, for example, [3]): (1) Pre-image resistance: given $h(x)$, it should be difficult to find x, that is, these hash functions are one-way functions; (2) Second pre-image resistance: given x_1, it should be difficult to find an x_2, such that $h(x_1) = h(x_2)$; (3) Collision resistance: it should be difficult to find any pair of distinct x_1, x_2, such that $h(x_1) = h(x_2)$. Note, that there are no one-way functions that are known to be provably hard to invert, the security of cryptographic hash functions is "computationally conditional".

In [4], [5] we defined a notion of quantum hashing as a quantum counterpart of classical hashing and presented approach for constructing quantum hash functions. It appeared that the quantum digital signature by Gottesman and Chuang is based on quantum functions which are actually quantum hash functions. Those quantum functions have "unconditionally one-way" property based on Holevo Theorem [6]. We have also shown that quantum hashing can be useful for constructing efficient quantum algorithms [7] and quantum communication protocols [8].

For a quantum hash function, which is a mapping that creates a quantum state from classical information, we require the following properties:

- It can be effectively computed given classical input;
- It is impossible to extract this input from the quantum hash;
- Hashes of different inputs can be distinguished with high probability, which also implies a reliable equality test.

In this paper we investigate the connection between the last two properties. We show that there is a trade-off between them and introduce a notion of the balanced quantum hash function that has both in a good combination.

2. Quantum Hashing

In this section we briefly recall the notion of the quantum hash function, formalize its properties and give several examples.

In the paper [9] we defined a notion of (ϵ, δ)-hash function where values ϵ and δ are numerical characteristics of the above two properties: (i) one-way resistance and (ii) collision resistance properties. The notion of the (ϵ, δ)-hash function is a generalization of the quantum hash function defined in [4], [5].

We present formal definitions now. For $s \geq 1$ let $(\mathcal{H}^2)^{\otimes s}$ be the 2^s-dimensional Hilbert space, describing the states of s qubits. Let $\mathbb{X}$ be a finite set of size $K = |\mathbb{X}|$. We define a $(K; s)$ quantum function ψ to be a unitary transformation (determined by an element $w \in \mathbb{X}$) of the initial state $|\psi_0\rangle^{\otimes s} \in (\mathcal{H}^2)^{\otimes s}$ to a quantum state $|\psi(w)\rangle \in (\mathcal{H}^2)^{\otimes s}$:

$$\psi : \{|\psi_0\rangle\} \times \mathbb{X} \to (\mathcal{H}^2)^{\otimes s}, |\psi(w)\rangle = U(w)|\psi_0\rangle,$$

where $U(w)$ is a unitary matrix. We will also use a shorter notation $\psi : \mathbb{X} \to (\mathcal{H}^2)^{\otimes s}$.

One-way Resistant Function. We present the following definition of quantum ϵ-resistant one-way function. Let decoding ("information extracting mechanism") $\mathcal{M}$ be a function $\mathcal{M} : (\mathcal{H}^2)^{\otimes s} \to \mathbb{X}$. Informally: $\mathcal{M}$ makes some measurement of the state $|\psi\rangle \in (\mathcal{H}^2)^{\otimes s}$ and decodes the result to $\mathbb{X}$.

Definition 2.1 *Let X be a random variable uniformly distributed over $\mathbb{X}$. Let $\psi : \mathbb{X} \to (\mathcal{H}^2)^{\otimes s}$ be a quantum function. Let Y is a random variable over $\mathbb{X}$ obtained by some decoding $\mathcal{M}$, i.e. $Y = \mathcal{M}(X)$. Let $\epsilon > 0$. We call a quantum function ψ a one-way ϵ-resistant function if for any decoding $\mathcal{M}$, the probability $Pr[Y = X]$ that $\mathcal{M}$ successfully decodes Y is bounded by ϵ*

$$Pr[Y = X] \leq \epsilon.$$

We will use here the following fact [10]. Let X be random variable uniformly distributed over a k-bit binary words $\{0,1\}^k$. Let $\psi : \{0,1\}^k \to (\mathcal{H}^2)^{\otimes s}$ be a $(2^k; s)$ quantum function. Let Y be a random variable over $\{0,1\}^k$ obtained by some decoding $\mathcal{M}$ of $|\psi(X)\rangle$ to $\{0,1\}^k$. Then the probability of correct decoding is bounded by

$$Pr[Y = X] \leq \frac{2^s}{2^k}.$$

That is, we should pick s as small as possible to make a quantum hash function one-way resistant.

Journal of Physics: Conference Series **681** (2016) 012019 doi:10.1088/1742-6596/681/1/012019

Collision Resistant Function. As we have noted in [4] there might be no collisions in the classical sense: since quantum hashes are quantum states they can store arbitrary amount of data and can be different for unequal messages. But the procedure of comparing those quantum states implies measurement, which can lead to collision-type errors.

That is, in order to make a quantum hash function resistant to quantum collisions, we must guarantee the distinguishability of quantum hashes for different inputs. Therefore, the pairwise inner product of the quantum hash function values should be bounded. This is formalized by the following definition.

Definition 2.2 *Let $\delta > 0$. Following [9] we call a quantum function $\psi : \mathbb{X} \to (\mathcal{H}^2)^{\otimes s}$ a collision δ-resistant function if for any pair w, w' of different elements,*

$$\left|\langle\psi(w)\,|\,\psi(w')\rangle\right| \leq \delta.$$

There is a known lower bound by Buhrman et al. [11] for the size of the sets of pairwise-distinguishable states: to construct a set of 2^k quantum states with pairwise inner products below δ we will need at least $\Omega(\log(k/\delta))$ qubits. Using the notation above this implies the bound $s = \Omega(\log\log K - \log\delta))$. The similar lower bound of $\log\log K - c(\delta)$ was proved by a different method in [5].

One-way Resistance and Collision Resistance. The above two definitions and considerations lead to the following formalization of the quantum cryptographic (one-way and collision resistant) function

Definition 2.3 *Let $K = |\mathbb{X}|$ and $s \geq 1$. Let $\epsilon > 0$ and $\delta > 0$. We call a function $\psi : \mathbb{X} \to (\mathcal{H}^2)^{\otimes s}$ a quantum (ϵ, δ)-Resistant $(K; s)$-hash function iff ψ is one-way ϵ-resistant and collision δ-resistant function.*

The trade-off between one-way resistance and collision resistance The following examples show that one-way resistance and collision resistance lead to the contradictory requirements on the size of the quantum hash and the "more" a quantum function is one-way resistant the "less" it is collision resistant and vice versa.

Example 2.1 *We encode a word $w \in \{0,1\}^k$ into one qubit:*

$$|\psi(w)\rangle = \cos\left(\frac{\pi w}{2^k}\right)|0\rangle + \sin\left(\frac{\pi w}{2^k}\right)|1\rangle.$$

This function has good one-way property with $\epsilon = \frac{2}{2^k}$, but also has poor collision resistance of $\delta = \cos\left(\pi/2^k\right)$.

Example 2.2 *We encode a word $w \in \{0,1\}^k$ into k qubits:*

$$|\psi(w)\rangle = |w\rangle.$$

This function has one-way resistance $\epsilon = 1$ (no resistance) and collision resistance with $\delta = 0$ (perfect resistance).

Example 2.3 *This example is based on the quantum fingerprinting by Buhrman et al. [11].*

Let $E : \{0,1\}^k \to \{0,1\}^n$ be an error-correcting code with Hamming distance $d \geq n - \delta n$ and $E_i(w)$ is the i-th bit of the codeword $E(w)$.

Quantum hash function $\psi_E : \{0,1\}^k \to (\mathcal{H}^{2n})$ *is defined as following:*

$$|\psi_E(w)\rangle = \frac{1}{\sqrt{n}}\sum_{i=1}^{n}|i\rangle|E_i(w)\rangle = \frac{1}{\sqrt{n}}\sum_{i=1}^{n}|i\rangle\left(\cos\frac{\pi E_i(w)}{2}|0\rangle + \sin\frac{\pi E_i(w)}{2}|1\rangle\right).$$

The properties of this function depend on the underlying error-correcting code. For instance, the original work [11] used Justesen code $E : \{0,1\}^k \mapsto \{0,1\}^n$ *with* $n = ck$ *and Hamming distance* $(1-\delta)n$ *(for* $c > 2$ *and* $\delta < 9/10 + 1/(15c)$*). For that code the corresponding function has one-way resistance* $\epsilon = 2ck/2^k$ *and collision resistance* $\delta < 9/10 + 1/(15c)$.

Example 2.4 *This example is based on the quantum fingerprinting by Ablayev and Vasiliev [12].*

Let $B = \{b_j : b_j \in \{0,\dots,2^k-1\}\}$, $\delta \in (0,1)$.

Quantum function $\psi_H : \{0,1\}^k \to (\mathcal{H}^{2|B|})$ *is defined as following*

$$|\psi_B(w)\rangle = \frac{1}{\sqrt{|B|}}\sum_{j=1}^{|B|}|j\rangle\left(\cos\frac{2\pi b_j w}{2^k}|0\rangle + \sin\frac{2\pi b_j w}{2^k}|1\rangle\right).$$

The properties of this function depend on the set of numeric parameters B. *We have proved the existence of such a set* B *with* $|B| = \lceil (2/\delta^2)\ln(2^{k+1})\rceil = O(k/\delta^2)$, *that the corresponding function has one-way resistance* $\epsilon = 2|B|/2^k$ *and collision resistance equal to the* δ *above (for proof see [12]).*

3. "Balanced" Quantum Hash Functions

The above considerations lead to the notion of a "balanced" quantum hash function. Informally, if we need to hash elements w from a domain $\mathbb{X}$, $|\mathbb{X}| = K$ and if one can build for $\delta > 0$ a collision δ-resistant $(K;s)$ hash function ψ with $s = O(\log\log K - \log\delta))$ qubits then the function f will be one-way ϵ-resistant with $\epsilon = O(\log K/(\delta * K))$.

The functions from Examples 2.3 and 2.4 are exactly such functions.

In [9], we have defined the concept of a quantum hash generator and offered design, which allows to build different quantum hash functions based on the composition of function from Example 2.4 and an arbitrary classical ϵ-universal hash family [13]. This construction allows to build a large family of balanced quantum hash functions [9]. In particular the following construction explicitly presented:

- Using the relationship between ϵ-universal hash families and Freivalds fingerprinting schemas we present explicit balanced quantum hash function and prove that this construction is optimal with respect to of number of qubits needed for construction.
- Using the relationship between ϵ-universal hash families and error correcting codes (see for example [13]) we present explicit balanced hash function based on Reed-Solomon codes and prove that this construction is optimal with respect to the number of qubits needed for construction.

Acknowledgments

The work is performed according to the Russian Government Program of Competitive Growth of Kazan Federal University. Work was in part supported by the Russian Foundation for Basic Research (under the grants 14-07-00878, 14-07-00557, 15-37-21160).

References

[1] Gottesman D and Chuang I 2001 Quantum digital signatures Tech. Rep. arXiv:quant-ph/0105032 Cornell University Library URL `http://arxiv.org/abs/quant-ph/0105032`
[2] Gavinsky D and Ito T 2013 *Quantum Information & Computation* **13** 583–606 ISSN 1533-7146 URL `http://dl.acm.org/citation.cfm?id=2535649.2535652`
[3] Amiri R and Andersson E 2015 *Entropy* **17** 5635–5659 ISSN 1099-4300 (*Preprint* 1508.01893) URL `http://arxiv.org/abs/1508.01893`
[4] Ablayev F M and Vasiliev A V 2014 *Laser Physics Letters* **11** 025202 URL `http://stacks.iop.org/1612-202X/11/i=2/a=025202`
[5] Ablayev F and Ablayev M 2015 *Lobachevskii Journal of Mathematics* **36** 89–96 ISSN 1995-0802 URL `http://link.springer.com/10.1134/S199508021502002X`
[6] Holevo A S 1973 *Probl. Pered. Inform. [Probl. Inf. Transm.]* **9** 311
[7] Ablayev F and Vasiliev A 2014 *Computing with New Resources* Lecture Notes in Computer Science ed Calude C S, Freivalds R and Kazuo I (Springer International Publishing) pp 149–160 ISBN 978-3-319-13349-2 URL `http://dx.doi.org/10.1007/978-3-319-13350-8_11`
[8] Vasiliev A 2015 *International Journal of Applied Engineering Research* **10** 31415–31426
[9] Ablayev F and Ablayev M 2015 (*Preprint* 1509.01268) URL `http://arxiv.org/abs/1509.01268`
[10] Nayak A 1999 *Foundations of Computer Science, 1999. 40th Annual Symposium on* pp 369–376 ISSN 0272-5428
[11] Buhrman H, Cleve R, Watrous J and de Wolf R 2001 *Phys. Rev. Lett.* **87** 167902 URL `www.arXiv.org/quant-ph/0102001v1`
[12] Ablayev F and Vasiliev A 2009 *Electronic Proceedings in Theoretical Computer Science* **9** 1–11 URL `http://arxiv.org/abs/0911.2317`
[13] Stinson D R 1996 *In Proc. Congressus Numerantium 114* pp 7–27

International Conference on Computer Simulation in Physics and Beyond 2015 IOP Publishing
Journal of Physics: Conference Series **681** (2016) 012020 doi:10.1088/1742-6596/681/1/012020

A model of quantum communication device for quantum hashing

A Vasiliev

Kazan Federal University, 18 Kremlyovskaya St., Kazan 420008, Russian Federation

E-mail: Alexander.KSU@gmail.com

Abstract. In this paper we consider a model of quantum communications between classical computers aided with quantum processors, connected by a classical and a quantum channel. This type of communications implying both classical and quantum messages with moderate use of quantum processing is implicitly used in many quantum protocols, such as quantum key distribution or quantum digital signature. We show that using the model of a quantum processor on multiatomic ensembles in the common QED cavity we can speed up quantum hashing, which can be the basis of quantum digital signature and other communication protocols.

1. Introduction

Nowadays the area of quantum computing technology is in active experimental phase, but a large-scale fully functional quantum computer has a long way to come. However, quantum communications are already here and further development of communication devices and protocols may give rise to new commercial products.

In this research we propose a model for one such device, which we have called a Quantum Communication Device and which can be seen as a part of quantum communication system. The main purpose of developing quantum communications is to implement quantum cryptographic protocols that outperform their classical counterparts. Among them is the well-known Quantum Digital Signature protocol [1].

In [2] we have proposed a quantum hashing technique that can be used in this protocol, and it also proved useful for constructing efficient quantum algorithms [3] and quantum communication protocols [4].

The model of the quantum processor on multiatomic ensembles in the common QED cavity we consider here was investigated in [5]. Using the logical encoding of qubits and the properties of the basic operations of this architecture we can speed up quantum hashing and thus performing communication protocols that are based on it.

2. A model of the quantum communication device

Due to severe limits of existing physical implementations of quantum computer it is natural to consider the restricted models of quantum computations. The one we use here is a hybrid classical-quantum model, where small quantum processor is controlled by a classical computer. To support this idea, we may note that such protocols as Quantum Key Distribution and Quantum Digital Signature (both demonstrated experimentally) imply both classical and

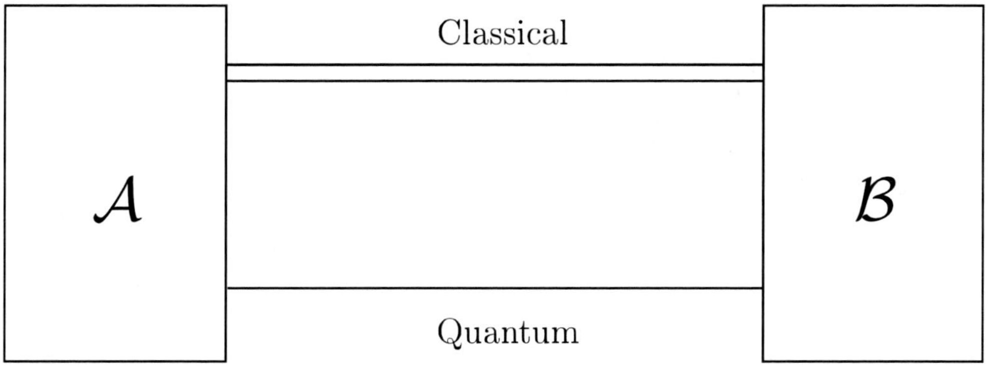

Figure 1. A simple abstraction of quantum communications over classical and quantum channels. $\mathcal{A}$ and $\mathcal{B}$ are classical computers aided with quantum processors.

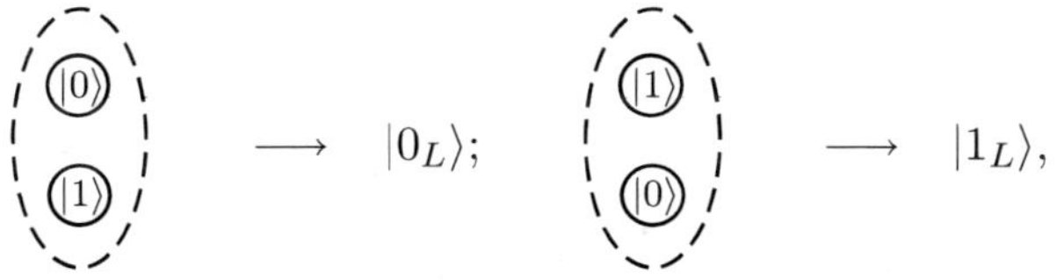

Figure 2. Pairwise qubit encoding. Small circles denote the processing nodes in the indicated quantum states.

quantum messages between communicating parties and a very moderate use of quantum processing device.

Based on this considerations we use the scheme of quantum communications (depicted in Figure 1), where communicating parties are classical computers aided with quantum processors on multiatomic ensembles in the common QED cavity [5], connected by a classical and a quantum channel.

In [5] we have investigated computational capabilities of such a quantum processor under the logical encoding of the states $|0_L\rangle$, $|1_L\rangle$ by pairs of physical qubits in states $|01\rangle$, $|10\rangle$ (see Figure 2).

This encoding eliminates the need for implementing single qubit gates, since our two-qubit gates are logical single-qubit in this encoding.

2.1. Basic operations of the quantum processor

The basic operations of the quantum processor are given by the underlying physical model and include the following.

- A Quantum Excitation Transfer between two ensembles (QET(θ)) is defined by

$$\begin{array}{l} |0\rangle|1\rangle \longrightarrow \cos\frac{\theta}{2}|0\rangle|1\rangle - i\sin\frac{\theta}{2}|1\rangle|0\rangle \\ |1\rangle|0\rangle \longrightarrow -i\sin\frac{\theta}{2}|0\rangle|1\rangle + \cos\frac{\theta}{2}|1\rangle|0\rangle \end{array}.$$

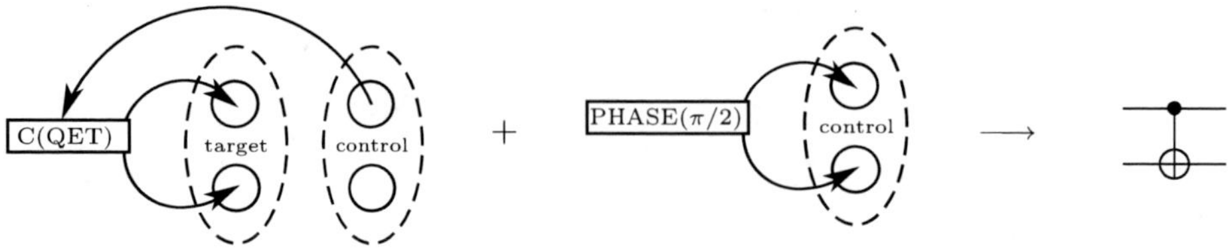

Figure 3. Logical CNOT gate implemented by the physical C(QET) gate coupled with PHASE ($\pi/2$).

In the encoded setting QET(θ) gate is actually a single-qubit rotation by the angle $-\theta$ about the $\hat{x}$ axis.

- PHASE (χ) operation is given by

$$\begin{array}{l} |0\rangle|1\rangle \longrightarrow e^{-i\chi/2}|0\rangle|1\rangle \\ |1\rangle|0\rangle \longrightarrow e^{i\chi/2}|1\rangle|0\rangle \end{array}.$$

In the logical encoding PHASE (χ) turns our composite qubit around the axis $\hat{z}$.

Based on the model of atomic transistor [6] it is possible to perform a Controlled-QET operation (denoted by C(QET)), that corresponds to logical Controlled NOT gate (up to the relative phase shift, that can be made global by adding PHASE operation, see Figure 3).

We have proved that the set of quantum gates $\{\mathrm{C(QET)}, \mathrm{QET}(\theta), \mathrm{PHASE}(\chi)\}$ is universal for the Hilbert subspace spanned by encoded states $|0_L\rangle = |01\rangle$, $|1_L\rangle = |10\rangle$ (for proof see [5]).

Though the architecture provides a universal set of quantum gates it is possible to speed up the usual implementation of some class of useful operations, in particular those used in the algorithm of quantum hashing.

3. Quantum hashing

In this section we recall the construction of quantum hash function from [2].

Let $q = 2^n$ and $B = \{b_1, b_2, \ldots, b_d\} \subset \mathbb{Z}_q$. We define a quantum hash function $\psi_{q,B} : \{0,1\}^n \to (\mathcal{H}^2)^{\otimes(\log d+1)}$ as follows. For an input $x \in \{0,1\}^n$ we let

$$|\psi_{q,B}(x)\rangle = \frac{1}{\sqrt{d}} \sum_{i=1}^{d} |i\rangle \left(\cos\frac{2\pi b_i x}{q}|0\rangle + \sin\frac{2\pi b_i x}{q}|1\rangle \right) . \tag{1}$$

It follows from this definition that the quantum hash $|\psi_{q,B}(x)\rangle$ of an n-bit string x consists of $\log d + 1$ qubits. We have shown in [2] that d can be about $O(n)$ without loosing the quality of hashing. Thus, this function maps n-bit inputs to quantum states of $O(\log n)$ qubits.

4. Algorithmic implementation of quantum hashing

In order to describe the implementation of the quantum hashing we use the following notations from [3].

We define a *Compound Controlled Rotation* operator (CCR):

$$CCR_{q,B}(\theta) = CCR_{q,B,1}(\theta) \cdot CCR_{q,B,2}(\theta) \cdots CCR_{q,B,d}(\theta), \tag{2}$$

Journal of Physics: Conference Series **681** (2016) 012020 doi:10.1088/1742-6596/681/1/012020

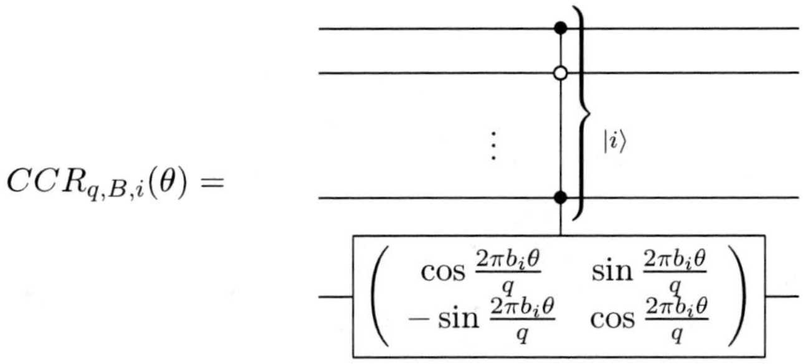

Figure 4. A circuit for the operator $CCR_{q,B,i}(\theta)$, that rotates the target qubit by the angle 2θ if the control qubits were in the state $|i\rangle$. Here, the single-qubit rotation is made around the $\hat{y}$ axis of the Bloch sphere.

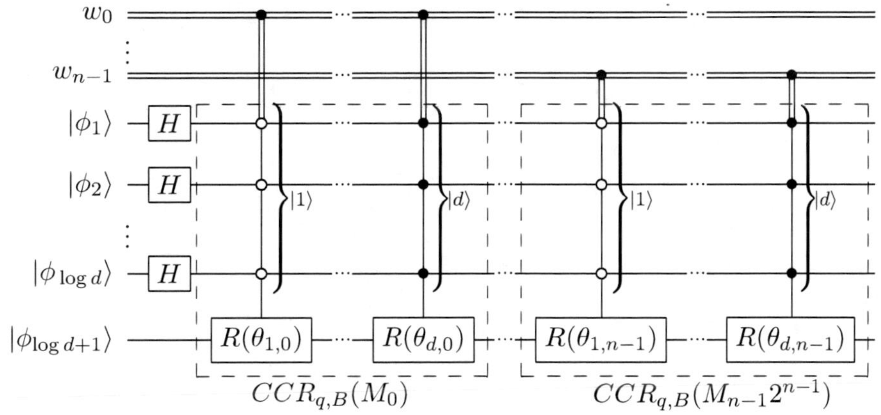

Figure 5. An algorithm, that hashes an n-bit input $w = w_0 \dots w_{n-1}$ into the state $|\psi_{q,B}(M)\rangle$ of $O(\log n)$ qubits. $R(\theta_{i,j})$ denotes a rotation by an angle $\frac{4\pi b_i 2^j}{q}$ around the $\hat{y}$ axis of the Bloch sphere.

where operator $CCR_{q,B,i}(\theta)$ rotates the target qubit by the angle θ if the control qubits were in the state $|i\rangle$ (its circuit representation is given in Figure 4).

In [3] we have given an efficient algorithm for computing quantum hash function in the model of quantum branching programs . This procedure (illustrated in Figure 5) consists of the following steps:

0. Initialization of the $\log d + 1$ qubits in the state $|0 \dots 0\rangle|0\rangle$.
1. Application of Hadamard transform to the first $\log d$ qubits:

$$\frac{1}{\sqrt{d}} \sum_{i=1}^{d} |i\rangle|0\rangle = |\psi_{q,B}(0)\rangle \ . \tag{3}$$

2. Application of $CCR_{q,B}(w)$ creates the quantum hash of the input bit string $w = w_0 \dots w_{n-1}$, which is also treated as a number $w = w_0 + w_1 2^1 + \dots + w_{n-1}2^{n-1}$:

$$CCR_{q,B}(w)|\psi_{q,B}(0)\rangle = |\psi_{q,B}(w)\rangle \ . \tag{4}$$

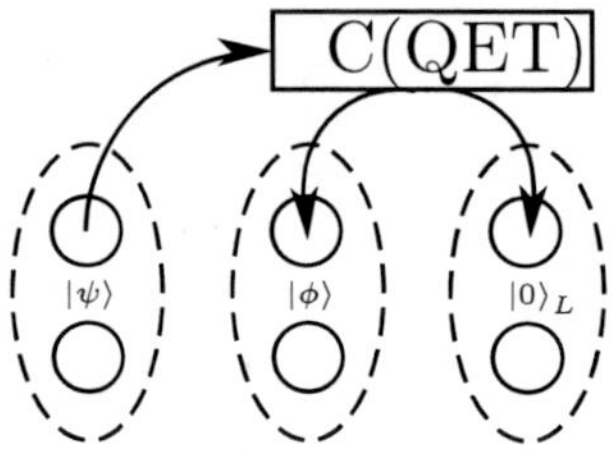

Figure 6. Implementing logical AND operation using Controlled-QET.

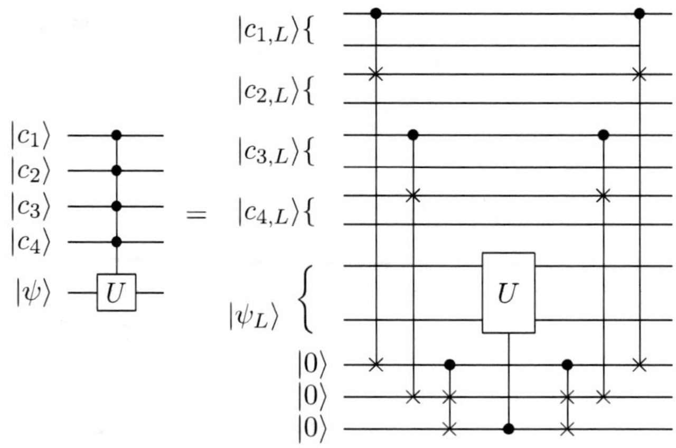

Figure 7. An example of improved implementation of the gate U controlled by 4 qubits.

5. Physical implementation of quantum hashing

In [5] we have proposed an effective physical implementation of compound multiply controlled operators for the model of quantum processor on multiatomic ensembles in the QED cavity. $CCR_{q,B,i}(\theta)$ is exactly such operator and thus can be accelerated in this architecture.

The key idea is to use the C(QET) operation to compute the logical AND operation in a manner depicted in Figure 6.

Let $C^t(U)$ be the general controlled gate, defined by the following equation in [7]:

$$C^t(U)|c_1c_2\dots c_t\rangle|\psi\rangle = |c_1c_2\dots c_t\rangle U^{c_1\cdot c_2\cdots c_t}|\psi\rangle. \quad (5)$$

Figure 7 demonstrates how the logical $C^t(U)$ gate can be reduced to the sequence of $2(t-1)$ C(QET) operations (which are implemented by two elementary gates in our model) and a Controlled-U gate. This scheme requires $t-1$ ancillary processing nodes instead of full logical qubits.

Finally, using the known construction [7] Controlled-U gate can decomposed into two CNOT gates (implemented here by C(QET) gate) and five single qubit rotations (corresponding to QET and PHASE operations).

Acknowledgments
The work is performed according to the Russian Government Program of Competitive Growth of Kazan Federal University. Work was in part supported by the Russian Foundation for Basic Research (under the grants 14-07-00878, 14-07-00557, 15-37-21160).

References

[1] Gottesman D and Chuang I 2001 Quantum digital signatures Tech. Rep. arXiv:quant-ph/0105032 Cornell University Library URL http://arxiv.org/abs/quant-ph/0105032
[2] Ablayev F M and Vasiliev A V 2014 *Laser Physics Letters* **11** 025202 URL http://stacks.iop.org/1612-202X/11/i=2/a=025202
[3] Ablayev F and Vasiliev A 2014 *Computing with New Resources* Lecture Notes in Computer Science ed Calude C S, Freivalds R and Kazuo I (Springer International Publishing) pp 149–160 ISBN 978-3-319-13349-2 URL http://dx.doi.org/10.1007/978-3-319-13350-8_11
[4] Vasiliev A 2015 *International Journal of Applied Engineering Research* **10** 31415–31426
[5] Ablayev F, Andrianov S, Moiseev S and Vasiliev A 2013 *Lobachevskii Journal of Mathematics* **34** 291–303 ISSN 1995-0802 (*Preprint* http://arxiv.org/abs/1301.2201) URL http://dx.doi.org/10.1134/S1995080213040094
[6] Moiseev S A, Andrianov S N and Moiseev E S 2013 *Optics and Spectroscopy* **115** 356–362 ISSN 0030-400X URL http://link.springer.com/10.1134/S0030400X13090166
[7] Nielsen M A and Chuang I L 2000 *Quantum Computation and Quantum Information* 1st ed (Cambridge University Press) ISBN 0521635039 URL http://www.worldcat.org/isbn/521635039

Dynamical systems for modeling the evolution of the magnetic field of stars and Earth

H Popova
Moscow State University, Moscow, Russia
E-mail: popovaelp@mail.ru

Abstract.
The cycles of solar magnetic activity are connected with a solar dynamo that operates in the convective zone. Solar dynamo mechanism is based on the combined action of the differential rotation and the alpha-effect. Application of these concepts allows us to get an oscillating solution as a wave of the toroidal field propagating from middle latitudes to the equator. We investigated the dynamo model with the meridional circulation by the low-mode approach. This approach is based on an assumption that the solar magnetic field can be described by non-linear dynamical systems with a relatively small number of parameters. Such non-linear dynamical systems are based on the equations of dynamo models. With this method dynamical systems have been built for media which contains the meridional flow and thickness of the convection zone of the star. It was shown the possibility of coexistence of quiasi-biennial and 22-year cycle. We obtained the different regimes (oscillations, vacillations, dynamo-bursts) depending on the value of the dynamo-number, the meridional circulation, and thickness of the convection zone. We discuss the features of these regimes and compare them with the observed features of evolution of the solar and geo magnetic fields. We built theoretical paleomagnetic time scale and butterfly-diagrams for the helicity and toroidal magnetic field for different regimes.

1. Introduction

It is generally accepted that the 11-year cycle of solar activity is connected with the action of the magnetic dynamo, the mechanism of which is based on the mutual work of the α-effect and the differential rotation [1]. It is assumed that Sun magnetic field has two components: the poloidal and the toroidal. The toroidal magnetic field originates from the poloidal one owing to the action of the differential rotation, which is situated inside the convective zone of the Sun. The reverse process of the transformation of the toroidal magnetic field into the poloidal one is realized as a result of the violation of the mirror symmetry of convection in a rotating body. The Coriolis force acting on the arising and diverging (descending and converging) vortices leads to the prevalence of right vortices in the northern hemisphere and left vortices in the southern one. The electromotive force, appearing as a result of the action of the Faraday's electromagnetic induction, after the averaging by velocity pulsations gain the component which is parallel to the mean magnetic field. It is this component that make close the self-excitation circuit in the Parker dynamo.

The 11-year cycle of solar activity is connected with its spot-producing activity. At the beginning of the solar activity cycle the spots appear at the mean latitudes and during approximately 11 years approach to the equator. Such latitudinal-temporal sunspot distribution looks like butterflies and is called butterfly-diagram or Maunder butterflies.

According to conceptions of dynamo theory, the butterfly appearance is connected with the motion of the toroidal component of the magnetic field (dynamo-wave) from higher latitudes to the equator in each hemisphere. The latitudinal-temporal diagram of the level lines of the magnetic field toroidal component obtained even using simplest dynamo models, qualitatively reproduces observational data of the butterfly-diagram. The form of the butterfly-diagram obtained in models is in essential dependence from the controlling model parameters. In [2], [8], [10], [11] it was shown numerically and analytically that meridional circulation directed against the dynamo-wave propagation can essentially slow down its motion. Besides, intensive meridional circulation "blows away" the dynamo-wave to the poles. In [6] it was shown that the turbulent diffusivity coefficient also has ability to influence the duration of solar activity cycle. Also there was shown that the α-effect latitudinal profile has influence upon the form of the Maunder butterfly.

Thus, the characteristics of the theoretical latitudinal-temporal diagrams for the magnetic fields of the Sun and other stars significantly depend on the controlling parameters in the dynamo models. On the other hand, the attempt to reproduce the observational latitudinal-temporal distributions for the stars magnetic fields by the choice of the appropriate values and the controlling processes depending on the coordinates and time could give the information about the physics of the investigated process. However, it is should be remembered that the physics of the process will be limited by the bounds of the chosen model.

2. Dynamo Model

Here we use a simple dynamo model which is a straightforward generalization of the initial Parker [7] migratory dynamo. The governing dynamo equations are derived by averaging the magnetic field over the radius within the thin shell, where the dynamo mechanism operates, and neglecting terms describing the effects of curvature near the pole. In this case the dynamo equations have the form:

$$\frac{\partial A}{\partial t} = R_\alpha \alpha B + \beta \frac{\partial^2 A}{\partial \theta^2} - V\frac{\partial A}{\partial \theta} - \eta^2 \beta A, \tag{1}$$

$$\frac{\partial B}{\partial t} = R_\omega \sin\theta \frac{\partial A}{\partial \theta} + \beta \frac{\partial^2 B}{\partial \theta^2} - \frac{\partial (VB)}{\partial \theta} - \eta^2 \beta B. \tag{2}$$

Here B represents the toroidal magnetic field, A is the azimuthal component of the vector potential of the poloidal magnetic field, and θ is the latitude measured from the pole, η is characterizes the thickness of the convective zone, t is the time. The distances are measured in units of the star of planet radius R and the time is in units of the diffusion time R^2/β. Baliunas (2006) showed that the terms proportional to η^2 describe the radial magnetic field diffusion. Since the thickness of the convective zone in the Sun is 1/3 of the radius, $\eta \approx 3$ for the Sun. The factor $\sin\theta$ is the decrease in the differential rotation at higher latitudes [4]. The second equation neglects the small contribution of the α-effect, i.e. we use so-called α-Ω-approximation. Curvature effects are absent in the diffusion terms. In Eq. 1–2 the parameters R_α and R_ω describe intensity of the α-effect and the differential rotation, respectively, the parameter $D = R_\alpha R_\omega$ is the dimensionless dynamo-number (a measure of the intensity of dynamo action), and β is the turbulent diffusivity. We used a simple scheme for the stabilization of the magnetic field growth, namely, the algebraic quenching of the helicity. This scheme assumes that $\alpha = \alpha_0(\theta)/(1+\xi^2 B^2) \approx \alpha_0(\theta)(1-\xi^2 B^2)$, where $\alpha_0(\theta) = \sin\theta$ is the helicity in the unmagnetized medium and $B_0 = \xi^{-1}$ is the magnetic field for which the α-effect is considerably suppressed.

In this paper we restrict ourselves to the consideration of dipole symmetry of the magnetic field with boundary conditions at the poles: $A(0,t) = B(0,t) = A(\pi,t) = B(\pi,t) = 0$, because we find only the solutions with the dipole symmetry.

Following [8], we consider latitudinal profile of the meridional circulation: $V(\theta) = v \sin 2\theta$. Since in our model the latitude is measured from the pole, the value with a positive sign corresponds to the meridional circulation directed along the dynamo-wave propagation.

The basic idea of the low-mode approach is that the mean-field dynamo equations are projected onto a minimum set of several first eigenfunctions of the problem describing the decay of magnetic fields without any generation sources. In this case it is necessary to choose the minimum set of functions in such a way that the solution which is a set of several first time-dependent Fourier coefficients taking into account the generation sources will describe the general behavior of the magnetic field of a given object and will not describe this field with any lower set of functions. Substituting the chosen set of components of the magnetic field into the dynamo equations one can obtain a dynamical system of equations containing the selected modes.

According to [13], we describe the magnetic field as

$$B(\theta,t) = b_1(t)\sin 2\theta + b_2(t)\sin 4\theta + b_3(t)\sin 6\theta,$$

$$A(\theta,t) = a_1(t)\sin\theta + a_2(t)\sin 3\theta + a_3(t)\sin 5\theta.$$

Substituting the chosen set of components of the magnetic field into the dynamo equations and collecting the coefficients of the sines of similar arguments, we can obtain a dynamical system of six equations containing the six modes [12], [13].

3. Results

As shown in [9], [13] dynamical system reproduces regimes of oscillations, vascillations (oscillations relative to a non-zero value of the magnetic field amplitude), dynamo-bursts and a special kind of oscillation, which is similar to the generation of quasi-biennial oscillation at the background of 22-year cycle. In [8], [12] [13] regimes of generation and their dependence on the control parameters v, D, β, η are discussed. However, in these papers the butterfly-diagrams of the helicity for these regimes and butterfly-diagrams of toroidal magnetic field for vascillations, dynamo-bursts and oscillations with two peaks which is similar to the dependence of Wolf number on time have not been built. Here we represent these butterfly-diagrams.

In Fig. 1a dependence of first component of the toroidal field on time is represented at $D = -2000$, $\beta = 1$, $v = 0$, and $\eta = 3$. Amplitude of the magnetic field and time are given in model units. Such dependence has oscillations with two peaks at maximum of the magnetic activity cycle which is similar to the dependence of Wolf number on time for solar cycles 22 and 23. For these dynamo parameters in Fig. 1b and Fig. 1c the butterfly-diagrams for the toroidal field and the helicity are represented. In Fig. 1d the butterfly-diagram for the helicity at $D = -2300$, $\beta = 1$, $v = 0$, and $\eta = 3$ is presented for combination of quasi-biennial and 11-years cycles. Theoretical butterfly-diagram for helicity represents Hale polarity rule similarly butterfly-diagram for observable current helicity in [5].

In Fig. 2a and Fig. 2b the butterfly-diagrams for the toroidal field and helicity are presented at $D = 90$, $\beta = 1$, $v = 0$, and $\eta = 0$ for the regime of vascillations. In Fig. 2c and Fig. 2d the butterfly-diagrams for the toroidal field and helicity are presented at $D = 128$, $\beta = 1$, $v = 0$, and $\eta = 0$ for the regime of dynamo-bursts. Fig.2a,b,c,d are built for fully convective stars.

It is suggested that generation of Earth magnetic field is carried out in outer core which is composed of iron and nickel. We use our dynamo model for modeling reversals of the geomagnetic poles. In [14] it was suggested that fluctuations in the α-effect are cause of geomagnetic reversals. It has been indicated that the reversal scale obtained in the scope of this model is rather close to the observed scale in its properties. In our model we take into account meridional circulation and consider magnetic field as

$$B(\theta, t) = b_1(t)\sin 2\theta + b_2(t)\sin 4\theta,$$

$$A(\theta, t) = a_1(t)\sin\theta + a_2(t)\sin 3\theta,$$

$\beta = 1$ and $\eta = 0$.

We obtained regimes similar to the geomagnetic reversals at $D \sim 500 \div 600$ and $v \sim -3 \div 0$. In Fig. 2e the theoretical dependence of first component amplitude of the poloidal field on time (top plot), theoretical paleomagnetic time scale (middle plot) and observable paleomagnetic time scale (bottom plot) are represented at $D = 500$, $v = -1$. Here time and amplitude of the magnetic field are given in model units. In Fig.3e for theoretical and observable paleomagnetic time scales we can see various periods of reversals which have a chaotic sequence.

4. Conclusion

The simple non-linear dynamic systems for the dynamo equations reproduce main features of solar and geo magnetic fields. We obtained that theoretical butterfly-diagrams for helicity represent Hale polarity rule. Our dynamic system is able to reproduce combination of quasi-biennial and 11-years cycles, moreover, we found regime of oscillations with two peaks at maximum of the magnetic activity cycle which is similar to the dependence of Wolf number on time for solar cycles 22 and 23. We obtained the regime similar to the geomagnetic reversals at constant governing parameters of the dynamo model.

References

[1] Brandenburg A and Subramanian K 2005 *Physics Reports* **417** 1
[2] Dikpati M and Gilman P A 2001 *Astrophys.J.* **559** 428
[3] Kleeorin N, Kuzanyan K, Moss D, et al. 2003 *Astron. and Astrophys.* **409** 1097
[4] Kuzanyan K M and Sokoloff D D 1995 *Geophys. Astrophys. Fluid Dynam.* **81** 113
[5] Baliunas S, Frick P, Moss D, et al. 2006 *Mon. Not. R. Astron. Soc.* **365** 181.
[6] Munoz-Jaramillo A, Nandy D, and Martens P C H 2011 *The Astrophysical Journal Letters* **727** 23
[7] Parker E N 1955 *Astrophys.J.* **122** 293
[8] Popova E P 2009 *Astronomy Reports* **53** 863
[9] Popova H 2013 *Magnetohydrodynamics* **49** 59
[10] Popova E P, Reshetnyak M Yu, and Sokoloff D 2008 *Astronomy Reports* **52** 157
[11] Popova H and Sokoloff D 2008 *Astron.Nachr.* **329** 766
[12] Popova H and Potemina K 2013 *Geomagnetism and Aeronomy* **53** 941
[13] Popova H and Yukhina N 2013 *Astronomy Letters* **39** 729
[14] Sobko G S, Zadkov V N, Sokoloff D D, and Trukhin V I 2012 *Geomagnetism and Aeronomy* **52** 254

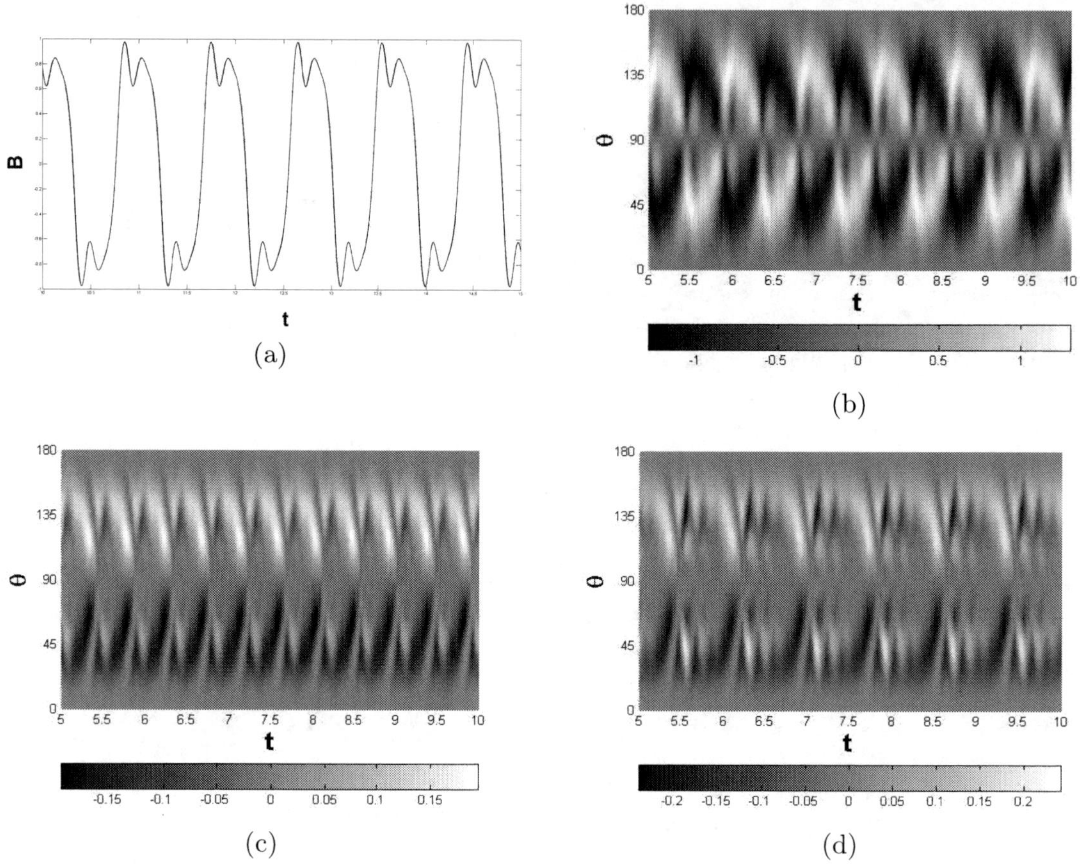

Figure 1. The dependence between the magnetic field amplitude and time (a), the butterfly-diagram for the toroidal field (b), the butterfly-diagram for the helicity (c) at $D = -2000$, $\beta = 1$, $v = 0$, and $\eta = 3$; the butterfly-diagram for the helicity at $D = -2300$, $\beta = 1$, $v = 0$, and $\eta = 3$

Journal of Physics: Conference Series **681** (2016) 012021 doi:10.1088/1742-6596/681/1/012021

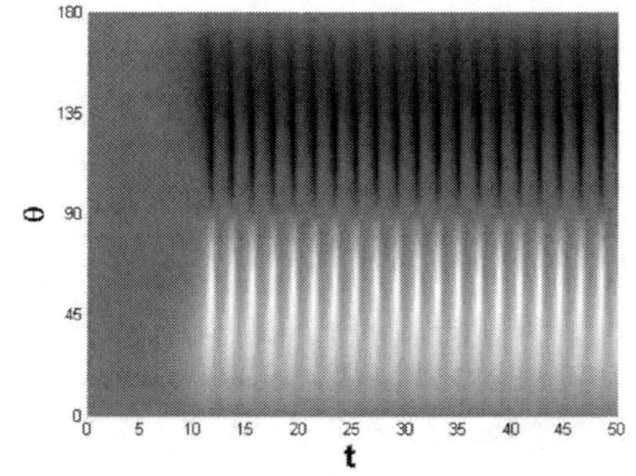

(a)

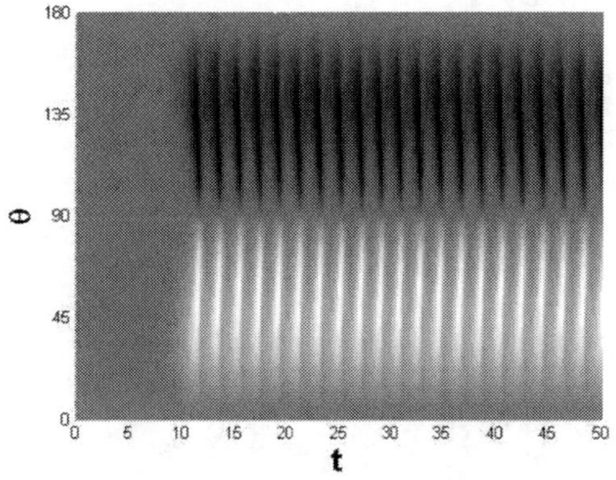

(b)

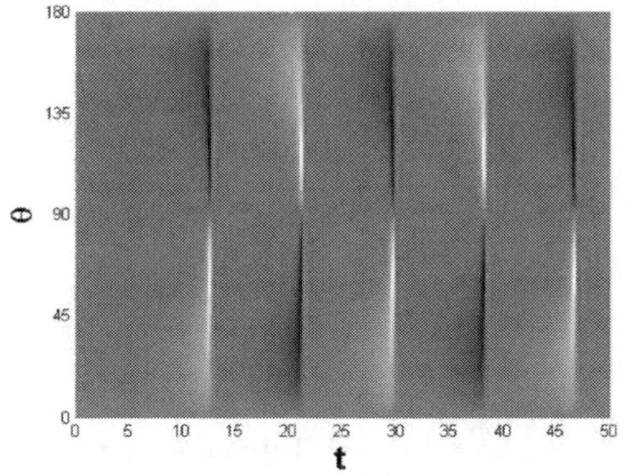

(c)

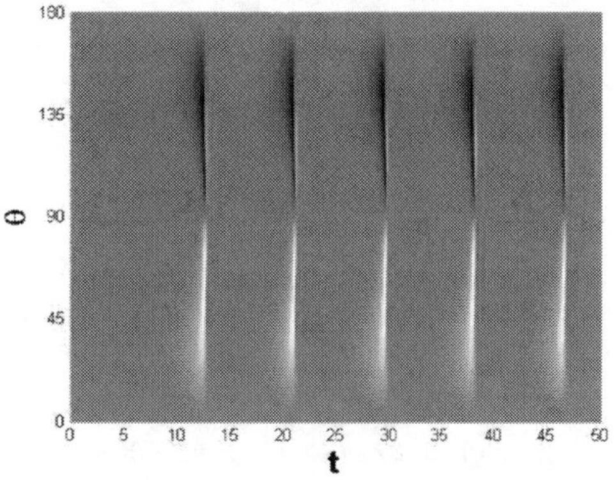

(d)

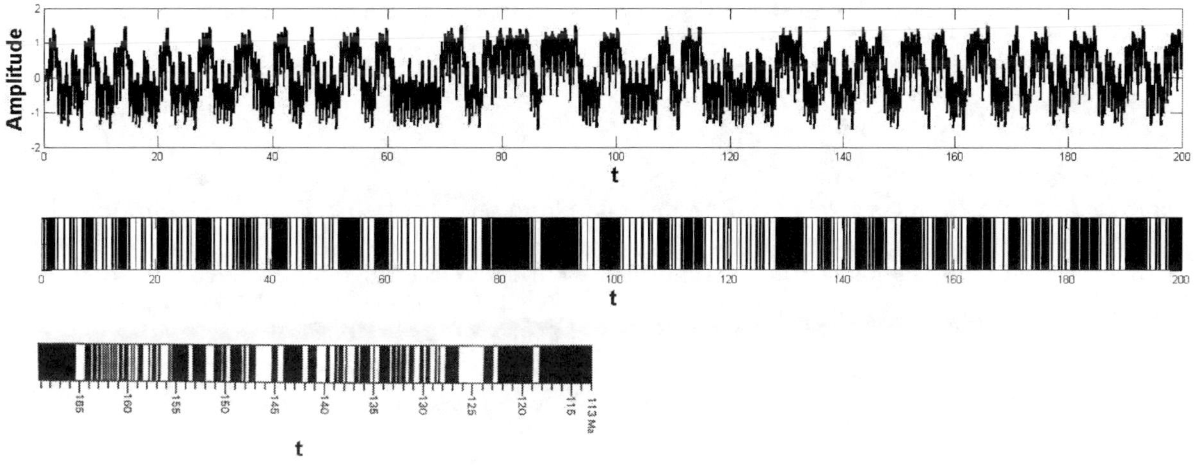

(e)

Figure 2. The butterfly-diagram for the toroidal field (a) and helicity (b) at $D = 90$, $\beta = 1$, $v = 0$, and $\eta = 0$; the butterfly-diagram for the toroidal field (c) and helicity (d) at $D = 128$, $\beta = 1$, $v = 0$, and $\eta = 0$; the theoretical dependence of first component amplitude of the poloidal field on time (top plot of (e)), theoretical paleomagnetic time scale (middle plot of (e)) and observable paleomagnetic time scale (bottom plot of (e)) at $D = 500$, $v = -1$, $\beta = 1$, and $\eta = 0$.

Journal of Physics: Conference Series **681** (2016) 012022 doi:10.1088/1742-6596/681/1/012022

High-Performance Computing in Astrophysical Simulations

Viktor Protasov[1,2], Alexander Serenko[2], Vladislav Nenashev[2], Igor Kulikov[1,2], and Igor Chernykh[1]

[1] Institute of Computational Mathematics and Mathematical Geophysics SB RAS, 6-th Lavrentyev avenue, Novosibirsk, Russian Federation

[2] Novosibirsk State Technical University, 20-th Karl Marks avenue, Novosibirsk, Russian Federation

E-mail: kulikov@ssd.sscc.ru

Abstract. The author's approach for simulating of multiscale astrophysical objects with using of supercomputers is described in the paper. Astrophysical objects consists of several components with different nature, and as a result are described with different mathematical models. This fact leads us to need of formulation of mathematical model and numerical method for each component. The two-phase model (gas + particles) was used in case of simulation of protoplanetary disks. The numerical method and details of parallel implementation for that model were disclosed. The mathematical model for galactic objects, describing stellar component and dark matter, based on the first momenta of Boltzmann equation was built. Such approach allows us to use unified numerical method to describe collisionless and gas component of galaxies.

1. Introduction

Observational astrophysics plays an important role in the study of the universe, formation of multiscale objects and different flows. However, observational data often just makes a question about formation of one or another configuration, but it couldn't give an answer about how any particular galaxy or planetary system was formed. Therefore, the main, and sometimes the only possible way to do that is to use mathematical modeling.

In recent years, with discovering of a big amount of planets, the problem of planet formation has become popular. Todays hypotheses about planet formation are based on the Kant-Laplace theory. It descibes the structure of the Solar system very well, but for some of other planetary systems it faltering. For example, many Hot Jupiters (gaseous giant planets with a very short orbit) were found. The question about how much time is needed to form the planet systems is opened too. As expected, it takes a hundreds of millions of years, but in fact the value is ten times less. Besides, it's not clear how multiple systems forms. Nowadays a couple of theories exist, described in [1], but they should be proved with using of mathematical modeling.

Also, the variety of observed galaxies is of great interest. An ordinary galaxy has up to a ten of collisions for a period of the Hubble time, that leads to formation of complex structures of galaxies. The observed variety of galaxies is directly related to collisions between them [2]. These observations led to the hypothesis that was proved with mathematical modeling [3].

The main problem in modeling of protoplanetary systems and galaxies is that it is necessery to model two different components – collisional (gas and dust in case of galaxies, and gas in case of protoplanetary disks) and collisionless (stellar component and dark matter in galaxies, and the dust component in protoplanetary disks). The gas component is described with system of gravitational gasdynamics equations, but the collisionless component is described with N-body problem, that leads to a couple of difficulties in its joint modeling.

The alternative to N-body problem is model based on the first momenta of the Boltzmann equation [4]. The ability of using such approach for describing of the collisionless component of colliding galaxies was experimentally proved [5]. This model isn't universal of course. It could be used only if: we are interested in the behaviour of cluster of collisionless component (it is true for the dynamics of stars in galactic scales); the velocity of cluster, as usually, has a directional movement and low dispersion of speed (it is also true for interacting galaxies, moreover, the very dispersion of speed of stellar component is observed in galaxies); there is no heat conduction (it is typical for all astrophysical processes).

The obvious advantage of the model is based on the first momenta of Boltzmann equation is ability of thermodynamically coherent phase transition between stellar and gas components. The mass and momentum conservation laws are satisfied but the entropy decreases in case of the N-body model. For example, in star formation process it is possible to save internal energy of the system while transition from gas component to collisionless component, but the mass and momentum are lost, that leads in fact to increasing of the gas entropy in the medium. But there is no contribution to the internal energy of the gas in case of supernovae feedback while transition from stellar component to gas with using of N-body model. The limitation on the description of small number of bodies, that holds in the case of protoplanetary disks, is the main disadvantage of this model. Nevertheless, such problem could be solved in two dimensional formulation, as opposed to simulation of colliding galaxies. So it need less computational resources and, consequently, is less demanding on the numerical methods in terms of performance.

The complexity and need of high detail of model leads to necessity of using the most powerful of the available supercomputers. Today the most powerful supercomputers are based on hybrid architecture with graphics accelerators and Intel Xeon Phi accelerators. For example, in the july 2015 version of Top500 the two first supercomputers (and four of Top10) are built with using of these technologies. It is quite clear, that first exaflop supercomputer wil be built on the basis of hybrid architecture. The using of the unified numerical method for describing of two components will allow us to use such hybrid computational resources [6].

2. Mathematical model

The equations of gravitational gasdynamics are used to describe gaseous component of galaxies and protoplanetary disks. The stellar component of galaxies is described with using of the first momenta of Boltzmann equation, and the dust component of protoplanetary disks is discribed with N-body model in 2D. The used model was substantiated in introduction, so let us give a formal description of it.

2.1. Equations

The gravitational gasdynamics equations could be written in redefined form.

$$\frac{\partial \rho}{\partial t} + \nabla \cdot (\rho \vec{u}) = 0 \qquad \frac{\partial \rho \vec{u}}{\partial t} + \nabla \cdot (\rho \vec{u}\vec{u}) = - \nabla (p) - \rho \nabla (\Phi)$$

$$\frac{\partial \rho \varepsilon}{\partial t} + \nabla \cdot (\rho \varepsilon \vec{u}) = -(\gamma - 1)\rho\varepsilon \nabla \cdot (\vec{u}) \qquad \frac{\partial \rho E}{\partial t} + \nabla \cdot (\rho E \vec{u}) = - \nabla \cdot (p\vec{u}) - (\rho\vec{u}, \nabla (\Phi))$$

where ρ – is a density, $\vec{u}$ – velocity, p – pressure, ε – internal energy, E – full energy, Φ – gravitational potential, γ – adiabatic index. The full energy is connected with internal energy by means of equation $\rho E = \rho\varepsilon + \frac{1}{2}\rho\vec{u}^2$. Internal energy and pressure are connected by equation of state for ideal gas $p = (\gamma - 1)\rho\varepsilon$.

Expressions for the first momenta of Boltzmann equation with using of diagonal tensor of velocity dispersion could be written in redefined form:

$$\frac{\partial n}{\partial t} + \nabla \cdot (n\vec{v}) = 0 \qquad \frac{\partial n\vec{v}}{\partial t} + \nabla \cdot (n\vec{v}\vec{v}) = -\nabla(\pi_{\xi\xi}) - n\nabla(\Phi)$$

$$\frac{\partial \pi_{\xi\xi}}{\partial t} + \nabla \cdot (\pi_{\xi\xi}\vec{v}) = -2\pi_{\xi\xi}\nabla \cdot (\vec{v}) \qquad \frac{\partial \rho W}{\partial t} + \nabla \cdot (\rho W\vec{v}) = -\nabla \cdot (\pi_{\xi\xi}\vec{v}) - (n\vec{v}, \nabla(\Phi))$$

where n – density, $\vec{v}$ – velocity, $\pi_{\xi\xi}$ – elements of diagonal tensor of velocity dispersion, W – full mechanical energy. The full energy is connected with a trace of diagonal tensor of velocity dispersion with relation: $\rho W = \frac{1}{2}\pi_{\xi\xi} + \frac{1}{2}\rho\vec{u}^2$.

Equations of particles movement is:

$$\frac{d^2x}{dt^2} = \frac{\nabla(\Phi)}{m}$$

where m – particle mass.

Poisson equation generally written as

$$\triangle\Phi = 4\pi G(\rho + n + \sigma)$$

where σ – density of particles, defined by an ansamble of particles using kernel.

2.2. Numerical method

To solve the equations of gasdynamics and first momenta of the Boltzmann equation, the combination of Fluids-in-Cells method (operator-splitting approach), Godunov method and piecewise-parabolic method on local stencil for achievement of high order of accuracy are used. The special modification of the Roe averaging scheme is used to average the quantity on the borders of cells. To recompute internal energy (or tensor of velocity dispersion in case of Boltzmann equation) in the region with high density and to correct the magnitude of velocity vector on the gas-vacuum border area and in the area rarefied space the redefined system of equations is used.

It is very important to know the movement trajectories of each particle with high accuracy in the case of simulation of planet formation process. The Particle-Mesh method with Clouds-in-Cells approach is used to simulate the dynamics of dust component of protoplanetary disk. According to it, the particles are discrete bodies translated into the same grid as used for gas density. Herewith, the mass of each particle is spreaded by some finite area, called "cloud", wich make a contribution to the distribution of density in gravitational potential and gravitational interaction between gas and particles without loosing of the ability of tracking each particle separately [7].

2.3. Parallel implementation

Geometrical decomposition of the computational area is in the heart of the parallel implementation of interacting galaxies model. Unique feature of the mathematical model allows us to formulate unified numerical method and, consequently, parallel algorithm and it's implementation. The implementation on heterogeneous supercomputer RSC PetaStream equipped with Intel Xeon Phi had achieved 134x acceleration on one Intel Xeon Phi, and 92% of efficiency was shown on 64 Intel Xeon Phi accelerators.

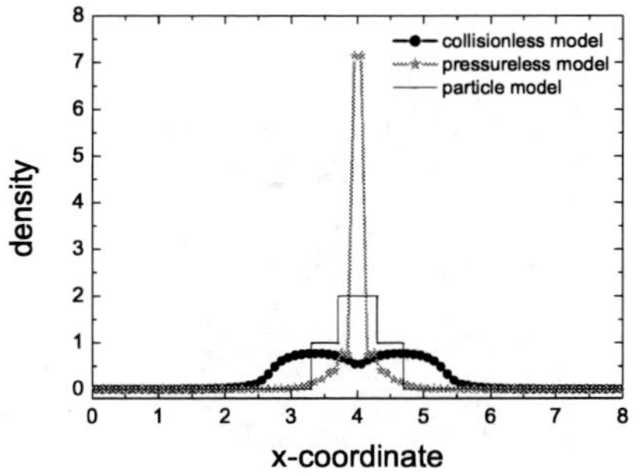

Figure 1. Collision of two waves with using of different models, $t = 1.2$

Graphics accelerators with Nvidia CUDA technology were used for parallel implementation of model of protoplanetary disks dynamics. The most expensive stage of simulation is solving of Poisson equation. The Fast Fourier Transform is used to solve the equation and it is implemented on GPU in cuFFT library. The computation of border conditions and potential in harmonic space are ported on GPU to improve performance . In this way the data transfering from host to device and back is minimal. So, almost 10x acceleration was achieved on Nvidia GeForce G210M opposed to one core of Intel Core2Duo. The particles simulation was implemented on GPU too, that had increased performance in 4 times. Simulation of gasdynamics is computed on CPU in the same time as particles dynamics computed on GPU.

3. Numerical experiments

The main focus of the paper is on demonstration of various models for describing of collisionless component. So let us make an experiment of one-dimension impact of density waves with using of different models. Also the late stage of protoplanetary disk evolution and collision of galaxies with differnet masses will be shown.

3.1. One-dimensional impact of density waves

We use three models to simulate the collision between two waves: models with particles with some initial speed and without collision; pressureless hydrodynamics model; model with first momenta of Boltzmann equation. The last one is equal to gasdynamics model with $\gamma = 3$ in 1D. Density distribution with using of these models is shown on the figure 1. In case of pressureless model the solution is δ-function, that makes it useless for describing of collisionless model. One can see the scattering of waves after collision in case of the third model, that could be used in galaxy collision. This fact was proved in paper [5].

3.2. The late stage of evolution of protoplanetary disk

Lets look at the process of evolution of protoplanetary disk from the moment when a couple of protoplanets formed. Let us make a gas disk with radius $R = 4$, density $\rho(r) = 3.5\frac{(R-r)(R+r)}{R^2}$, pressure $p(r) = 0.8\frac{(R-r)(R+r)}{R^2}$ and angular velocity $\omega = 1$ in the area $[-5, 5] \times [-5, 5]$. The protoplanets are 10 particles that distributed uniformly within a disk with radius $R \in [1.0, 2.5]$ and has masses $m \in [9e - 3,\ 9e - 2]$ and angular velocity $\omega = 2.5 + \epsilon$, $\epsilon \sim N(1.0, 0.5)$. The result is shown on figure 2. Formation of three gaseous objects with high density, which could be interpreted as possible stars, is shown on it. Also, there is a rarefying space in the left of central star. It was formed as a result of stellar wind that threw matter away to the outter part

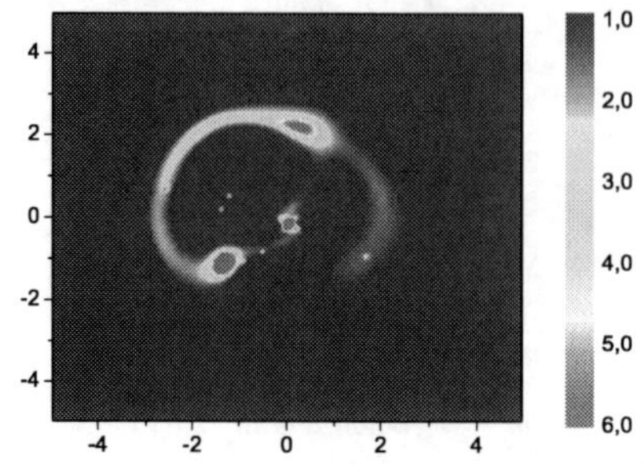

Figure 2. Simulation of the late stage of evolution of protoplanetary disk, $t = 9.5$

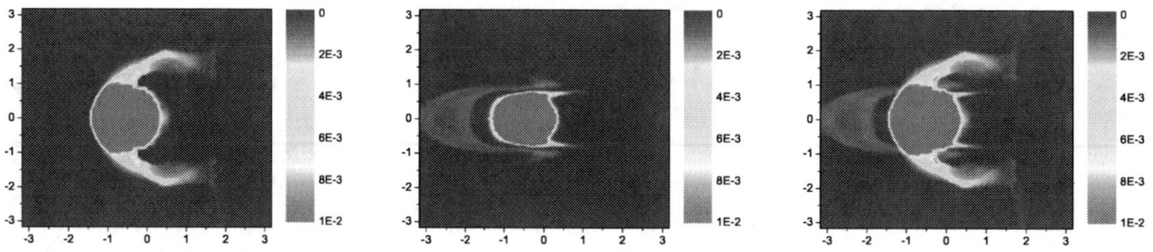

Figure 3. Simulation of galaxies collision. Dimensionless distribution of gas density (left), stellar component and dark matter (center), total density (right)

of the disk. In particular, it is one of reasons why the gas giant couldn't be formed near the sun.

3.3. Simulation of collision of galaxies with different masses

As a model problem, let us compute a central collision of galaxies with mass fraction $1 : 20$ with total mass of "big" one $10^{13} M_{\odot}$. The fraction of halo mass to disk mass is $10 : 1$ in both galaxies. Each component is given by the equilibrium configuration. The results (dimensionless density distribution) are in figure 3. The instabilities after the massive galaxy because of ram-pressure mechanism made by the "small" galaxy are shown on the figure 3.

4. Conclusion

The author's approach for modeling of multiscale astrophysical objects with supercomputers is described in this paper. The two-phase model (gas + particles) was used to simulate the dynamics of protoplanetary disks. Also the numerical method and details of parallel implementation were described for it. The mathematical model based on the first momenta of Boltzmann equation for describing of stellar component and the dark matter of galactic objects is given. Such approach allows us to use unified numerical method to simulate collisionless and gaseous component of galaxies and also get high efficiency of the parallel implementation. The process of collision of two galaxies with different masses and the late stage of evolution of protoplanetary disk were simulated.

Acknowledgments

Acknowledgements This work was partially supported by Russian Foundation for Basic Research (grants 15-31-20150, 15-01-00508, 13-07-00589, and 14-01-31199) and by Grant of the President of Russian Federation for the support of young scientists number MK – 6648.2015.9. This project was partially supported by the Russian Ministry of Education and Science.

References

[1] Tutukov A.V., Fedorova A.V. Formation of Planets During Evolution of Single and Binary Stars // Astronomy Reports. 2012. 54, 305-314

[2] Tutukov A., Lazareva G., Kulikov I. Gas Dynamics of a Central Collision of Two Galaxies: Merger, Disruption, Passage, and the Formation of a New Galaxy // Astronomy Reports. 2011. 55, 9, 770-783

[3] Vshivkov V., Lazareva G., Snytnikov A., Kulikov I., Tutukov A. Hydrodynamical code for numerical simulation of the gas components of colliding galaxies // The Astrophysical Journal Supplement Series. 2011. 194, 47, 1-12

[4] Mitchell N., Vorobyov E., Hensler G. Collisionless Stellar Hydrodynamics as an Efficient Alternative to N-body Methods // Monthly Notices of the Royal Astronomical Society. 2013. 428, 2674-2687

[5] Kulikov I. GPUPEGAS: A New GPU-accelerated Hydrodynamic Code for Numerical Simulations of Interacting Galaxies // The Astrophysical Journal Supplements Series. 2014. 214, 12, 1-12

[6] Kulikov I.M., Chernykh I.G., Snytnikov A.V., Glinskiy B.M., Tutukov A.V. AstroPhi: A code for complex simulation of dynamics of astrophysical objects using hybrid supercomputers // Computer Physics Communications. 2015. 186, 71-80

[7] Protasov V.A., Kulikov I.M. PADME - A New Code for Modeling Planet Georesources Formation on Heterogeneous Computing Systems // Bulletin of the Tomsk Polytechnic University. 2015. 326, 8, 61-70

Spatially inhomogeneous structures in the solution of Fisher-Kolmogorov equation with delay

S V Aleshin[1,2], **S D Glyzin**[1,2] **and S A Kaschenko**[1,3]

[1] P.G. Demidov Yaroslavl State University, Sovetskaya str., 14, Yaroslavl, 150000, Russia
[2] Scientific Center in Chernogolovka RAS, Lesnaya str., 9, Chernogolovka, Moscow region, 142432, Russia
[3] National Research Nuclear University MEPhI, Kashirskoye shosse 31, Moscow, 115409, Russia

E-mail: fktiby@yandex.ru, glyzin@uniyar.ac.ru, kasch@uniyar.ac.ru

Abstract. We consider the problem of density wave propagation in a logistic equation with delay and diffusion (Fisher–Kolmogorov equation with delay). A Ginzburg–Landau equation was constructed in order to study the qualitative behavior of the solution near the equilibrium state. The numerical analysis of wave propagation shows that for a sufficiently small delay this equation has a solution similar to the solution of a classical Fisher–Kolmogorov equation. The delay increasing leads to existence of the oscillatory component in spatial distribution of solutions. A further increase of delay leads to destruction of the traveling wave. That is expressed in the fact that undamped spatio-temporal fluctuations exist in a neighborhood of the initial perturbation. These fluctuations are close to the solution of the corresponding boundary value problem with periodic boundary conditions. Finally, when the delay is sufficiently large we observe intensive spatio-temporal fluctuations in the whole area of wave propagation.

1. Introduction

A. N. Kolmogorov, I. G. Petrovskii and N. S. Piskunov proposed the logistic equation with diffusion [1] to simulate the genetic wave propagation

$$\frac{\partial u}{\partial t} = \frac{\partial^2 u}{\partial x^2} + u[1 - u]. \tag{1}$$

We will use abbreviation KPP for this equation.

Here $u(t, x)$ is a density distribution of the individuals number which have a dominant gene, $t \geq 0$ is a time variable and $x \in (-\infty, \infty)$ is a spatial variable. We consider the population density wave propagation problem from some nonzero initial conditions. Article by R. A. Fisher [2], which devoted to the same problem, was published almost simultaneously. Therefore the equation (1) is sometimes called the Fisher equation or the Fisher-Kolmogorov equation.

The equation (1) is used in a wide range of applications related to the space distribution of different waves — from the concentration waves to the population density waves. Among the large number of publications on the topic, we note a books [3, 4, 5], which contains problem summarizing results and we note the extensive bibliography of [5].

For any of classical boundary conditions (Neumman boundary conditions: $\frac{\partial u}{\partial x}\big|_{x=a} = \frac{\partial u}{\partial x}\big|_{x=b} = 0$; periodic boundary conditions $u(t, x+T) \equiv u(t, x)$ and some other) the equation (1) has only one attractor — a homogeneous equilibrium state $u_0 \equiv 1$. All other equilibrium states are unstable.

Let's consider the generalization of the KPP equation with delay $h > 0$

$$\frac{\partial u}{\partial t} = \frac{\partial^2 u}{\partial x^2} + u[1 - u(t-h,x)]. \tag{2}$$

Under certain assumptions this equation with diffusion and delay can be derived from the equations of population density dynamics which are studied in a series of articles by Britton N., Kuang Y., Gourley S., So J., Wu J. [6, 7, 8] (also refer the bibliography of [9]). This equation is one of the simplest problems of "reaction-diffusion" type and it was seen in various formulations [10, 11, 12, 13, 14, 15, 16, 17].

For the equation (2) we investigate the behavior of solutions with the initial conditions of a sufficiently small neighborhood (in $C[-\infty,\infty] \times C[-h,0]$) of the equilibrium state $u_0 \equiv 1$. It is shown that the behavior of solutions (2) in the neighborhood of u_0 is determined by a special non-linear parabolic equations of Ginzburg-Landau in the case of parameter h is close to $\pi/2$. Then we construct the equation of wave propagation for (2) and prove some statements about their dynamic properties. At the end of the article we presented the results of numerical modeling of the density wave in the equation (2) in the case of an unbounded x region.

2. Problem statement

The logistic equation with delay

$$\dot{u} = u[1 - u(t-h)] \tag{3}$$

has been studied in [18, 19, 20, 21, 22, 23, 24, 25, 26]. Solution of (3) monotonically tends to 1 for $h < e^{-1}$ and oscillatory tends to 1 for $e^{-1} < h < \pi/2$. In the article [18] it was shown that all positive solutions (3) tend to 1 for $h \leq 37/24$ and $t \to +\infty$. The equilibrium state u_0 is asymptotically stable for $h \leq \pi/2$. Apparently ([21]) u_0 is globally stable for $h \leq \pi/2$. For $h > \pi/2$ we have slowly oscillating stable periodic solution $u_*(t)$ (the distance between a peaks is greater than h). In the case of $h - \pi/2 \ll 1$ and $h \gg 1$ asymptotic behavior of solution is given in [26].

Situation for the equation (2) with classical boundary conditions is more complicated. Suppose, for example, we have a periodic boundary conditions

$$u(t, x+T) \equiv u(t,x). \tag{4}$$

Of course, the boundary value problem (2), (4), as well as the equation (3), have spatially homogeneous periodic solution $u_*(t)$ at $h > \pi/2$. If parameter h is close to $\pi/2$ and $h \gg 1$ then this solution is stable. If h and T are sufficiently large then it is possible that the solution $u_*(t)$ loses its stability and complex spatially inhomogeneous structures appear [24, 25, 26, 27, 28].

Lets us consider the case of delay h proximity to $\pi/2$ in detail and suppose the following relation holds:

$$h = \pi/2 + \varepsilon h_1 \quad \text{and} \quad 0 < \varepsilon \ll 1. \tag{5}$$

If we additionally suppose that the parameter T of (4) satisfies

$$T \gg 1, \tag{6}$$

then the dynamic of boundary value problem (2), (4) becomes more complicated. In the main the behavior of solutions in a small neighborhood of the equilibrium state $u_0 \equiv 1$ is determined by nonlocal behavior of solutions of the normalized complex equation (Ginzburg-Landau equation)

$$\frac{\partial \xi}{\partial \tau} = \sigma\delta \frac{\partial^2 \xi}{\partial y^2} + h_1 \delta \xi + d|\xi|^2 \xi, \tag{7}$$

$$\xi(\tau, y+1) \equiv \xi(\tau, y). \tag{8}$$

Here $\tau = \varepsilon t$, $y = T^{-1}x$ — new time and spatial variables,
$\sigma = T^{-2}\varepsilon^{-1}$ is of order of 1,

$$\delta = (4 - 2\pi i)(4+\pi^2)^{-1}, \quad d = -2(3\pi - 2 + i(\pi+6))\big(5(4+\pi^2)\big)^{-1}. \tag{9}$$

Solution of (2), (4) and solutions of (7), (8) are related by the formula

$$u(t,x) = 1 + \varepsilon^{1/2}[\xi(\tau,y)\exp(it) + \overline{\xi}(\tau,y)\exp(-it)] + O(\varepsilon). \tag{10}$$

The dynamic properties (7) – (8) depend essentially on the parameter σ. For example, for sufficiently small σ all simple periodic solution of the form

$$\rho_m \exp(2\pi imy + i\varphi_m\tau) \quad (m = 0, \pm1, \pm2, \dots) \tag{11}$$

are unstable [29].

In the articles [27, 30] for problem (7) with periodic boundary conditions (8) and Neumann boundary conditions a numerical experiment was performed. As a result of numerical experiment we can say that if we decrease the parameter σ then oscillations modes of solution became disordered and have more complex structure of the spatial variable. It should be noted that the approval of the correspondence between the solutions of quasinormal form (7), (8) and initial boundary value problem can be substantiated only in the case when problem (7), (8) has hyperbolic attractor. Therefore, in an article [31] for the equation (2) with Neumann boundary conditions and the boundary conditions (8) authors make the transition to a special difference analog and then performed a numerical analysis of the discrete system.

3. The normalized equation construction

Consider the behavior of equation solution near the equilibrium state $u_0 \equiv 1$

$$\frac{\partial u}{\partial t} = \frac{\partial^2 u}{\partial x^2} + u\Big[1 - u\Big(t - \frac{\pi}{2} - \varepsilon, x\Big)\Big]. \tag{12}$$

The characteristic (wave) equation for the linearized on u_0 equation (12) have a form

$$\lambda = -k^2 - \exp\Big(-\Big(\frac{\pi}{2} + \varepsilon\Big)\lambda\Big), \ k \in (-\infty, \infty). \tag{13}$$

With sufficiently small k^2 equation (13) has a root $\lambda(k,\varepsilon)$. This root has sufficiently small real part. Let us state this conclusion more precisely. Let's $k = \varepsilon^{1/2}m$.

Then equation (13) has a root $\lambda_m(\varepsilon) = \lambda(\varepsilon^{1/2}m, \varepsilon)$ and the following asymptotic equality holds for $\varepsilon \to 0$

$$\lambda_m(\varepsilon) = i + 2\varepsilon(1-m^2)(2 - i\pi)(4+\pi^2)^{-1} + O(\varepsilon). \tag{14}$$

Thus for each $|m| < 1$ a real part of $\lambda_m(\varepsilon)$ is close to zero and imaginary part is close to 1.

Let us consider the formal series

$$u(t,x) = 1 + \varepsilon^{1/2}\big[\xi(\tau,y)\exp(it) + \overline{\xi}(\tau,y)\exp(-it)\big] + \varepsilon u_2(\tau,y,t) + \varepsilon^{3/2}u_3(\tau,y,t) + \dots, \tag{15}$$

and $\tau = \varepsilon t$, $y = \varepsilon^{1/2}x$, $\xi(\tau,y)$ — unknown "amplitude" to be defined, and function $u_j(\tau,y,t)$ has 2π-periodic dependence of t. Substitute (15) in (12) and equate the coefficients of corresponding powers of $\sqrt{\varepsilon}$. At the second step we find $u_2(\tau,y,t)$. At the third step we collect the coefficients for $\varepsilon^{3/2}$ and obtain the equation with respect to u_3. From the condition of solvability of this equation

Journal of Physics: Conference Series **681** (2016) 012023 doi:10.1088/1742-6596/681/1/012023

in a specified class of functions we got the Ginzburg-Landau equation (as in the preceding section) to find $\xi(\tau, y)$:

$$\frac{\partial \xi}{\partial \tau} = \delta \frac{\partial^2 \xi}{\partial y^2} + \delta \xi + d|\xi|^2 \xi, \tag{16}$$

here δ and d are determined by formulas (9). Note that $\mathrm{Re}\,\delta > 0$ and $\mathrm{Re}\,d < 0$.

The main result of this section is following:

Theorem 1. *Let the equation (16) have a bounded solution $\xi_0(\tau, y)$ for $\tau \to +\infty$ and $y \to \pm\infty$. Then equation (12) has asymptotic solution $u(t, x, \varepsilon)$ with residual $O(\varepsilon^{3/2})$:*

$$u(t, x, \varepsilon) = 1 + \varepsilon^{1/2}[\xi_0(\tau, y)\exp(it) + \bar{\xi}_0(\tau, y)\exp(-it)] + \\ 0.2\varepsilon\big[(2 - i)\xi^2(\tau, y)\exp(2it) + (2 + i)\bar{\xi}^2(\tau, y)\exp(-2it)\big] + O(\varepsilon^{3/2}).$$

Equation (16) has one-parameter family of the simplest periodic solutions

$$u_m(\tau, y) = \rho_m \exp(imy + i\omega_m \tau), \tag{17}$$

where $\rho_m^2 = (m^2 - 1)\mathrm{Re}\,\delta(\mathrm{Re}\,d)^{-1}$, $|m| < 1$, $\omega_m = \Big[\mathrm{Im}\,\delta - \mathrm{Im}\,d\mathrm{Re}\,\delta(\mathrm{Re}\,d)^{-1}\Big](1 - m^2) - \mathrm{Im}\,\delta$. Since we have $1 + \mathrm{Im}\,\alpha(\mathrm{Re}\,\alpha)^{-1} \cdot \mathrm{Im}\,d(\mathrm{Re}\,d)^{-1} = -(\pi^2 + 4)/(6\pi - 4) < 0$, then from [29] we conclude that all solutions of (17) are unstable. Taking into account this result, in the next section we consider some properties of the wave solutions of (2).

4. Some properties of the wave equation

Consider a substitution in the equation (2) in the form of a traveling wave $u(t, x) = w(2t \pm x)$ and introduce a new time $s = 2t \pm x$, then for $w(s)$ we have the following second order equations with delay

$$w'' - 2w' + w[1 - w(s - 2h)] = 0, \tag{18}$$

here $'$ is a derivative with respect to s. The properties of stability of the zero solution of (18) do not depend on h. This solutions is unstable. It corresponds to a multiple root equal to 1. The stability properties of equilibrium state 1 are determined by the location of the roots of the characteristic quasipolinom

$$P(\lambda) \equiv \lambda^2 - 2\lambda - \exp(-2h\lambda). \tag{19}$$

We prove the following lemmas.

Lemma 1. *Quasipolinom $P(\lambda)$ has three real roots at $0 < h < h_1^*$. One root is positive and two roots are negative. $P(\lambda)$ has only one positive real root at $h > h_1^*$. Here $h_* = (\lambda_* - 1)/(\lambda_*^2 - 2\lambda_*) \approx 0.5608$, and λ_* is a root of transcendental equation $\lambda^2 - 2\lambda - \exp((2 - 2\lambda)/(\lambda - 2)) = 0$.*

Lemma 2. *All roots of quasipolinom $P(\lambda)$ lie in the left half-plane for $0 < h < h_2^*$, except for one real positive root. Here $h_2^* = (\sqrt{5} - 2)^{-1/2} \cdot \arccos(-\sqrt{5} + 2)/2 \approx 1.8617$. The pair $\lambda = \pm i\omega_0$ of pure imaginary roots goes to the imaginary axis at $h = h_2^*$ and $\omega_0 = (\sqrt{5} - 2)^{-1/2} \approx 0.4859$.*

In a neighborhood of solution $w(s) \equiv 1$ we find the asymptotic of the cycle, branching from this solution at $h = h_2^* + \mu$, $0 < \mu \ll 1$. In this case, the following statement holds.

Lemma 3. *There exists $\mu_0 > 0$ such that for all $0 < \mu < \mu_0$ equation (18) has dichotomous cycle which one-dimensional unstable manifold and following asymptotic*

$$w(s, \mu) = 1 + 2\sqrt{-\mu\varphi_0/d_0}\cos((\omega_0 + \mu(\psi_0 - c_0\varphi_0/d_0))s + \gamma) + O(\mu), \tag{20}$$

here $\varphi_0 + i\psi_0 = 2\omega_0^2(-1 + i\omega_0)(P'(i\omega_0))^{-1}$, $d_0 + ic_0 = \Big(2\omega_0^2(1 - \omega_0^2 - 2i\omega_0) + b\big((\omega_0^2 + 2i\omega_0)^2 - (\omega_0^2 + 2i\omega_0)^{-1}\big)\Big)(P'(i\omega_0))^{-1}$, $b = (\omega_0^2 + 2i\omega_0)(4\omega_0^2 + 4i\omega_0 + (\omega_0^2 + 2i\omega_0)^2)^{-1}$ and γ — is an arbitrary constant, which determines the phase shift along the cycle. Approximate values of the coefficients are $\varphi_0 + i\psi_0 \approx 0.1368 - 0.2066i$ and $d_0 + ic_0 \approx -0.0443 - 0.0366i$.

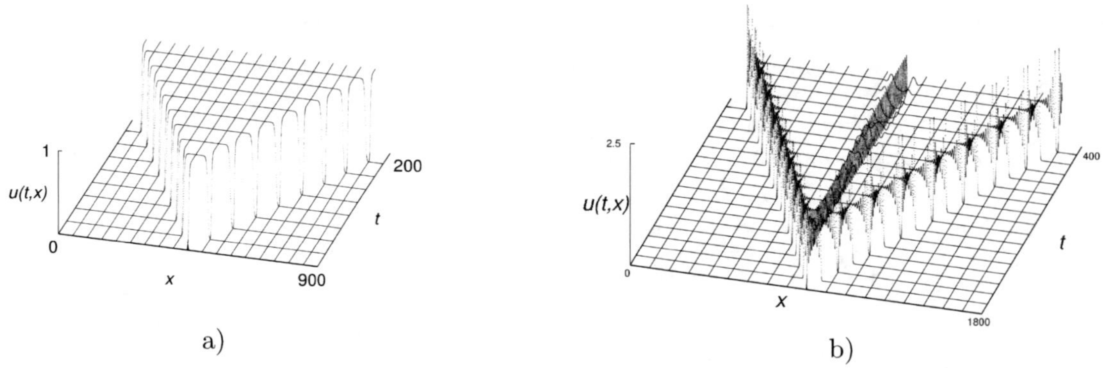

Figure 1. Wave propagation in KPP with delay a) $h = 0$; b) $h = 1.6$

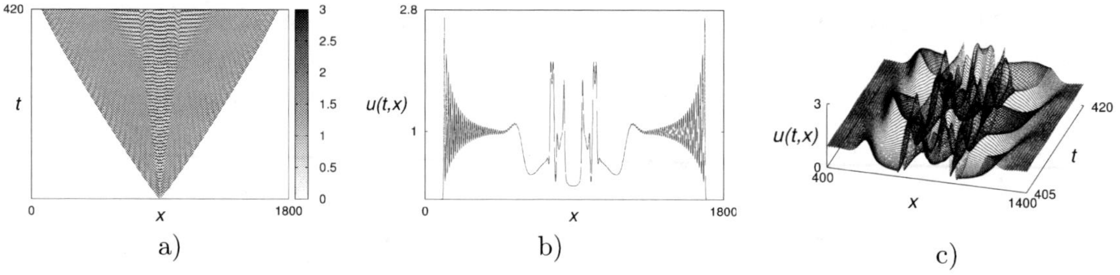

Figure 2. The solution of (2) at $h = 1.8$ a) density distribution $u(t, x)$ in a grayscale; b) cross-section $t = 400$; c) fragment at $t \in [405 \times 420]$ and $x \in [400 \times 1400]$;

5. Description of the numerical experiment

Described in the above sections analytical results have allowed us to make a numerical experiment. In this experiment we distinguish the qualitative features of the structure of the concentration wave of (2) from spatially localized initial disturbance. We considered (2) on the interval $[a, b]$ and $|a-b|$ was a sufficiently large. The boundary values were zero. Initial condition at $-h \leq t \leq 0$ were localized near central part of the space variable pulse of 0.1 amplitude.

Below we a present the results of numerical experiments obtained for different values of delay (0, 1.6, 1.8 and 2) that are the most revealing to study of the structure of the wave front. The figures 1 – 3 show an image of solutions density distribution of the KPP equation with delay and some cuts in different planes, allowing to clarify the process of emergence and propagation of the oscillations on the border of the wave front and on the central part of it.

6. Main results

In order to clarify some of the features of the qualitative behavior of solutions of KPP with delay we built Ginzburg-Landau equation that describes the dynamic properties of the equations near equilibrium state $u_0 \equiv 1$.

We studied the wave propagation equation (18) and found the critical parameter of delay for which changes in the structure of the spatial distribution of the solution appear.

Numerical analysis, based on analytical results, allowed us to distinguish the following intervals of delay: 1) the behavior of solutions of problem with delay close to the behavior of the problem without delay; 2) in the center part of the wave solutions area appears a region with complex spatial distribution; 3) in the spatial distribution of solutions there is preserved a region where the solution tends to 1; 4) the whole area of wave propagation is filled with the

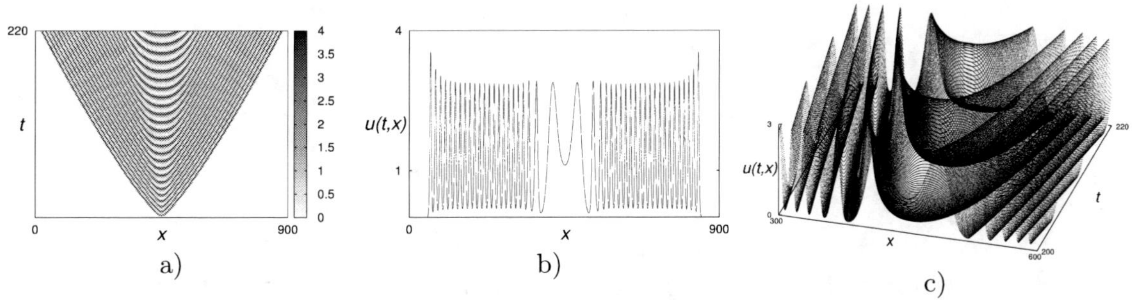

Figure 3. The solution of (2) at $h = 2$ a) density distribution $u(t, x)$ in a grayscale; b) cross-section $t = 200$; c) fragment at $t \in [200 \times 220]$ and $x \in [300 \times 600]$;

intensive oscillations in space and time variables.

Acknowledgments
This work was supported by the Russian Science Foundation (project nos. 14-21-00158).

References

[1] Kolmogorov A, Petrovsky I and Piscounov N 1937 *Moscou Univ. Bull. Math.* **1** 1–25
[2] Fisher R 1937 *Annals of Eugenics* **7** 355–69
[3] Murray J 2001 *Mathematical Biology. I. An Introduction* Third Edition (Berlin)
[4] Danilov V, Maslov V and Volosov K 1995 *Mathematical Modelling of Heat and Mass Transfer Processes* (Kluwer: Dordrecht)
[5] Volpert A, Volpert V and Volpert V 2000 *American Mathematical Society*
[6] Britton N 1990 *SIAM J. Appl. Math.* **50:6** 1663–88
[7] Gourley S and Britton N 1993 *IMA Journal of Applied Mathematics* **51** 229–310
[8] Gourley S and Kuang Y 2003 *Proc. R. Soc. Lond. A* **459** 1563–79
[9] Gourley S, So J and Wu J 2004 *Journal of Mathematical Sciences* **124:4** 5119–53
[10] Chow S, Lin X and Mallet-Paret J 1989 *Journal of Dynamics and Differential Equations* **1:1** 3–43
[11] Rottschäfer V and Wayne C 2001 *Journal of Differential Equations* **176** 532–60
[12] Ashwin P, Bartuccelli M, Bridges T and Gourley S 2002 *Z. angew. Math. Phys.* **53** 103–22
[13] Danumjaya P and Pani A 2006 *International Journal of Numerical Analysis and Modeling* **3:2** 186–210
[14] Wang Z, Li W and Ruan S 2006 *J. Differential Equations* **222** 185–232
[15] Ai S 2007 *J. Differential Equations* **232** 104–33
[16] Wang H 2009 *J. Differential Equations* **247** 887–905
[17] Pan S 2014 *Electronic Journal of Differential Equations* **2014:65** 1–10
[18] Kakutani S., Markus L. 1958 *Ann. Math. Stud.* Princenton University Press Princenton **IV** 1–18
[19] Kuang Y 1993 *Delay Differential Equations. With Applications in Population Dynamics* (Academic Press)
[20] Wright E 1955 *J. Reine Angew. Math.* **194:1–4** 66–87
[21] Kashchenko S 1985 *Nelineynyye kolebaniya v zadachakh ekologii* Yaroslavl YarGU 55–62
[22] Jones G 1962 *T. Math. Anal. and Appl.* **5** 435–50
[23] Kolesov A 1985 *Matematicheskie modeli v biologii i medicine* **1** Vilnus: In-t. math. AN Lit. SSR 93–103
[24] Kashchenko S 1987 *DAN SSSR* **292:2** 327–30
[25] Kashchenko S 1990 *Matem. mod.* **2:9** 49–69
[26] Kashchenko S 2013 *Automatic Control and Computer Science* **47:7** 470–94
[27] Glyzin S 2009 *Modeling and Analysis of Information Systems* **16:3** 96–116
[28] Wu J 1996 *Theory and Applications of Partial Functional Differential Equations Theory and Applications of Partial Functional Differential Equations* (New York: Springer-Verlag)
[29] Kashchenko A 2011 *Modeling and Analysis of Information Systems* **18:3** 58–62
[30] Glyzin S 2013 *Automatic Control and Computer Sciences* **47:7** 452–69
[31] Glyzin S, Kolesov A and Rozov N 2010 *Computational Mathematics and Mathematical Physics* **50:5** 816–30

On Behavior of Stochastic Synchronization Models

Anatoly Manita

Lomonosov Moscow State University, Faculty of Mechanics and Mathematics, 119991, GSP-1, Moscow, Russia

Higher School of Economics (HSE) Moscow Institute of Electronics and Mathematics, Russia

E-mail: manita@mech.math.msu.su, amanita@hse.ru

Abstract. We consider N-component synchronization models defined in terms of stochastic particle systems with special interaction. For general (nonsymmetric) Markov models we discuss phenomenon of the long time stochastic synchronization. We study behavior of the system in different limit situations related to appropriate changes of variables and scalings. For $N = 2$ limit distributions are found explicitly.

1. Model

Synchronization models studied in the present paper are motivated by some computer science applications (parallel computations [1], wireless sensor networks [2, 3] etc.). These models surprisingly have a lot in common with interacting particle systems in physics. Recall the definition of N-component stochastic synchronization system. Let $x_j \in \mathbb{R}^d$ represent the state of a component j, $j = \overline{1,N}$. The dynamics of the system is a continuous time stochastic process $x(t) = (x_1(t), \ldots, x_N(t))$, $t \geq 0$, which evolution is composed of two parts called respectively a free dynamics and a synchronizing interaction. The interaction between components is happened only at some random epochs $0 = T_0 < T_1 < T_2 < \cdots$ and has a form of instantaneous synchronizing jumps $x(T_n) \mapsto x(T_n + 0)$ to be precised later on. The free dynamics means that between successive epochs of interaction the components $x_j(t)$ evolve independently. Namely there is a family of mutually independent stochastic processes $x_j^\circ(t)$, $j = \overline{1,N}$, such that

$$x_j(t) - x_j(T_{n-1} + 0) = x_j^\circ(t) - x_j^\circ(T_{n-1} + 0), \qquad T_{n-1} < t \leq T_n, \qquad n \geq 1, \quad j = \overline{1,N}.$$

Before going into further details, let us reformulate the model in terms of particle systems. Instead of components we will speak of N particles with coordinates x_j.

To introduce a pair-wise synchronizing interaction between particles it is convenient to assume that particles can share information about each other by sending and receiving messages. Imagine, for example, that at time T_q the particle $j_1^{(q)}$ sends a message to some another particle $j_2^{(q)}$. For breavity notation let us write $j_1 = j_1^{(q)}$ and $j_2 = j_2^{(q)}$. The message contains information on the current value of x_{j_1}. Assume that messages reach their destinations instantly. After receiving the message from j_1 the particle j_2 ajusts its coordinate to the value x_{j_1}: $x_{j_2}(T_q + 0) = x_{j_1}(T_q)$. This is the only jump in the system at the time T_q: $x_j(T_q + 0) = x_j(T_q)$

for all $j \neq j_2$. Define a (linear) map $M_{j_1 j_2} : (\mathbb{R}^d)^N \to (\mathbb{R}^d)^N$ such that x and $y = M_{j_1 j_2} x$ differ only in the j_2-component:

$$y_{j_2} = x_{j_1}, \quad y_k = x_k, \quad k \neq j_2, \qquad x = (x_1, \ldots, x_N), \quad y = (y_1, \ldots, y_N).$$

Thus the synchronizing jump can be expressed as $x(T_q + 0) = M_{j_1 j_2} x(T_q)$.

The key idea of this synchronization model is that the sequence $\mathcal{H} = \left\{ \left(T_n;\, j_1^{(n)}, j_2^{(n)} \right) \right\}_{n \geq 1}$ should be random. Consider point processes

$$\boldsymbol{T}^{j\to k} := \{T_n : (T_n; j, k) \in \mathcal{H}\}, \quad (j \neq k), \qquad \boldsymbol{T}^{j\to} := \bigcup\nolimits_k \boldsymbol{T}^{j\to k}, \qquad \boldsymbol{T} := \bigcup\nolimits_{j,k} \boldsymbol{T}^{j\to k}$$

where, for brevity, we have put $\boldsymbol{T}^{j\to j} = \varnothing$. Evidently, $\boldsymbol{T}$ is the sequence of synchronization epochs $\{T_n\}_{n\geq 1}$. For any interval $[t_1, t_2] \subset \mathbb{R}_+$ and any route (j_1, j_2) denote

$$\nu_{j_1 j_2}([t_1, t_2]) := \mathrm{card}\left(\{(T_n;\, j_1, j_2) : T_n \in [a, b]\}\right) = \mathrm{card}\left(\boldsymbol{T}^{j_1\to j_2} \cap [a, b]\right),$$

the random number of messages sent from j_1 to j_2 during the time interval $[a, b]$. Reasonable assumptions covering a wide field of applications are the following.

M1) Message flows $\boldsymbol{T}^{j\to}$ generated by different particles j are independent.

M2) The particle j sends messages to another particle k with the rate $\beta_{jk} \geq 0$, i.e., for any j and $t, h \geq 0$

$$\begin{aligned} \mathsf{P}\left(\nu_{jk}([t, t+h]) = 1\right) &= \beta_{jk} h + o(h), \qquad k \neq j, \\ \mathsf{P}\Big(\sum_{k:\, k\neq j} \nu_{jk}([t, t+h]) = 0\Big) &= 1 - \beta_j h + o(h) \quad \text{as } h \to 0 \qquad \text{where } \beta_j := \sum_{k:\, k\neq j} \beta_{jk}. \end{aligned}$$

(If $\beta_{jk} = 0$ then $\boldsymbol{T}^{j\to k} = \varnothing$, that is, there is no message flows on the route (j, k).)

M3) For any j the message flow $\boldsymbol{T}^{j\to}$ generated by the particle j is nonzero, i.e., $\sum_{k:\, k\neq j} \beta_{jk} > 0$.

With the set of parameters $\{\beta_{jk},\, k \neq j\}$ we associate a directed graph $\mathcal{G}$ with vertices $1, \ldots, N$ and directed arcs (j, k) such that $\beta_{jk} > 0$.

M4) The graph $\mathcal{G}$ is strongly connected, i.e., any pair of vertices can be connected by a directed path.

Assumptions M1–M2 implies that

$$\begin{aligned} \mathsf{P}\left(\mathrm{card}\,(\boldsymbol{T} \cap [a, b]) \geq 2\right) &= \mathsf{P}\Big(\sum_{j,k:\, k\neq j} \nu_{jk}([t, t+h]) \geq 2\Big) = o(h), \\ \mathsf{P}\left(\mathrm{card}\,(\boldsymbol{T} \cap [a, b]) = 0\right) &= 1 - \beta h + o(h) \qquad \text{where } \beta := \sum_j \beta_j = \sum_{j\neq k} \beta_{jk}. \end{aligned}$$

Put $X_0 = 0$. A point process $\boldsymbol{X} = \{X_1, X_2, \ldots, X_n, \ldots\} \subset (0, +\infty)$ is said to be a Poisson process of rate $\gamma > 0$ if $\{X_n - X_{n-1}\}_{n\geq 1}$ is a sequence of independent and identically distributed exponential random variables with the mean γ^{-1} that is $\mathsf{P}\left(X_n - X_{n-1} > s\right) = \exp(-\gamma s)$ for $s \geq 0$. Differences $X_n - X_{n-1}$ are called inter-event intervals. For further details the interested reader is referred to [4].

From Assumptions M1–M3 and the classic point processes theory it follows that

- any message flow $\boldsymbol{T}^{j_1\to j_2}$ is a Poisson process of rate $\beta_{j_1 j_2}$;
- the flows $\boldsymbol{T}^{j\to k}$ corresponding to different routes (j, k) are mutually independent;

Journal of Physics: Conference Series **681** (2016) 012024 doi:10.1088/1742-6596/681/1/012024

- the point processes $\boldsymbol{T}^{j\to}$ and $\boldsymbol{T} = \{T_n\}_{n\geq 1}$ are Poisson with rates β_j and β respectively.

So the mechanism of *synchronizing interaction* is very transparent: at epochs of the message flow $\boldsymbol{T}^{j_1\to j_2}$ we apply the map $M_{j_1 j_2}$.

Before specifying the free dynamics we fix some notation. Given a random vector $Z \in \mathbb{R}^d$, let P_Z and ψ_Z denote its probability distribution and characteristic function respectively, i.e.,

$$P_Z(A) := \mathsf{P}\,(Z \in A)\,, \quad A \subset \mathbb{R}^d, \qquad \psi_Z(a) := \int_{\mathbb{R}^d} e^{\mathrm{i}\,a\cdot q}\, P_Z(dq) = \mathsf{E}\,\exp\left(\mathrm{i}\,a\cdot Z\,\right), \quad a \in \mathbb{R}^d,$$

where E is the expectation with respect to P and $a \cdot q$ denotes the Euclidian scalar product of two vectors a and q in $\mathbb{R}^d$. It is readily seen that the function ψ_Z is the Fourrier transform of the probability measure P_Z.

Recall [5, 6] that an $\mathbb{R}^d$-valued stochastic process $z(t)$, $t \geq 0$, is called a Lévy process if 1) it has independent and stationary increments; 2) it is stochastically continuous; 3) $z(0) = 0$. A remarkable property of Lévy processes is a representation for $\psi_{z(t)}$,

$$\psi_{z(t)}(a) = \psi_{z(s+t)-z(s)}(a) = \exp\left(t\phi_z(a)\right), \qquad t \geq 0, \quad s > 0,$$

where ϕ_z has the Lévy-Khintchine form [5]. The function $\phi_z : \mathbb{R}^d \to \mathbb{C}$ is called a characteristic exponent of the Lévy process $z = z(t)$. It is easy to see that $\phi_z(a)$ is continuous, $\phi_z(0) = 0$ and $\mathrm{Re}\,\phi_z(a) \leq 0$. Our main assumption about the *free dynamics* is the following.

F) The free dynamics $x_j^\circ(t)$, $j = \overline{1,N}$, are independent Lévy processes. For further convenience, the characteristic exponent for the particle j will be denoted as $\phi_{x_j^\circ}(a) = -\eta_j(a)$, i.e.,

$$\mathsf{E}\,\exp\left(\mathrm{i}\,\lambda_j \cdot x_j^\circ(t)\,\right) = \exp\left(-t\,\eta_j(\lambda_j)\right), \qquad \lambda_j \in \mathbb{R}^d.$$

Many interesting examples of one-particle dynamics satisfy Assumption F. In particular, if $x_j^\circ(t)$ is a Brownian motion with diffusion matrix Σ and drift vector $v_j \in \mathbb{R}^d$ then $\eta_j(\lambda_j) = -\mathrm{i}\,v_j \cdot \lambda_j + \frac{1}{2}\,(\Sigma\Sigma^\top\lambda_j)\cdot\lambda_j$. Sometimes we will need stronger assumptions.

F0) Assumption F holds and the functions η_j are such that $\mathrm{Re}\,\eta_j(\lambda_j) > 0$ for all $\lambda_j \neq 0$.

FS$\boldsymbol{\alpha}$) Assumption F0 holds and each $x_j^\circ(t)$ is an α-stable Lévy process. In other words, the characteristic exponents $-\eta_j(\lambda_j)$, $j = \overline{1,N}$, satisfy the following condition: there exists $\alpha \in (0, 2]$ such that

$$\eta_j(s\lambda_j) = s^\alpha \eta_j(\lambda_j) \qquad \forall s > 0, \quad \lambda_j \in \mathbb{R}^d. \tag{1}$$

The number α is called the stability index. The classical examples of stable Lévy dynamics are Brownian motions with zero drift ($\eta_j(\lambda_j) = \frac{1}{2}\,(\Sigma\Sigma^\top\lambda_j)\cdot\lambda_j$, $\alpha = 2$). More information about stable laws can be found in [7,8].

The only assumption about *initial configuration* $x(0)$ is that it is independent of the free dynamics and all message flows $\boldsymbol{T}^{j\to k}$.

Under above assumptions $x(t)$ is a Markov process. Synchronization model is called symmetric if $\eta_j = \eta$ for all j and $\beta_{jk} = c/(N-1)$ for all $j \neq k$.

2. Stochastic synchronization

It is clear from definition of $x(t)$ that collective behavior of particle systems with synchronization is a superposition of two tendencies. Indeed, the processes $x_j^\circ(t)$, $j = \overline{1,N}$, are mutually independent and have independent increments. Thus, due to the free dynamics, the spread of the particle system has the tendency to increase with the course of time. An opposite tendency is produced by synchronizing jumps which try to decrease the spread of the particle configuration.

In order to state precise results on a long-time behavior of $x(t)$ it is convenient to use the concept of convergence in distribution. We don't touch here all nuances of this notion and refer

Journal of Physics: Conference Series **681** (2016) 012024 doi:10.1088/1742-6596/681/1/012024

interested readers to [9]. We need only the following useful statement which can be found in most probability textbooks: an $\mathbb{R}^d$-valued random sequence $\{Z_n\}_{n=1}^{\infty}$ converges in distribution to a random vectors Z iff $\psi_{Z_n}(a) \to \psi_Z(a)$ for all $a \in \mathbb{R}^d$.

The long time behavior of different *symmetric* synchronization models was studied in a series of papers [3, 10–14]. In particular, it is known that

a) components $x_j(t)$ have no limits in distribution as $t \to \infty$,

b) any difference $x_j(t) - x_k(t)$ converge in distribution to a nontrivial probability law,

c) an *improved process* $y(t) = (y_1(t), \ldots, y_N(t)) \in \left(\mathbb{R}^d\right)^N$ such that $y_i(t) = x_i(t) - M(t)$ and $M(t) := N^{-1} \sum_{m=1}^{N} x_m(t)$ *is ergodic.*

In particular, there exists a limit in distribution of $y(t)$ as $t \to \infty$. Evidently, the process $y(t)$ describes the particle system $x(t)$ viewed by an observer placed in the center of mass $M(t)$.

In the presence of b) we say that for the system $x(t)$ a *long time stochastic synchronization* takes place. Note that statement c) is stronger than statement b).

The goal of this and subsequent papers is to obtains results similar to a)–c) for *non-symmetric* synchronization models. We are interested in the joint distribution of the total system $x(t) = (x_1(t), \ldots x_N(t)) \in \left(\mathbb{R}^d\right)^N$ and, hence, in the joint characteristic function (c.f.)

$$\chi_{1,2,\ldots,N}(t;\, \lambda) = \mathsf{E} \exp\Big(\mathrm{i} \sum_j \lambda_j \cdot x_j(t)\Big), \qquad \lambda_j \in \mathbb{R}^d, \quad \lambda = (\lambda_1, \ldots, \lambda_N).$$

It contains all information about distribution of the system and, in particular, about its marginal distributions. For instance, $\psi_{x_1(t)}(\lambda_1) = \chi_{1,2,\ldots,N}(t;\, \lambda_1, 0, \ldots, 0)$ etc. But the function $\chi_{1,2,\ldots,N}$ appears to be very complicated. Due to the synchronizing interaction the particles $x_1(t), \ldots x_N(t)$ are dependent. Hence $\chi_{1,2,\ldots,N}(t;\, \lambda_1, \ldots, \lambda_N) \neq \psi_{x_1(t)}(\lambda_1) \cdots \psi_{x_N(t)}(\lambda_N)$ for any finite $t > 0$ and, as we will see later, this feature survives also in the asymptotic limit $t \to \infty$. Nevertheless, we will get a nice description of the long time behavior of $x(t)$ by making an appropriate change of variables. Consider $y(t)$ and $M(t)$ as defined in item *c*). Let $\psi_{y(t),M(t)}$ be their joint c.f.,

$$\psi_{y(t),M(t)}(u, \rho) = \mathsf{E} \exp\Big(\mathrm{i} \sum_j u_j \cdot y_j(t) + \mathrm{i}\, \rho \cdot M(t)\Big), \quad u_j, \rho \in \mathbb{R}^d, \quad u \in \left(\mathbb{R}^d\right)^N,$$

$\psi_{y(t)}(u) = \psi_{y(t),M(t)}(u, 0)$ and $\psi_{M(t)}(\rho) = \psi_{y(t),M(t)}(0, \rho)$ be c.f.s of $y(t)$ and $M(t)$. The results on their long time behavior are summarized in next two theorems.

Theorem 1. *Let Assumptions M1-M4 and F0 hold.*

i) None of the vectors $M(t)$, $x_j(t)$, $j = \overline{1, N}$, *has a limit in distribution as* $t \to \infty$.

ii) For any $u \in \left(\mathbb{R}^d\right)^N$ $\psi_{y(t)}(u) \to \psi_{y(\infty)}(u)$ *as* $t \to \infty$ *where* $\psi_{y(\infty)}(u)$ *is a c.f. of some probability law on* $\left(\mathbb{R}^d\right)^N$.

Item *ii* implies that $y(t)$ has a limit in distribution. Remark that $\sum_j y_j(t) \equiv 0$, hence all distributions $P_{y(t)}$ and $P_{y(\infty)}$ are supported on the linear subspace $\left\{y : \sum_j y_j = 0\right\}$.

Theorem 2. *Let Assumptions M1-M4 and FSα hold. Denote* $m(t) := M(t)/t^{1/\alpha}$. *Then*

$$\forall u, \rho \qquad \psi_{m(t)}(\rho) \to \psi_{m(\infty)}(\rho), \qquad \psi_{y(t),m(t)}(u, \rho) \to \psi_{y(\infty)}(u)\psi_{m(\infty)}(\rho) \qquad (t \to \infty)$$

where $\psi_{m(\infty)}(\rho)$ *is a c.f. of some probability law on* $\mathbb{R}^d$.

This theorem implies that the joint distribution of $(y(t), m(t)) = (y(t), M(t)/t^{1/\alpha})$ has a limit as $t \to \infty$ and, after passage to the limit, the components y and m become independent.

Below we give a detailed study of the case $N = 2$. For $N > 2$ complete proofs of Theorems 1 and 2 demand some additional constructions which might exceed the size of the present publication. So the case $N \geq 3$ will be considered in another paper.

Journal of Physics: Conference Series **681** (2016) 012024 doi:10.1088/1742-6596/681/1/012024

3. Complete treatment for $N = 2$

Fix $\lambda_1, \lambda_2 \in \mathbb{R}^d$ and denote $\varepsilon_{12} = \eta_1(\lambda_1) + \eta_2(\lambda_2)$, $\varepsilon_1 = \eta_1(\lambda_1 + \lambda_2)$, $\varepsilon_2 = \eta_2(\lambda_1 + \lambda_2)$,

$$f_{12}(t) = \chi_{1,2}(t;\, \lambda_1, \lambda_2), \qquad f_1(t) = \chi_{1,2}(t,;\, \lambda_1 + \lambda_2, 0), \qquad f_2(t) = \chi_{1,2}(t;\, 0, \lambda_1 + \lambda_2).$$

By using Markov property of $x(t)$ it is straightforward to check that the functions f_{12}, f_1, f_2 satisfies the following system of linear differential equations

$$\frac{d}{dt}\begin{pmatrix} f_{12}(t) \\ f_1(t) \\ f_2(t) \end{pmatrix} = \begin{pmatrix} -\varepsilon_{12} - \beta_{12} - \beta_{21} & \beta_{12} & \beta_{21} \\ 0 & -\varepsilon_1 - \beta_{21} & \beta_{21} \\ 0 & \beta_{12} & -\varepsilon_2 - \beta_{12} \end{pmatrix} \begin{pmatrix} f_{12}(t) \\ f_1(t) \\ f_2(t) \end{pmatrix} \tag{2}$$

or in short notation $\frac{d}{dt}\vec{f} = \mathcal{A}(\boldsymbol{\varepsilon})\,\vec{f}$, $\boldsymbol{\varepsilon} := (\varepsilon_{12}, \varepsilon_1, \varepsilon_2)$. Hence $\vec{f}(t) = \exp(t\mathcal{A}(\boldsymbol{\varepsilon}))\,\vec{f}(0)$ is the solution. To study behavior of $\vec{f}(t)$ as $t \to \infty$ let us find an eigensystem of $\mathcal{A}(\boldsymbol{\varepsilon})$: $\mu\vec{v} = \mathcal{A}(\boldsymbol{\varepsilon})\,\vec{v}$. The simplest pair is $\mu_1(\boldsymbol{\varepsilon}) = -\varepsilon_{12} - \beta_{12} - \beta_{21}$ and $\vec{v}_1(\boldsymbol{\varepsilon}) = (1,0,0)^\top$. The next two eigenvalues can be found in an explicit form containing radical fuctions. After some algebra we get their expansions as $\boldsymbol{\varepsilon} \to 0$, namely,

$$\mu_2(\boldsymbol{\varepsilon}) = -\frac{\beta_{12}\varepsilon_1 + \beta_{21}\varepsilon_2}{\beta_{12} + \beta_{21}} + o(\|\boldsymbol{\varepsilon}\|), \qquad \mu_3(\boldsymbol{\varepsilon}) = -\beta_{12} - \beta_{21} + o(1). \tag{3}$$

The corresponding eigenvectors $\vec{v}_2(\boldsymbol{\varepsilon})$ and $\vec{v}_3(\boldsymbol{\varepsilon})$ approaches as $\boldsymbol{\varepsilon} \to 0$ to the column vectors $(1,1,1)^\top$ and $(0, -\beta_{21}, \beta_{12})^\top$ respectively. It is important to remark that $\mu_2(\boldsymbol{\varepsilon})$ and $\mu_3(\boldsymbol{\varepsilon})$ do not depend on ε_{12}. Note that $\boldsymbol{\varepsilon} \to 0$ as $(\lambda_1, \lambda_2) \to 0$. Denote $S = \big\{(\lambda_1, \lambda_2) \in \mathbb{R}^{2d} : \lambda_1 + \lambda_2 \neq 0\big\}$. If the pair $(\lambda_1, \lambda_2) \in S$ is sufficiently small then all eigenvalues have strictly negative real parts. This follows from Assumption F0. Thus $\vec{f}(t) \to 0$ as $t \to +\infty$ for small $(\lambda_1, \lambda_2) \in S$. Since $\chi_{1,2}(t;\, 0,0) \equiv 1$ the limit of $\chi_{1,2}(t;\, \lambda_1, \lambda_2)$ is discontinuous at $(\lambda_1, \lambda_2) = 0$. By the Lévy continuity theorem the stochastic process $(x_1(t), x_2(t))$ has no limit in distribution as $t \to \infty$. Now item i of Theorem 1 follows.

Note that if $N = 2$ then $y = (y_1, y_2) = \frac{1}{2}(x_1 - x_2, x_2 - x_1)$. Therefore it is enough to consider c.f. of $r(t) := x_1(t) - x_2(t)$, i.e., $\psi_{r(t)}(\lambda) = \mathsf{E}\, e^{\mathrm{i}\lambda\cdot r(t)} = \chi_{1,2}(t;\, \lambda, -\lambda)$. We can use system of equations (2) with substitution $(\lambda_1, \lambda_2) = (\lambda, -\lambda)$. Since now $f_1(t) = f_2(t) = 1$ for all t we have to solve only the equation for $\chi_{1,2}$:

$$\frac{d}{dt}\psi_{r(t)}(\lambda) = (-\eta_1(\lambda) - \eta_2(-\lambda) - \beta_{12} - \beta_{21})\,\psi_{r(t)}(\lambda) + \beta_{21} + \beta_{12}.$$

Solving it and letting $t \to +\infty$ we get the proof of item ii of Theorem 1

$$\psi_{r(t)}(\lambda) \to \psi_{r(\infty)}(\lambda) = \frac{\beta_{12} + \beta_{21}}{\beta_{12} + \beta_{21} + \eta_1(\lambda) + \eta_2(-\lambda)}. \tag{4}$$

Consider now a joint distribution of $(r(t), m(t)) = \left(x_1(t) - x_2(t),\, \dfrac{x_1(t) + x_2(t)}{2t^{1/\alpha}}\right)$ where α is the stability index from Assumption **FS$\boldsymbol{\alpha}$)**. The joint c.f. is

$$\psi_{r(t),m(t)}(\lambda, \rho) = \mathsf{E}\, e^{\mathrm{i}\lambda\cdot r(t) + \mathrm{i}\rho\cdot m(t)} = \chi_{1,2}\Big(t;\, \lambda + \frac{\rho}{2t^{1/\alpha}}, -\lambda + \frac{\rho}{2t^{1/\alpha}}\Big), \qquad \lambda, \rho \in \mathbb{R}^d.$$

Using eigenvectors of $\mathcal{A}(\boldsymbol{\varepsilon})$ we decompose $\vec{f}(0) = \sum_{k=1}^{3} c_{f,k}(\boldsymbol{\varepsilon})\vec{v}_k(\boldsymbol{\varepsilon})$ and rewrite solution of (2) as

$$\vec{f}(t) = \sum_{k=1}^{3} e^{t\mu_k(\boldsymbol{\varepsilon})} c_{f,k}(\boldsymbol{\varepsilon})\vec{v}_k(\boldsymbol{\varepsilon}). \tag{5}$$

Since this representation holds for any fixed ε and t to obtain $\psi_{r(t),m(t)}(\lambda,\rho)$ we substitute

$$\varepsilon=\varepsilon(t)=\Big(\eta_1\Big(\lambda+\frac{\rho}{2t^{1/\alpha}}\Big)+\eta_2\Big(-\lambda+\frac{\rho}{2t^{1/\alpha}}\Big),\eta_1\Big(\frac{\rho}{t^{1/\alpha}}\Big),\eta_2\Big(\frac{\rho}{t^{1/\alpha}}\Big)\Big)$$

into (5). Note that $\varepsilon(t)\to\varepsilon(\infty):=(\eta_1(\lambda)+\eta_2(-\lambda),0,0)$ as $t\to\infty$. It follows from the above analysis that the first and the third terms in (5) vanish as $t\to\infty$. The second term is a product of two multipliers. The first multiplier, $e^{t\mu_2(\varepsilon)}|_{\varepsilon=\varepsilon(t)}$ depends on ρ and does not depend on λ. The second one, $c_{f,2}(\varepsilon)\vec{v}_2(\varepsilon)|_{\varepsilon=\varepsilon(t)}$, depends on the both variables ρ and λ. By continuity of $c_{f,2}(\varepsilon)$ and $\vec{v}_2(\varepsilon)$ we obtain that $c_{f,2}(\varepsilon)\vec{v}_2(\varepsilon)|_{\varepsilon=\varepsilon(t)}\to c_{f,2}(\varepsilon)\vec{v}_2(\varepsilon)|_{\varepsilon=\varepsilon(\infty)}$ as $t\to\infty$. Recalling definition of $\varepsilon(\infty)$ we conclude that $c_{f,2}(\varepsilon)\vec{v}_2(\varepsilon)|_{\varepsilon=\varepsilon(\infty)}$ depends only on λ and its value is the same as the value of $\psi_{r(\infty)}(\lambda)$ calculated already in (4). Let us evaluate the limit of $t\mu_2(\varepsilon(t))$ as $t\to\infty$. By the stability property (1) we have $\eta_j\left(\rho t^{-1/\alpha}\right)=t^{-1}\eta_j(\rho)$. Using (3) we get

$$t\mu_2(\varepsilon(t))=-\frac{\beta_{12}\eta_1(\rho)+\beta_{21}\eta_2(\rho)}{\beta_{12}+\beta_{21}}+o(1),\qquad t\to\infty.$$

So we have proved that

$$\psi_{r(t),m(t)}(\lambda,\rho)\longrightarrow\frac{\beta_{12}+\beta_{21}}{\beta_{12}+\beta_{21}+\eta_1(\lambda)+\eta_2(-\lambda)}\exp\Big(-\frac{\beta_{12}\eta_1(\rho)+\beta_{21}\eta_2(\rho)}{\beta_{12}+\beta_{21}}\Big).$$

This is even more than it was stated in Theorem 2. Indeed, putting $(\lambda,\rho)=(0,\rho)$ we obtain not only convergence of $\psi_{m(t)}(\rho)$ to some limit $\psi_{m(\infty)}(\rho)$ but also an explicit formula for the limiting c.f., namely,

$$\psi_{m(\infty)}(\rho)=\exp\Big(-\frac{\beta_{12}\eta_1(\rho)+\beta_{21}\eta_2(\rho)}{\beta_{12}+\beta_{21}}\Big).$$

It is readily seen that distribution of $m(\infty)$ is α-stable.

Distributions similar to (4) were previously obtained in context of synchronization models in [13, 14]. They are related to the class of Linnik distributions.

References

[1] Bertsekas D P and Tsitsiklis J N 1987 *Parallel and Distributed Computation: Numerical Methods* (Belmont: Athena Scientific)

[2] Sundararaman B, Buy U, Kshemkalyani A D 2005 Clock synchronization for wireless sensor networks: a survey *Ad Hoc Networks* **3** pp 281–323

[3] Manita A 2014 Clock synchronization in symmetric stochastic networks *Queueing Systems* V **76** 2 pp 149–180

[4] Kingman J F C 1993 *Poisson Processes* (Oxford: Clarendon Press)

[5] Applebaum D 2004 *Lévy Processes and Stochastic Calculus* (Cambridge: Cambridge University Press)

[6] 1995 *Flights and Related Topics in Physics* Lecture Notes in Physics **450** ed M F Shlesinger, G M Zaslavsky and U Frisch

[7] Uchaikin V V and Zolotarev V M 1999 *Chance and Stability. Stable Distributions and their Applications* (Utrecht: VSP)

[8] Samorodnitsky G and Taqqu S 1994 *Stable Non-Gaussian Random Processes: Stochastic Models with Infinite Variance* (Taylor & Francis)

[9] Billingsley P 1999 *Convergence of Probability Measures* (Wiley)

[10] Malyshev V and Manita A 2006 Phase Transitions in the Time Synchronization Model *Theory of Probability and its Applications* **50** 1 pp 134–141

[11] Manita A 2009 Stochastic Synchronization in a Large System of Identical Particles *Theory of Probability and its Applications* **53** 1 pp 155–171

[12] Manita A 2011 Brownian Particles Interacting via Synchronizations *Communications in Statistics - Theory and Methods* **40** 19–20 pp 3440–3451

[13] Manita A 2014 Intrinsic space scales for multidimensional stochastic synchronization models *New Perspectives on Stochastic Modeling and Data Analysis* ed J R Bozeman, V Girardin *et al* (Athens: ISAST) pp 271–282

[14] Manita A 2014 Intrinsic scales for high-dimensional Lévy-driven models with non-Markovian synchronizing updates *Preprint* ArXiv 1409.2919 pp 1–50

Resonance regions of extended Mathieu equation

V P Semyonov[1,2], **A V Timofeev**[2,1]

[1] Moscow Institute of Electronics and Mathematics, National Research University Higher School of Economics, Talinskaya 34, Moscow 123458, Russia
[2] Joint Institute for High Temperatures of the Russian Academy of Sciences, Izhorskaya 13 Bldg 2, Moscow 125412, Russia

VPSemenov@gmail.com

Abstract. One of the mechanisms of energy transfer between degrees of freedom of dusty plasma system is based on parametric resonance. Initial stage of this process can de described by equation similar to Mathieu equation. Such equation is studied by analytical and numerical approach. The numerical solution of the extended Mathieu equation is obtained for a wide range of parameter values. Boundaries of resonance regions, growth rates of amplitudes and times of onset are obtained. The energy transfer between the degrees of freedom of dusty plasma system can occur over a wide range of frequencies.

1. Introduction

Dusty plasma is an ionized gas containing micrometer-sized charged dust particles. In a gas-discharge plasma such particles obtain high negative charge and can have abnormally high temperatures, even higher then electron and ion temperatures [1,2]. Mean kinetic energy of dust particles vertical and horizontal motion may be significantly different [3,4]. Phenomenon of energy transfer between different degrees of freedom of a dusty plasma system, which causes such heating, are of great interest. One of such mechanisms is based on parametric resonance and can be described by an extended Mathieu equation [3,4]:

$$\ddot{x} + 2\lambda\dot{x} + \omega_0^2\left(1 + h\cos\omega_p t\right)x = \eta(t), \tag{1}$$

where λ is a friction coefficient, h is an amplitude of modulation, ω_0 is an eigenfrequency of system, ω_p is a frequency of parameter and $\eta(t)$ is a stochastic force with zero mean value.

Mathieu equation [5] is used to describe dust particles motion in various articles [6,7]. In [6,7] the equation is derived by assuming, that the dust particle charge fluctuates harmonically. In this article extended Mathieu equation is derived by considering features of the near-electrode sheath in gas-discharge plasma and random fluctuations of the dust particle charge [3,4]. The parameters of equation (1) in such case have values significantly different from the ones considered in [6,7].

Classical Mathieu equation is studied for $h \ll 1$, $\lambda \ll 1$ and $\left|\omega_p - 2\omega_0/n\right| \ll \omega_0$, approximations of third and lower orders of accuracy is obtained in [8]. This range of parameters is too small for dusty plasma, since there can be significant friction and values of h may be above 3.

Connection of energy transfer between different degrees of freedom in dusty plasma and extended Mathieu equation is discussed in the second section. Analytical solution of the extended Mathieu equation is derived in the third section. Method of numerical solution is presented in the fourth section and the results of numerical simulation are presented and discussed in the fifth section.

2. Energy transfer between different degrees of freedom

The motion of dust particles is described by model presented in [3]. Since x and y axes are symmetrical in such system, only fluctuations along x and z axes are considered:

$$\begin{cases} m\ddot{x} = F^{inter} + F^{trap} + F^{fr}, \\ m\ddot{z} = F^{inter} + F^{fr} + F^{grav} + F^{el}, \end{cases} \tag{2}$$

where F^{inter} is the force of interaction between dust particles defined by Yukawa potential, F^{fr} is the friction force, F^{grav} is the gravity force, F^{el} is electrostatic force, which balances gravity force, and F^{trap} is the force of potential trap, which prevents negatively charged particles from scattering horizontally.

The amplitude of oscillations is small, comparing to inter-particle distance before the resonance starts, so the Tailor series expansion can be used:

$$\begin{cases} \ddot{x} = -a_1 x + a_2 xz + a_3 x^3 + a_4 xz^2 + \ldots - 2\lambda\dot{x}, \\ \ddot{z} = -b_1 z + b_2 z^2 + b_3 z^3 + b_4 zx^2 + \ldots - 2\lambda\dot{z} + g\delta q(t). \end{cases} \tag{3}$$

First and fourth terms is much larger than the others in the conditions of standard laboratory dusty plasma experiment, so:

$$\ddot{x} \approx -\left(a_1 x - a_4 xz^2\right) - 2\lambda\dot{x}. \tag{4}$$

Substitution of one harmonic of vertical motion $z \approx A_z \cos(\omega_z t)$ in equation (4) leads to extended Mathieu equation (1), where $\omega_0^2 = a_1 - a_4 A_z^2/2 + ..$, $\omega_p = 2\omega_z$ and $h = -a_4 A_z^2/\omega_0^2$.

Since resonance leads to growth of the amplitude, impact of other terms in (3) becomes greater. So equation (1) can't be used to describe late stages of the energy transfer between degrees of freedom but it can be used to describe the first stages of energy transfer.

3. Analytical solution

Approximations of boundaries of the resonance regions can be derived analytically acting by analogy with [8] and using averaging over an ensemble of distributions of $\eta(t)$. This approach works only for $h \ll 1$, $|\varepsilon| = |\omega_p - 2\omega_0/n| \ll \omega_0$, where n is number of the resonance region.

The first resonance region boundaries for $\lambda = 0$ are obtained with fourth-order accuracy by a method similar to one in [8].

$$-\omega_0\varepsilon \pm \frac{\omega_0^2 h}{2} = 0, \tag{9}$$

$$-\omega_0\varepsilon - \frac{\varepsilon^2}{4} \pm \frac{\omega_0^2 h}{2} + \frac{\omega_0^2 h^2}{32} = 0, \tag{10}$$

$$-\omega_0\varepsilon - \frac{\varepsilon^2}{4} \pm \frac{\omega_0^2 h}{2} + \frac{\omega_0^2 h^2}{32} - \frac{9\omega_0 h^2 \varepsilon}{2^8} = 0, \tag{11}$$

$$-\omega_0\varepsilon - \frac{\varepsilon^2}{4} \pm \frac{\omega_0^2 h}{2} + \frac{\omega_0^2 h^2}{32} - \frac{9\omega_0 h^2 \varepsilon}{2^8} + \frac{h^4}{3\cdot 2^{15}} + \frac{9\varepsilon^2 h^2}{2^{13}} = 0. \tag{12}$$

And next three equations are approximations of second, fourth and sixth orders of accuracy for second resonance region derived for the same conditions.

$$\begin{cases} -2\omega_0\varepsilon + \dfrac{\omega_0^2 h^2}{12} = 0, \\ -2\omega_0\varepsilon - \dfrac{5\omega_0^2 h^2}{12} = 0; \end{cases} \tag{13}$$

$$\begin{cases} -2\omega_0\varepsilon + \dfrac{\omega_0^2 h^2}{12} - \varepsilon^2 - \dfrac{2\omega_0\varepsilon h^2}{9} + \dfrac{\omega_0^2 h^4}{9\cdot 2^7} = 0, \\ -2\omega_0\varepsilon - \dfrac{5\omega_0^2 h^2}{12} - \varepsilon^2 - \dfrac{2\omega_0\varepsilon h^2}{9} + \dfrac{\omega_0^2 h^4}{9\cdot 2^7} = 0; \end{cases} \tag{14}$$

$$\begin{cases} -2\omega_0\varepsilon + \dfrac{\omega_0^2 h^2}{12} - \varepsilon^2 - \dfrac{2\omega_0\varepsilon h^2}{9} + \dfrac{\omega_0^2 h^4}{9\cdot 2^7} - \dfrac{\varepsilon^2 h^2}{9} - \dfrac{\omega_0\varepsilon h^4}{2^{10}} + \dfrac{h^6}{3\cdot 5\cdot 2^{12}} = 0, \\ -2\omega_0\varepsilon - \dfrac{5\omega_0^2 h^2}{12} - \varepsilon^2 - \dfrac{2\omega_0\varepsilon h^2}{9} + \dfrac{\omega_0^2 h^4}{9\cdot 2^7} - \dfrac{\varepsilon^2 h^2}{9} - \dfrac{\omega_0\varepsilon h^4}{2^{10}} + \dfrac{h^6}{3\cdot 5\cdot 2^{12}} = 0. \end{cases} \tag{15}$$

Approximations of third and fourth orders of accuracy of boundaries of the first resonance region are almost the same even for $h > 1$. Equations (14) and (15) are almost the same as well. These approximations are closer to data obtained numerically then approximations of lower orders of accuracy. However none of this approximations are close to data obtained numerically for $h > 1$.

It is usually assumed that the friction causes the amplitude to decay as $e^{-\lambda t}$ [8]. Such assumption leads to

$$\left(\frac{h\omega_0}{2}\right)^2 - \varepsilon^2 - 4\lambda^2 = 0. \tag{16}$$

This approximation is in serious disagreement with numerical solution. This is due to the fact that this approach takes into account only terms of zero-order of accuracy with λ. So this approach is applicable only for $\lambda \ll 1$ and for solution of the first order of accuracy.

The other way to describe resonance region in the presence of friction is solving equation (1) by an analogy with the case of zero friction without any additional assumptions. Such approach gives results closer to data obtained numerically. It also explains the shift of the bottom point of resonance region in terms of amplitude of modulation h. For first resonance region in approximation of the first order of accuracy this approach gives

$$\frac{\omega_0^4 h^2}{4} - \omega_0^2\varepsilon^2 - 4\lambda^2\left(\omega_0 + \frac{\varepsilon}{2}\right)^2 = 0 \tag{17}$$

instead of (16).

In equation (16) $h^{min} = 4\lambda/\omega_0$ when $\varepsilon = 0$, in equation (17) h is minimum for

$$\varepsilon_{fr} = -\frac{2\omega_0\lambda^2}{\omega_0^2 + \lambda^2} \tag{18}$$

and equals

$$h_{fr}^{min} = \frac{2\left(\varepsilon^2\lambda^2 + 4\varepsilon\lambda^2\omega_0 + \varepsilon^2\omega_0^2 + 4\lambda^2\omega_0^2\right)^{1/2}}{\omega_0^2}. \tag{19}$$

Comparison of the boundaries specified by equations (16) and (17) and data obtained numerically is presented on figure 1.

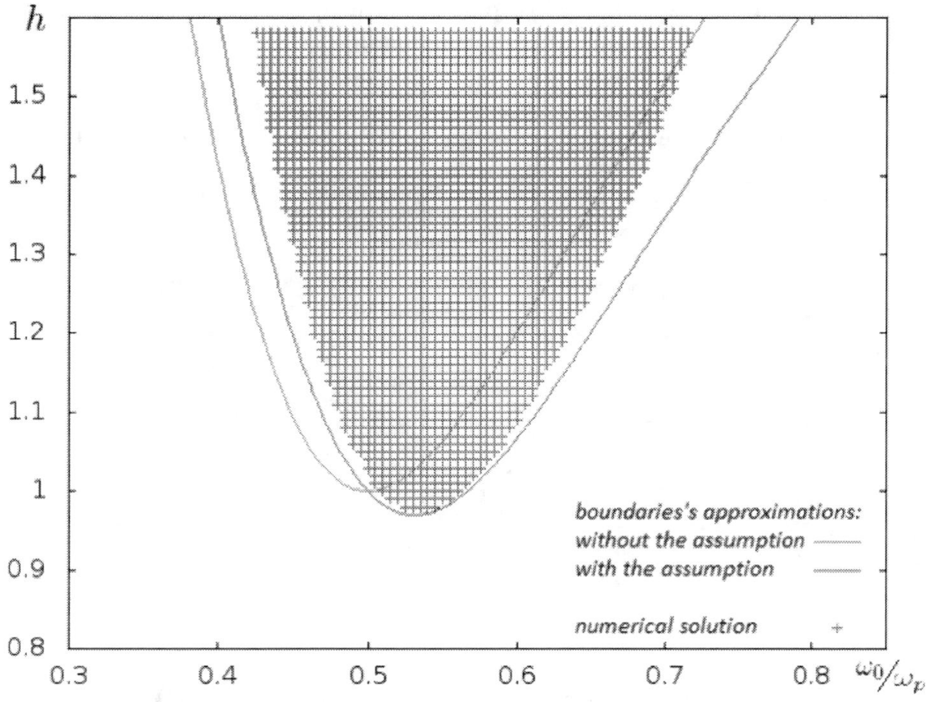

Figure 1. Boundaries of the resonance region obtained analytically with the assumption of exponential decay (blue line) and without this assumption (green line) and the resonance region obtained numerically (red crosses).

4. Numerical method

In case of numerical detection of resonance regions, growth rates of the amplitude and the time of onset (to be explained below) equation (1) is solved by numerical integration for each fixed set of parameters. The velocity Verlet algorithm is used. Initial values are $x(0)=0$ and $\dot{x}(0)=1$ and aren't affecting the results. Time step is $\Delta t = 10^{-5}$ sec.

Criterion of the resonance occurrence is exponential growth of the amplitude. Moment of the amplitude value from which the growth starts is considered to be the time of onset. Thus the time of onset is obtained with an accuracy of quarter period.

5. Results of the numerical simulation

Resonance regions for equations (1) is found for $h<4$ and $\omega_0/\omega_p<4$. Resonance regions for fixed $\omega_0=50$ and $\lambda=10$ are shown on figure 2 and ones for fixed $\omega_p=70$ and $\lambda=10$ is shown on figure 3.

Figure 2 shows that resonance regions are very close to each other for $h>2$. So resonance occurs at almost any value of $\omega_0/\omega_p>0.35$. Since this result is obtained for fixed ω_0 (corresponding to horizontal oscillations), it shows that fixed harmonic of horizontal oscillations is heated by a wide spectrum of vertical oscillations, not only those which is close to $2\omega_0/n$. Figure 3 shows the same result for fixed ω_p. So each harmonic of vertical oscillations warms up a wide spectrum of horizontal oscillations with frequencies above $0.35\omega_p$.

A wide spectrum of dust particles oscillations is also observed in a standard laboratory dusty plasma experiment. The possibility of participation of a wide range of oscillation frequencies in the resonant energy transfer between the degrees of freedom, allows to explain a series of experiments [9] and makes this result very important for the description of the energy transfer in dusty plasmas.

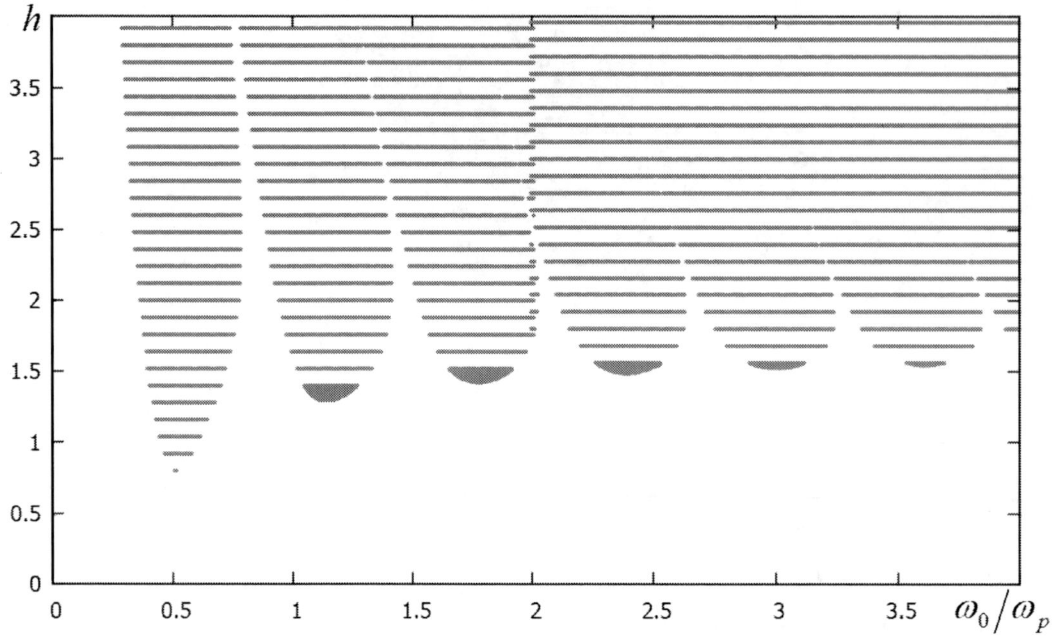

Figure 2. Resonance regions obtained numerically for $\omega_0 = 50$ and $\lambda = 10$.

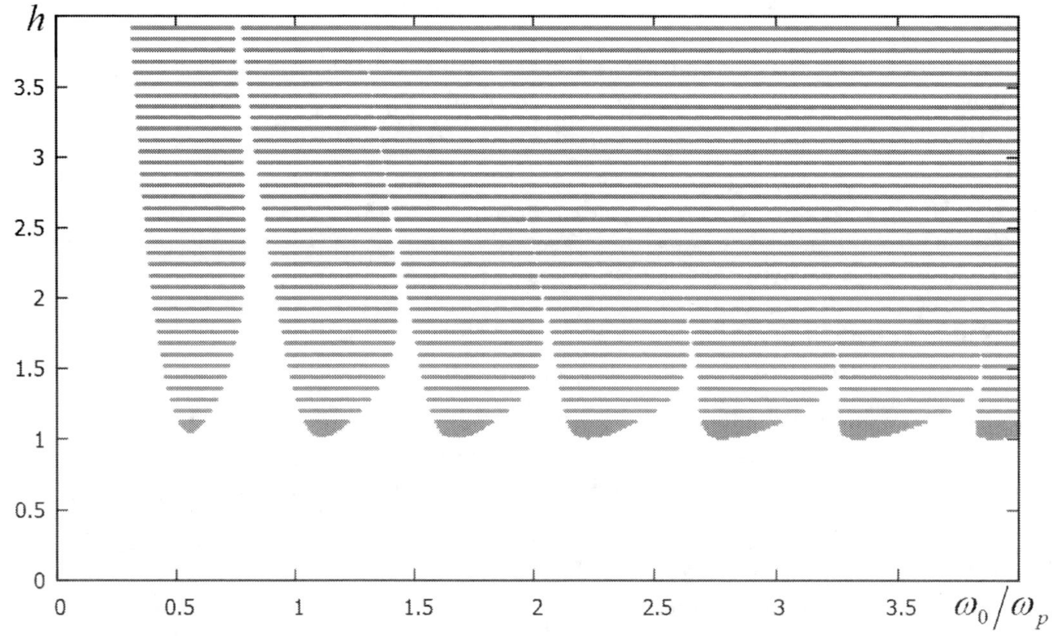

Figure 3. Resonance regions obtained numerically for $\omega_p = 70$ и $\lambda = 10$.

Stochastic force $\eta(t)$ isn't affecting previous results, however it is affecting the time of onset. The time of onset is stochastic and has normal distribution (figure 4) because of the stochastic force.

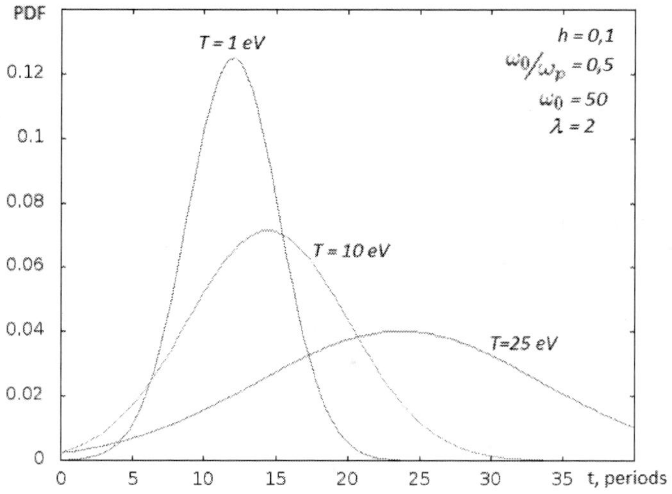

Figure 4. Probability density function of the time of onset obtained numerically for various mean kinetic energies T of the dust particle.

The more amplitude of stochastic force (corresponding to mean value of kinetic energy of dust particle) the more mean value and dispersion of the time of onset. However even for large values of the kinetic energy (25eV) the time of onset hardly can be over 40 periods of oscillations.

6. Conclusions

Connection between energy transfer between degrees of freedom in dusty plasma and extended Mathieu equation is shown. Approximations of boundaries of the resonance regions of various orders of accuracy are derived analytically. The shift of the point in which resonance occurs with a minimum value of h is shown by simulation and described analytically. Resonance regions, growth rates of the amplitude and the times of onset are obtained using numerical simulation. It is shown that the resonance regions obtained this way are close to each other for $h > 2$. This shows that resonance can occur at almost any values of $\omega_0/\omega_p > 0.35$. And since spectral peaks of vertical motion in dusty plasma are wide, each harmonic of vertical oscillations are exciting a wide spectrum of horizontal oscillations and single harmonic of horizontal oscillations are excited by a wide spectrum of vertical oscillations. Stochastic force shifts the timing of the resonance onset upward and increases the width of the onset time distribution.

Acknowledgments
This work is supported by the Russian Science Foundation (project № 14-19-01295).

References

[1] Vaulina O S, Petrov O F, Fortov V E, Khrapak S A and Khrapak A G 2009 *Dusty Plasmas: Experiment and Theory* (Moscow: FIZMATLIT)
[2] Tsytovich V N, Morfill G E and Thomas H M 2008 *Elementary Physics of Complex Plasmas, Lecture Notes in Physics* Vol. 731 (Springer, Berlin-Heidelberg)
[3] Norman G E and Timofeev A V 2011 *Phys. Rev.* E 84 056401
[4] Norman G E, Stegailov V V and Timofeev A V 2011 *JETP* 113 887–900
[5] Choudhury A G and Guha P 2014 *Appl. Math. Comput.* 229 85–93
[6] Momeni M, Kourakis I, Moslehi-Fard M and Shukla P K 2007 *J. Phys. A: Math. Theor.* 40 F473–F481
[7] Bora M P and Sarmah D 2013 *Pramana J. Phys.* 81 677–690
[8] Landau L D and Lifshitz E M 1959 *Theoretical Physics. Mechanics* (London: Pergamon)
[9] Ichiki R, Ivanov Y, Wolter M, Kawai Y and Melzer A 2004 *Phys. Rev.* E 70 066404.

Journal of Physics: Conference Series **681** (2016) 012026 doi:10.1088/1742-6596/681/1/012026

Computational problems in Arctic Research

I Petrov
Professor, Head of Computer Science and Computational Mathematics Department at MIPT,

E-mail: petrov@mipt.ru

Abstract. This article is to inform about main problems in the area of Arctic shelf seismic prospecting and exploitation of the Northern Sea Route: simulation of the interaction of different ice formations (icebergs, hummocks, and drifting ice floes) with fixed ice-resistant platforms; simulation of the interaction of icebreakers and ice-class vessels with ice formations; modeling of the impact of the ice formations on the underground pipelines; neutralization of damage for fixed and mobile offshore industrial structures from ice formations; calculation of the strength of the ground pipelines; transportation of hydrocarbons by pipeline; the problem of migration of large ice formations; modeling of the formation of ice hummocks on ice-resistant stationary platform; calculation the stability of fixed platforms; calculation dynamic processes in the water and air of the Arctic with the processing of data and its use to predict the dynamics of ice conditions; simulation of the formation of large icebergs, hummocks, large ice platforms; calculation of ridging in the dynamics of sea ice; direct and inverse problems of seismic prospecting in the Arctic; direct and inverse problems of electromagnetic prospecting of the Arctic. All these problems could be solved by up-to-date numerical methods, for example, using grid-characteristic method.

1. Introduction

Many natural impacts on industrial structures in the Arctic regions can be numerically simulated with a high degree of reliability by the models of continuum media, and by the advanced numerical methods for solving the corresponding systems of partial differential equations, and using correct problem definitions and high-performance multiprocessor computing algorithms.

Since the full-scale experiments in the study of Arctic processes are expensive and sometimes difficult to make, the appropriate computer modeling is the only realistic approach.

Attention to the development of the Arctic shelf of the Russian Federation explained the real need for a major exploration and development of oil-fields and gas-fields, as well as the real need for international use of the Northern Sea Route. On the Arctic shelf of Russian Federation there are eight fields discovered in 1983-1992 with estimates of reserves about 2.7 trillions m^3. Five of these eight fields are the largest objects related to the objects of federal significance: Ledovoe, Ludlovskoe, Murmanskoe in the Barents Sea, Pomorskoe, Gulyaevskoe in the Pechora Sea, Leningradskoe, Rusanovskoe in the Kara Sea. For oil production in the Pechora Sea the ice-resistant platform called Prirazlomnaja was installed. For gas production the project of the Shtokman field in the Barents Sea is designed. For these fields the improvement of made earlier estimates of hydrocarbon reserves volumes is required. A significant complicating factor in the production of hydrocarbons in the North Seas is the existence of ice formations. In the Kara Sea almost during all year there are drifting ice. In the Barents and in the Pechora Sea there are icebergs and ice hummocks. The depths of these North Seas in the industrial zones are reaching up to 300 m.

For the hydrocarbon transportation the pipelines are made in the northern regions of Russian Federation including the bottom zones. Obviously that for the long-term operation of industrial gas hydrotechnical facilities in difficult ice conditions it is necessary to calculate the impacts of different types of ice formations on these objects (platforms, pipelines, ships, icebreakers, etc.).

The creation of the Northern Sea Route for transporting cargo from Europe to America and to the Far East reduces the distance on 30-60 percent and reduces the transportation time on 10 days during comparison to traditional maritime routes (via the Panama and Suez canals). Now days the main cargo flow on this route is associated with delivery of nickel from the port Dudinka and its volume is equal to approximately 1.2 millions tons per year.

The main problems of development of the Northern Sea Route with the extent 2200-2900 nautical miles from the Novaya Zemlya to the Bering Strait are associated with difficult ice conditions and the use of powerful icebreakers that can move into the ice cover with thickness up to 2 m (now the track has 7 icebreakers and diesel-electric icebreakers and ice-class vessels).

Ice hummocks, icebergs, ice is often a serious impediment to these ships. Large ice formations are also a danger to the stationary ice-resistant platforms and sea bed pipelines [1-3].

The movement of ice masses occurs under the action of winds and currents. The ice cover is characterized by the presence of icebergs, ice hummocks, drifting ice, cracks in the ice, spring floods [4-6]. The presence of ice hummocks materially affects on the ice surface roughness and leads to the increase of the friction forces from wind and currents. The average distance between sails of ice hummocks in various regions of the Arctic is about 200-300 meters, the height of the hummock sail can reach several meters, the deep can reach a few tens of meters. The problems of ice ridging is discussed in papers [7-11]. The characteristic dimensions of icebergs are more than several times bigger than the characteristic dimensions of the ice hummocks.

Most time of the year, for example, the Pechora Sea and the Kara Sea are covered with drifting ices, the speed of which ices may exceed 5 m/s, the thickness of plane ice is up to 2 m, the thickness of draft ice hummocks is 20 m. Thus, the structure and parameters of the ice cover of the Northern seas are significant parameters determining the extreme loads at fixed and floating offshore oil and gas industrial structures.

In this regard, the importance of solving the problem of the modeling of dynamic processes in the air, water and soil of the Arctic with the processing of observation results, and prediction of the ice conditions, and the estimation of the further stability of stationary platforms, sea bed pipelines, the further security of icebreakers and ice-class vessels is clear [12,13].

The hydrocarbon exploration in the Arctic area has its own specificity. In particular, one of the layers, through which the seismic signals propagate, is the sea [14-17], the other layer is ice. The icebergs, ice hummocks, drifting ice, ice cover have also contributed to the measured or calculated responses in the seismic prospecting. During carrying out exploration work on the land one must take into account the effect from the permafrost. In addition to seismic technology the electrical exploration of hydrocarbons is an effective approach. The review of studies on this topic is given in [18,19].

We can highlight some of the major classes of problems at the Arctic area, which can be solved by numerical simulations on high-performance computing systems:

- Simulation of the interaction of different ice formations (icebergs, ice hummocks, and drifting ice floes) with fixed ice-resistant platforms;
- Simulation of the interaction of icebreakers and ice-class vessels with ice formations;
- Modeling of the impact of the ice formations on the ground pipelines;
- Neutralization of damage for fixed and mobile offshore industrial structures from ice formations;
- Calculation of the strength of the ground pipelines;
- Transportation of hydrocarbons by pipeline;

- The problem of migration of large ice formations;
- Modeling of the formation of ice hummocks on ice-resistant stationary platform;
- Calculation the stability of fixed platforms;
- Calculation dynamic processes in the water and air of the Arctic with the processing of data and its use to predict the dynamics of ice conditions;
- Simulation of the formation of large icebergs, ice hummocks, large ice platforms;
- Calculation of ridging in the dynamics of sea ice;
- Direct and inverse problems of seismic prospecting in the Arctic;
- Direct and inverse problems of electromagnetic prospecting of the Arctic.

The system of equations for modeling of these processes is the system of equations which describes continuum mechanics, in particular, solid state physics, acoustics, and fluid dynamics [20-53].
For the numerical solution of the relevant problems it is necessary to develop or adopt adequate modern computational methods and algorithms for high-performance computers [21-53].

2. Examples of numerical modeling

We can give some examples of numerical solutions to some up-to-date problems of oil and gas industry in the Arctic conditions.

Figure 1 shows the results of numerical modeling of the impact of the ice floes on the oil platform.

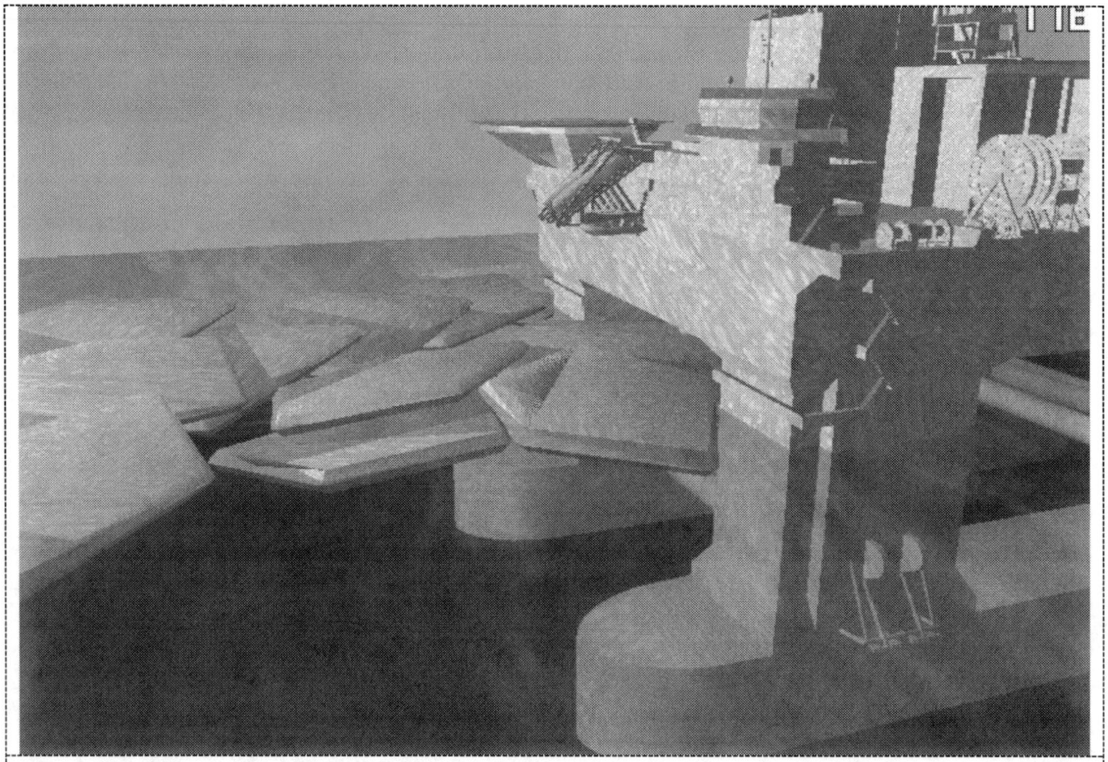

Figure 1. The impact of the ice floes on the oil platform.

Figure 2 shows the wave pattern that occurs under the influence of an iceberg, white color shows the image of the system of the cracks. The colors scale shows the velocity module, the red color corresponds to the maximum value, and the blue color corresponds to the zero value. The calculation was made using grid-characteristic method [40-52], the destruction was calculated on the basis of the criteria of main stress. Hereafter, we solve the systems of equations describing the state of the lincar-elastic body [21] and the system of equation describing acoustic waves [20].

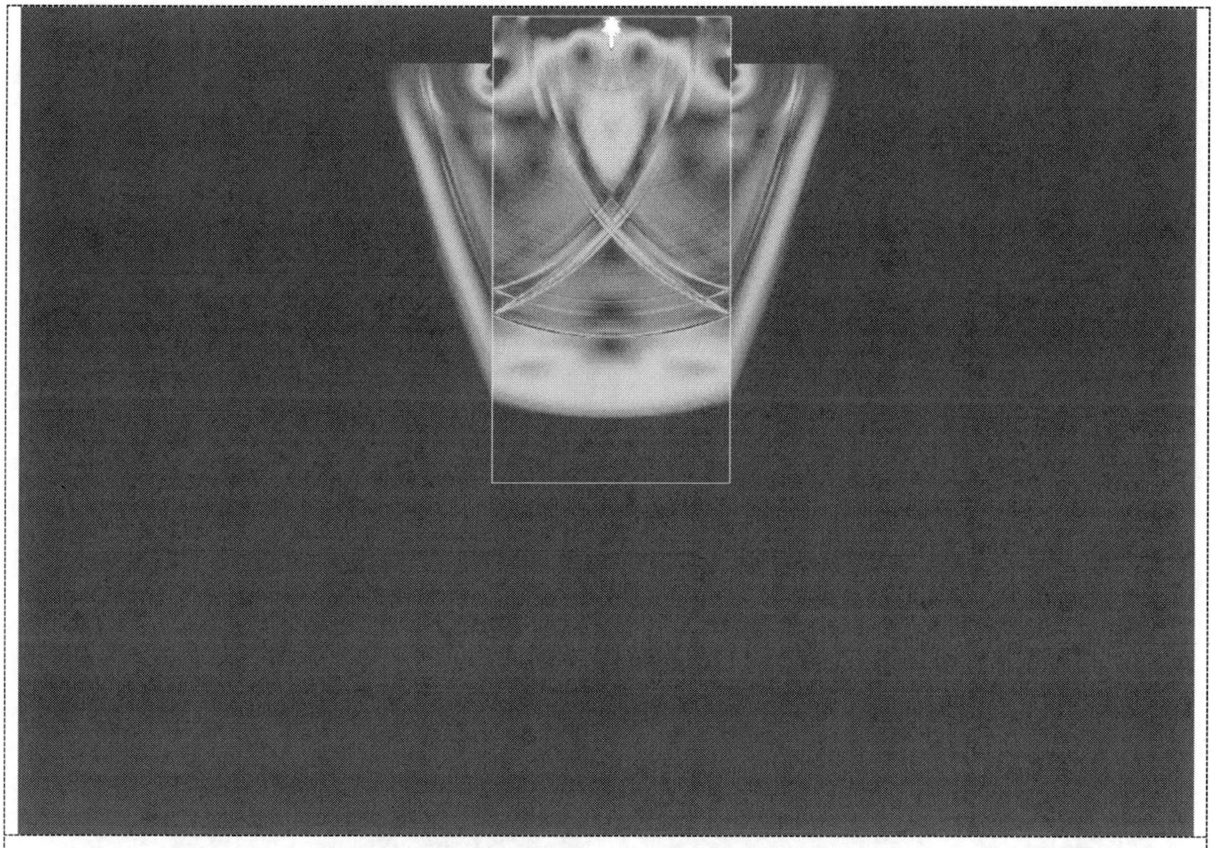

Figure 2. Impact on the iceberg. Wave pattern.

Figure 3 shows the computational grid and the wave pattern that occurs during the collision between the iceberg and fixed oil platform. The calculation was performed using a discontinuous Galerkin method [14-17,40], color shows the module of velocity, the scale is shown at figure 3.

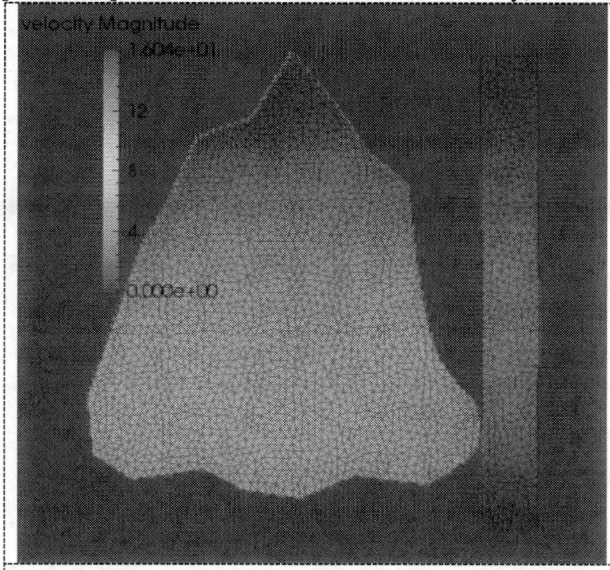

Figure 3. Collision between the iceberg and the fixed oil platform.

Figure 4 represented the wave patterns resulting from the numerical solution of problems of seismic exploration in the Arctic shelf. The colors shows the module of velocity. The red color corresponds to the maximum value, and blue color corresponds to the zero value. Figures 4, 5 demonstrates the source located in the ice, figures 6, 7 demonstrates the source located at the sea bed. Figures 5, 7 demonstrates the case without carbon reservoir. We consider the complex system composed from ice, water, soil and carbon reservoir and use grid-characteristic method [40-52].

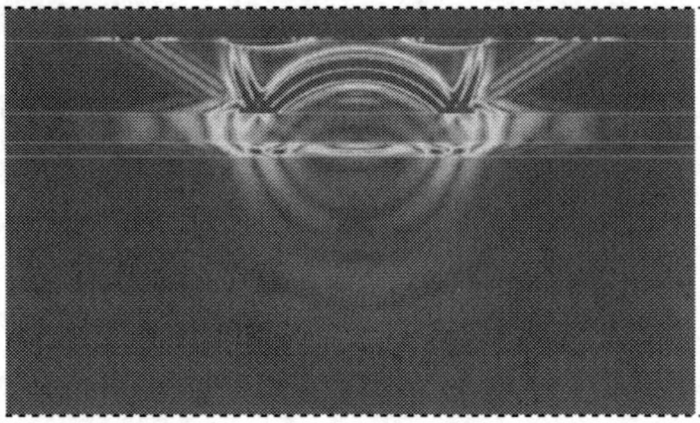

Figure 4. Wave pattern. Source in the ice. With reservoir.

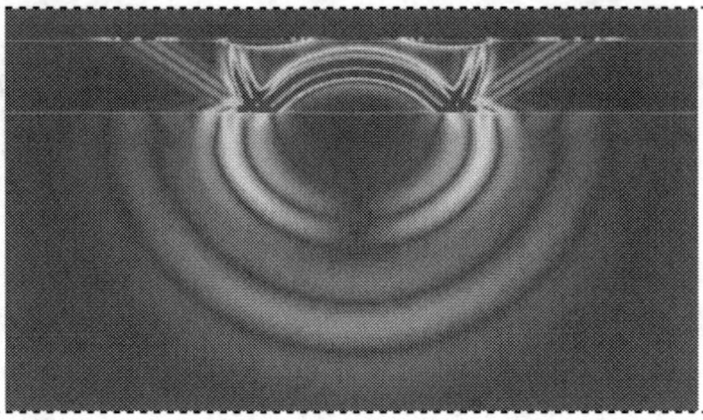

Figure 5. Wave pattern. Source in the ice. Without reservoir.

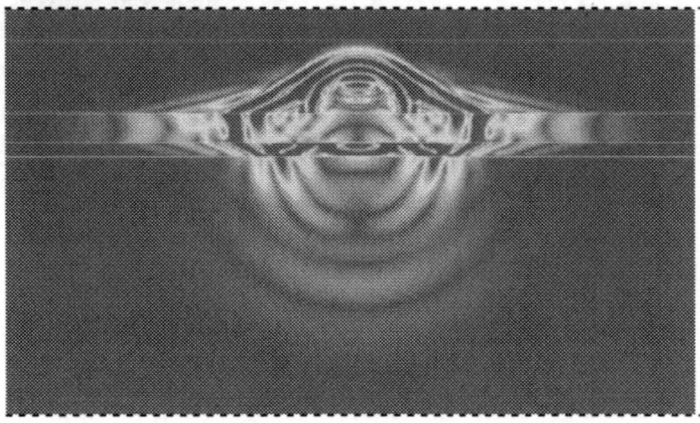

Figure 6. Wave pattern. Source at the sea bed. With reservoir.

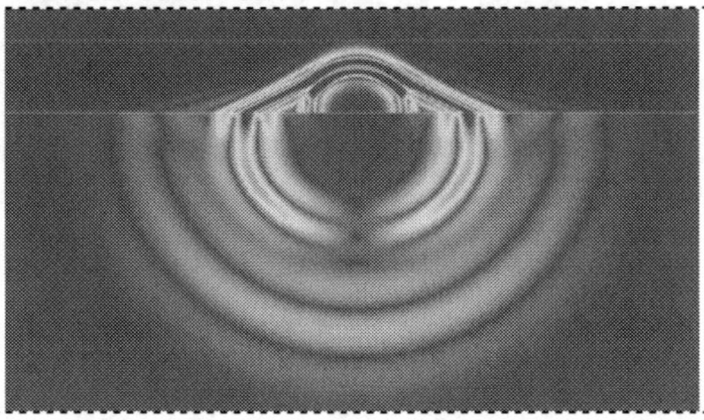

Figure 7. Wave pattern. Source at the sea bed. Without reservoir.

Figure 5 shows the seismograms corresponding to all the cases considered above, recorded on a receivers located on the ice and measuring the vertical component of velocity.

Figure 8. Seismogram. Source in the ice. With reservoir.

Figure 9. Seismogram. Source in the ice. Without reservoir.

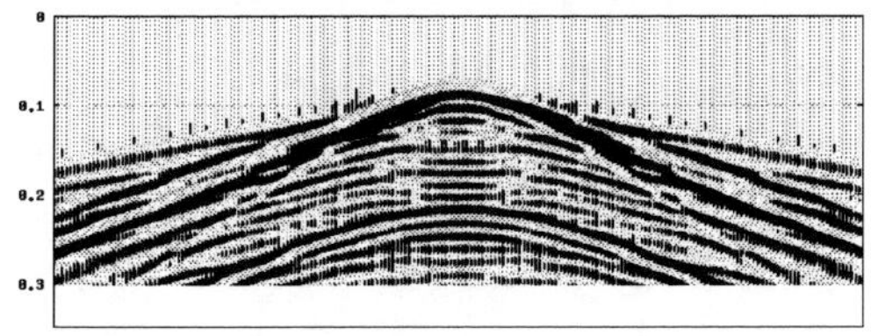

Figure 10. Seismogram. Source at the sea bed. With reservoir.

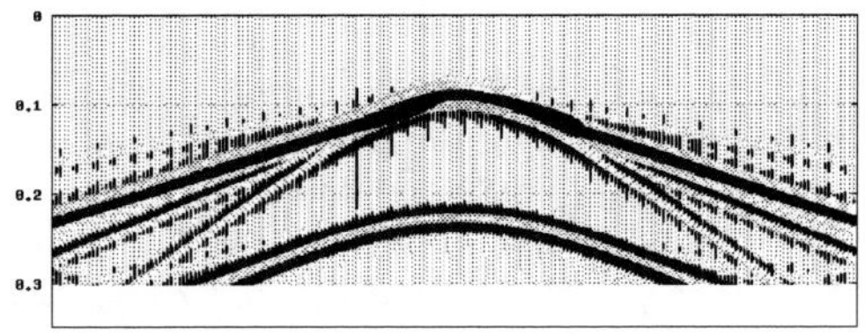

Figure 11. Seismogram. Source at the sea bed. Without reservoir.

Figure 12 shows the seismic and acoustic waves from a series of point sources in the complex system "water-soil-carbon reservoir". Grayscale shows the module of velocity. We use grid-characteristic method [40-52].

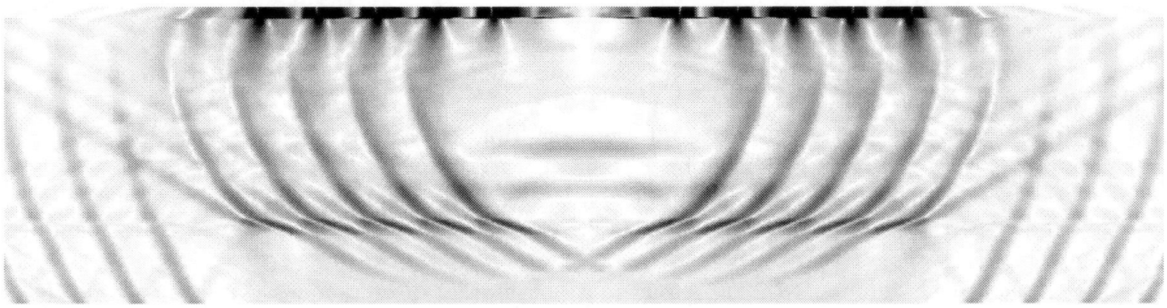

Figure 12. The wave pattern in the system "water-soil-carbon reservoir". Several sources.

3. Conclusions

In this article we discussed several kinds of Arctic problems, several kinds of numerical methods for computer study of these problems and several examples representing results of numerical solutions of these problems using grid-characteristic method [40-52] and discontinues Galerkin method [14-17,40].

The study was funded by Ministry of Education and Science of the Russian Federation under grant agreement No. 14.575.21.0084 on October 20, 2014 (the unique identifier PNI: RFMEFI57514X0084) in the Moscow Institute of Physics and Technology (State University).

I express my gratitude to my students adopted up-to-date numerical methods to solve these problems and obtained the numerical solution of these problems: Favorskaya A.V., Khokhlov N.I, Sannikov A.V., Miryaha V.A., and Petrov D.I.

4. References

[1] Gagnon R E and Wang J 2012 Numerical simulations of a tanker collision with a bergy bit incorporating hydrodynamics, a validated ice model and damage to the vessel *Cold regions. Science and Technology*

[2] Lee S G, Lun S H and Kong G Y 2013 Modeling and simulation system for marine accident cause investigation *Collision and Graunding of Ships and Offsore Structure* ed Amdahl, Ehlers and Leira (Taylor and France Group, London) pp 39–47

[3] Bekker A T, Sabobash O A, Seliverstov V I, Koff G I and Pipko E N 2009 Estimation of Lomit Loads on Engeneering Offshore Structures Proceeding of the Nineteenth *International Offshore and Polar Engeneering Conference* (Osaka, Japan) pp 574–579

[4] Weiss J 2013 Drift, Deformation and Fracture of Sea Ise. A perspective Across Scales. *Springer* p 83

[5] Pavlov V, Pavlova O and Korsnes R 2004 Sea ice fluxes and drift trajectories from potential pollution sources, computed with a statistical sea ice model of the Arctic Ocean *Journal of marine Systems* **48** pp 133–157
[6] Eik K 2009 Iseberg drift modeling and validation of applied metocean hindcust data *Cold Region Science and Technologhy* **57** pp. 67–90
[7] Garbrecht T, Luphes C, Augstein E and Wamser C 1999 Influence of a sea ice ridge on low-level airflow *J. Geophysics. Res.* **104** pp 2449–24507
[8] Shinohara Y 1999 A redistribution function applicable to a dynamic sea ice model *J. Geophysics. Res.* **95** pp 13423–13431
[9] Marchenko A 2008 Thermodynamic consolidation and melting of sea ice ridges *Cold regions. Science and Technology* **52 (3)**
[10] Shinohara Y 1990 A redistribution foundation ice model. *J. Geophysics. Res.* **95** pp 13423–14431
[11] Gray J and Killworth P D 1996 Sea ice ridging schemes *J. Physics. Oceanogr.* **26** pp 2420–2428
[12] Julev S K and Belyaev K 2012 Probility distribution characteristics for surface ari-sea turbulent heat Aluxes over the global ocean *Journal of Climate* **25 (1)** pp 184–206
[13] Belyaev K P, Tuchkova N P and Cubash U 2010 Response of a coupled ocean ice atmosphere model to data assimilation in the tropical zone of the tropical zone of the Pacific ocean *Oceanology* **50(3)**
[14] Käser M and Dumbser M 2006 An arbitrary high-order discontinuous Galerkin method for elastic waves on unstructured meshes – I. The two-dimensional isotropic case with external source terms *Geophys. J. Int.* **166(2)** pp 855–877
[15] Käser M and Dumbser M 2006 An arbitrary high-order discontinuous Galerkin method for elastic waves on unstructured meshes – II. The three-dimensional isotropic case *Geophys. J. Int.* 167(1) pp 319–336
[16] Käzer M and Dumbser M 2008 A highly accurate discontinuous Galerkin method for complex interfaces between solids and moving fluids *Geophysics* **73(3)** pp T723–T725
[17] Käser M and Igel H 2001 Numerical simulation of 2D wave propagation on unstructured grids using explicit differential operators *Geophysical Prospecting* 49(5) pp 607–619
[18] Zhdanov M S 2002 *Geophysical Inverse Theory and Regularization Problems* Elsevier
[19] Zhdanov M S 2015 *Inverse Theory and Applications in Geophysics* Elsevier
[20] Landau L D and Lifshitz E M 1959 *Fluid mechanics* (Pergamon Press)
[21] LeVeque R 2002 *Finite volume methods for hyperbolic problems* Cambridge University Press
[22] Aki K and Richards P G 2002 *Quantitative seismology, theory and methods*, second edition, University Science Books, Sausalito, California
[23] Bording R P and Lines L R 1997 Seismic Modeling and Imaging with the Complete Wave Equation *SEG proceedings*
[24] Carcione J M, Herman G C and Kroode A P E 2002 Seismic modeling *Geophysics* **67(4)** pp 1304−1325.
[25] Chapman C 2004 Fundamentals of Seismic Wave Propagation *Cambridge University Press*
[26] Di Bartolo L, Dors C and Mansur W J 2012 A new family of finite-difference schemes to solve the heterogeneous acoustic wave equation *Geophysics* **77(5)** pp T187−T199
[27] Etgen J T and O'Brien M J 2007 Computational methods for large-scale 3D acoustic finite-difference modeling: A tutorial *Geophysics* **72(5)** pp SM223−SM230
[28] Hestholm S 2009 Acoustic VTI modeling using high-order finite differences *Geophysics* **74(5)** pp T67−T73
[29] Hobro J W D, Chapman C H and Robertsson J O A 2014 A method for correcting acoustic finite-difference amplitudes for elastic effects *Geophysics* **79(4)** pp T243−T255

[30] Jianfeng Z 1997 Quadrangle-grid velocity-stress finite-difference method for elastic-wave-propagation simulation *Geophys. J. Int.* **131(1)** pp 127–134

[31] Levander A. 1988 Fourth-order finite-difference P-SV seismograms *Geophysics* **53(11)** pp 1425–1436.

[32] Peter D, Komatitsch D, Luo Y, Martin R, Le Goff N, Casarotti E, Le Loher P, Magnoni F, Liu Q, Blitz C, Nissen-Meyer T, Basini P. and Tromp J 2011 Forward and adjoint simulations of seismic wave propagation on fully unstructured hexahedral meshes *Geophys. J. Int.* **186(2)** pp 721–739.

[33] Sanyi Y, Shangxu W, Wenju S, Lina M and Zhenhua L 2014 Perfectly matched layer on curvilinear grid for the second-order seismic acoustic wave equation *Exploration Geophysics* **45(2)** pp 94–104

[34] Seriani G, Priolo E, Carcione J M and Padovani E 1992 High-order spectral element method for elastic wave modeling, *62nd Annual International Meeting and Exposition, SEG, Extended Abstracts* pp 1285–1288.

[35] Tong P, Yang D and Hua B 2011 High accuracy wave simulation revised derivation, numerical analysis and testing of a nearly analytic integration discrete method for solving acoustic wave equation, International Journal of Solids and Structures **48** pp 56–70

[36] Van Vossen R, Robertsson J O A and Chapman C H 2002 Finite-difference modeling of wave propagation in a fluid-solid configuration *Geophysics* **67(2)** pp 618–624

[37] Wang X and Liu X 2007 3-D acoustic wave equation forward modeling with topography *Appl. Geophys.* **4** pp 8–15

[38] Zeng Y Q and Liu Q H 2004 A multidomain PSTD method for 3D elastic wave equations *Bull. seism. Soc. Am.* **94** pp 1002–1015

[39] Zhang W, Zhang Z and Chen X 2012 Three-dimensional elastic wave numerical modelling in the presence of surface topography by a collocated-grid finite-difference method on curvilinear grids *Geophys. J. Int.* **190(1)** pp 358–378

[40] Favorskaya A V, Petrov I B, Khokhlov N I, Miryakha V A, Sannikov A V, Golubev V I 2015 Monitoring the State of the Moving Train by Use of High Performance Systems and Modern Computation Methods *Mathematical Models and Computer Simulations* **7(1)** pp 50–60

[41] Muratov M V and Petrov I B 2013 Estimation of wave responses from subvertical macrofracture systems using a grid characteristic method *Mathematical Models and Computer Simulations* 2013 **5(5)** pp 479-491

[42] Golubev V I, Petrov I B and Khokhlov N I 2013 Numerical simulation of seismic activity by the grid-characteristic method *Computational Mathematics and Mathematical Physics* **53(10)** pp 1523–1533.

[43] Magomedov K M and Kholodov A S 1988 *Grid characteristic methods* Nauka, Moscow

[44] Kholodov A S, Kholodov Ya A 2006 Monotonicity criteria for difference schemes designed for hyperbolic equations *Computational Mathematics and Mathematical Physics* **46(9)** pp 1560-1588

[45] Kvasov I E and Petrov I B 2012 High-performance computer simulation of wave processes in geological media in seismic exploration *Computational Mathematics and Mathematical Physics* **52(2)** pp 302–313

[46] Favorskaya A V, Petrov I B, Muratov M V and Sannikov A V 2014 Grid Characteristic Method on Unstructured Tetrahedral Meshes *Computational Mathematics and Mathematical Physics* . 54(5) pp 837–847

[47] Favorskaya A V, Petrov I B, Shevtsov A V, Vasyukov A V, Potapov A P, Ermakov A S 2014 Combined Grid Characteristic Method for the Numerical Solution of ThreeDimensional Dynamical Elastoplastic Problems *Computational Mathematics and Mathematical Physics* **54(7)** pp 1176 – 1189

[48] Favorskaya A V, Petrov I B, Muratov M V, Biryukov V A, Sannikov A V 2014 Grid-Characteristic Method on Unstructured Tetrahedral Grids *Doklady Mathematics* **90(3)** pp 781 – 783

[49] Favorskaya A V, Petrov I B, Shevtsov A V, Vasyukov A V, Potapov A P, Ermakov A S 2015 Combined Method for the Numerical Solution of Dynamic Three-Dimensional Elastoplastic Problems *Doklady Mathematics* **91(1)** pp 111–113

[50] Favorskaya A V, Petrov I B, Vasyukov A V, Ermakov A S, Beklemysheva K A, Kazakov A O, Novikov A V 2014 Numerical Simulation of Wave Propagation in Anisotropic Media *Doklady Mathematics* **90(3)** pp 778–780

[51] Petrov I B, Favorskaya A V, Sannikov A V, Kvasov I E 2013 Grid-Characteristic Method Using High Order Interpolation on Tetrahedral Hierarchical Meshes with a Multiple Time Step Mathematical *Models and Computer Simulations* 5(5) pp 409–415

[52] Petrov I B, Favorskaya A V, Beklemysheva K A. Numerical Simulation of Processes in Solid Deformable Media in the Presence of Dynamic Contacts Using the Grid-Characteristic Method *Mathematical Models and Computer Simulations* 2014 **6(3)** pp 294–304

[53] Bathe K-J *Finite Element Procedures* (Prentice Hall, Upper Saddle River, New Jersey)

Journal of Physics: Conference Series **681** (2016) 012027 doi:10.1088/1742-6596/681/1/012027

Numerical simulation of small-scale mixing processes in the upper ocean and atmospheric boundary layer

O Druzhinin[1] Yu Troitskaya[1] and S Zilitinkevich[2]

[1]Institute of Applied Physics RAS, Nizhny Novgorod, Russia

[2]Finnish Meteorological Institute, Helsinki, Finland

E-mail: druzhinin@hydro.appl.sci-nnov.ru

Abstract. The processes of turbulent mixing and momentum and heat exchange occur in the upper ocean at depths up to several dozens of meters and in the atmospheric boundary layer within interval of millimeters to dozens of meters and can not be resolved by known large-scale climate models. Thus small-scale processes need to be parameterized with respect to large scale fields. This parameterization involves the so-called bulk coefficients which relate turbulent fluxes with large-scale fields gradients. The bulk coefficients are dependent on the properties of the small-scale mixing processes which are affected by the upper-ocean stratification and characteristics of surface and internal waves. These dependencies are not well understood at present and need to be clarified. We employ Direct Numerical Simulation (DNS) as a research tool which resolves all relevant flow scales and does not require closure assumptions typical of Large-Eddy and Reynolds Averaged Navier-Stokes simulations (LES and RANS). Thus DNS provides a solid ground for correct parameterization of small-scale mixing processes and also can be used for improving LES and RANS closure models. In particular, we discuss the problems of the interaction between small-scale turbulence and internal gravity waves propagating in the pycnocline in the upper ocean as well as the impact of surface waves on the properties of atmospheric boundary layer over wavy water surface.

1. Introduction

The processes of small-scale mixing in the atmospheric boundary layer and the upper ocean are of primary importance for mass, momentum and heat exchange between atmosphere and hydrosphere. The mixing occurs due to turbulent motions coupled with the gravity waves at the sea surface as well as internal gravity waves in the oceanic pycnocline. These coupling dramatically affects the dynamics of turbulence and thus needs to be taken into account in climate prognostic models [1,2].

In the state of the art models, the turbulent fluxes of momentum, energy and matter are calculated via bulk formulae which to a large extent are based on empirical coefficients. The latter are obtained from field and laboratory experiments and exhibit essential variability whose nature in relation to turbulence-wave interaction processes is not yet fully understood [3]. Of special interest, from the point of view of the present study, is the influence of density (or temperature) stratification of the atmospheric surface layer on wind-wave interactions and the air-sea fluxes in the atmospheric boundary layer and the interaction between small-scale turbulence and internal gravity waves propagating in the seasonal pycnocline in the upper ocean.

Measurements in the atmospheric boundary layer above waved water surface require special efforts. Especially complicated are measurements in wave troughs and in a thin layer above the water surface covering the viscous sublayer and the adjacent buffer layer. The typical height of the region to be examined is of order of millimetres, which is usually much smaller than the surface wave amplitude. Here, detailed properties of air flow cannot be detected using contact measurement techniques, such as wave-following probes [4,5]. Methods based on particle image velocimetry (PIV) technique allow for measuring wind velocity at heights over the surface of order 1 mm. This is sufficient for investigating both viscous and buffer layers at not too strong winds, with friction velocities less than 10 cm/s [6,7]. Precise measurements of the air temperature in the viscous layer and buffer layer above waves are strongly needed but still remain difficult.

In laboratory studies of the interaction of small-scale turbulence with internal waves (IWs), in the absence of mean shear, turbulent motions are usually induced by an oscillating grid [2,8]. One of the most interesting and practically important aspects of the turbulence-IWs interaction in the absence of mean shear is the effect of damping of IWs by turbulence on the one hand, and the possibility of enhancement of small-scale turbulence by non-breaking IWs on the other hand. The phenomena of IWs damping by turbulence was observed in laboratory experiments [1, 9, 10]. Measurements of the IW amplitudes [10] showed an effective enhancement of the decay rate of IW under the effect of turbulence which was found to be in good agreement with the theoretical prediction [11] based on RANS-closure approach.

The effect of enhancement of small-scale turbulence by mechanically-generated, non-breaking internal wave propagating in the pycnocline was also observed experimentally [12]. The results show that a sufficiently strong (as compared to turbulent velocity fluctuations), non-breaking internal wave can significantly increase the kinetic energy of turbulence in the well-mixed layer above the pycnocline. However, the obtained experimental data did not provide enough detail to study the modification of turbulence kinetic energy spectra under the influence of IWs.

Numerical experiments make a complementary tool to laboratory and field measurements. Earlier numerical modelling of interaction between turbulent air flow and surface waves was based on the Reynolds (ensemble) Averaged, Navier-Stokes (RANS) equations for stationary, two-dimensional flow [13,14]. These RANS equations employed semi-empirical concept of turbulent viscosity to express unknown Reynolds (turbulent) stresses through known mean-field gradients. This approach is computationally cheap and provides a useful picture of the mean velocity, pressure and turbulent fluxes in the air flow. However, it is not immediately applicable to modelling air flow in the viscous sublayer and buffer region, which requires inclusion of the molecular transports [15].

More advanced (but more computationally expensive) is Large-Eddy Simulation (LES) based on numerical integration of three-dimensional, non-stationary, filtered Navier-Stokes equations and capable of resolving 3D large-scale turbulent structure of the flow [16,17]. However, LES do not resolve properties of the flow close to the water surface (or solid wall). Typically the first LES grid node in the vertical direction is located within the logarithmic region of the boundary layer. As a remedy, one can perform *wall-resolved* LES [18] and resolve the viscous sublayer; however, numerical cost strongly increases.

Direct Numerical Simulation (DNS) is free of this drawback and resolves the entire turbulence spectrum from large eddies down to the dissipation length scale. Early studies [19, 20] employed DNS to investigate turbulent wind flow over surface waves for bulk Reynolds number $\mathrm{Re} = (U_0\lambda/v)$ (where U_0 is the bulk wind flow velocity, λ the surface wave length, and v the air molecular viscosity) less than 10^4 and wave slope $ka = 0.1$ (where k is the wavenumber and a the wave amplitude). Later studies [21] considered a larger wave slope (*ka*

= 0.25) and nearly the same Re (≈ 10000). Although DNS provides full description of the turbulent flow at all physically significant scales, so far it has not been able to achieve Reynolds numbers of order 10^5 typical of the wind-wave interaction in laboratory experiments. Recent DNS considered Re = 15000 with maximum wave slope $ka = 0.2$ [22].

The interaction between internal waves (IWs) propagating through a pycnocline with a turbulent layer above the pycnocline was also studied with the use of DNS [23, 24]. The results show that if the ratio of IW amplitude vs. turbulent pulsations amplitude is sufficiently small (less than 0.5), turbulence strongly enhances IW damping. In this case, the damping rate obtained in DNS agrees well with the prediction of a semi-empirical closure approach based on the RANS equations [11]. However, for larger IW amplitude the effect of damping of IWs by turbulence is much weaker, and, in this case, the RANS-based theory overestimates the IW damping rate by the order of magnitude. DNS results also show that in the absence of IW turbulence decays, but its decay rate is reduced in the vicinity of the pycnocline where stratification effects are significant. In this case, at sufficiently late times most of turbulent energy is located in a layer close to the pycnocline center. Here turbulent eddies are collapsed in the vertical direction and acquire the "pancake" shape. Strong IWs modify turbulence dynamics, in that the turbulence kinetic energy (TKE) is significantly enhanced as compared to the TKE in the absence of IWs. As in the case without IW, most of turbulent energy is localized in the vicinity of the pycnocline center. Here the TKE spectrum is considerably enhanced in the entire wavenumber range as compared to the TKE spectrum in the absence of IWs.

In the present paper, we discuss how DNS can be employed to study the interaction between turbulence and surface and internal gravity waves in the upper ocean and marine boundary layer. In Section 2 the numerical algorithm used in DNS of turbulent, *stably-stratified* air flow over water waves is presented and some simulation results are discussed. Numerical algorithm and results of DNS of the dynamics of small-scale turbulence coupled with a strong internal gravity wave propagating in a pycnocline are discussed in Sec. 3. Conclusions are presented in Sec.4.

2. DNS of turbulent air-flow over waved water surface

2.1 Numerical method

We perform direct numerical simulation of turbulent stably stratified Couette flow above waved water surface (Figure 1). A Cartesian framework is considered where x-axis is oriented along the mean wind, z-axis is directed vertically upwards and y-axis is orthogonal to the mean flow and parallel to the wave front. We prescribe two-dimensional, x-periodic water wave with amplitude a, wavelength λ and phase velocity c. The maximum wave slope considered in our DNS is $ka = 2\pi a/\lambda = 0.2$. The rectangular computational domain has the sizes $L_x = 6\lambda$, $L_y = 4\lambda$ and $L_z = \lambda$ in the x-, y-, and z-directions, respectively, and the air flow is assumed to be periodical in the x- and y-directions. DNS is performed in a reference frame moving with the wave phase velocity, c, so that the horizontal coordinate in the moving framework is $x = x' - ct$, where x' is the coordinate in the laboratory reference frame. Then the lower boundary, representing the wave surface, is stationary in the moving reference frame. As in all previous DNS studies mentioned above, in the present DNS study we do not consider capillary ripples riding on the wave generally found in a realistic sea situation. The presence of ripples increases the effective wavy-surface slope and may cause local flow

Journal of Physics: Conference Series **681** (2016) 012027 doi:10.1088/1742-6596/681/1/012027

separation which significantly complicates numerical solution. The no-slip boundary condition is prescribed at the lower boundary, so that the air flow velocity here

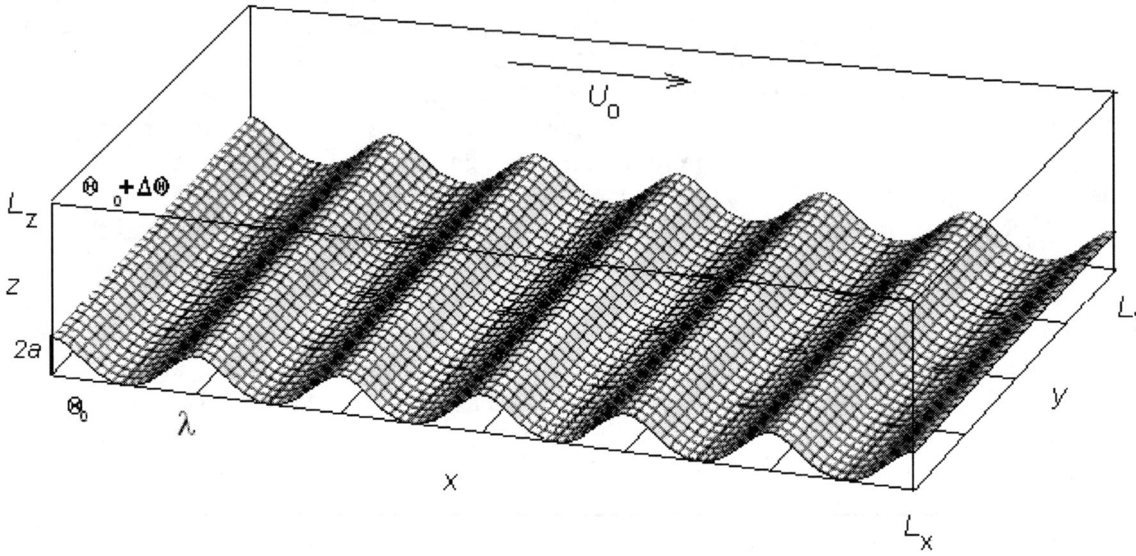

Fig. 1. Schematic of numerical experiment.

coincides with the velocity in the water wave. The no-slip boundary condition is also prescribed at the upper horizontal plane moving in *x*-direction with bulk velocity U_0. This condition provides the external source of momentum due to the viscous shear stress, which compensates the viscous dissipation in the boundary layer and makes the flow statistically stationary. The stable density stratification is specified by prescribing the potential temperature at the wavy surface as $\Theta = \Theta_0$ and the top boundary plane as $\Theta = \Theta_0 + \Delta\Theta$, where $\Delta\Theta > 0$.

Numerical algorithm is based on the integration of full, 3D Navier-Stokes equations for incompressible fluid under the Boussinesq approximation [25]:

$$\frac{\partial U_i}{\partial t} + \frac{\partial (U_i U_j)}{\partial x_j} = -\frac{\partial P}{\partial x_j} + \nu \frac{\partial^2 U_i}{\partial x_j \partial x_j} + \delta_{iz} \frac{g}{\Theta_0} (\Theta - \Theta_{ref}), \quad (1)$$

$$\frac{\partial U_j}{\partial x_j} = 0, \quad (2)$$

$$\frac{\partial \Theta}{\partial t} + \frac{\partial (U_j \Theta)}{\partial x_j} = \frac{\nu}{\Pr} \frac{\partial^2 \Theta}{\partial x_j \partial x_j}, \quad (3)$$

where $x_i = x, y, z$, $U_i (i = x, y, z)$ are the velocity components, *P* is pressure and Θ_{ref} is the reference temperature, ν is the kinematic air viscosity and *g* the gravity acceleration. The Prandtl number, $\Pr = \nu / \mu$ (where μ is thermal diffusivity).

The integration is performed in curvilinear coordinates (ξ, y, η) which are related to the Cartesian coordinates (x, y, z) by the mapping:

$$x = \xi - a \exp(-k\eta) \sin k\xi, \quad (4)$$

$$z = \eta + a \exp(-k\eta) \cos k\xi. \quad (5)$$

This mapping transforms the lower waved boundary $z_b(x) = a\cos k\xi(x)$ into a plane boundary at $\eta = 0$ [22]. It is easy to show that for small, finite wave amplitude a the shape of the boundary, $z_b(x)$, up to the terms of order $O(k^2a^3)$, coincides with asymptotic solution for the Stokes wave [13]:

$$z_b(x) = a\cos kx + \frac{1}{2}a^2k(\cos 2kx - 1)\,. \tag{6}$$

The governing equations (1)-(3) are rewritten in dimensionless variables normalized by the wavelength λ, bulk velocity U_0 and temperature difference $\Delta\Theta$, and pressure normalized by ρU_0^2 (where ρ is the air density). The integration is performed using dimensionless curvilinear coordinates, $\xi_1 = \xi/\lambda$, $\xi_2 = y/\lambda$, $\xi_3 = \eta/\lambda$, and Cartesian velocity components $U_1 = U_x/U_0$, $U_2 = U_y/U_0$, and $U_3 = U_z/U_0$. We also introduce a linear reference temperature profile, $\Theta_{ref}(\xi_2) = \Delta\Theta\eta/\lambda$, and perform the integration with respect to the dimensionless deviation of the temperature, $\widetilde{T} = (\Theta - \Theta_{ref})/\Delta\Theta$.

The governing parameters in DNS are the bulk Reynolds and Richardson numbers defined as:

$$\mathrm{Re} = \frac{U_0\lambda}{\nu}\,, \tag{7}$$

$$\mathrm{Ri} = g\frac{\Delta\Theta}{\Theta_0}\frac{\lambda}{U_0^2}\,. \tag{8}$$

The Prandtl number, $\Pr = \nu/\mu$ (where μ is thermal diffusivity), is prescribed as $\Pr = 0.7$.

The governing equations are discretized in a rectangular domain with sizes $0 < \xi_1 < 6$, $0 < \xi_2 < 4$, and $0 < \xi_3 < 1$ by employing a finite difference method of the second-order accuracy on a uniform staggered grid consisting of $360\times240\times180$ nodes. An additional mapping is employed to compress the grid in the vertical direction near the boundaries in order to resolve the viscous boundary layer. In present DNS we prescribe relatively high Reynolds number (up to 80000) which requires sufficiently high resolution of fine scales where dissipation of turbulence kinetic energy occurs in DNS. In the considered case of boundary-layer flow, the viscous scale is ν/u_*, where u_* is the friction velocity and ν the kinematic viscosity. This scale generally varies in DNS for different bulk Reynolds number and Richardson number. Our DNS employ the grid with mesh sizes $\Delta\xi_1$ = 1/60 in the streamwise and spanwise directions, ξ_1 and ξ_2, whereas in the vertical direction ξ_3 the mesh size increases from $\Delta\xi_3^w \approx 0.0008$ near the walls to $\Delta\xi_3^c \approx 0.009$ in the centre of computational domain. When normalized by the wall scale, the mesh sizes for different Re and Ri varies as $\Delta\xi_1^+ \approx 2 \div 10$ in the horizontal plane and from $\Delta\xi_3^{w+} \approx 0.2 \div 0.6$ near the boundaries to about $\Delta\xi_3^{c+} \approx 2 \div 5$ in the centre of the domain. These grid spacings are generally regarded as sufficient to resolve fine-scale turbulent motions in DNS of turbulent boundary-layer flows [26].

The velocity field is initialized as a weakly perturbed laminar Couette flow and the initial temperature deviation field is put to zero. After a transient, at sufficiently large times, a statistically stationary flow regime is established. Then sampling of the velocity and temperature fields is performed at discrete time moments t_k, $k = 1,\ldots,1000$, with interval

$t_{k+1} - t_k = 0.2$. The mean profiles are obtained by averaging over time and spanwise and streamwise coordinates.

2.2. Numerical results

DNS results show that there are two distinct flow regimes which are realized depending on the stratification strength measured by the bulk Richardson number, Ri. If Ri is small enough, the flow remains turbulent throughout the domain and is qualitatively similar to that in the non-stratified case (Fig. 2).

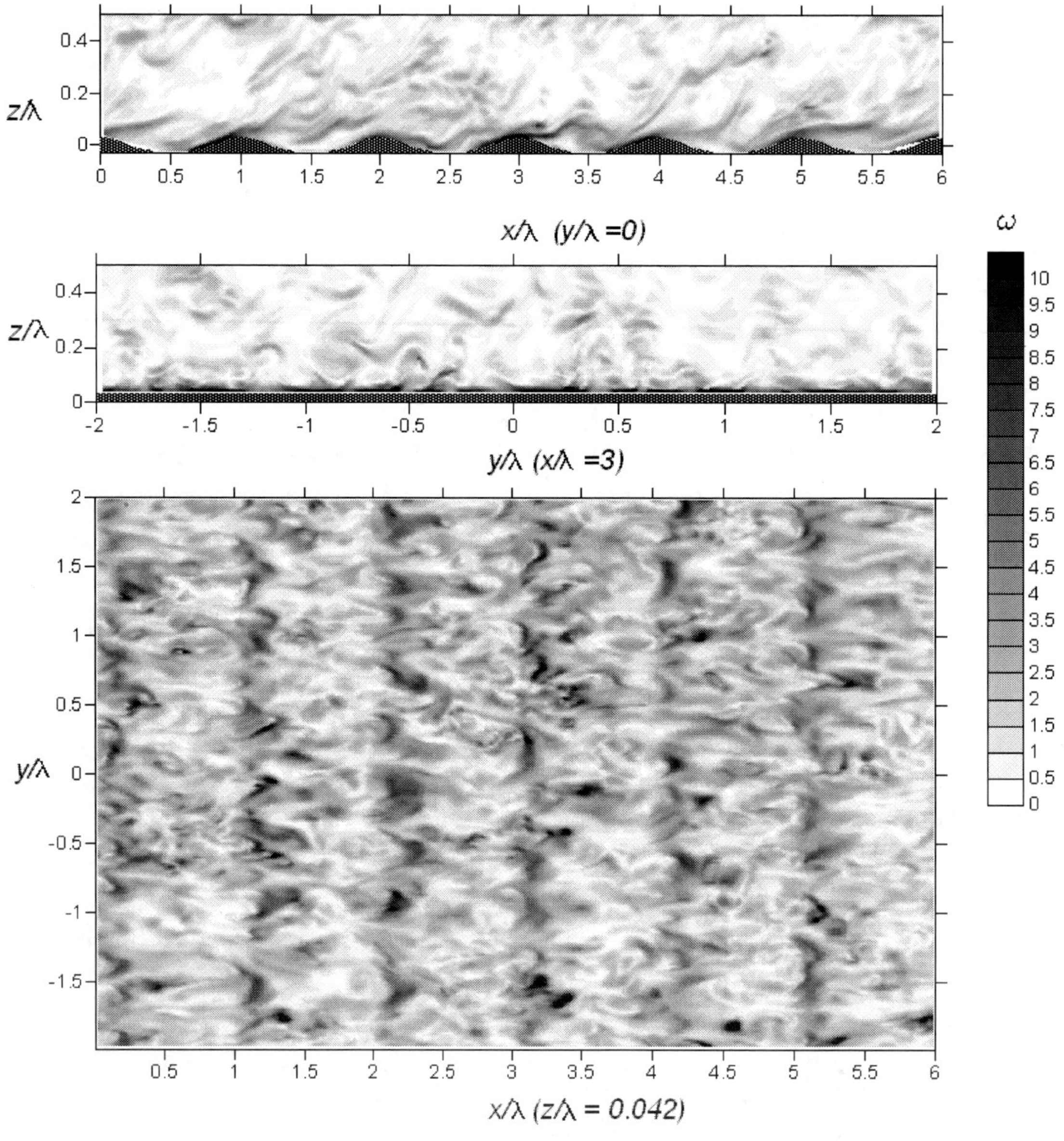

Fig. 2. Figure 4a. Instantaneous contours of the vorticity modulus in (x,z), (y,z) and (x,y) planes (top, middle and bottom, respectively) in statistically stationary, stably-stratified boundary-layer flow over waved surface with the wave slope $ka = 0.2$ and dimensionless phase velocity $c/U_0 = 0.05$, obtained in DNS for Re = 15000 and Ri = 0.04 at $t = 1000$.

Under this statistically-stationary, turbulent regime the mean velocity and temperature profiles are well predicted by the Monin & Obukhov asymptotic solutions [25].

On the other hand, if stratification is strong enough, and Ri is sufficiently large, turbulent fluctuations are not sustainable and die out in the bulk of the flow domain. However, under the influence of the surface wave, finite fluctuations are still present in the vicinity of the water surface (Fig. 3).

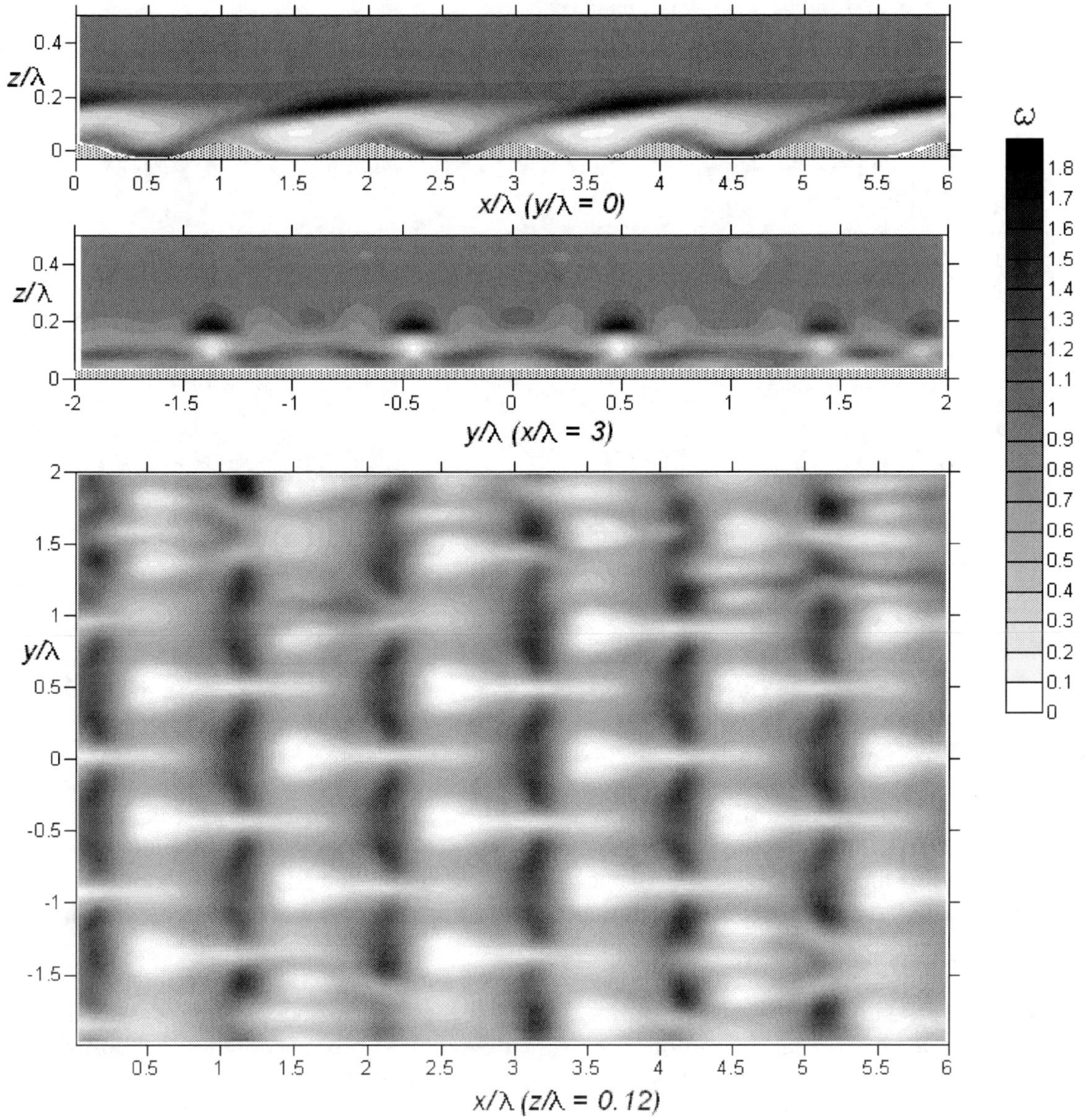

Fig. 3. Instantaneous field of the vorticity modulus in (x,z) plain at $y = 0$, (y,z) plain at $x = 3$ and (x,y) plane at $z = 0.12$, obtained from DNS of the stably stratified boundary-layer flows for the bulk Reynolds number Re = 15000, bulk Richardson number Ri = 0.08, wave slope $ka = 0.2$, and dimensionless phase velocity $c/U_0 = 0.05$, at time $t = 1000$.

The results in Figure 3 show that the fluctuations "survive" only for sufficiently large wave slope, and the amplitude of the fluctuations monotonically decreases as Ri increases.

In order to investigate the origin of complicated flow structure in the vicinity of the water surface in Figure 4 we evaluated the instantaneous power spectrum, $E(k_x, k_y)$, obtained by the Fourier transform of the vorticity field in the horizontal (x, y) plane in Fig. 3.

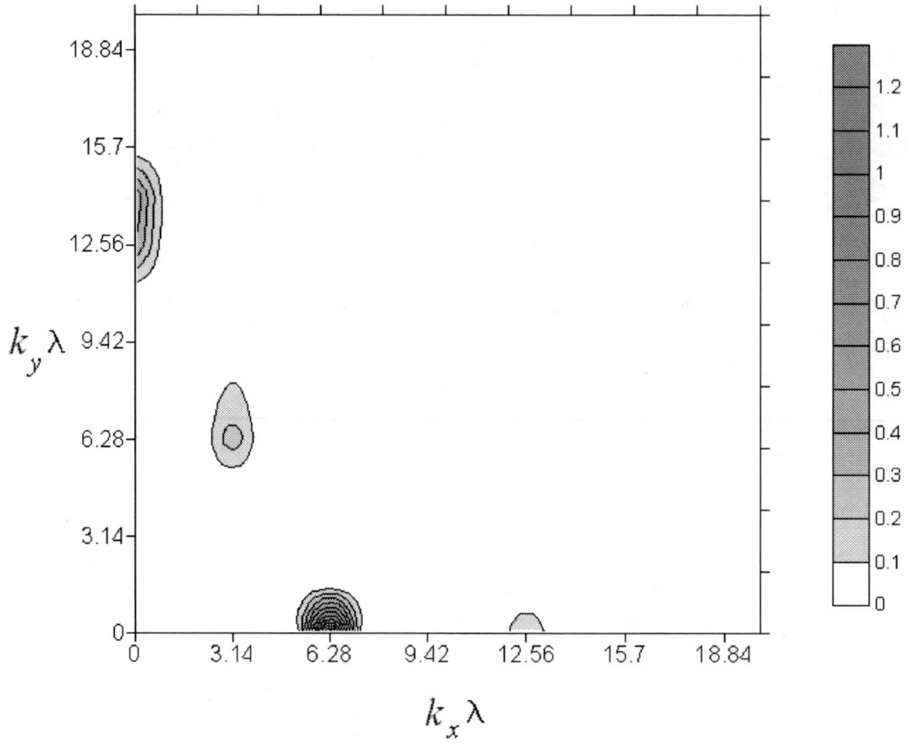

Fig. 4. Instantaneous power spectrum of the vorticity field in the air flow over the waved water surface for $c/U_0 = 0.05$, Re = 15000, Ri = 0.08 and $ka = 0.2$ [the case shown in Fig. 3: (x,y) plane at $z = 0.12$].

The spectrum exhibits peaks at the wavenumbers $\mathbf{k} = (2\pi, 0)$, $\mathbf{k}_1 = (\pi, 2\pi)$, $\mathbf{k}_2 = (4\pi, 0)$ and $\mathbf{k}_3 = (0, 4\pi)$. The peak at $\mathbf{k} = (2\pi, 0)$ corresponds to the two-dimensional (2D) disturbance directly induced in the air-flow by the surface wave. However, the 2D forcing alone would not be able to directly produce a three dimensional (3D) disturbance, heterogeneous in y-direction along the surface-wave front and characterized by the energy peaks at wavenumbers $\mathbf{k}_1$ and $\mathbf{k}_3$. A theoretical analysis (not presented here) shows that the 3D structure of the vorticity field in the horizontal (x,y)-plane in Fig. 3 and the spectral peak at the wavenumber $\mathbf{k}_1 = (\pi, 2\pi)$ can de regarded as a consequence of the development of a secondary instability of the 2D wave-induced disturbances, caused by the interaction of the former with the oblique waves via parametric resonance. Indeed, Figure 4 shows that the parametric-resonance condition

$$k_{1x} = \frac{1}{2}k = \pi \tag{9}$$

is satisfied. Additional peaks at wave numbers $\mathbf{k}_3$ and $\mathbf{k}_4$ in the spectrum $E(k_x, k_y)$ may result from non-linear generation of second harmonics.

3. DNS of interaction of small-scale turbulence with internal waves

3.1 Numerical method and initial conditions

We consider a stably stratified fluid with a pycnocline (Fig. 5).

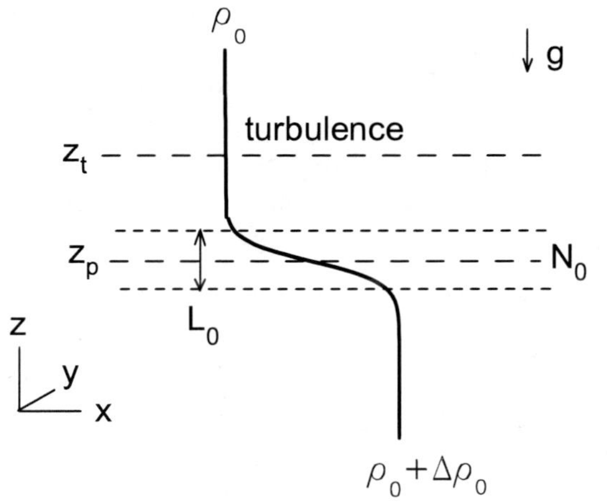

Fig. 5. Schematic of the numerical experiment: x,y,z are the Cartesian coordinates; ρ_0 is the density above the pycnocline; $\Delta\rho_0$ the density jump across the pycnocline; g the gravity acceleration, N_m the buoyancy frequency in the pycnocline center; L_0 the pycnocline thickness; z_p and z_t the locations of the pycnocline and the turbulent layer centers.

Initial turbulence field is localized in a layer at some distance above the pycnocline. The first mode of the internal wave propagating along the pycnocline from left to right is also prescribed as initial condition. Periodic boundary conditions in the horizontal, x and y, directions and Neumann (zero normal gradient) boundary condition in the vertical z direction are considered. The thickness of the pycnocline, L_0, and the buoyancy frequency in the middle of the pycnocline, $N_0 = \left(-\frac{g}{\rho_0}\frac{d\rho_0}{dz}\right)^{1/2}$ (where g is the gravity acceleration and $\rho_0(z)$ the fluid density), are chosen to define the characteristic length and time scales, L_0 and $T_0 = 1/N_0$, which are used further to write the governing equations in the dimensionless form.

As in the previous case (Sec. 2) DNS solves the Navier-Stokes equations for the fluid velocity written under the Boussinesq approximation where the coordinates, time and velocity are normalized by the length, time and velocity scales, L_0, T_0 and $U_0 = L_0/T_0$. The density deviation ρ is normalized by the density jump across the pycnocline, $\Delta\rho_0$ (Fig.1).

The Navier-Stokes equations for the fluid velocity and density are integrated in a cubic domain with sizes $0 \le x \le 40, -10 \le y \le 10$ and $0 \le z \le 20$ by employing a finite difference method of the second-order accuracy on a uniform rectangular staggered grid consisting of $400 \times 200 \times 200$ nodes in the x-, y- and z- directions, respectively. The integration is advanced in time using the Adams-Bashforth method with time step $\Delta t = 0.01$. The Poisson equation for the pressure is solved by FFT transform over x and y coordinates, and Gaussian elimination method over z coordinate [23, 24]. The Neumann (zero normal gradient) boundary condition is prescribed for all fields in the horizontal (x,y) planes at $z = 0$ and $z = 20$, and periodic boundary conditions are prescribed in the longitudinal (x) and transverse (y) directions. In DNS we prescribe the Reynolds number to be Re = 20000. This number is sufficiently large to render the viscous damping of IWs negligible.

The initial condition for the velocity and density fields is prescribed as a first mode of internal wave field with wavelength λ (and wavenumber $k = 2\pi / \lambda$) and frequency ω. The solution of the linearized Navier-Stokes equations for the progressive internal wave propagating from left to right in the x-direction [1]. DNS was performed with initial conditions corresponding to the IW fields with wavelength $\lambda = 10$ (i.e. about ten times larger as compared to pycnocline thickness) and amplitude $W_0 = 0.1$. Under these conditions, non-linear effects during the IW propagation in the pycnocline remain negligible. The mid-pycnocline level was prescribed at $z_p = 8$. The turbulence layer center was set at $z_t = 10$. The value of z_t was chosen to ensure that the effects of turbulent mixing and internal wave generation by turbulence in the pycnocline remained sufficiently small. Turbulent velocity field is initialized in DNS as a random, divergence-free field with a given power spectrum. The turbulence parameters were chosen so that the effects of turbulent mixing on the pycnocline structure and generation of internal waves by turbulence remain negligible during the considered time interval.

3.2 Numerical results

DNS was performed both in the freely-decaying turbulent-layer case (without initially induced IWs) and in the case where turbulence evolved in the presence of IWs propagating in the pycnocline. DNS results show that, in the former case (not shown here), there occurs a collapse of three-dimensional turbulence and formation of quasi-2D pancake vortex structures in the vicinity of the pycnocline. The horizontal spatial scale of these structures is considerably larger as compared to the characteristic size of 3D turbulent eddies which still survive in the non-stratified region sufficiently far from the pycnocline. In the latter region, the decay rate of the turbulent kinetic energy is enhanced as compared to the region in the vicinity of the pycnocline ($E(z = 11) \sim t^{-1.6}$ as compared to $E(z = 9) \sim t^{-0.9}$). As a result, the location of the kinetic energy maximum is shifted with time from the center of the turbulent layer at $z_t = 10$ (at $t = 0$) to the level $z = 9$, i.e. closer to the pycnocline. At sufficiently late times ($t > 400$) most of turbulent kinetic energy is located in a layer occupied by pancake large-scale eddies in the vicinity of the pycnocline.

Strong IWs propagating in the pycnocline modify turbulence evolution in that at sufficiently late times the turbulence intensity is significantly enhanced in a thin layer in the vicinity of the pycnocline center (Fig. 6). Evaluation of the kinetic energy power spectrum in Figure 7 shows significant amplification (by the order of magnitude) of the kinetic energy spectrum under the effect of IW in the entire wavenumber (k) range. The maximum peak in the IW spectrum (in the absence of turbulence) is due to the first harmonics at $k = 2\pi/10$, the second harmonics peak (at $k = 4\pi/10$) being less by two orders of magnitude (Fig.7, left panel).

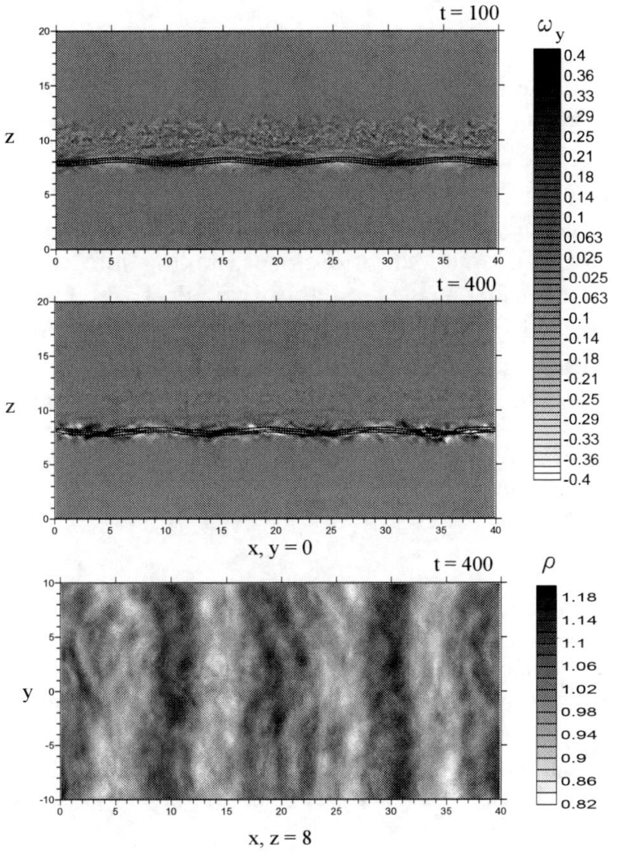

Fig. 6. Instantaneous distribution of the vorticity *y*-component ω_y with imposed density contours (1.3, 1.5, 1.7) in the central (x,z)-plane at times t = 100 and t = 400, and density distribution in the (x,y)-plane at the pycnocline level (z = 8, bottom panel) at t = 400 obtained in DNS of turbulence layer. IW wavelength λ = 10.

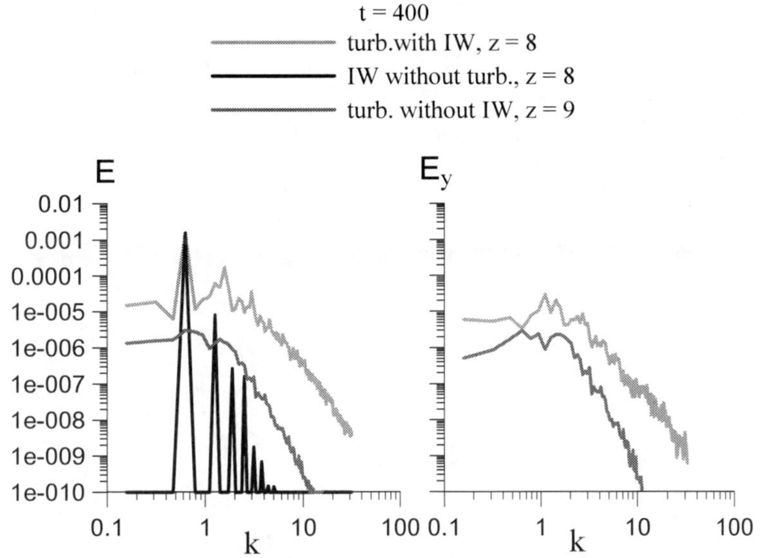

Fig. 7. The kinetic energy power spectrum, $E(k)$, (left) and the spectrum of the *y*-velocity component, $E_y(k)$, (right) obtained in DNS with initially excited IWs at the pycnocline center level (z = 8) at time t = 400.

4. Conclusions

We have performed direct numerical simulation (DNS) of turbulent atmospheric boundary layer over waved water surface and the interaction of small-scale turbulence with strong internal waves in the upper ocean. DNS resolves all physically important scales of the flow and thus does not require modeling of the subgrid-scale-flow effects on the large-scale, mean flow typical both of Large-Eddy Simulations (LES) and Reynolds-Averaged Navier-Stokes-equations simulations (RANS). Since subgrid closure taking into account the interaction of turbulence with internal and surface waves remains difficult, DNS provides benchmark results and solid ground for further LES/RANS modeling. DNS also reveals new physical effects related to the possibility of the enhancement of turbulent motions by the waves which affect mixing and thus need to be taken into account in large-scale prognostic models .

Acknowledgement: This work was supported by RFBR (Nos. 14-05-00367, 13-05-00865, 14-05-91767, 15-35-20953). Yu. Troitskaya and O. Druzhinin were partially supported by the Russian Science Foundation (Nos. 14-17-00667 and 15-17-20009, respectively).

References

[1] Phillips O 1977 *The dynamics of the upper ocean* (2nd ed., Cambridge).
[2] Thorpe S 2007 *An introduction to ocean turbulence* (Cambridge).
[3] Fairall C Bradley E Hare J 2003 *J. Climate* **16** 571-591.
[4] Donelan M Babanin A Young I et al. 2005 *J. Atmos. Oceanic Technol.* **22** 799-813.
[5] Hsu C Hsu E Street R 1981 *J. Fluid Mech.* **105** 87–117.

[6] Veron F Saxena G Misra S 2007 *Geophys. Res. Lett.* **34** L19603 doi: 10.1029/2007GL031242.
[7] Troitskaya Yu. Sergeev D Ermakova O Balandina G 2011 *J. Phys. Oceanogr.* **41** 1421-1454.
[8] Turner J 1973 *Buoyancy effects in fluids* Cambridge Univ. Press.
[9] Kantha L 1979 *Dynamics of Atmospheres and Oceans* **3** 39-46.
[10] Ostrovsky L Kazakov V Matusov P Zaborskikh D 1996 *J. Phys. Oceanogr.* **26** 398-405.
[11] Ostrovsky L Zaborskikh D 1996 *J. Phys. Oceanogr.* **26** 388-397.
[12] Matusov P Ostrovsky L Tsimring L Dokl. 1989 Akad. Nauk USSR **307** 979-984.
[13] Gent P Taylor P 1976 *J. Fluid Mech.* **77** 105-128.
[14] Chalikov D 1986*Boundary Layer Meteorol.* **34** 63-98.
[15] L'vov V Pomyalov A Procaccia I Zilitinkevich S 2006 *Phys. Rev. E.* **73** 016303.

[16] Sullivan P Edson J Hristov T McWilliams J 2008 *J. Atmos. Sci.* **65** 1225-1245.
[17] Sullivan P McWilliams J Patton E 2014 *J. Atmos. Sci.* **68** 2395–2415.
[18] Pope S 2000 *Turbulent Flows.* Cambridge University Press.
[19] Sullivan P McWilliams J Moeng C 2000 *J. Fluid Mech.* **404**: 47-85.
[20] Sullivan P McWilliams J 2002 *Phys. Fluids* **14** 1182-1195.
[21] Yang D Shen L 2010 *J. Fluid Mech.* **650** 131-180.
[22] Druzhinin O Troitskaya Y Zilitinkevish S 2012 *J. Geophys. Res.* **117** C00J05.
[23] Druzhinin O Ostrovsky L Zilitinkevich S 2013 *Nonlin. Processes Geophys.* **20** 977-986.

[24] Druzhinin O Ostrovsky L 2015 *Nonlin. Processes Geophys.* **22** 337–348.

[25] Monin A Yaglom A 1971 *Statistical Fluid Mechanics: Mechanics of Turbulence*. V. 1. MIT Press: Cambridge, Massachusetts, and London, England; 769 pp.

[26] Moin P Manesh K 1998 *Annu. Rev. Fluid Mech.* **30** 539–78.

Journal of Physics: Conference Series **681** (2016) 012028 doi:10.1088/1742-6596/681/1/012028

New compact equation for numerical simulation of freak waves on deep water

A I Dyachenko[1,2], D I Kachulin[1] and V E Zakharov[1,2,3,4]

[1] Novosibirsk State University, Pirogova 2, Novosibirsk, 630090, Russia
[2] Landau Institute for Theoretical Physics RAS, Semenova pr. 1-a, Chernogolovka, Moscow Region, 142432, Russia
[3] Lebedev Physical Institute of RAS, Leninskiy pr. 53, Moscow, 119991, Russia
[4] Department of Mathematics, University of Arizona, Tucson, AZ, 857201, USA

E-mail: d.kachulin@gmail.com

Abstract. Considering surface gravity waves which propagate in same direction we applied canonical transformation to a water wave equation and drastically simplify the Hamiltonian. After this transformation, corresponding equation of motion is written in x-space in a compact form. This new equation is suitable for analytical studies and numerical simulations. Localized in space breather-type solution was found numerically by using iterative Petviashvili method. Numerical simulation of breathers collision shows the stability of such solutions. We observed the freak wave formation in numerical simulations of sea surface waving in the framework of new equation.

1. Compact equation

A one dimensional potential flow of an ideal incompressible fluid with a free surface in a gravity field is considered. The system is Hamiltonian system and Hamiltonian variables are the free surface elevation $\eta(x,t)$ and the velocity potential at the free boundary $\psi(x,t)$ [1]. Considering not very steep surface waves (small parameter is a steepness of the waves $\mu \sim \eta'_x$), the Hamiltonian can be represented by an infinite power series expansion on its natural variables. The Hamiltonian expansion up to 4th order as follows:

$$H = \frac{1}{2}\int g\eta^2 + \psi\hat{k}\psi dx - \frac{1}{2}\int\{(\hat{k}\psi)^2 - (\psi_x)^2\}\eta dx + \frac{1}{2}\int\{\psi_{xx}\eta^2\hat{k}\psi + \psi\hat{k}(\eta\hat{k}(\eta\hat{k}\psi))\}dx \tag{1}$$

Here $\hat{k}$ is modulus operator ($|k|$) in Fourier space and g is a gravitational acceleration. It is well known that third order terms correspond to nonresonant three-wave processes and can be excluded by using nonlinear canonical transformation.

Considering waves which propagate in same direction we applied canonical transformation not only to remove cubic nonlinear terms but to simplify drastically fourth order terms in Hamiltonian. Unlike in [2],[3] the four-wave interaction coefficient is chosen as follows:

$$\begin{aligned} \tilde{T}^{kk_1}_{k_2k_3} &= \frac{(kk_1k_2k_3)^{\frac{1}{2}}}{2\pi} min(k,k_1,k_2,k_3)\theta(kk_1k_2k_3) = \qquad (2)\\ &= \frac{(kk_1k_2k_3)^{\frac{1}{2}}}{8\pi}\left(k + k_1 + k_2 + k_3 - |k-k_2| - |k-k_3| - |k_1-k_2| - |k_1-k_3|\right)\theta(kk_1k_2k_3) \end{aligned}$$

Journal of Physics: Conference Series **681** (2016) 012028 doi:10.1088/1742-6596/681/1/012028

Transformation from $b(x,t)$ to physical variables $\eta(x,t)$ and $\psi(x,t)$ up to the second order has the form:

$$\begin{aligned}
\eta(x) &= \frac{1}{\sqrt{2}g^{\frac{1}{4}}}(\hat{k}^{-\frac{1}{4}}c(x)+\hat{k}^{-\frac{1}{4}}c(x)^*)+\frac{\hat{k}}{4\sqrt{g}}[\hat{k}^{-\frac{1}{4}}c(x)-\hat{k}^{-\frac{1}{4}}c^*(x)]^2,\\
\psi(x) &= -i\frac{g^{\frac{1}{4}}}{\sqrt{2}}(\hat{k}^{-\frac{3}{4}}c(x)-\hat{k}^{-\frac{3}{4}}c(x)^*)+\frac{i}{2}[\hat{k}^{-\frac{1}{4}}c^*(x)\hat{k}^{\frac{1}{4}}c^*(x)-\hat{k}^{-\frac{1}{4}}c(x)\hat{k}^{\frac{1}{4}}c(x)]+\\
&+ \frac{1}{2}\hat{H}[\hat{k}^{-\frac{1}{4}}c(x)\hat{k}^{\frac{1}{4}}c^*(x)+\hat{k}^{-\frac{1}{4}}c^*(x)\hat{k}^{\frac{1}{4}}c(x)] \qquad (3)
\end{aligned}$$

The third order terms are very cumbersome, so we don't give them here. After the transformation the Hamiltonian take the compact form:

$$H=\int c^*\hat{V}c\,dx+\frac{1}{2}\int\left[\frac{i}{4}(c^2\frac{\partial}{\partial x}c^{*2}-c^{*2}\frac{\partial}{\partial x}c^2)-|c|^2\hat{k}(|c|^2)\right]dx \qquad (4)$$

Here operator $\hat{V}$ in k-space is so that $V_k=\frac{\omega_k}{k}$. $\omega_k=\sqrt{gk}$ – dispersion law for surface gravity waves on deep water. Equation of motion is the following:

$$\frac{\partial c(x,t)}{\partial t}+i\hat{\omega}_k c(x,t)-i\hat{P}^+\frac{\partial}{\partial x}\left(|c(x,t)|^2\frac{\partial c(x,t)}{\partial x}\right)=\hat{P}^+\frac{\partial}{\partial x}\left(\hat{k}|c(x,t)|^2c(x,t)\right) \qquad (5)$$

here $\hat{P}^+$ is projection operator to the upper half-plane.

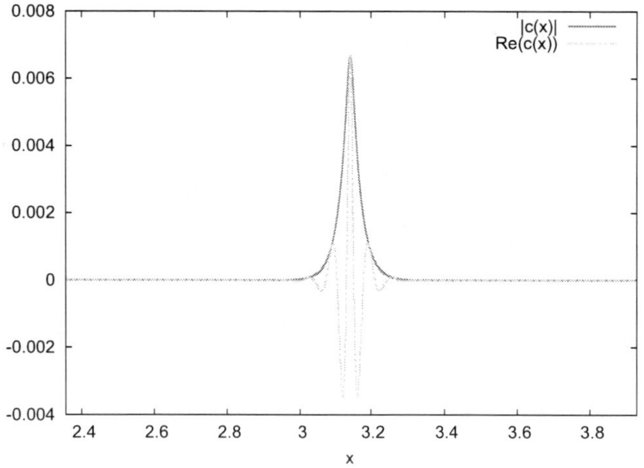

Figure 1. Narrow breather with $V=5\cdot10^{-2}$, $\Omega=5.2$ (wavenumber of the carrier wave $k_0\sim100$). Green dashed curve corresponds to the real part ($Re(c(x,0))$); red solid curve – modulus ($|c(x,0)|$)

A monochromatic wave

$$c(x,t)=C_0e^{i(k_0x-\omega_0t)} \qquad (6)$$

is the simplest solution of (5). Here C_0 is arbitrary complex constant. Substituting (6) into (5) yields the following relation for frequency shift:

$$\omega_0=\omega_{k_0}+k_0^2|C_0|^2. \qquad (7)$$

Recalling the transformation (3), one can see that relation (7) coincides with the well-known Stokes correction to the frequency due to finite wave amplitude:

$$\omega_0=\omega_{k_0}(1+\frac{1}{2}k_0^2|\eta_0|^2),\quad |C_0|^2=\frac{1}{2}\omega_{k_0}\eta_0^2 \qquad (8)$$

2. Breathers

Breather is the localized solution (5) following type:

$$c(x,t) = C(x - Vt)e^{i(k_0 x - \omega_0 t)} \tag{9}$$

Here k_0 is the wavenumber of the carrier wave. In the Fourier space breather can be written as follow:

$$c_k = e^{i(\Omega + Vk)t}\phi_k \tag{10}$$

Such solutions are completely determined by two real parameters: V – group velocity and Ω. Functions ϕ_k satisfies the equation:

$$(\Omega + Vk - \omega_k)\phi_k = \frac{1}{2}\int \tilde{T}_{kk_1}^{k_2k_3}\phi_{k_1}^*\phi_{k_2}\phi_{k_3}\delta_{k+k_1-k_2-k_3}dk_1dk_2dk_3 \tag{11}$$

Breather (9) can be found by Petviashvili method

$$\begin{aligned}\phi_k^{n+1} &= \frac{NL_k^n}{M_k}\left[\frac{<\phi^n \cdot NL(\phi^n)>}{\phi^n \cdot M\phi^n}\right]^\gamma, \qquad M_k = \Omega + Vk - \omega_k,\\ NL(\phi^n) &= -P^+\frac{\partial}{\partial x}\left(|\phi^n|^2\frac{\partial\phi^n}{\partial x}\right) + iP^+\frac{\partial}{\partial x}\left(\hat{k}\left(|\phi^n|^2\right)\phi^n\right)\end{aligned}$$

Breather solution of this equation in the periodic domain 2π with $V = 5 \cdot 10^{-2}$, $\Omega = 5.2$ (wavenumber of the carrier wave $k_0 \sim 100$) is shown in Fig.1. Figure 2 shows the spectrum of the breather in log scale. These breathers can be considered as a simple model of the freak waves.

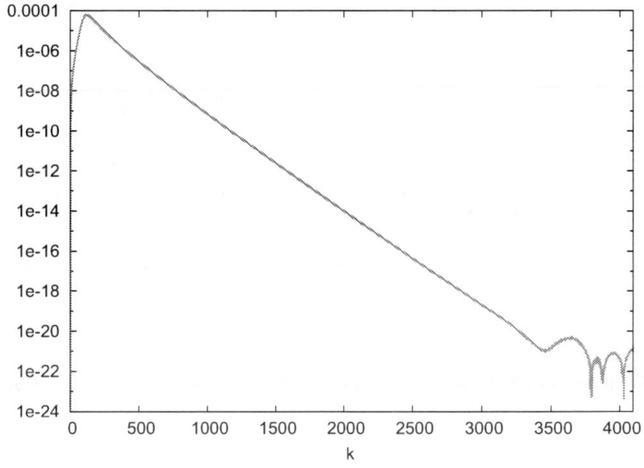

Figure 2. Spectrum ($|c_k(0)|$) of narrow breather with $V = 5 \cdot 10^{-2}$, $\Omega = 5.2$ (wavenumber of the carrier wave $k_0 \sim 100$) in logariphmic scale.

3. Interaction of breathers

Breather is very stable structure. We performed numerical simulation of breathers collision to prove this. The pseudospectral Fourier algorithm was applied for simulation the equation (5). Numerical integration of the equation (5) were carried out on the base of Runge-Kutta method 4th order accurate in time. Simulation was performed in the center-of-momentum frame.

- As the initial condition we have used two breathers separated in space (distance was equal to π.)

Journal of Physics: Conference Series **681** (2016) 012028 doi:10.1088/1742-6596/681/1/012028

- The first breather has the following parameters: $V_1 = 5 \cdot 10^{-2}$, $\Omega_1 = 5.005$ (wavenumber of the carrier wave $k_0 \sim 100$)
- For the second breather $V_2 \simeq 3.535 \cdot 10^{-2}$, $\Omega_2 \simeq 7.081$ (wavenumber of the carrier wave $k_0 \sim 200$)

Collision of two breathers is shown in Fig.3. Fourier spectrum ($|c_k|$) in log scale for different time moments is shown in Fig. 4. One can see that spectrum returns to it's initial state and the breathers profiles almost have not changed.

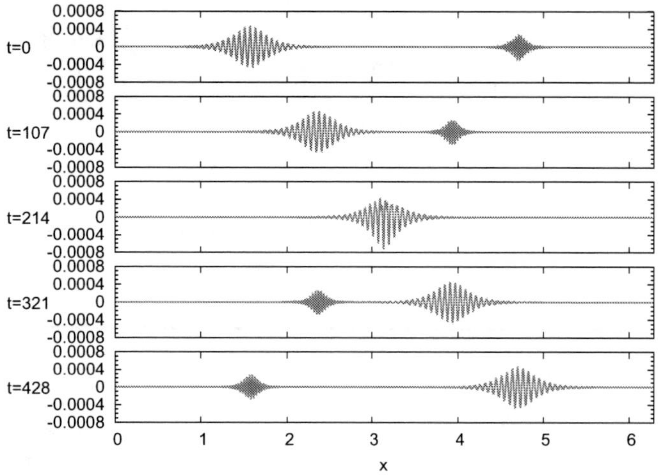

Figure 3. Snapshots of breather collision. Free surface for different time moments $t = 0, 107, 214, 321, 428$.

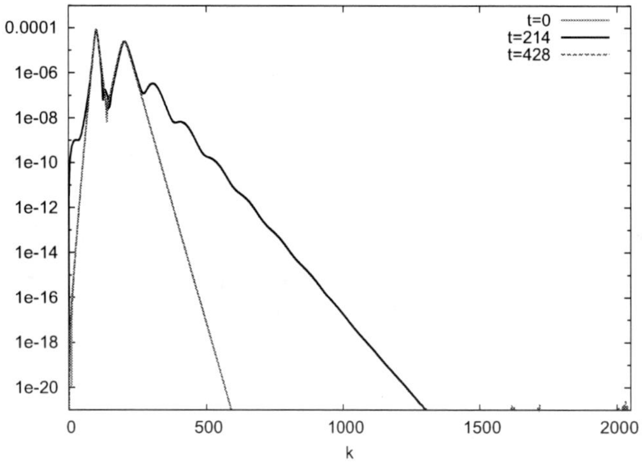

Figure 4. Fourier spectrum ($|c_k|$) in log scale for different time moments $t = 0, 214, 428$. Red solid curve corresponds to initial statement ($t = 0$); black solid curve corresponds to the collision moment ($t = 214$); blue dashed curve – spectrum after breathers collision ($t = 428$)

4. Modulational Instability of Monochromatic wave

The numerical simulation of the modulational instability of the homogeneous wave train in the framework of compact equation (5) was performed by pseudospectral Fourier method. Initial condition was chosen as slightly perturbed monochromatic wave with the steepness $\mu \sim 0.82$ and wavenumber $k_0 = 100$. After ~ 1000 wave periods on a free surface the freak wave was formed. The lifetime of this wave was about 10 periods, after which it disappeared. Snapshots of free surface for different times ($t = 0, 615, 635, 655$) is shown on figure 5. Figure 6 shows the zoomed freak wave profile at $t = 635$. One can see that the freak wave's heigth is more than 2 times greater than the heigth of neigbour waves.

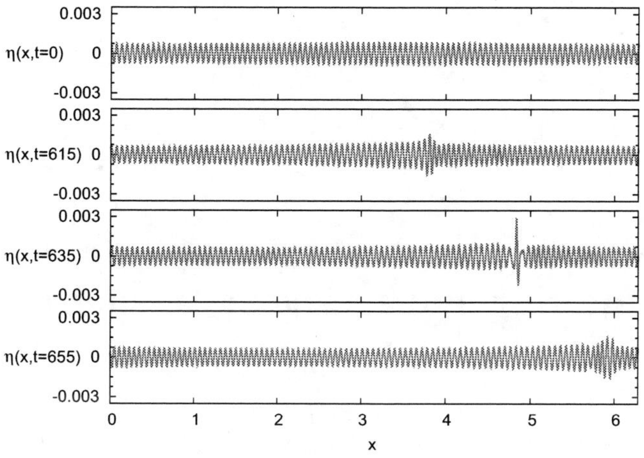

Figure 5. Freak wave formation. Snapshots of free surface for different times ($t = 0, 615, 635, 655$) is shown

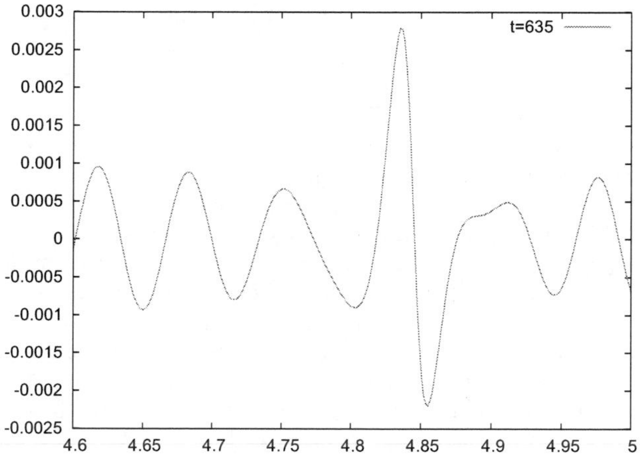

Figure 6. Freak wave on the free surface at $t = 635$.

5. Conclusion

We have performed numerical simulations of nonlinear stage of modulational instability up to the freak-wave formation in the framework of equation (5). The new compact equation (5) can be generalized for the almost 2D waves i.e. waves slightly inhomogeneous in the transverse direction y. In this case frequency $\hat{\omega}$ depends on both k_x and k_y as $\hat{\omega}_{k_x,k_y}$, while nonlinear terms not depend on y, and c now depends on both x and y:

$$\frac{\partial c(x,y,t)}{\partial t} + i\hat{\omega}_{k_x,k_y} c(x,y,t) - i\hat{P}_x^+ \frac{\partial}{\partial x}\left(|c(x,y,t)|^2 \frac{\partial c(x,y,t)}{\partial x}\right) = \hat{P}_x^+ \frac{\partial}{\partial x}\left(\hat{k}_x |c(x,y,t)|^2 c(x,y,t)\right) \tag{12}$$

Due to specific structure of nonlinearity the equation (12) can be effectively solved numerically.

Acknowledgments

This work was supported by the Grant "Wave turbulence: theory, numerical simulation, experiment" #14-22-00174 of Russian Science Foundation.

References

[1] Zakharov V E 1968 *J. Appl. Mech. Tech. Phys* **9** 190–4
[2] Dyachenko A I and Zakharov V E 2011 *JETP Lett.* **93(12)** 701–5
[3] Dyachenko A I and Zakharov V E 2012 *Europ. J. Mech. B* **32** 17–21

International Conference on Computer Simulation in Physics and Beyond 2015 IOP Publishing
Journal of Physics: Conference Series **681** (2016) 012029 doi:10.1088/1742-6596/681/1/012029

Data-intensive multispectral remote sensing of the nighttime Earth for environmental monitoring and emergency response

M Zhizhin[1,2], A Poyda[3], V Velikhov[3], A Novikov[3], A Polyakov[3]
[1]Space Research Institute of RAS, 84/32 Profsoyuznaya Str, Moscow, Russia.
[2]University of Colorado Boulder, Broadway, Boulder, CO, USA.
[3]National Research Centre "Kurchatov institute", 1 Akademika Kurchatova pl., Moscow, Russia.

E-mail: mikhail.zhizhin@gmail.com, poyda_aa@nrcki.ru

Abstract. All Most of the remote sensing applications rely on the daytime visible and infrared images of the Earth surface. Increase in the number of satellites, their spatial resolution as well as the number of the simultaneously observed spectral bands ensure a steady growth of the data volumes and computational complexity in the remote sensing sciences. Recent advance in the night time remote sensing is related to the enhanced sensitivity of the on-board instruments and to the unique opportunity to observe "pure" emitters in visible infrared spectra without contamination from solar heat and reflected light. A candidate set of the night-time emitters observable from the low-orbiting and geostationary satellites include steady state and temporal changes in the city and traffic electric lights, fishing boats, high-temperature industrial objects such as steel mills, oil cracking refineries and power plants, forest and agricultural fires, gas flares, volcanic eruptions and similar catastrophic events. Current satellite instruments can detect at night 10 times more of such objects compared to daytime. We will present a new data-intensive workflow of the night time remote sensing algorithms for map-reduce processing of visible and infrared images from the multispectral radiometers flown by the modern NOAA/NASA Suomi NPP and the USGS Landsat 8 satellites. Similar radiometers are installed on the new generation of the US geostationary GOES-R satellite to be launched in 2016. The new set of algorithms allows us to detect with confidence and track the abrupt changes and long-term trends in the energy of city lights, number of fishing boats, as well as the size, geometry, temperature of gas flares and to estimate monthly and early flared gas volumes by site or by country. For real-time analysis of the night time multispectral satellite images with global coverage we need gigabit network, petabyte data storage and parallel compute cluster with more than 20 nodes. To meet the processing requirements, we have used the supercomputer at the Kurchatov Institute in Moscow.

1. Introduction

When an emergency it is important to promptly and as fully as possible to assess the situation and the scale of the disaster to make the right decisions about the first respond to the disaster. Recent disasters, observed in different regions of the Earth, prove the urgency of the task. For example: 1) Hurricane Sandy [1], which led to large losses in the US in late October 2012 and in particular in New York on 29 October 2012, 2) War 2012 - 2015 in Syria and Iraq, which led to numerous casualties and partial destruction of infrastructure, and 3) an earthquake in Nepal on 25 April 2015, which led to more than 8,000 casualties [2]. In all these cases, the satellite observations of rapid changes and trends in the

Journal of Physics: Conference Series **681** (2016) 012029 doi:10.1088/1742-6596/681/1/012029

brightness of the nighttime lights in different spectral bands allow to evaluate the nature and the scale of the consequences of emergency situations such as power outages and disruption of production processes.

Currently, only two US satellites are sensitive enough to provide data from the night side of the Earth in visible and infrared bands: DMSP F18 and Suomi NPP [3]. However, the number of satellites in the near future will grow. In 2017 NOAA and NASA will launch another satellite, JPSS-1, which will provide multispectral data at night, similar Suomi NPP, and later satellite JPSS-2. That means a possibility of continuous satellite monitoring of the nighttime lights of the Earth for decades.

Using time-series of pixel brightness instead of instant images increases the accuracy of assessing of the emergency extent, as it accounts for natural variability associated with the seasons, cloud cover, the moon light, and makes it possible to monitor the dynamics of the disasters and their socio-economic and demographic consequences.

The scale and complexity of the problem arise from the need for statistical analysis of a large spatio-temporal dataset (collection of images), geolocated on a latitude-latitude grid of high resolution (from 15 to 30 arc seconds), especially the time series of brightness obtained for all pixels in the study region (administrative unit, the country or continent). The flow of "raw" data from the satellite Suomi NPP is about 3 terabytes per month. The volume of the raster database to be analyzed is more than 200 terabytes. With the launch in the 2016-2017 of the new satellites JPSS-1 and GOES-R, which will also provide the imagery of the Earth at night, suitable for detection of emergencies, the amount available for the data analysis will get close to 1 petabyte.

2. Data Model and Algorithms

To study rapid changes in the brightness of the nighttime lights and to assess the scale and dynamics of an emergency or a disaster, we use the following approaches and methods.

1. A geospatial database PostGIS that contains the images from the nighttime side of the Earth received from the Suomi NPP satellite with global coverage of the Earth's surface every night at spatial resolution of about 700 meters per pixel in the visible, near, mid and thermal IR bands (9 spectral channels).
   In the course of the creation of raster base we have to correct the satellite images for the terrain and to change the projection of the images from a satellite view into a regular latitude-longitude grid [4].
   In addition to the values of brightness in different spectral channels, the database contains information on the cloud cover, including the type and the optical thickness of the clouds, the orbital parameters and the orientation of the satellite, and other flags, which can be used to assess the quality of data for all images pixel by pixel.
   Data access interface can mosaic images from adjacent orbits and extract time series with pixel brightness in every spectral channel.
2. A new Nightfire algorithm is used to detect combustion sources at the night time surface of the Earth [5]. In many cases for the detected source we can estimate its temperature and size. The essence of the Nightfire is a non-linear fitting of the temperature and area of a combustion source using the simplex method. Combustion source observed by satellite is modeled as a black body. The size and temperature of the subpixel fire are found by minimizing the sum of squared residuals between the observed radiances in the IR bands and the Planck curve model. The area of the combustion source may be substantially less (up to 1 m square, the sensitivity depends on the temperature of combustion) of surface area observed in the satellite image one pixel (about 1 km square).
   Compared to the previous fire detection algorithms developed by NASA for the MODIS sensors, the new algorithm at night detects an order of magnitude more sources in the same area of observation. Moreover, the Nightfire results provide not only a place, but also the temperature and the area of fire, explosion, or gas flare.

Journal of Physics: Conference Series **681** (2016) 012029 doi:10.1088/1742-6596/681/1/012029

3. The task of detecting blackouts (or reduce power) by rapid changes of nighttime lights in the visible band is the most complex. It depends on many factors, among which the natural variation of the brightness of the nighttime lights of the villages and small towns, small transparency of atmosphere and clouds in the visible range as compared to the IR, additional illumination by moonlight (moon phases), the sun (summer in the polar latitudes) and air glow, as well as by fires when they occur during emergencies.

3. High Performance Computing Platform

To process the large volumes of the remote sensing data we have developed a parallel conveyor cloud-system. The system includes the MapReduce computing model, workflows, job queues, virtual computing nodes and cluster file system.

The main features of the parallel cloud conveyor are:

- The system monitors the state of the environment and starts the processing task only when there are all the necessary conditions. The system allows to track receipt and availability of all the files required for a certain stage of processing, updating some of the files processed before, user queries, the completion of the previous required steps of the processing flow, timer, etc.
- Using of composite tasks (workflows), in which any hosted applications hosted on operating system and called via the command line can be used. Workflows used in the system correspond to a MapReduce computational model. Using the MapReduce model is justified by the fact that the majority of workflows' atomic tasks (jobs) can be parallelized on the data.
- Using templates to describe the workflows. Templates allow to define depending on names of input and output parameters, time and order of execution of jobs, the conditions of the beginning of each task/job, etc.
- Each workflow can be performed in conveyor mode on a potentially endless stream files: treating another dataset, the application takes the next one.
- The system allows control of the number of concurrent workflows and the volume of allocated computing resources.
- Using cloud computing platform and special program module for dynamic balancing resources by increasing and decreasing of the number and configuration of running virtual machines.
- Ensuring system reliability is guaranteed by locking states - the presence of restore points, queues, messaging and objectives, as well as a distributed file system allows you to increase the resiliency of the system.

Figure 1 shows the general structure of workflow management. The files from remote storage of satellite data pre-loaded into the buffer storage system. Incoming data are recorded in the database. Then the manager creates the conveyors instances according to templates. The conveyor instances take the input files from the input stream and activate the jobs. The jobs are sent to job distribution module that sends them to cluster nodes for execution under certain circumstances (presence of a specific software, available memory, processor cores and free disk space). We use SLURM [6] as the job distribution module.

Information module logs the workload of the job queue, the amount of RAM and processor cores, compute nodes, running and using virtual machines, their resources, and the total quota of the cloud provider.

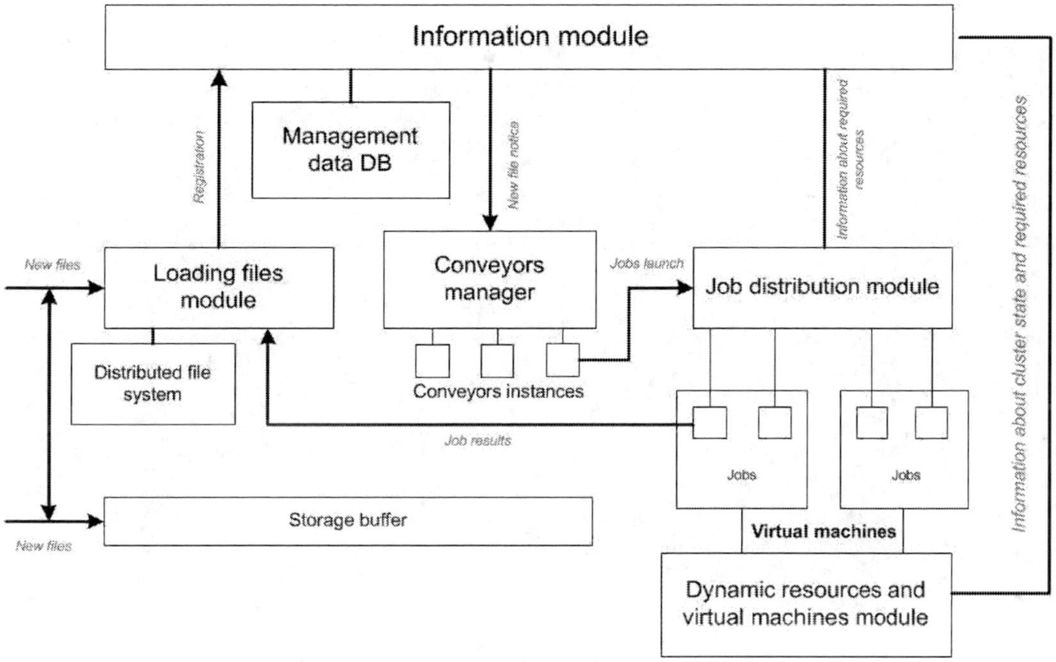

Figure 1. General structure of workflow management system

Depending on the load the system automatically changes the composition of the compute cluster by starting or stopping virtual machines. Machines are configured so that when they start they automatically register in the SLURM and from that moment they are ready to receive and process the jobs.

4. Results

An example of the detection by algorithm Nightfire from the weather satellite Suomi NPP is the explosion in Chapaevsk, Russia, 18 June 2013 is shown in figure 2. Location of the explosion is shown in blue pushpin on the Google Earth map. Detected parameters of the combustion source at the time of flight of the satellite were:

Time = 18-Jun-2013 22:14:24 UTC Latitude = 52.991741 deg North Longitude = 49.800602 deg East
Temperature = 1079 degrees K Energy combustion = 2.63 MW Area combustion = 34.22 sq m

The combustion source was detected through the cloud cover. The figures below show the explosion photo by witnesses from the scene and the result of the Planck curve fitting to the brightness recorded in the near infrared satellite channels.

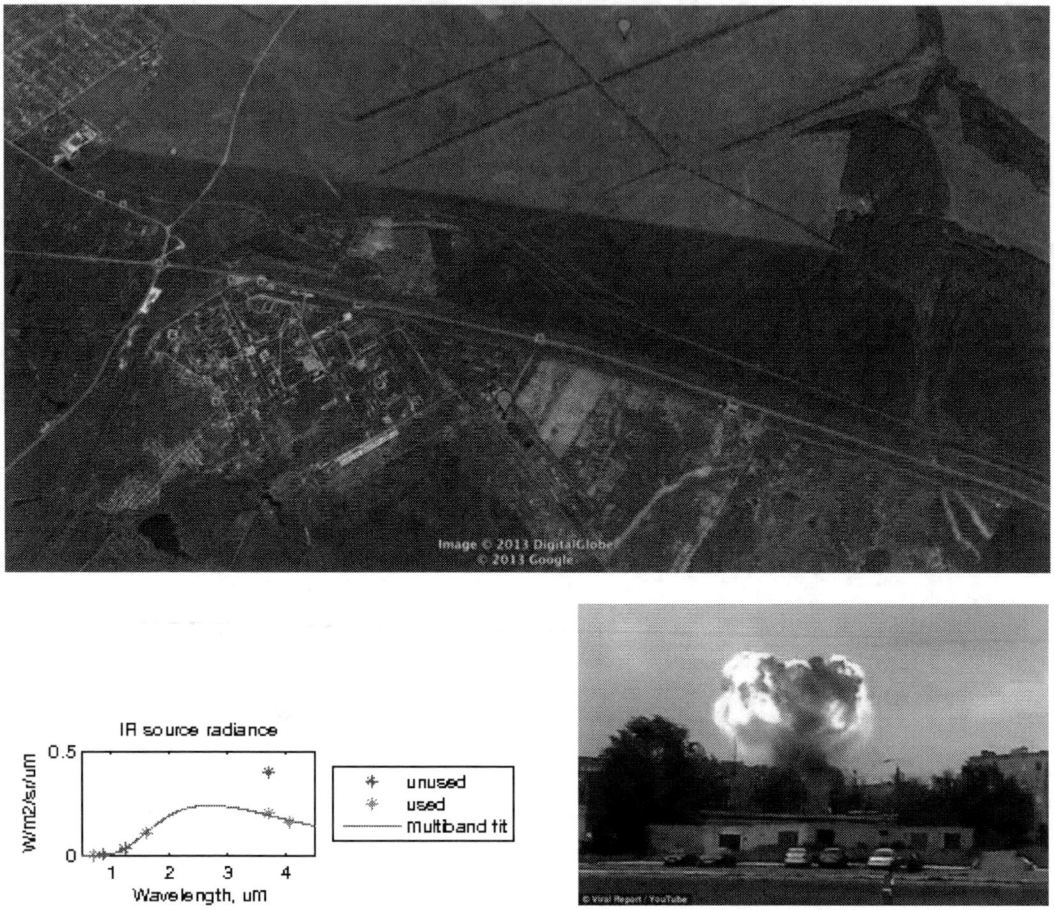

Figure 2. Nightfire detection of the Chapaevsk explosion

The Figure 3 illustrates the preliminary results of the analysis of time series of the brightness of nighttime lights in the visible (400-800 nm) and near infrared (SWIR 1.6 m) bands from the satellite Suomi NPP of oil refinery in Syria controlled by ISIS for the time period from April 2012 to April 2015. From the spring of 2013 this area is located in a war zone. The site coordinates are 35.122955 deg North and 40.543999 deg East.

Joint analysis of the time series in the infrared and visible bands, taking into account the results of the Nightfire algorithm (temperature time series of the gas flare at the refinery) allows us to conclude that the cracking at the site was suspended in February 2013, but the site was illuminated with electric lights almost all the time during the observation period, possibly except for time periods from December 2013 to May 2014 and from October 2014 to January 2015.

5. Conclusion and Future Work

Development of the automated system to monitor changes in night time lights after disasters and in war zones is still a work in progress. We have most of the components in hand, including the parallel raw data processing conveyor, the geospatial database for data mining, multispectral detector of fires and flares. However, we still analyse visible night time light changes "by hand", but this task is close toe be automated in the nearest future.

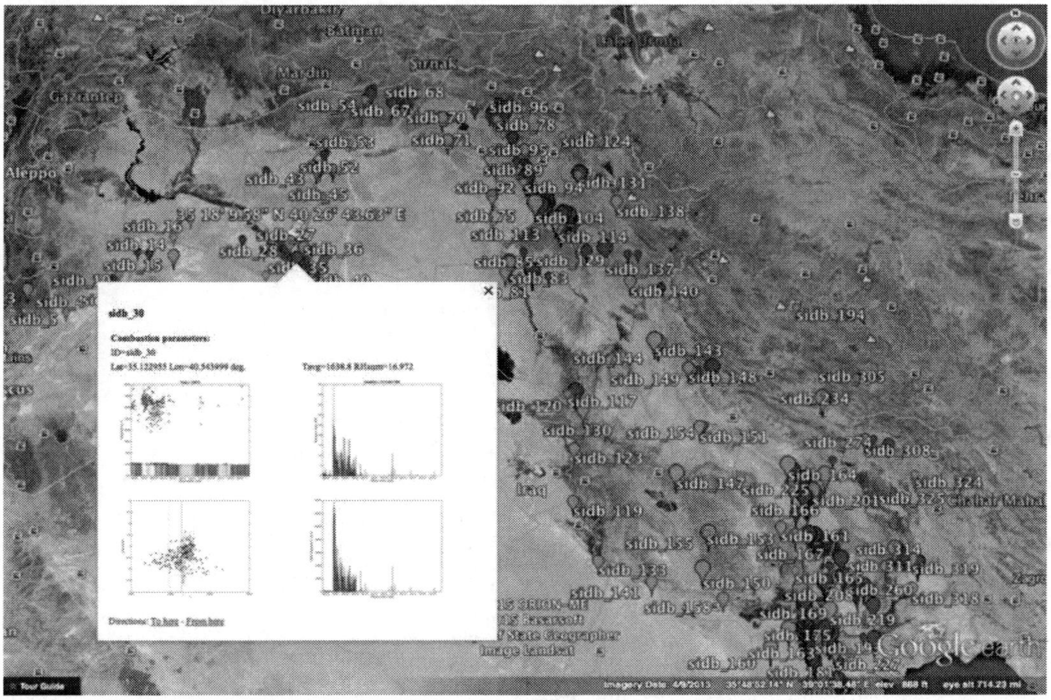

Figure 3. Time series of flare temperature and visible brightness for an oil production site in Syria

Acknowledgments

The study was partially funded by RFBR according to the research project № 15-07-07710 a. The satellite images were kindly provided by the Earth Observation Group (EOG) of the NOAA National Geophysical Data Center in Boulder, Colorado, USA. The Nightfire algorithm was developed by the EOG with contributions from M. Zhizhin and A. Poyda.

References

[1] Changyong C 2013 Detecting Light Outages After Severe Storms Using the S-NPP/VIIRS Day/Night Band Radiances, *Geoscience and Remote Sensing Letters, IEEE*, **10**, 1582–86. http://ieeexplore.ieee.org/xpl/articleDetails.jsp?arnumber=6553428

[2] NASA 2015 NASA blog entry on the night time lights change after the Nepal earthquake on April 25, http://earthobservatory.nasa.gov/IOTD/view.php?id=85816

[3] Elvidge C D, Baugh K, Zhizhin, M and Hsu F-C 2013 Why VIIRS data are superior to DMSP for mapping nighttime lights, *Proceedings of the Asia-Pacific Advanced Network*, **35**, 62-69. http://dx.doi.org/10.7125/APAN.35.7

[4] Baugh K, Hsu F-C, Elvidge C D and Zhizhin M 2013 Nighttime Lights Compositing Using the VIIRS Day-Night Band: Preliminary Results, *Proceedings of the Asia-Pacific Advanced Network*, **35**, 70-86. http://dx.doi.org/10.7125/APAN.35.8

[5] Elvidge C D, Zhizhin M, Hsu F-C and Baugh K E 2013 VIIRS Nightfire: Satellite Pyrometry at Night. *Remote Sensing*, **5(9)**, 4423-49. http://www.mdpi.com/2072-4292/5/9/4423

[6] Lawrence Livermore National Laboratory 2008 SLURM: A highly scalable resource manager. Project site: https://computing.llnl.gov/linux/slurm/

Modelling of stratified flows in the problem of the morphological behaviour of a sandpit

Ya N Parshakova[1], T P Lyubimova[1, 2], A O Ivantsov[1, 2]
[1] Institute of Continuous Media Mechanics UB RAS, Academika Koroleva 1, Perm 614013, Russia
[2] Perm State University, Bukireva 15, Perm 614990, Russia

E-mail: parshakova@icmm.ru

Abstract. The problem of the removal of brines from underwater sand pits is studied. The intensity of the removal of brines from the pits due to the changes in the hydrological regime of the river is calculated in three-dimensional approach. It is found that for the flow rates typical for the summer, the removal of brines collected in underwater pits does not occur. At flow rates typical for the spring, there is a fairly intensive removal of accumulated brines. Numerical experiments have shown that the underwater pits can serve as brines batteries, becoming an additional source of river pollution under certain hydrological conditions.

1. Introduction

The production of non-metallic building materials (NBM) from the river channel pits is one of the most widely used methods of producing NBM. In spite of the large number of investigations, associated with the evaluation of the impact of such developments on rivers, many important aspects remain open. As a rule, evaluation of the impact of mining is focused primarily on the analysis of changes in the velocity regime of the river. In most cases, these indicators adequately define the parameters governing the use of the river. The studies of changes in water level and velocity regimes due to the large-scale mining are carried out quite successfully using the one-dimensional (1D) [1-5] and two-dimensional (2D) [6-8] models. However, for areas of high anthropogenic impact on rivers it is necessary to evaluate not only the hydraulic flow parameter changes, but also hydrochemical regimes of the river. To correctly solve these problems the use of three-dimensional models (3D) is needed.

Until now 3D hydrodynamic models did not receive wide distribution in river hydraulics because of their complexity. One an example of application of such models for the solution of problems of increasing the efficiency of water supply and regulating wastewater can be found in [9]. In this study, one of the most powerful software packages CFD Fluent was used for the development of 3D-models. 3D calculations were performed on a supercomputer «URAN» of the Institute of Mathematics and Mechanics, Ural Branch of the Russian Academy of Sciences. Characteristics of the model according to the tasks of river hydraulics can be found in [10].

The present study is dedicated to the simulation of dilution and transport of highly mineralized brines in rivers taking into account density stratification effects. This problem is of high significance for ecological safety due to the growth of the exploitation of salt depositions. The production of one million tones of fertilizer is accompanied with the same quantity of highly mineralized brines demanding utilization. The desalination of such quantity of highly mineralized brines under current conditions requires extremely high power inputs and is practically impossible. That is why in Russia and abroad the main method of brine utilization is the dumping into surface water bodies or underground. Numerical investigation of the dilution and transport of highly mineralized brines in rivers in the framework of two-dimensional models is sufficient only when there is no significant density stratification, in this case it is applicable to use depth-averaged characteristics.

The problem configuration for the investigation of underwater pit washing is shown in Fig. 1.

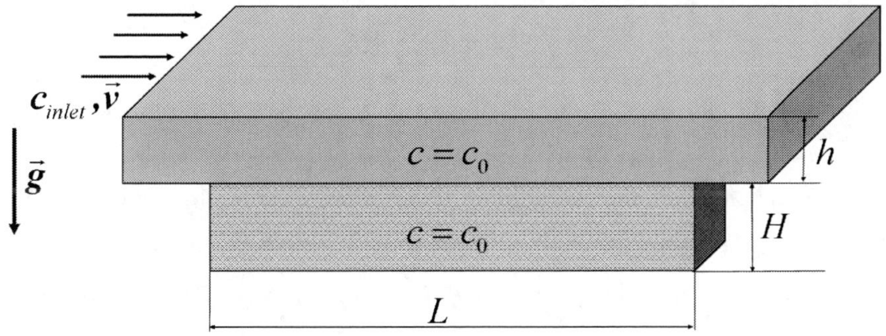

Figure 1. Problem configuration for the investigation of underwater pit washing: h – the thickness of the water layer above the pit; H & L are the depth and length of the pit

Input hydrodynamical characteristics needed to implement 3D model, were calculated on the basis of 2D-hydraulic model, realized for the whole Kama reservoir: Kama river (Tyulkino village) – the backwater of Kama hydroelectric power station. Now even with the use of powerful computational tools, such as the cluster "URAN" is not possible to build 3D models for the considered site because of the very large area of the river system under consideration.

2. Governing equations and boundary conditions.

We implement the Reynolds-averaged Navier-Stokes equations:

$$\frac{\partial \rho}{\partial t} + \frac{\partial}{\partial x_i}(\rho u_i) = 0 \quad (1)$$

$$\frac{\partial}{\partial t}(\rho u_i) + \frac{\partial}{\partial x_j}(\rho u_i u_j) = -\frac{\partial p}{\partial x_i} + \frac{\partial}{\partial x_j}\left[\mu\left(\frac{\partial u_i}{\partial x_j} + \frac{\partial u_j}{\partial x_i} - \frac{2}{3}\delta_{ij}\frac{\partial u_l}{\partial x_l}\right)\right] +$$
$$+\frac{\partial}{\partial x_j}\left[\mu_t\left(\frac{\partial u_i}{\partial x_j} + \frac{\partial u_j}{\partial x_i}\right) - \frac{2}{3}\left(\rho k + \mu_t \frac{\partial u_l}{\partial x_l}\right)\delta_{ij}\right] + \rho g_i. \quad (2)$$

,

Here, ρ is the density, x_i are coordinates (we use Cartesian coordinate system), u_i are the velocity components, μ is the kinematic viscosity, μ_t is the turbulent viscosity.

The turbulence kinetic energy k and rate of its dissipation ε are obtained from the following transport equations:

$$\frac{\partial}{\partial t}(\rho k) + \frac{\partial}{\partial x_i}(\rho k u_i) = \frac{\partial}{\partial x_j}\left[\left(\mu + \frac{\mu_t}{\sigma_k}\right)\frac{\partial k}{\partial x_j}\right] + G_k + G_b - \rho\varepsilon, \quad (3)$$

$$\frac{\partial}{\partial t}(\rho \varepsilon) + \frac{\partial}{\partial x_i}(\rho \varepsilon u_i) = \frac{\partial}{\partial x_j}\left[\left(\mu + \frac{\mu_t}{\sigma_\varepsilon}\right)\frac{\partial \varepsilon}{\partial x_j}\right] + C_{1\varepsilon}\frac{\varepsilon}{k}(G_k + C_{3\varepsilon}G_b) - C_{2\varepsilon}\rho\frac{\varepsilon^2}{k}. \quad (4)$$

In equations (3)-(4), G_k represents the generation of turbulence kinetic energy due to the mean velocity gradients $G_k = \mu_t S^2$ where S is the modulus of the mean strain rate tensor, defined as $S = \sqrt{2S_{ij}S_{ij}}$, $S_{ij} = 0.5\left(\partial u_j / \partial x_i + \partial u_i / \partial u_i\right)$, G_b is the turbulence kinetic energy due to buoyancy, which is calculated as

$$G_b = g_i \frac{1}{\rho}\left(\frac{\partial \rho}{\partial c}\right)\mu_{st}\frac{\partial c}{\partial x_i}. \tag{5}$$

i.e. taking into account the dependence of the density on concentration, β is the concentration expansion coefficient, μ_{st} is the parameter of density stratification due to the concentration, $\Pr_t$ is the turbulent Prandtl number, g_i is the component of the gravity vector in *i*-th direction; $G_{1\varepsilon}$, $C_{2\varepsilon}$, and $C_{3\varepsilon}$ are constants, σ_k and σ_ε are the turbulent Prandtl numbers for k and ε respectively. The turbulent viscosity μ_t is computed as follows: $\mu_t = \rho C_\mu k^2 / \varepsilon$, where C_μ is a constant.

The model constants $\Pr_t$, $G_{1\varepsilon}$, $C_{2\varepsilon}$, C_μ, σ_k and σ_ε were taken to have the following values: $\Pr_t = 0.85$, $C_{1\varepsilon} = 1.44$, $C_{2\varepsilon} = 1.92$, $C_\mu = 0.09$, $\sigma_k = 1.0$, $\sigma_\varepsilon = 1.3$.
The equation of mass transfer has the form

$$\frac{\partial}{\partial t}(\rho c) + \nabla \cdot (\rho \vec{v} c) = -\nabla \cdot \vec{J}, \tag{6}$$

where $\vec{J} = -(\rho D + \mu_t / \mathrm{Sc}_t)\nabla c$, is the mass flux, $Sc_t = \mu_t / \rho D_t$ is the turbulent Schmidt number, and the turbulent diffusion parameter D_t describes turbulent mass transfer.

The dependence of density on concentration is defined according to the linear law $\rho = \rho_0 - \rho_0 \beta_c (c - c_0)$, where ρ_0, c_0 are the density and concentration of the water in the absence of brines.
We set the no-slip condition, zero mass flux condition on the rigid boundaries (bottom of the river):

$$u_x = u_y = u_z = 0, \quad \frac{\partial c}{\partial n} = 0 \tag{7}$$

At the entrance of computational domain we impose the velocity of river, in which a component (perpendicular to the outfall axis) is the only non-zero constant $\mathrm{v} = \{\mathrm{v}, 0, 0\}$; the concentration is set equal to the background concentration of pollutant in the river water.

$$x = 0: \; u_x = V, \; u_y = u_z = 0, \; c = c_0 \tag{8}$$

The conditions on the other boundaries of the computational domain are:

$$\frac{\partial u_i}{\partial n} = 0 \quad (i = x, y, z), \quad \frac{\partial c}{\partial n} = 0 \tag{9}$$

At the beginning of the calculation we impose the constant or linear distribution of brines concentration:

$$c_{\max} = a + b\,z. \tag{10}$$

The computations were made using the ANSYS Fluent package of computational fluid dynamics and the k-ε model describing the turbulent pulsations. The problem was solved by applying the non-stationary isothermal approach. The second-order accuracy scheme was used to perform spatial discretization of equations. The temporal evolution was simulated using the explicit scheme of the second order. To study the effect of the computational mesh size we have performed test calculations using four meshes: with 163000, 269000, 639000 and 1258000 nodes. On the basis of the results of these test calculations, in main calculations we used the mesh with 269000 nodes. The number of grid nodes throughout the depth of the computational domain was taken equal to 30.

3. Results of modelling

The calculations included two stages. During the first stage the accumulation of brines ejected from the rectangular outlet device of the dimension 1 m x 1 m, located near the bottom at a distance of fifty meters from the front edge of the pit was simulated. The discharge of brines with the concentration c_{inlet} 300 g/l (0.3 wt%) was considered. It lasts 2 hours then stops. When the discharge occurs the accumulation of brines in the pit takes place. During the second stage of calculations the removal of accumulated brines from the pit was simulated.

The pit has the following dimensions: 100 m width, 400 m length (in the direction of flow), 6 m depth, the depth of the river h 6 meters.

The simulations were carried out for two velocities of the river according to the summer period 0,3 m/s and to the spring period 1.2 m/s. In the first case the velocity in the river during the second stage was taken to be the same as in the stage of brines accumulation in the pit $v = 0.3\ m/s$, the results of calculation for this case are presented in Figures 2 and 3. In Figure 2 the concentration fields in the middle of the vertical section of the computational domain are presented. Figure 3 presents the vertical distributions of the brines concentration at a distance 40 meters from the pit for different time moments.

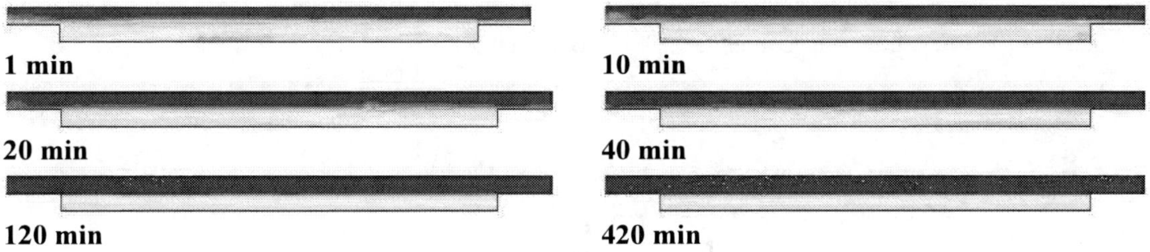

Figure 2. The intensity of washing the underwater pit in summer $v = 0.3\ m/s$. Mineralization of water: blue – less than 0.5 g/l; yellow – 10 g/l; green – transition zone

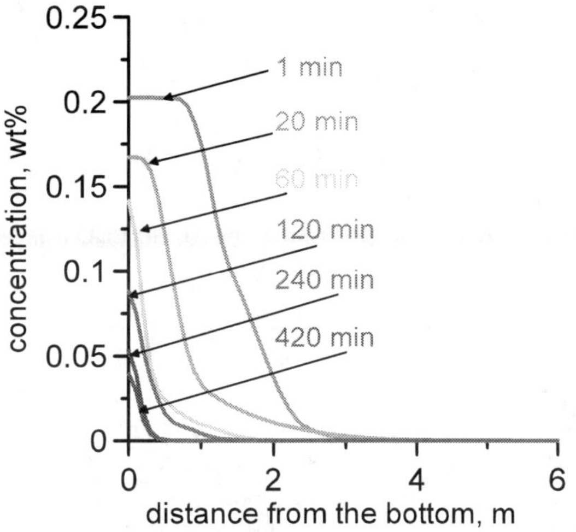

Figure 3. Vertical distribution of brine concentration in the middle of computational domain at a distance of 40 m from the pit at different time moments. Case $v = 0.3\ m/s$

As one can see from Figures 2 and 3, during the first stage the brines are carried away from the river bottom along the river, while the brines in the pit are not washed away with time. The stage of quick changes in the concentration in the river is observed within 2 hours and then the concentration field becomes stationary, brines concentration in the computational domain is not varied.

In simulations for the spring season the velocity of flow in the river was taken to be four times larger than in the calculations for summer ($v = 1.2\ m/s$). In this case, the most part of the brines is washed out from the pit (figure 4 shows the concentration of brines in the pit). The washing time is reduced with the growth of the river velocity: at the same period of time greater amount of brines is removed from the pit in spring than in the summer (see figures 3 and 5 for comparison).

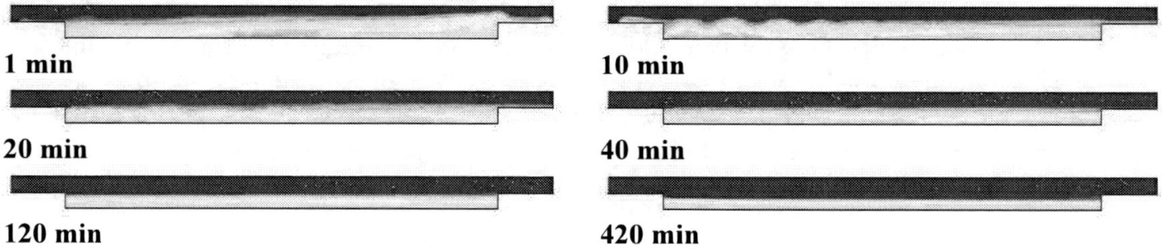

Figure 4. The intensity of washing underwater pits during the spring. Case $v = 1.2\ m/s$. Mineralization of water: blue – less than 0.5 g/l; yellow – 10 g/l; green – transition zone.

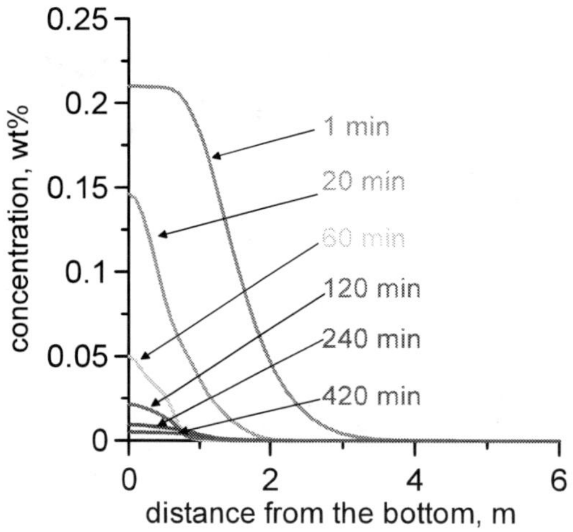

Figure 5. Vertical distribution of brine concentration in the middle of computational domain at a distance of 40m from the pit at different time moments. Case $v = 1.2\ m/s$

4. Conclusion

Thus, at flow rates typical for the summer period, there is no removal of brines accumulated in the underwater pits. At flow rates typical for the spring season there is a fairly intensive discharge of accumulated pollutants. Numerical experiments have shown that the underwater pits created in the process of large-scale production of non-metallic building materials, can serve as original batteries of brines becoming an additional tense source of river pollution under certain hydrological conditions.

Acknowledgments
This work was supported by RFBR and Perm Region Government (grant 13-01-96040) and by President of Russian Federation (grant 4310.2014.1 for the support of young PhD scientists and grant 4022.2014.1 for the support of Leading Scientific Schools).

References
[1] Fredse J, Andersen O H, Silberg S 1985 *J.Waterway, Port, Coast and Ocean Eng.* **111** (6) 1041
[2] Villaret C, Davies A G, Frey J M 2000 *Proceedings of 27th International Conference on Coastal Engineering* 2561
[3] Davies A G, Li Z 1997 *Continental Shelf Research* **17** (5) 555
[4] Ribberink J S, Al Salem A A 1995 *Coastal Engineering* **25** 205
[5] Dohmen-Janssen C M, Hassan W N, Ribberink J S 2001 *J. Geophys. Research* **106** (C11) 27103
[6] Andersen K H, Fredse J 1999 *Proceedings of Coastal Sediments* 78
[7] Brors B 1999 *J. Hydraulic Eng.* **125** (5) 511
[8] Utnes T, Meling T S 1998 *Comp. Methods in Applied Mech. and Engineer.* **169** 123
[9] Lepikhin A P, Lyubimova T P, Parshakova Y N, Tiunov A A 2012 *Journal of Mining Science* **48** (2) 390
[10] Lyubimova T, Lepikhin A, Konovalov V, Parshakova Ya, Tiunov A 2014 *Journal of Hydrology* **508** 328

International Conference on Computer Simulation in Physics and Beyond 2015 IOP Publishing
Journal of Physics: Conference Series **681** (2016) 012031 doi:10.1088/1742-6596/681/1/012031

On Boiling of Crude Oil under Elevated Pressure

Anastasiya V Pimenova[1] **and Denis S Goldobin**[1,2,3]

[1]Institute of Continuous Media Mechanics, UB RAS, Perm 614013, Russia
[2]Department of Theoretical Physics, Perm State University, Perm 614990, Russia
[3]Department of Mathematics, University of Leicester, Leicester LE1 7RH, UK

E-mail: Anastasiya.Pimenova@gmail.com, Denis.Goldobin@gmail.com

Abstract. We construct a thermodynamic model for theoretical calculation of the boiling process of multicomponent mixtures of hydrocarbons (*e.g.*, crude oil). The model governs kinetics of the mixture composition in the course of the distillation process along with the boiling temperature increase. The model heavily relies on the theory of dilute solutions of gases in liquids. Importantly, our results are applicable for modelling the process under elevated pressure (while the empiric models for oil cracking are not scalable to the case of extreme pressure), such as in an oil field heated by lava intrusions.

The problem of boiling of crude oil fields by lava intrusions is an important geophysical problem. The impact of the tectonic activity on oil fields not only controls the composition and geological characteristics of the deposits experienced this impact, but may have also an important influence on the climate. For instance, the Great Permian Extinction contemporary to the formation of the Ural mountains is hypothetically associated with massive release of light hydrocarbon gases into the atmosphere. The today's assessments of the amount of methane hydrates on the Earth are essentially more modest than the earlier ones and suggest that the destabilisation of methane hydrates could not be a sufficient source of these gases. Meanwhile, the oil fields boiled by lava intrusions in the region of forming Ural mountains could be this source. Indeed, the oil deposits in this region and the adjoin part of the contemporary basin of the Volga river are features by a nearly absent light fraction.

For modelling this process, one needs a thermodynamic model for crude oil boiling at elevated pressure. The empiric data on the oil cracking are insufficient here, since the technological process of the oil cracking is performed at atmospheric or lower pressure. In what follows we suggest the physical model for boiling of multicomponent mixtures of hydrocarbons, derive mathematical model governing the kinetics of the boiling process and evaluate its physical parameters. The derived model can be employed for modelling of the boiling process under pressure not significantly exceeding 100 atm, which corresponds to 1 km depth in sediments.

1. Boiling of multicomponent mixture of hydrocarbons

1.1. Basic physical model

As a first step, we need to determine thermodynamic equilibrium composition of the vapour phase above the liquid multicomponent mixture of hydrocarbons at its boiling temperature. This can be achieved on the basis of the following theses:

• The components are indexed by the number m of C-atoms in their molecules, and the difference in thermodynamic characteristics of the interaction of the molecules with the liquid phase between various hydrocarbons with the same number m is neglected. In support to this model reduction, let us notice that the boiling temperatures of aromatic hydrocarbons, cycloalkanes, and isomers of alkanes with m C-atoms are between the boiling points of n-alkanes with $(m-1)$ and $(m+1)$ with rare exceptions for isomers which are untypical for oil up to $m \approx 12$.

• Within the framework of our model reduction, the oil composition is described by the molar fraction $X(m)$ (or $X^{[m]}$ in our notations below) of the molecules with m C-atoms.

• As the interaction between hydrocarbon molecules is short-range and, therefore, a molecule in liquid oil interacts dominantly with the short nearest segments of nearby passing chains of other molecules, the interaction potential energy will be similar, *e.g.*, for segments of the octane's chain and the dodecane's chain. Hence, as a rough approximation, the potential energy of interaction of a molecule with the oil liquid is assumed to be independent of the oil composition.

• The molar fraction of each component in oil is small; therefore, one can consider the relation between the vapour pressure of this component above the liquid oil and its concentration in oil adopting the theory of solubility of gases in liquids for infinitely dilute solutions—specifically, the scaled particle theory [1].

• The vapour phase is considered as a perfect gas, which is reasonable as it is mainly composed by light hydrocarbons and the reference pressure is 100 atm, which corresponds to the depth of oil formation about 1 km. (In appendix of [2], one can see that the van der Waals corrections to the gas solubility are small for pressure about 100 atm.)

According to the scaled particle theory the saturated vapour pressure of specie m dilutely dissolved in the liquid is

$$P_{\mathrm{vap}}^{[m]} = \frac{RT}{v_{\mathrm{liq}}} X^{[m]} \exp\left[\frac{G_c^{[m]} + G_i^{[m]}}{k_B T}\right], \tag{1}$$

where $R = 8.314\,\mathrm{J/(mol \cdot K)}$ is the universal gas constant, T is temperature, v_{liq} is the volume of 1 mol of the liquid matter, $k_B = 1.38 \cdot 10^{-23}\,\mathrm{J/K}$ is the Boltzmann constant, G_c is the cavity formation work for the guest molecule in the solvent, G_i is the internal energy of interaction of the solute molecule with the ambient solvent molecules. The cavity formation work reads [1]

$$\frac{G_c^{[m]}}{k_B T} = -\ln(1-y) + \frac{3y}{1-y}\frac{\sigma^{[m]}}{\sigma_{\mathrm{liq}}} + \left[\frac{3y}{1-y} + \frac{9}{2}\left(\frac{y}{1-y}\right)^2\right]\left(\frac{\sigma^{[m]}}{\sigma_{\mathrm{liq}}}\right)^2 + \frac{yP}{n_{\mathrm{liq}} k_B T}\left(\frac{\sigma^{[m]}}{\sigma_{\mathrm{liq}}}\right)^3, \tag{2}$$

where $\sigma^{[m]}$ is the effective diameter of the solute molecules, σ_{liq} is the effective diameter of the solvent molecules, y is the fraction of the volume occupied by the solvent molecules in the liquid (*i.e.*, $(1-y)$ is the fraction of the volume in between the liquid molecules), $y = (\pi/6) n_{\mathrm{liq}} \sigma_{\mathrm{liq}}^3$, n_{liq} is the solvent particle number density. In what follows we take into account that, typically, the last term in equation (2) is negligible until pressure is as high as 10^8 Pa, which is not the case for our study. Energy $G_i^{[m]}$ is yet-to-be-determined.

The condition of oil boiling at pressure P is

$$P = \sum_m P_{\mathrm{vap}}^{[m]}. \tag{3}$$

At the boiling temperature (which depends on the oil composition), the equilibrium between the vapour and liquid phases is determined by the system of relations

$$Y^{[m]} P = \frac{RT}{v_{\mathrm{liq}}} X^{[m]} \exp\left[\frac{G_c^{[m]} + G_i^{[m]}}{k_B T}\right], \tag{4}$$

where $Y^{[m]}$ is the molar fraction of specie m in the vapour phase. Notice obvious relations $\sum_m X^{[m]} = 1$ and $\sum_m Y^{[m]} = 1$.

Journal of Physics: Conference Series **681** (2016) 012031 doi:10.1088/1742-6596/681/1/012031

1.2. Change of the mixture composition in the course of boiling

During boiling of the mixture, the volatile components are evaporated faster than the heavy ones. Hence, the boiling temperature increases as a part of mixture is boiled-out. Within the framework of the above-described physical model, one can derive the equation system for the change of oil composition with temperature increase.

Let us assume that under pressure P the boiling temperature T_b of the mixture with composition $X^{[m]}$ was attained. One can calculate what number δM of liquid molecules has to pass into the vapour phase for to raise the boiling point from T_b to $T_b + \delta T$;

$$(M - \delta M)\overline{X}^{[m]} + \delta M \overline{Y}^{[m]} = M X^{[m]}\,, \tag{5}$$

where M is the number of molecules in the liquid phase and the bar marks the new values of $X^{[m]}$ and $Y^{[m]}$. Equation (4) determines relation between the changed values $\overline{X}^{[m]}$ and $\overline{Y}^{[m]}$;

$$\overline{Y}^{[m]} P = \frac{R(T_b + \delta T)}{v_{\text{liq}} + \delta v_{\text{liq}}} \overline{X}^{[m]} \exp\left[\frac{G_c^{[m]}(T_b + \delta T) + G_i^{[m]}}{k_B (T_b + \delta T)}\right]. \tag{6}$$

Here $v_{\text{liq}} = \rho_{\text{liq}}^{-1} \sum_m \mu_m X^{[m]}$, where μ_m is the effective molar mass of the specie m. Neglecting variation of ρ_{liq}, one finds $\delta v_{\text{liq}} = \rho_{\text{liq}}^{-1} \sum_m \mu_m \delta X^{[m]}$.

After straightforward mathematical manipulations, one can rewrite equation (5) for small variations $\delta X^{[m]} = \overline{X}^{[m]} - X^{[m]}$ and $\delta Y^{[m]} = \overline{Y}^{[m]} - Y^{[m]}$;

$$\delta \ln X^{[m]} = \left(1 - \frac{Y^{[m]}}{X^{[m]}}\right) \delta\omega\,, \tag{7}$$

where $\delta\omega = \delta M / M$, which means that the current number of molecules in the liquid phase is related to the initial number of particles M_0 as $M = M_0 \exp(-\omega)$ and the initial value of ω is 0. Equation (6) can be rewritten as

$$\delta Y^{[m]} = Y^{[m]} \delta \ln X^{[m]} + Y^{[m]} \left(1 - \frac{G_i^{[m]}}{k_B T_b}\right) \delta \ln T_b - Y^{[m]} \frac{\sum_s \mu_s \delta X^{[s]}}{\sum_s \mu_s X^{[s]}}\,. \tag{8}$$

Substituting equation (7) into (8) and summing over m, one can obtain

$$\delta \ln T_b = \frac{\displaystyle\sum_m \frac{(Y^{[m]})^2}{X^{[m]}} - \frac{\sum_m \mu_m Y^{[m]}}{\sum_m \mu_m X^{[m]}}}{\displaystyle 1 + \sum_m \frac{-G_i^{[m]}}{k_B T_b} Y^{[m]}} \delta\omega\,.$$

Substituting the ratio $(Y^{[m]}/X^{[m]})$ from equation (4), one finds

$$\delta \ln T_b = \frac{\displaystyle\sum_m \left(\frac{R T_b}{P v_{\text{liq}}} \exp \frac{G_c^{[m]} + G_i^{[m]}}{k_B T_b}\right)^2 X^{[m]} - \frac{\displaystyle\sum_m \mu_m \frac{R T_b}{P v_{\text{liq}}} \exp \frac{G_c^{[m]} + G_i^{[m]}}{k_B T_b} X^{[m]}}{\displaystyle\sum_m \mu_m X^{[m]}}}{\displaystyle 1 + \sum_m \frac{R}{P v_{\text{liq}}} \frac{-G_i^{[m]}}{k_B} X^{[m]} \exp \frac{G_c^{[m]} + G_i^{[m]}}{k_B T_b}} \delta\omega\,. \tag{9}$$

Simultaneously, equation (7) reads

$$\delta\ln X^{[m]}=\left(1-\frac{RT_b}{Pv_{\mathrm{liq}}}\exp\frac{G_c^{[m]}+G_i^{[m]}}{k_BT_b}\right)\delta\omega\,. \tag{10}$$

Equations (9) and (10) form a closed equation system for calculation of the change of $X^{[m]}$ and the relative change of the number of molecules in the liquid phase, $M/M_0=\exp(-\omega)$, as the boiling temperature increases.

2. Evaluation of physical parameters of the system

In this section we evaluate the physical parameters of the system from indirect empiric data.

2.1. Saturated vapour pressure of liquid n-alkanes

From the data on the saturated vapour pressure (including the boiling temperature) of long-chain n-alkanes, one can assess the potential energy of the interaction of the n-alkanes with the “broth” of the crude oil $G_i^{[\mathrm{C_mH_{2m+2}}]}$. The saturated vapour pressure at 293 or 298 K for the substances from n-pentane to n-hexadecane and the boiling temperatures of the substances from n-hexane to n-tetradecane are well resembled by the expression

$$P_{\mathrm{vap}}^{[\mathrm{C_mH_{2m+2}}]}=\frac{RT}{v_{\mathrm{liq}}^{[\mathrm{C_mH_{2m+2}}]}}\exp\left[\gamma+\gamma_{\ln}\ln(m)-\frac{q(m)}{T}\right] \tag{11}$$

with $\gamma=4.178$, $\gamma_{\ln}=1.237$,

$$q(m)=\alpha m,\qquad \alpha=500.3\,\mathrm{K}; \tag{12}$$

for liquid n-alkanes $v_{\mathrm{liq}}^{[\mathrm{C_mH_{2m+2}}]}=(16.18\,m+33.00)\,\mathrm{cm^3/mol}$ at $T=273\,\mathrm{K}$ and atmospheric pressure. Here $\gamma+\gamma_{\ln}\ln(m)$ represents entropic part of the enthalpy of molecules in the liquid phase (implies that the specific entropy is a logarithm of a power-law function of the chain length m), while

$$q(m)=-G_i^{[\mathrm{C_mH_{2m+2}}]}/k_B\approx -G_i^{[m]}/k_B$$

does the potential energy of the interaction of the molecule with the ambient liquid.

2.2. Solubility of light gaseous n-alkanes in long-chain n-alkanes

Solubility of gaseous n-alkanes in long-chain n-alkanes yields information on the effective diameter $\sigma^{[m]}$ of n-alkane molecules. On the basis of the scaled particle theory the gas solubility in liquid alkane is

$$X_{(0)}^{[\mathrm{C_mH_{2m+2}},\mathrm{C_lH_{2l+2}}]}=\frac{P\,v_{\mathrm{liq}}^{[\mathrm{C_lH_{2l+2}}]}}{RT}\exp\left[-\frac{G_c^{[\mathrm{C_mH_{2m+2}},\mathrm{C_lH_{2l+2}}]}+G_i^{[\mathrm{C_mH_{2m+2}},\mathrm{C_lH_{2l+2}}]}}{k_BT}\right]. \tag{13}$$

The cavity formation work (2) reads

$$\frac{G_c^{[\mathrm{C_mH_{2m+2}},\mathrm{C_lH_{2l+2}}]}}{k_BT}=-\ln(1-y^{[\mathrm{C_lH_{2l+2}}]})+\frac{3y^{[\mathrm{C_lH_{2l+2}}]}}{1-y^{[\mathrm{C_lH_{2l+2}}]}}\frac{\sigma^{[\mathrm{C_mH_{2m+2}}]}}{\sigma^{[\mathrm{C_lH_{2l+2}}]}}$$
$$+\left[\frac{3y^{[\mathrm{C_lH_{2l+2}}]}}{1-y^{[\mathrm{C_lH_{2l+2}}]}}+\frac{9}{2}\left(\frac{y^{[\mathrm{C_lH_{2l+2}}]}}{1-y^{[\mathrm{C_lH_{2l+2}}]}}\right)^2\right]\left(\frac{\sigma^{[\mathrm{C_mH_{2m+2}}]}}{\sigma^{[\mathrm{C_lH_{2l+2}}]}}\right)^2. \tag{14}$$

Fitting the empiric data on the solubility of ethane and propane in long-chain alkanes [4, 5] at various temperatures, one finds

$$\sigma^{[m]}\approx\sigma^{[\mathrm{C_mH_{2m+2}}]}=\sigma_0 m^{1/3},\qquad \sigma_0=2.74\,\text{Å}\,. \tag{15}$$

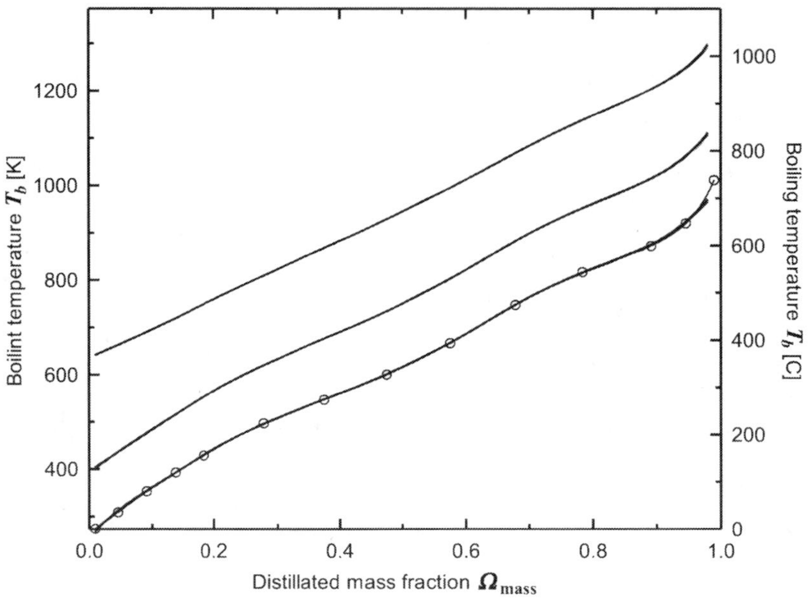

Figure 1. Temperature–distillation chart of typical crude oil: Experimental distillation chart for atmospheric pressure is plotted with circles. Bold lines represent the distillation charts calculated with the model (9)–(10) for a reconstructed composition of crude oil at $P = 1\,\text{atm}$, $1\,\text{MPa}$ and $10\,\text{MPa}$ (from bottom to top).

2.3. Solubility of gaseous n-alkanes in crude oil

The important-for-entropy characteristics of the oil "broth" as a solvent, y and σ_{liq}, can be assessed from the empiric data on the solubility of light gases (methane, ethane) in crude oil. From empiric data for crude oils [6], with equations (4), (2), (12), (15), one can find

$$y = 0.401\,, \qquad \sigma_{\text{liq}} = 6.20\,\text{Å}\,. \tag{16}$$

With these parameters we calculate solubility determined as the number of the solute molecules per the unit volume of solvent (X/v_{liq}), since this quantity (but not the molar fraction) varies insignificantly for solutions of a light gas in long-chain hydrocarbons or crude oils. For calculations with the molar fraction X and equations (1) or (4), one has to use

$$v_{\text{liq}} = \frac{\sum_m \mu_m X^{[m]}}{\rho_{\text{liq}}}\,,$$

where $\mu_m = 14m\,\text{g/mol}$ is the effective molar mass of the specie m.

2.4. Crude oil composition and temperature–distillation chart

Up to the authors' knowledge, the comprehensive data on the composition of crude oil (even for some sort of it), similar to data of chromatographic analysis, are not available in the literature. However, with known parameters (12), (15) and (16), one can use the model (9)–(10) to reconstruct the initial composition $X^{[m]}$ from the temperature-distillation chart—the dependence of the boiling temperature on the evaporated volume fraction of liquid oil. The volume fraction Ω_{vol} first needs to be recalculated into the mass fraction Ω_{mass};

$$1 - \Omega_{\text{mass}} = (1 - \Omega_{\text{vol}})\frac{\rho_{\text{liq}}(\Omega_{\text{vol}})}{\rho_{\text{liq}}(\Omega_{\text{vol}} = 0)}\,.$$

The value ω is related to Ω_{mass} as follows;

$$1 - \Omega_{\text{mass}} = \frac{e^{-\omega} \sum_m \mu_m X^{[m]}}{\sum_m \mu_m X^{[m]}_{\text{init.}}},$$

where $X^{[m]}_{\text{init.}}$ is the oil composition prior to distillation.

In figure 1, one can see the results of reconstruction of the crude oil composition for a typical temperature–distillation chart. Notice, the temperature–distillation charts for elevated pressure are not results of mere rescaling or simple transformation of the chart for atmospheric pressure; the shift of the chart along temperature neither correspond to nearly logarithmic nor nearly polynomial dependence of the boiling temperature on pressure. For higher pressures the chart becomes more linear.

One should treat the results on oil composition for high values of m with care. The derived composition provides correct behaviour for the early and middle stages of the distillation process; however, for the late stage, when only few components are remaining in the liquid phase, their molar fractions become non-small and one cannot rely on the theory of dilute solutions anymore. Additionally, for extremely high temperatures, chemical transformations of species become important for the distillation process. Their role increases for elevated pressure, since the boiling temperatures increases as pressure increases. Nonetheless, the subject of our primary interest is the early stages of oil distillation, where it looses light fractions. Thus, the discussed restrictions on the theory applicability are not relevant for our task even for the case of elevated pressure.

3. Conclusion

We have constructed the theoretical model of boiling of multicomponent mixtures of hydrocarbons (equations (9)–(10)), calculated physical parameters of this model (equations (12), (15) and (16)) and employed them for evaluation of the crude oil composition from the temperature–distillation chart. With these data evaluated, one can numerically simulate the kinetics of crude oil boiling under elevated pressure. In particular, one can quantitatively assess the role of boiling of oil fields by lava intrusions for the light hydrocarbon gases release into the atmosphere, which is considered as a potential driving force of the Great Permean Extinction.

Acknowledgments

Authors acknowledge financial support by the Government of Perm Region (Contract C-26/0004.3) and the Russian Foundation for Basic Research (project no. 14-01-31380_mol_a).

References

[1] Pierotti R A 1976 *Chem. Rev.* **76** 717–726
[2] Goldobin D S and Brilliantov N V 2011 *Phys. Rev.* E **84** 056328
[3] Goff J A 1957 *Trans. Am. Soc. Heat. Ventilat. Eng.* **63** 347–354
[4] Bufkin B A, Robinson R L, Esirera S S and Luks K D 1980 *J. Chem. Eng. Data* **31** 421–423
[5] Hayduk W, Walter E B and Simpson Ph 1972 *J. Chem. Eng. Data* **17** 59–61
[6] Hammer I P, Hessel F A, Martin J W and Rust J B 1959 Gaseous hydrocarbon and carbon dioxide solutions in hydrocarbons, US Patent 2875832 A

Implicit scheme for Maxwell equations solution in case of flat 3D domains

Marina Boronina, Vitaly Vshivkov

Institute of Computational Mathematics and Mathematical Geophysics SB RAS, Novosibirsk, Russia

E-mail: boronina@ssd.sscc.ru

Abstract. We present a new finite-difference scheme for Maxwell's equations solution for three-dimensional domains with different scales in different directions. The stability condition of the standard leap-frog scheme requires decreasing of the time-step with decreasing of the minimal spatial step, which depends on the minimal domain size. We overcome the conditional stability by modifying the standard scheme adding implicitness in the direction of the smallest size. The new scheme satisfies the Gauss law for the electric and magnetic fields in the final-differences. The approximation order, the maintenance of the wave amplitude and propagation speed, the invariance of the wave propagation on angle with the coordinate axes are analyzed.

1. Introduction

A standard scheme for Maxwell equation solution in plasma physics is the leap-frog scheme. The advantages of the scheme are second order of accuracy, time reversibility, simplicity, however the stability condition is $c\tau/h < 1$, where $h = min\{hx; hy; hz\}$, c the light speed. In case of flat ultrarelativistic beams in supercolliders the beam sizes ratio $\sigma_x : \sigma_y : \sigma_z$ may be $1 : 200 : 60000$. The high ratio of vertical beam sizes provides higher luminosity, and the purpose of the experiments is to increase the maximum available luminosity. The high relativistic factor requires the domain boundaries to be close to the beam. The three-dimensional problem of ultrarelativistic beam dynamics in supercolliders is complicated [1], and the present effective parallel load-balanced algorithms based on the leap-frog scheme limit the beam sizes ratio to $\sim 1 : 50 : 500$. However, the stability condition in case of such flat domains (but not the accuracy condition) requires decreasing of the time-step with decreasing of the minimal spatial step, and decreasing of the number of times-steps in order to achieve a specified time moment [2].

We suggest the new scheme to be implicit in the direction of the smallest beam size and the domain. We present the study of the scheme approximation, convergence, behaviour of the wave amplitude and propagation speed in the calculations using the scheme.

2. The scheme study

The leap-frog scheme for Maxwell's equations is defined on the grids, which are shifted by half steps in time and space:

$$\frac{H^{m+1/2} - H^{m-1/2}}{\tau} = -rot_h E^m,$$

$$\frac{E^{m+1}-E^{m}}{\tau}=j^{m+1/2}+rot_h H^{m+1/2},$$

where

$$rot_h H=\left|\begin{array}{c}\dfrac{Hz_{i,k,l-1/2}-Hz_{i,k-1,l-1/2}}{h_y}-\dfrac{Hy_{i,k-1/2,l}-Hy_{i,k-1/2,l-1}}{h_z}\\ \dfrac{Hx_{i-1/2,k,l}-Hx_{i-1/2,k,l-1}}{h_z}-\dfrac{Hz_{i,k,l-1/2}-Hz_{i-1,k,l-1/2}}{h_x}\\ \dfrac{Hy_{i,k-1/2,l}-Hy_{i-1,k-1/2,l}}{h_x}-\dfrac{Hx_{i-1/2,k,l}-Hx_{i-1/2,k-1,l}}{h_y}\end{array}\right|,$$

$$rot_h E=\left|\begin{array}{c}\dfrac{Ez_{i-1/2,k+1/2,l}-Ez_{i-1/2,k-1/2,l}}{h_y}-\dfrac{Ey_{i-1/2,k,l+1/2}-Ey_{i-1/2,k,l-1/2}}{h_z}\\ \dfrac{Ex_{i,k-1/2,l+1/2}-Ex_{i,k-1/2,l-1/2}}{h_z}-\dfrac{Ez_{i+1/2,k-1/2,l}-Ez_{i-1/2,k-1/2,l}}{h_x}\\ \dfrac{Ey_{i+1/2,k,l-1/2}-Ey_{i-1/2,k,l-1/2}}{h_x}-\dfrac{Ex_{i,k+1/2,l-1/2}-Ex_{i,k-1/2,l-1/2}}{h_y}\end{array}\right|.$$

The scheme provides the second order of accuracy in space and time because of the "central" differences. Let us consider a component of an electric field in a system without currents:

$$\frac{Ex^{m+1}_{i,l-1/2,k-1/2}-Ex^{m}_{i,l-1/2,k-1/2}}{\tau}=$$
$$\frac{Hz^{m+1/2}_{i,l,k-1/2}-Hz^{m+1/2}_{i,l-1,k-1/2}}{h_y}-\frac{Hy^{m+1/2}_{i,l-1/2,k}-Hy^{m+1/2}_{i,l-1/2,k-1}}{h_z}$$

The magnetic field components we can exclude subtracting the values on the previous time step:

$$\frac{Ex^{m+1}_{i,l-1/2,k-1/2}-2Ex^{m}_{i,l-1/2,k-1/2}+Ex^{m-1}_{i,l-1/2,k-1/2}}{\tau^2}=$$
$$\frac{Ex^{m}_{i+1,l-1/2,k-1/2}-2Ex^{m}_{i,l-1/2,k-1/2}+Ex^{m}_{i-1,l-1/2,k-1/2}}{h_x^2}+$$
$$\frac{Ex^{m}_{i,l+1/2,k-1/2}-2Ex^{m}_{i,l-1/2,k-1/2}+Ex^{m}_{i,l-3/2,k-1/2}}{h_y^2}+$$
$$\frac{Ex^{m}_{i,l-1/2,k+1/2}-2Ex^{m}_{i,l-1/2,k-1/2}+Ex^{m}_{i,l-1/2,k-3/2}}{h_z^2}$$

The equation is hyperbolic, and looks similar for any other component of the electric and magnetic fields. The scheme for the one-dimensional case is following:

$$\frac{y_i^{m+1}-2y_i^m+y_i^{m-1}}{\tau^2}=\frac{y_{i+1}^m-2y_i^m+y_{i-1}^m}{h^2}$$

Let us analyze it. The solution $y_l^m=\lambda^m e^{i\alpha l}$ provides the following relation:

$$\lambda-2-\frac{1}{\lambda}=\frac{\tau^2}{h^2}(e^{i\alpha}-2+e^{i\alpha})$$

and the roots:

$$\lambda = 1 - 2\frac{\tau^2}{h^2}\sin^2\frac{\alpha}{2} \pm 2\frac{\tau}{h}\sin\frac{\alpha}{2}\sqrt{\frac{\tau^2}{h^2}\sin^2\frac{\alpha}{2} - 1}$$

The production of the roots is 1, and if the roots are real and different, then the absolute value of one root is grater then 1, and the scheme is unstable. If the roots are complex, then the absolute values of the both roots are equal to 1, what means the maintenance of the wave amplitude and the stability of the scheme. Hence, the stability condition is following:

$$\frac{\tau^2}{h^2}\sin^2\frac{\alpha}{2} \le \frac{\tau^2}{h^2} \le 1 \tag{1}$$

Considering the solution as $y_l^m = A\exp(-i(\omega tm - klh))$ we obtain the dispersion relation:

$$\sin\frac{\omega\tau}{2} = \frac{\tau}{h}\sin\frac{kh}{2},$$

and, using the Taylor series for τ and h for

$$\omega = \frac{2}{\tau}\arcsin\left(\frac{\tau}{h}\sin\frac{kh}{2}\right),$$

we obtain the relation for the speed of the wave propagation:

$$u = \frac{\omega}{k} = 1 + k^2\left(\frac{\tau^2 - h^2}{24}\right) + O(h^3, \tau^3, h\tau^2, h^2\tau)$$

This implies, that if the stability criterion (1) is satisfied, the speed of the wave propagation is smaller then the speed of the light. The speed decreases with the increasing of the wavenumber k. The maximum value of the wavenumber for the domain length L depends on the number of the grid nodes n for the wave description, thus $k_{max} = 2\pi n/L = 2\pi/h$, and the speed of the wave propagation can be negative when $1+4\pi^2(\tau^2-h^2)/24h^2 < 0$, that is when $\tau^2/h^2 < 1-6/\pi^2 \sim 0.4$. So the leap-frog scheme has the restrictions for the steps usage for the short-wave modes.

We add the values from $m+1$ time step to the right hand side in order to increase the scheme stability. Having a need to keep the symmetry of the scheme in time we suggest the following scheme:

$$\frac{y_i^{m+1} - 2y_i^m + y_i^{m-1}}{\tau^2} = \delta\Delta_{xx}y_i^{m+1} + (1-2\delta)\Delta_{xx}y_i^m) + \delta\Delta_{xx}y_i^{m-1}$$

where

$$\Delta_{xx}y_i^m = \frac{y_{i+1}^m - 2y_i^m + y_{i-1}^m}{h^2},$$

and $0 \le \delta \le 1/2$. For $\delta = 0$ we obtain the original leap-frog scheme. Making similar deductions we analyze the scheme. The corresponding equation for the roots is following:

$$(1+4\delta A)\lambda^2 - 2(1-2(1-2\delta)A)\lambda + 1 + 4\delta A = 0,$$

where $A = \frac{\tau^2}{h^2}\sin^2\frac{\alpha}{2}$. The stability condition turns to the condition of the negativity of the discriminant $(1-2(1-2\delta)A)^2 - (1+4\delta A)$, that is $\delta \ge 1/4$. Hence, for $\delta = 1/4$ the scheme is absolutely stable [3] and maintains the wave amplitude. Similarly to the previous case, the dispersion relation provides the results for the propagation speed:

$$u = \frac{\omega}{k} = 1 + k^2 \left(\frac{\tau^2 - h^2}{24} + \frac{\delta \tau^2}{2} \right) + O(h^3, \tau^3, h\tau^2, h^2\tau)$$

and the speed of the wave propagation depends on the wavenumber k, the oscillations with a high wavenumber may have negative propagation speed for a certain ratio of the spatial and the time steps. For example, for $\delta = 1/4$ $k = k_{max}$ with $\tau^2/h^2 < (\pi^2 - 6)/4\pi^2 \sim 0.09$ the waves travel in the opposite direction. Note, that the optimal step ration for the scheme is $\tau/h = 1$, for the implicit scheme the optimal ratio is $\tau/h = 0.5$, in this case the approximation order arises. The fig. 1 demonstrates the solution for $k = 10$. The graphs for both schemes coincide, as the wavenumber is not high. The fig. 2 demonstrates the solutions for $k = 100$, the difference in the wave speeds is the result of the difference $k^2\tau^2/8$ between the two schemes.

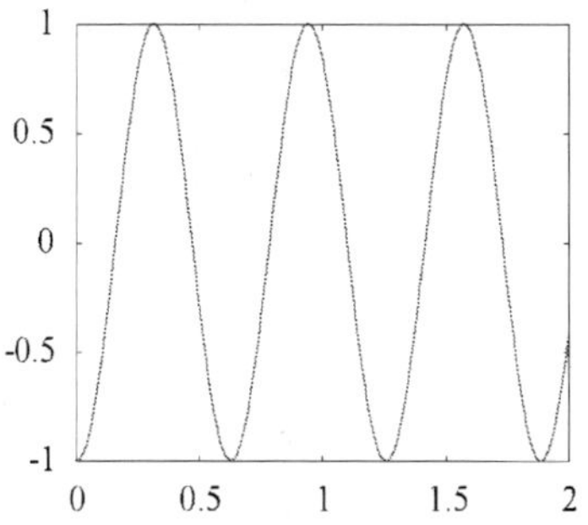

Figure 1. Figure caption for first of two sided figures.

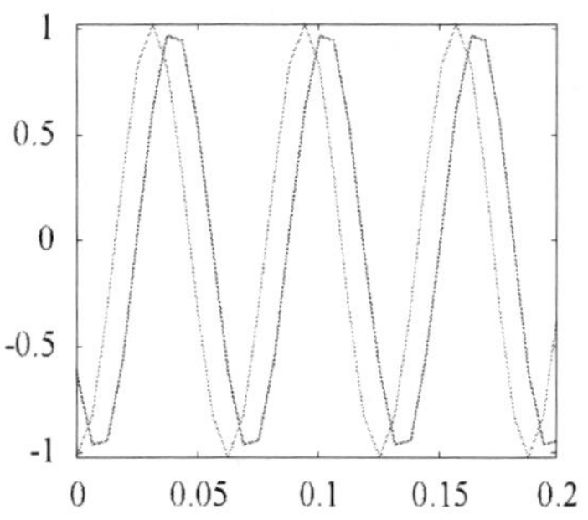

Figure 2. Figure caption for second of two sided figures.

So, both schemes change the propagation speed in cases of high wavenumbers. But the implicit scheme allows using a big time step in the calculations.

Let us use the following scheme for the three-dimensional case:

$$\frac{Ex^{m+1}_{i,l,k} - 2Ex^m_{i,l,k} + Ex^{m-1}_{i,l,k}}{\tau^2} =$$

$$\delta \Delta_{xx} Ex^{m+1}_{i,l,k} + (1 - 2\delta) \Delta_{xx} Ex^m_{i,l,k} + \delta \Delta_{xx} Ex^{m-1}_{i,l,k}$$

$$+ \delta \Delta_{yy} Ex^m_{i,l,k} + \delta \Delta_{zz} Ex^m_{i,l,k}.$$

Taking the solution as $Ex^m_{j,l,k} = \lambda^m \exp(i(\alpha j + \beta l + \gamma k))$, we obtain, that contrary to the stability condition of the explicit scheme $\tau^2/h_x^2 + \tau^2/h_y^2 + \tau^2/h_z^2 \leq 1$ the stability condition of the scheme with implicitness does not depend on h_x: $\tau^2/h_y^2 + \tau^2/h_z^2 \leq 1$.

The analysis above was was made only for one component of the electric field. However, the similar results can be obtains for any other components.

So, the final three-dimensional scheme is following:

$$\frac{Hx^{m+1/2} - Hx^{m-1/2}}{\tau} = \Delta_z Ey^m - \Delta_y Ez^m$$

$$\frac{Hy^{m+1/2} - 2Hy^{m-1/2} + Hy^{m-3/2}}{\tau^2} = \Delta_{yy} Hy^{m-1/2} + \Delta_{zz} Hy^{m-1/2} +$$
$$+\frac{1}{4}\Delta_{xx} Hy^{m+1/2} + \frac{1}{2}\Delta_{xx} Hy^{m-1/2} + \frac{1}{4}\Delta_{xx} Hy^{m-3/2}$$

$$\frac{Hz^{m+1/2} - 2Hz^{m-1/2} + Hz^{m-3/2}}{\tau^2} = \Delta_{yy} Hz^{m-1/2} + \Delta_{zz} Hz^{m-1/2} +$$
$$+\frac{1}{4}\Delta_{xx} Hz^{m+1/2} + \frac{1}{2}\Delta_{xx} Hz^{m-1/2} + \frac{1}{4}\Delta_{xx} Hz^{m-3/2}$$

$$\frac{Ex^{m+1} - Ex^m}{\tau} = \Delta_y Hz^{m+1/2} - \Delta_z Hy^{m+1/2}$$

$$\frac{Ey^{m+1} - Ey^m}{\tau} = \Delta_z Hx^{m+1/2} - \Delta_x Hz^{m+1/2}$$

$$\frac{Ez^{m+1} - Ez^m}{\tau} = \Delta_x Hy^{m+1/2} - \Delta_y Hx^{m+1/2}$$

The Gauss laws in finite differences are following:

$$\frac{1}{4}\Delta_x \left(Hx^{m+1/2} + 2Hx^{m-1/2} + Hx^{m-3/2}\right) + \Delta_y Hy^{m-1/2} + \Delta_z Hz^{m-1/2} = 0$$

for $divH = 0$ and

$$\Delta_x Ex^m + \Delta_y Ey^m + \Delta_z Ez^m = 4\pi\rho^m$$

for $divE = 4\pi\rho$.

Acknowledgments
The support of the Russian Foundation for Basic Research grant 14-01-00392 "Development of the effective parallel algorithms for modeling the processes in the relativistic plasma physics" is gratefully acknowledged.

3. Conclusion

A new scheme for solution of Maxwell equations in flat 3D domains is presented. The scheme is based on the standard leap-frog scheme and is implicit in the direction of the smallest domain size. The approximation order and the stability of the scheme in one-dimensional and three-dimensional cases are analyzed. The study of the wave propagation in different directions demonstrated that the both schemes are non-invariant on the angle with the coordinate axes. It is shown, that the scheme maintains the amplitude of the wave and maintains the wave propagation speed with the second order in space and time. The difference between the two schemes results are insignificant, however the new scheme allows using bigger time step in order to decrease of the calculations time.

References

[1] Boronina M. and Vshivkov V. 2015 3D computer sumulations of the ultrarelativistic beam dynamics in supercolliders *Proceedings of IPAC2015, Richmond, VA, USA*

[2] Boronina M and Korneev V. 2013 The parallel three-dimensional PIC code for the numerical modeling of ultrarelativistivc beams *Bull. Nov. Comp. Center* **16** 15–22.

[3] Samarskiy A. and Gulin A. 1989 *Numerical methods* (Moscow: Nauka)

International Conference on Computer Simulation in Physics and Beyond 2015 IOP Publishing
Journal of Physics: Conference Series **681** (2016) 012033 doi:10.1088/1742-6596/681/1/012033

Mass transfer during drying of colloidal film beneath a patterned mask that contains a hexagonal array of holes

Yu Yu Tarasevich, I V Vodolazskaya
Astrakhan State University, Russia
E-mail: tarasevich@asu.edu.ru

Abstract. We simulated an experiment in which a thin colloidal sessile droplet is allowed to dry out on a horizontal hydrophilic surface when a mask just above the droplet predominantly allows evaporation from the droplet free surface directly beneath the holes in the mask [Harris D J, Hu H, Conrad J C and Lewis J A 2007 *Phys. Rev. Lett.* **98** 148301]. We considered one particular case when centre-to-centre spacing between the holes is much less than the drop diameter. In our model, advection, diffusion, and sedimentation were taken into account. FlexPDE was utilized to solve an advection-diffusion equation using the finite element method. The simulation demonstrated that the colloidal particles accumulate below the holes as the solvent evaporates. Diffusion can reduce this accumulation.

1. Introduction

In our research, we try to simulate the results of the actual experiments [1, 2, 3]. When a colloidal film dries under a mask, pattern formation can be observed. The mask induces periodic variations between regions of free and of hindered evaporation. The inhomogeneous evaporation induces flows in the film. The flows carry the colloidal particles towards regions of higher evaporative flux. The particles therefore accumulate under the holes in the mask. Completely dried films exhibit patterns. These patterns can be regulated by the mask geometry, the distance between the mask and the underlying film, and the initial concentration of the colloidal particles. In fact, this pattern formation is a kind of well-known coffee-ring effect [4]. Our model is based on the use of an advection-diffusion equation [5, 6, 7, 8, 9].

2. Model

In the experiments [1, 2, 3], the diameter of the droplet, d, was about 15 mm and initial height, h_0, was about 100 μm, i.e. $h_0/d \simeq 0.01$. The patterned mask contained a hexagonal array of holes with diameter $d_h = 250\mu$m, i.e. $d_h/d \simeq 0.01$. Centre-to-centre spacing between the holes, P, varied from $2d_h$ to $10d_h$, i.e. $P/d \simeq 0.1$. As a first approximation, the film and the mask can be treated as infinitely large. In this case, only one hexagonal cell is needed for consideration due to translational symmetry. The behaviour of the such cell is expected to be similar with a drop covered by a lid that had only a small hole over the centre of the drop through which the vapor could escape. The lid restricted the evaporation from the perimeter. The resulting deposit was uniform rather than being concentrated at the edge [10]. This situation has been

Journal of Physics: Conference Series **681** (2016) 012033 doi:10.1088/1742-6596/681/1/012033

simulated using lubrication approximation [11]. Recently, effect of diffusion and hole radius on the component redistribution inside the droplet has been investigated numerically [12]. The opposite limiting case, when the distance between the holes is comparable with the diameter of the drop ($P \approx d$) and the drop itself can be treated as a spherical cup, has been considered [13].

We considered a thin colloidal film on a horizontal substrate (Fig. 1). The substrate was assumed to be solid, hydrophilic, impermeable, and indissoluble by the solvent. We supposed that the suspended particles have no effect on the hydrodynamics. Moreover, we assumed that volume fraction of the colloidal particles, Φ, is small in any internal volume of the film at any time ($\Phi \ll 1$). For the coffee-ring effect, a model accounting finite volume of the solute particles in evaporating sessile drops of a colloidal solution on a plane substrate was proposed in [14].

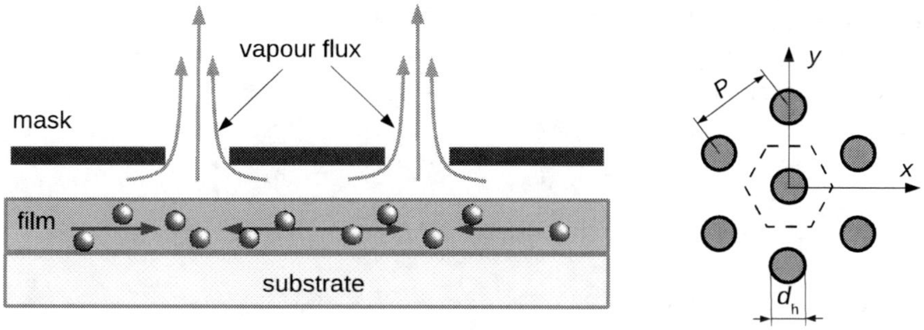

Figure 1. Sketches of a colloidal film beneath a mask (left) and hexagonal array of holes in the mask (right)

Under the normal ambient conditions (room temperature, normal atmospheric pressure, room humidity, unheated substrate), evaporation of the solvent can be considered as slow and the flows inside the droplets can be treated as steady state [14]. We ignored any thermal effects and suppose that the surface tension is independent of the particle concentration, thus Marangoni convection cannot occur. We assumed that the diffusivity is constant. Within continual approaches, any effects connected with the particles are ignored. Among these effects are jamming [15, 16] and shape-dependent capillary interactions [17].

We assumed that the liquid is incompressible. Mass conservation yields $\nabla \mathbf{v} = 0$. The boundary conditions are

- at the substrate ($z = 0$) $\partial_z u = 0$, where $\mathbf{v} = -\nabla u$,
- at the imaginary walls of the hexagonal cell $\partial_{\mathbf{n}} u = 0$, where $\mathbf{n}$ is the outward normal unit vector at the wall,
- at the free surface (Σ) $-\partial_{\mathbf{n}} u = \mathbf{v}_s \mathbf{n} + \frac{J}{\rho}$ or $(\mathbf{v} - \mathbf{v}_s)\mathbf{n} = \frac{J}{\rho}$, where $\mathbf{n}$ is the outward normal unit vector at the liquid-air interface, J is the density of the vapor flux above the free surface and ρ is the density of liquid.

Mass transfer inside the film can be described using the advection–diffusion equation [5, 7]

$$\partial_t c + \mathbf{v} \nabla c = D \nabla^2 c, \tag{1}$$

where c is the concentration of the suspended particles, D is their diffusivity, $\mathbf{v}$ is the velocity of the flow inside the film. Eq. (1) should be solved in the region with a moving boundary, because the free surface of the film goes down due to evaporation.

The advection–diffusion equation (1) was solved subject to the boundary conditions

- at the free surface (Σ) $c\mathbf{v}\mathbf{n} - D\partial_{\mathbf{n}} c = c\mathbf{v}_s \mathbf{n}$, where $\mathbf{n}$ is the outward normal unit vector at the liquid-air interface, $\mathbf{v}_s$ is the velocity of the free surface,

- at the substrate ($z = 0$) $\partial_z c = -\tilde{k}_d/c$, where $\tilde{k}_d$ is the deposition rate constant [5].
- at the imaginary walls of the hexagonal cell $c\mathbf{v}\mathbf{n} - D\partial_{\mathbf{n}}c = 0$, where $\mathbf{n}$ is the outward normal unit vector at the wall.

To mimic the influence of a mask above the film, the speculative evaporative flux function was utilized

$$J(x,y) = \begin{cases} J_0, & \text{if } \sqrt{x^2+y^2} < r, \\ 0, & \text{otherwise,} \end{cases} \tag{2}$$

where x, y are the coordinates, J_0 is the constant, and $r = d_h/2$. This formula qualitatively correctly reproduces the features of the evaporative landscape above the drying films calculated using finite-element modeling [1].

The free surface of the film was assumed to be flat. Due to evaporation, the film height, $h = h(t)$, changes with time as

$$\frac{dh}{dt} = -\frac{1}{\rho A_6}\int_{A_6} J(x,y)\,dA = -\frac{aJ_0}{\rho},$$

where $a = \frac{\pi r^2}{A_6}$ is the ratio of a hole area to the area of the base of a hexagonal cell $A_6 = 2\sqrt{3}(P/2)^2$.

All coordinates were scaled by the $P/2$. The time taken for the complete evaporation of the film, $t_0 = \frac{\rho h(0)}{aJ_0}$, is the natural characteristic time. The characteristic velocity is $v_0 = \frac{aJ_0}{\rho}$. We defined the dimensionless time as $\tau = \frac{t}{t_0} = \frac{aJ_0 t}{\rho h(0)}$, the dimensionless film height as $l = \frac{2h(t)}{P}$, the dimensionless vapor flux as $j = \frac{J}{J_0}$, and the dimensionless velocity as $\mathbf{w} = \frac{\mathbf{v}}{v_0}$. $l = \varepsilon(1-\tau)$, $dl/d\tau = -\varepsilon$.

The dimensionless Péclet number is defined as the ratio of the rate of advection of mass by the flow to the rate of diffusion of the mass driven by the diffusion. In our case, the Péclet number may be written as $\mathrm{Pe} = \frac{Pv_0}{2D} = \frac{aJ_0P}{2D\rho}$. Concentration can conveniently be measured with respect to the initial concentration. We denote $k_d = \tilde{k}_d P/2$.

In thin-film limit ($\varepsilon = 2h(0)/P \ll 1$), variations of the concentration and velocity along the z-axis is assumed to be negligible. In this case, the problem is quasi-two-dimensional, and therefore the boundary conditions on the free surface and on the substrate can be included in the governing equations as the source or sink terms. The conservation of the solute is written as

$$\frac{dl}{d\tau} + \varepsilon l \nabla \mathbf{w} = -\frac{\varepsilon j}{a} \tag{3}$$

or

$$\nabla \mathbf{w} = \frac{1 - j/a}{\varepsilon(1-\tau)}$$

and its solution should satisfy the boundary condition at the walls of the hexagonal cell $\mathbf{w_n} = 0$.

The conservation of the dispersed particles is written as

$$\partial_\tau(cl) + \varepsilon l \nabla(c\mathbf{w}) = \varepsilon \mathrm{Pe}^{-1}\left(l\nabla^2 c - k_d c\right) \tag{4}$$

or

$$\partial_\tau c + \varepsilon\nabla(c\mathbf{w}) = \mathrm{Pe}^{-1}\varepsilon\nabla^2 c + \frac{1 - k_d\mathrm{Pe}^{-1}}{1-\tau}c$$

and its solution should satisfy the boundary condition

$$\partial_{\mathbf{n}} c = 0. \tag{5}$$

In our approach, Eqs. (3) and (4) are independent and may be solved separately. To solve the partial differential equations (3) and (4), we utilized finite element methods using FlexPDE Professional 6.37.

3. Simulation results and discussion

When a colloidal film dries under a mask, the liquid flows towards regions with the highest evaporation. Such behaviour is intuitively clear, similar to the results obtained for a droplet [12], and follows the mass conservation (Fig. 2).

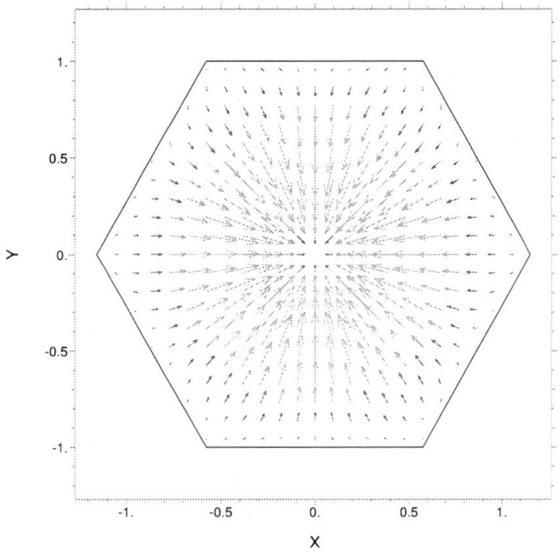

Figure 2. Velocity field inside a hexagonal cell ($\varepsilon = 0.01$, $P = 2d_h$, $\tau = 0.5$).

Inside a hexagonal cell the liquid flows towards the centre. The highest velocity corresponds to an edge of the hole (Fig. 3).

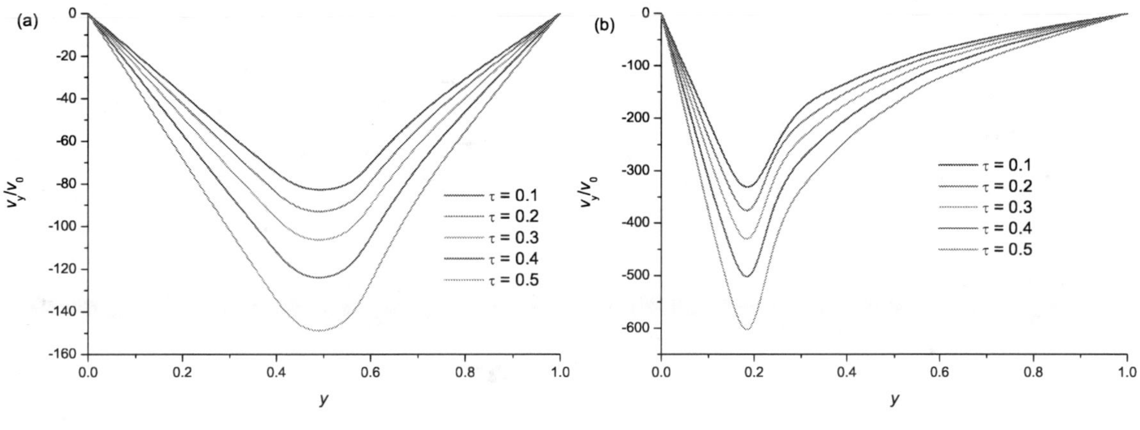

Figure 3. Velocity along y-axis for different times ($\varepsilon = 0.01$) (a) $P = 2d_h$, (b) $P = 6d_h$.

Having the velocity field, we can simulate the transfer of particles. To simplify our consideration, we start with the situation when both the diffusion and sedimentation are negligible ($\mathrm{Pe}^{-1} = 0$, $k_d = 0$), i.e. the right hand side of Eq. 4 is equal to zero. It is natural to expect that the particles will accumulate under the holes. Our computations demonstrate that the concentration of the particles under the holes increases by several times during half of the total time taken for evaporation (Fig. 4).

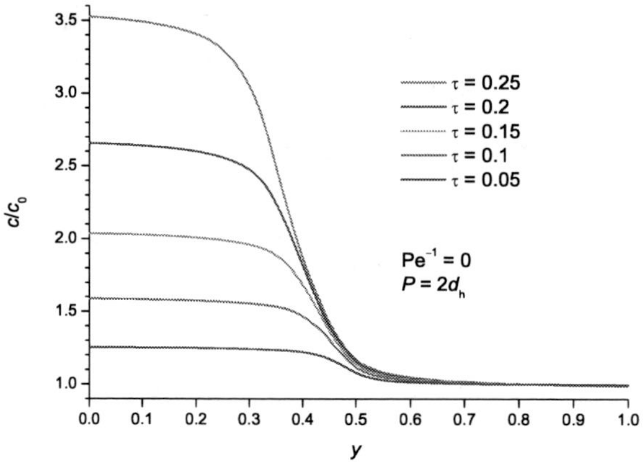

Figure 4. Concentration of the dispersed particles inside the thin film for different times ($\varepsilon = 0.01$, $P = 2d_h$) ($k_d = 0$).

We examined the effect of diffusion for different values of the Péclet number (Fig 5). When the dispersed particles have high diffusivity, advective transfer could be hindered by diffusion. Nevertheless, this effect is visible only in a narrow region near the hole edge, where the concentration gradient is large.

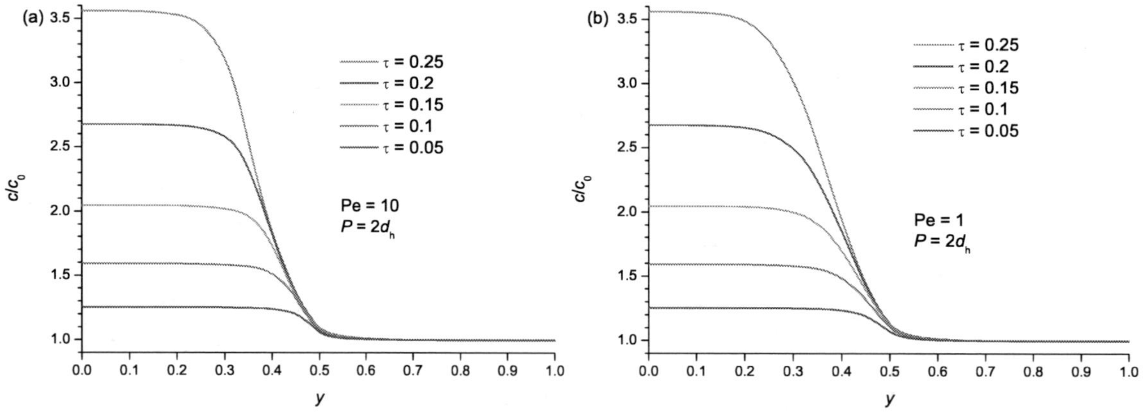

Figure 5. Concentration of the dispersed particles inside the thin film for different times ($\varepsilon = 0.01$, $P = 2d_h$); (a) Pe = 10, (b) Pe = 1. The sedimentation is absent ($k_d = 0$).

4. Conclusion

The proposed model allows a qualitative description of the real system. Nevertheless, it has some obvious shortcomings originating from the assumptions stated in Section 2. There are different ways to improve the model. First of all, one should take into account that the free surface is not perfectly flat. The flow inside the film is actually caused by the curvature of the surface shape. Moreover, the liquid should be considered as viscous. In this case, the velocity will vary from the bottom of the droplet to its apex, hence the quasi-two-dimensional approach would hardly be acceptable. The viscosity, the surface tension, and the diffusivity depend on the particle concentration. These dependencies should be taken into account. The dependence of the surface tension on concentration may produce Marangoni flow. The thermal effects ignored in the present investigation may be the cause of convective flows. The lubrication-approximation [18] may be used as one of the possible ways to incorporate these effects into our consideration. This approach has been successfully applied to desiccated colloidal droplets with axial symmetry [8, 11, 19]. In further studies, the speculative evaporative flux (2) should be replaced with a calculated one. Furthermore, the solute particles occupy finite volume. When we take into account this effect, the governing equations are no longer independent and should be solved together. Taking account of such additional effects significantly complicates the model, but in reality leads to little a significant change in the results. Thus, although we can identify a number of different ways to improve the proposed model, we believe that this model is useful even in the present simplest form.

Acknowledgements
The reported research was supported by the Ministry of Education and Science of the Russian Federation through Project No. 643.

References
[1] Harris D J, Hu H, Conrad J C and Lewis J A 2007 *Phys. Rev. Lett.* **98** 148301
[2] Harris D J and Lewis J A 2008 *Langmuir* **24** 3681–3685
[3] Harris D J, Conrad J C and Lewis J A 2009 *Philos. T. R. Soc. A* **367** 5157–5165
[4] Deegan R D, Bakajin O, Dupont T F, Huber G, Nagel S R and Witten T A 1997 *Nature* **389** 827–829
[5] Widjaja E and Harris M 2008 *AIChE J.* **54** 2250–2260
[6] Okuzono T, Kobayashi M and Doi M 2009 *Phys. Rev. E* **80** 021603
[7] Maki K L and Kumar S 2011 *Langmuir* **27** 11347–11363
[8] Tarasevich Y Y, Vodolazskaya I V and Bondarenko O P 2013 *Colloid Surface A* **432** 99 – 103
[9] Eales A D, Routh A F, Dartnell N and Simon G 2015 *AIChE J.* **61** 1759–1767
[10] Deegan R D, Bakajin O, Dupont T F, Huber G, Nagel S R and Witten T A 2000 *Phys. Rev. E* **62** 756–765
[11] Fischer B J 2002 *Langmuir* **18** 60–67
[12] Vodolazskaya I V and Dyakova V M 2014 *Bull. PFUR. Ser. Math. Inform. Sciences. Phys.* 239243 in Russian
[13] Tarasevich Y Y, Vodolazskaya I V and Sakharova L V 2015 *Eur. Phys. J. E* In press
[14] Popov Y O 2005 *Phys. Rev. E* **71** 036313
[15] Marín A G, Gelderblom H, Lohse D and Snoeijer J H 2011 *Phys. Rev. Lett.* **107**(8) 085502
[16] Shao F, Huynh T, Somers A, Liu B, Fu J, Muradoglu M and Ng T W 2014 *Physica B* **441** 21–27
[17] Yunker P J, Still T, Lohr M A and Yodh A G 2011 *Nature* **476** 308–311
[18] Anderson D M and Davis S H 1995 *Phys. Fluids* **7** 248–265
[19] Ozawa K, Nishitani E and Doi M 2005 *Jpn. J. Appl. Phys.* **44**(6A) 4229–4234

International Conference on Computer Simulation in Physics and Beyond 2015 IOP Publishing
Journal of Physics: Conference Series **681** (2016) 012034 doi:10.1088/1742-6596/681/1/012034

Free-molecular gas flow through the high-frequency oscillating membrane

V L Kovalev[1,2], A N Yakunchikov[1,2] and V V Kosiantchouk[1,2]

[1] Institute of Mechanics, M.V. Lomonosov Moscow State University, Michurinskii prosp. 1, Moscow 119192, Russian Federation

[2] Department of Mechanics and Mathematics, M.V. Lomonosov Moscow State University, Leninskie Gory 1, Moscow 119991, Russian Federation

E-mail: vasiliy.kosyanchuk@gmail.com, art-ya@mail.ru

Abstract. The possibility of using a high frequency oscillating track membranes as diffusion membranes for gas separation was studied. High frequency forced oscillation of the membrane was considered because of assumption that the membrane conductivity for a given gas can be controlled by varying the frequency and amplitude of oscillation. The problem about free-molecular gas flow through a oscillating in its plane membrane was stated and the possibility of separation of gases using such a device was investigated. Also, optimal values of membrane oscillation parameters for most efficient gas separation have been found.

1. Introduction

Membrane technologies are widely used in many areas of space industry nowadays. Applications for membranes in space include: lunar and planetary habitats, RF reflectors and waveguides, optical and IR imaging, solar concentrators for solar power and propulsion, sun shades, solar sails and many others [1, 2]. Other important application is the construction of novel separation devices working in microgravity [3]. The problem with modern membrane separation technologies is that they use membranes "as is" and the performance of any membrane separator is based only on the properties (physical or chemical) of membrane material.

In recent years, many researches discovered novel physical effects caused by external impacts such as optical radiation [4], temperature gradient [5] and induced oscillations [6] on gas flow in microstructures where gas is flowing in rarefied regime. The influence of all these effects on the flow of multicomponent gas mixtures was shown to lead to mixture separation.

In authors previous papers [7, 8] a novel effect of gas separation using track membranes oscillating with high frequency was discovered; it was shown, that varying the amplitude and frequency of oscillations one can control the conductivity of the membrane for the given gas, and as a result, achieve the effect of gas separation. The purpose of the present work is to determine optimal values of membrane oscillation parameters for the most efficient gas separation.

2. Problem Statement

Multicomponent gas flow through the oscillating in its plane membrane, consisting of straight cylindrical channels of length L and radius R is considered. Membrane connects two volumes with constant pressures — p_1, p_2, and constant temperatures — T_1, T_2 (Fig.1) Temperatures of the gas in volumes are equal and coincide with the temperature of the membrane ($T_1 = T_2 = T_w$, where T_w — is the temperature of membrane). Membrane moves as absolutely rigid body and the law of motion of each channel axis is as follows: $(x, y) = (x_c, y_c + Asin(\omega t + \psi))$, where x_c, y_c — initial coordinates of channel axis (which is directed along z), A and ω — amplitude and frequency of oscillations. It is assumed that the influence of gas flow on the motion of membrane can be neglected. Temperature of channel wall is assumed to be constant in time and along channel axis. Molecules are modelled as material points, mass forces and internal degrees of freedom are neglected. Gas flow in channel is assumed to be free-molecular and the velocity distribution of molecules entering from volumes is equilibrium corresponding to volume temperature. Interaction of gas molecules with channel surface is described in terms of scattering kernel. For specific combination of gas and membrane material, the parameters of the kernel (and the scattering model itself) can be obtained from molecular dynamics trajectories computations [9, 10, 11]. The choice of the theoretical model to describe scattering in a number of problem appears to be fundamental. For example, the use of specular-diffusion Maxwell model in the problem of thermal transpiration leads to results qualitatively different from experiments [12], and the Cercignani-Lampis scattering kernel does not allow the dependence of the energy and momentum conservation coefficients on the energy of incident gas molecules [13]. In the present work, calculations were conducted using Maxwell scattering kernel with full energy and impulse accommodation which is the extreme case for all scattering models.

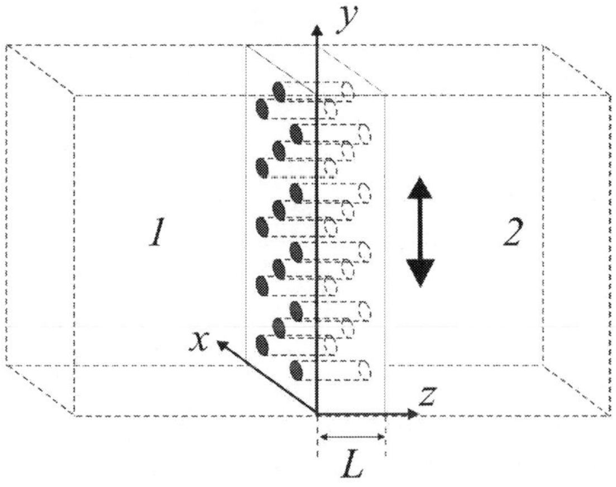

Figure 1. Problem scheme: *1*, *2* — tanks, separated by track membrane, oscillation direction shown by arrow (y-axis).

3. Computations

Problem was studied using event-driven molecular dynamics method [14]. For detailed description, please refer to [7]. Assumption that all membrane channels are equal and membrane moves as absolutely rigid body implies that flow can be studied only in one channel. Assumption about the free-molecular flow regime allows to simplify modelling to sequential computations of trajectories independently for each molecule. The computation was as follows. The same way as in [7], coordinate and velocity of a new gas molecule at the inlet section of the channel were calculated according to given distributions. Then, the point of collision between gas molecule and channel wall was calculated. After that, a new velocity of molecule was calculated according to the scattering kernel and then a new point of collision was calculated. This continued up until molecule leaved the channel.

4. Results

Studied model has five defining parameters R, L, A — radius, length and amplitude of channel oscillations, ω — oscillation frequency, $v_m = \sqrt{2kT/m}$ — most probable thermal speed of gas molecules (where, k — boltzmann constant, T — volume temperature, m — mass of gas molecule). The problem was solved using dimensionless variables for simplicity of computations and analysis of results. Presented model has only two dimensions — length and time. Thus, parameters R, v_m were chosen for nondimensionalization:

$$\mathbf{r}' = \frac{\mathbf{r}}{R}, \qquad A' = \frac{A}{R}, \qquad L' = \frac{L}{R}, \qquad \mathbf{u}' = \frac{\mathbf{u}}{v_m}, \qquad t' = \frac{tv_m}{R}, \qquad \omega' = \frac{\omega R}{v_m}.$$

where $\mathbf{r}, \mathbf{u}, t$ — coordinate, velocity and time, $\mathbf{r}', \mathbf{u}', t'$ — their dimensionless analogues. According to π-theorem, passing probability P is a function of three dimensionless combinations, composed of five defining parameters. These combinations were chosen as follows. Dimensionless channel length, dimensionless oscillation amplitude and the ratio of characteristic oscillations speed to characteristic thermal speed of molecules:

$$L' = \frac{L}{R}, \qquad A' = \frac{A}{R}, \qquad c = \frac{A\omega}{v_m},$$

During the simulations, these combinations were changed in the ranges (5–500; 0,1–10; 0–7). Further the stroke indices are omitted and the use of dimensionless variables is assumed.

4.1. Effect of gas separation

The values of the parameter c depend on the characteristic speed of the molecules, and, therefore, are not equal for different gases. Thus, the stronger probability P depends on c, the more will be the difference between values of membrane conductivity for different gases. To study the influence of oscillations on membrane conductivity P, its value in the figures is normalized by the calculated probability of passing a static channel $P_0(L)$ (as we are primarily interested in how much the passing probability decreases in oscillating channel in comparison with the static one), which is in a good agreement with Clausing's empirical formula [15, 16]:

$$P_0(L) = \frac{1 + 0.4L}{1 + 0.95L + 0.15L^2}.$$

Further, we deal with already normalized dependency $P(L, A, c)$.

Fig.2 shows dependencies $P(c)$ at $L = 5$ for different values of A. The decreasing behaviour of $P(c)$ represent the effect of gas separation. Indeed, consider two species of gas with masses of molecules m_1, m_2. Given parameters of membrane oscillations, two values of parameters c_1, c_2 and different passing probabilities $P(L, A, c_1), P(L, A, c_2)$, correspond to these masses. As passing probabilities are different for two gases, the molecular flow through membrane will differ too which will result in gas separation.

4.2. Determination of optimal parameters

Note that values c_1, c_2 mentioned above are related as follows

$$k = \frac{c_1}{c_2} = \frac{v_{m_2}}{v_{m_1}} = \sqrt{\frac{m_1}{m_2}}.$$

Without loss of generality, assume that $P(L, A, kc) > P(L, A, c)$. Then for a given k the values of parameters L, A, c will be optimal for gas separation if the ratio of membrane conductivity for two gases — $P(L, A, kc)/P(L, A, c)$ reaches its maximum.

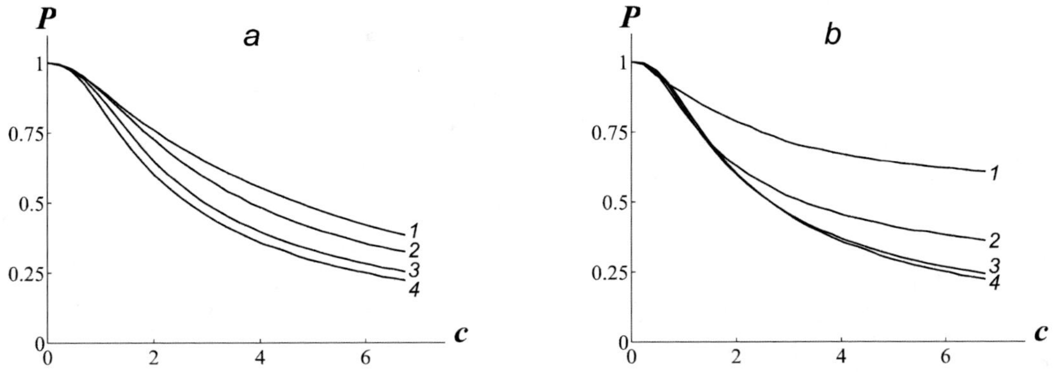

Figure 2. Dependency of normalized passing probability P/P_0 on c for $L = 5$: a — curve *1* corresponds to $A = 10$, *2* — 5, *3* — 2, *4* — 1; b — curve *1* corresponds to $A = 0.2$, *2* — 0.5 , *3* — 0.8 , *4* — 1.

As one can see from Fig.2,a at $A > 1$ probability $P(c)$ monotonically decreases as A decreases from 10 to 1. Thus, the conductivity decreases most rapidly with increase of c at $A = 1$. As is seen from Fig.2,b at $A < 1$ and small values of c ($c < 0,4$) probability $P(c)$ decreases most rapidly at the smallest value of A ($A = 0,2$). However, when c increases, the monotony on A breaks and at big values of c the minimal value of $P(c)$ is again reached at $A = 1$.

Thus, the value of amplitude $A = 1$ is optimal for gas separation since membrane conductivity at this value of A decreases most as c increases. Calculations showed that this conclusion is also valid for other values of L. Fig.3,a shows dependencies $P(c)$ at $A = 1$ for values $L = 5, 10, 20, 50$. Note, that curves for different values of channels length are identical to the stretching along the c axis, i.e. dependency $P(L, c)$ has the form $P(f(L) \cdot c)$. it can be shown that in this case variation of channel length does not increase the effectiveness of separation. Indeed, if for length L maximum ratio of membrane conductivities for two gases is reached at some value c_0

$$c_0 = \arg\max \frac{P(f(L) \cdot kc)}{P(f(L) \cdot c)} \qquad (L, k - \text{const})$$

then for any other length L' one can achieve the same ratio setting $c_0' = (f(L) \cdot c_0)/f(L')$. At the same time, there is no such value c_1' for length L', which would give greater ratio than c_0'. Otherwise, turning back to L, and setting $c_1 = (f(L') \cdot c_1')/f(L)$, we get the ratio of conductivities, greater than at c, which stands in contradiction with the original assumption.

It is seen from Fig.3,a that the value c required to achieve the equivalent level of conductivity decreases as L grows. Hence, the value of oscillation speed required to obtain maximal effect of gas separation also decreases. Fig.3,b shows points of the curves from Fig.3,a , after stretching of each of them with its own coefficient $f(L)$ along c axis. One can obtain this coefficient from Fig.3,a , for instance: $f(5) = 1$; $f(10) = 1.21$; $f(20) = 1.35$; $f(50) = 1.47$. As seen from Fig.3,b, all the points fit to one curve. Equation of the approximation curve is

$$P(c) = \frac{1 + 0.4c^{3.75}}{1 + 0.41c^{2.76} + 0.25c^{4.76}}. \qquad (1)$$

Using dependency 1, one can find optimal value of parameter c for any L as an argument of maximum of the function

$$g(c) = \frac{P(f(L) \cdot kc)}{P(f(L) \cdot c)}.$$

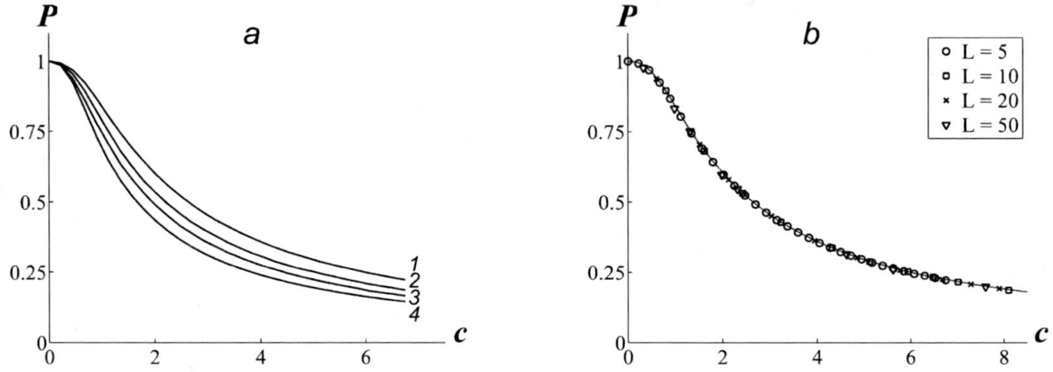

Figure 3. Dependency of normalized passing probability P/P_0 on c for $A = 1$: a — curve *1* corresponds to $L = 5$, *2* — 10, *3* — 20, *4* — 50; b — approximating curve *(1)*.

5. Conclusions

Problem was solved numerically by event-driven molecular dynamics method. Passing probability was obtained in dependence on dimensionless parameters. It was shown that varying the frequency and amplitude of oscillation can lead to difference of membrane conductivity for gases with different molecular mass. This means that one can control the conductivity of the membrane for a given gas by changing the parameters of its forced oscillations.

Also, for separation of binary gas mixtures in free molecular flow by the means of an oscillating membrane:

(i) The optimal amplitude of membrane oscillation was found.

(ii) Approximation for the dependence of membrane conductivity on the dimensionless oscillation speed was constructed, which allows to determine the optimal value of oscillation speed.

(iii) It was shown that variation of the channel length does not increase efficiency of gas separation but allows to reduce the required speed of membrane oscillation.

Acknowledgments

Work was supported by Russian Foundation of Basic Research (RFBR), grant 14-01-00310a. Computation was done on SKIF-MSU "CHEBYSHEV" supercomputer.

References

[1] Jenkins C H 2001 *Progress In Astronautics and Aeronautics: Gossamer Spacecraft: Membrane and Inflatable Structures Technology for Space Applications* vol 191 (AIAA)

[2] Roberts W T 2005 Optical membrane technology for deep space optical communications filters *Aerospace Conference, 2005 IEEE* (IEEE) pp 1991–2000

[3] Witt D 1993 *Membrane Techno* **1993** 6–9

[4] Chermyaninov I and Chernyak V 2014 Thermo-optical piston in gases *Proceedings of the 29th International Symposium on Rarefied Gas Dynamics* vol 1628 (AIP Publishing) pp 653–660

[5] Nakaye S, Sugimoto H, Gupta N K and Gianchandani Y B 2015 *European Journal of Mechanics - B/Fluids* **49, Part A** 36 – 49 ISSN 0997-7546

[6] Kovalev V, Yakunchikov A and Kosiantchouk V 2015 *Acta Astronautica* **116** 282 – 285 ISSN 0094-5765 URL `http://www.sciencedirect.com/science/article/pii/S009457651500291X`

[7] Yakunchikov A, Kovalev V and Kosiantchouk V 2015 *Microfluid Nanofluidics* **18** 1039–1043

[8] Kovalev V L, Kosyanchuk V V and Yakunchikov A N 2014 *Fluid Dyn* **49** 524–529

[9] Yamaguchi H, Shobatake K and Niimi T 2008 Molecular dynamics study on rare gas-graphite (0001) surface scattering *Proceedings of the 26th International Symposium on Rarefied Gas Dynamics* vol 1084 (AIP Publishing) pp 647–652
[10] Yamamoto K, Takeuchi H and Hyakutake T 2007 *Phys Fluids* **19** 087102
[11] Kovalev V and Yakunchikov A 2011 Accommodation coefficients for hydrogen molecules on graphite surface *Proceedings of the 27th International Symposium on Rarefied Gas Dynamics* vol 1333 (AIP Publishing) pp 481–485
[12] Sharipov F 2002 *Eur J Mech B/Fluids* **21** 113–123
[13] Yakunchikov A, Kovalev V and Utyuzhnikov S 2012 *Chem Phys Lett* **554** 225–230
[14] Donev A, Garcia A L and Alder B J 2008 *J Comput Phys* **227** 2644–2665
[15] Knudsen M 1950 *The kinetic theory of gases: some modern aspects* (Methuen & Company)
[16] Shidlovskiy V P 1965 *Dynamica razrezhennogo gaza* (Moscow: Nauka) ISBN 200000931461

3D simulation and analytical model of chemical heating during silicon wet etching in microchannels

S A Konakov [1], V V Krzhizhanovskaya [1,2,3]

[1] Peter the Great St. Petersburg Polytechnic University, 195251, St. Petersburg, Russia
[2] University of Amsterdam, 1098 XH Amsterdam, The Netherlands
[3] ITMO University, 197101, St. Petersburg, Russia

E-mail: konakov.st@gmail.com; Valeria.Krzhizhanovskaya@gmail.com

Abstract. We investigate chemical heating of a Silicon-on-Glass (SOG) chip during a highly exothermic reaction of silicon etching in potassium hydroxide (KOH) solution in a microchannel of 100-micron width inside a 1x1 cm SOG chip. Two modeling approaches have been developed, implemented and compared. (1) A detailed 3D model is based on unsteady Navier-Stokes equations, heat and mass transfer equations of a laminar flow of viscous incompressible fluid in microchannel, coupled to the heat transfer equation in the solid chip. 3D simulation results predicted temperature distributions for different KOH flow rates and silicon etching areas. Microchannels of a small diameter do not heat the chip due to the insufficient chemical heating of the cold fluid, whereas large-area etching (large channel diameter and/or length) leads to local overheating that may have negative effects on the device performance and durability. (2) A simplified analytical model solves a thermal balance equation describing the heating by chemical reactions inside the microchannel and energy loss by free convection of air around the chip. Analytical results compare well with the 3D simulations of a single straight microchannel, therefore the analytical model is suitable for quick estimation of process parameters. For complex microstructures, this simplified approach may be used as the first approximation.

1. Introduction in technology of etching in microchannel

Microtechnology is a rapidly developing area of research with many promising applications in semiconductor, optoelectronic, microfluidic devices, MEMS and other microsystems [1]. There are some problems and challenges in technologies, and novel methods of microfabrication should be investigated to provide a continuous progress. Microfluidics is one of the most promising parts of microtechnology. Lab-on-a-chip for DNA analysis and biomedical research, microreactors for nanoparticle synthesis and many other applications attract attention of scientists and engineers to microfluidics [2].

Microchannels for microfluidic devices are produced by etching through the mask. The groove may be of a different cross-section: circular, rectangle, trapezoidal, triangle, etc. To fabricate an enclosed microchannel, the second wafer is bonded to the grooved one. The top wafer may have a matching groove or be flat. This way, a microchannel of complex cross-section may be produced [3]. But there is a problem of misalignment while bonding the two wafers, which leads to uncontrollable changes of the cross-section shape and of hydraulic properties of the microchannel [4]. Another serious problem of this technology is the impossibility to change the channel after production. The technology applied in our research solves both problems. The concept of etching-in-microchannel technology is illustrated in Figure 1 by an example of producing a microstructure in silicon-on-glass (SOG) wafer.

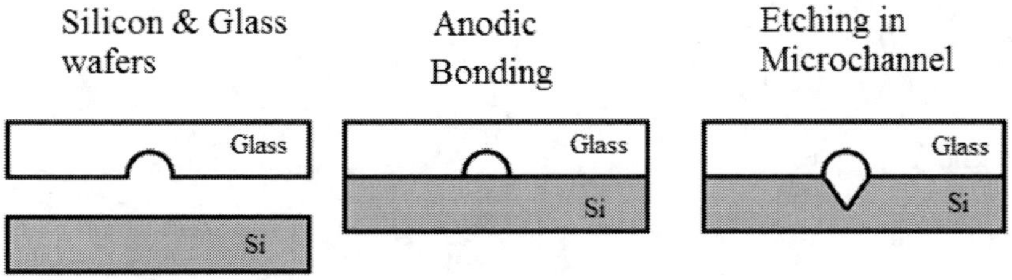

Figure 1. Scheme of etching-in-microchannel technology

The technology includes three steps. At the first, microchannel is etched in glass. At the second, silicon with (100) orientation and glass wafers are bonded together. The third step is supplying potassium hydroxide (KOH) solution into microchannel. KOH reacts with silicon, as in typical process of silicon anisotropic wet etching, and forms trapezoidal or triangle profile [5]. Edges of microchannel in glass play a role of mask for silicon wet etching and it leads to perfect alignment of microstructure. Set a time of KOH supplying into microchannel we can control geometry with high accuracy. Moreover it is possible to measure hydraulic properties of system in-situ and change them during the etching process, reaching the required parameters. Figures 2 and 3 show profiles before and after etching of silicon in microchannel by KOH solution.

Figure 2. Microchannel in SOG ship before silicon etching

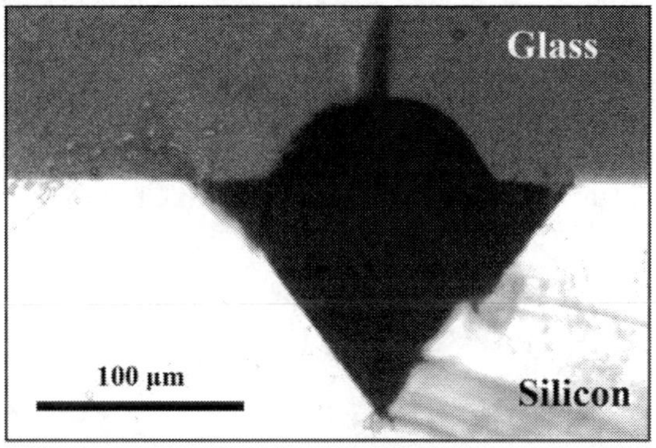

Figure 3. Microchannel in SOG ship after silicon wet etching by KOH solution

Our global goal is comprehensive investigation of technology of etching-in-microchannel. In this paper we present results of our work where problem of heat transfer during silicon etching in the microchannel was considered in detail. Figure 4 shows a scheme of SOG ship with microchannel which was investigated in the work. The chip with 10x10x1 mm dimensions consist of silicon and glass wafer. 100 µm channel have been etched in the glass before bonding of two wafers. The reaction (R1) of silicon dissolution by KOH is highly exothermal and this is a heat source leading to temperature increase.

$$Si + 2KOH + 4H_2O => K_2[Si(OH)_6] + H_2 + P_R \qquad \textbf{(R1)}$$

where P_R is heat by chemical reaction of silicon dissolution in KOH.

The chip located in air and its surfaces are cooling by free convection. In this situation we can write a govern heat balance equation (1) for the process of silicon etching in microchannel during the time when KOH solution is being supplied in it.

$P_R+P_e+P_L=0$ (1)

where: $P_R>0$ is heat by chemical reaction; $P_e<0$ is power which heats KOH from initial temperature (T_i) to temperature at outlet ($T_{out}>T_i$); $P_L<0$ is heat loss by free convection in air with temperature T_a from top (P_{L1}) and bottom (P_{L2}) surfaces of chip. For this case we should obtain an equilibrium temperature of the chip.

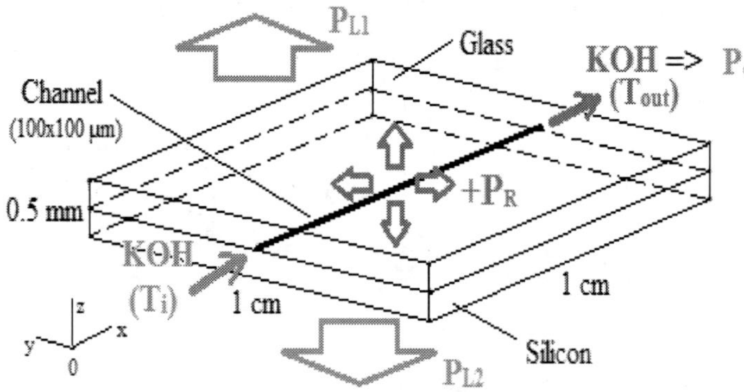

Figure 4. Scheme of SOG chip with microchannel

2. Mathematical model

2.1 Equations

We selected a continuum modeling approach to simulate a liquid flow in microchannel. Calculated Reynolds number Re=0.15 corresponds to a laminar regime. In model mass conservation law (2) and Navier-Stokes equations for a flow of KOH solution (3) describe the fluid dynamics in the channel:

$$\frac{\partial\rho}{\partial t} + \nabla\cdot\rho V = 0 \quad (2)$$

where ρ is density, t is time, V is velocity vector

$$\frac{\partial(\rho u_i)}{\partial t} + \nabla\cdot(\rho V u_i) = -\frac{\partial p_i}{\partial x_i} + \nabla\cdot(\mu\nabla u_i),\ i = x,y,z \quad (3)$$

where u_i is projection of velocity vector on i axis, μ is the kinematic viscosity

Heat transfer equation can be written as (4)

$$\frac{\partial(\rho h_o)}{\partial t} + \nabla\cdot(\rho V h_o) = \nabla\cdot(\lambda\nabla T) + \frac{\partial p}{\partial t} + P_R \quad (4)$$

where h_o is enthalpy, λ is thermal conductivity, T is temperature, p is pressure, P_R is the heat source due to chemical reactions.

2.2 Boundary conditions

For calculation of the flow in the microchannel we applied conditions of constant velocity at the inlet and outlet. The expression (5) gives us a value of the velocity.

$$U(x=0,x=10\ mm)=R/S_{ch} \quad (5)$$

Where U is flow velocity (x direction) [m/s], R is flow rate of etchant [m^3/s], S_{ch} is cross section of channel [m^2]. Velocity at the walls of microchannel was 0 [m/s]. Temperature of KOH solution was set at 300 [K] at the inlet and outlet.

Another temperature condition was applied to silicon surface in microchannel, because the etching process generates some quantities of heat. In general form this conditions expressed by (6)

$$P_R = F_{ch}\cdot P'_R \quad (6)$$

where $P_R>0$ is heat by chemical reaction of silicon dissolution in KOH [W], F_{ch} is area of Si etching [m^2], P'_R is surface heating power by Si dissolution [W/m^2]. Surface heating power can be obtained by equation (7)

$$P'_R = -\Delta H_R\cdot\rho_{Si}R_{etch}(T)/M_{si} \quad (7)$$

where $\Delta H_R = -247 \pm 6$ [kJ/mol] is enthalpy of reaction (R1) [6], ρ_{Si} is density [kg/m^3], M_{Si} is Si molar weight [kg/mol], R_{etch} is etching rate of (100) plane [m/s]. This rate is a function of temperature, described by (8)

$$R_{etch} = R_0 C_{H_2O}^4 C_{KOH}^{0.25}\exp(-E_A/K_bT) \quad (8)$$

where $R_0 = 689\cdot10^{-9}$ [m/s] is constant, C_{H2O}, C_{KOH} is molar concentration [mol/L], $E_a = 0.595$ [eV] is activation energy, $K_b = 8.617\cdot10^{-5}$ [eV/K] is Boltzmann constant, T is temperature [K].

Journal of Physics: Conference Series **681** (2016) 012035 doi:10.1088/1742-6596/681/1/012035

Substituting expression (8) into (6) we obtain (9)

$$P_R = F_{ch} \cdot (-\Delta H_R \rho_{Si}/M_{Si}) \cdot R_0 C_{H_2O}^4 C_{KOH}^{0.25} \exp(-E_A/K_b T) \quad (9)$$

This is the boundary condition for silicon wall in microchannel in general form. Simplifying it for our case for constant concentration of KOH we can write (10).

$$P_R = F_{ch} \cdot K_e \cdot (-E_A/K_b T) \quad (10)$$

where K_e=1.233·10^{11} [W/m^2] is constant.

For boundary conditions, which applying to outer wall of the chip, we have (11)

$$P_L = P_{L1}'F_{s1} + P_{L2}'F_{s2} \quad (12)$$

where P_L' is surface heat loss by convection [W/m^2], F_s – area of chip surface [m^2]. Index 1 and 2 denote top and bottom surfaces of the chip respectively. In our model for side surfaces of the chip we consider adiabatic conditions because its area is small. To calculate P_l' we use equation (13)

$$P_l' = \alpha(T_a - T_s) \quad (13)$$

where α is heat transfer coefficient [W/m^2·K]. Expression (14) couple α and Nusslt number

$$Nu = \alpha \cdot X/\lambda_a \quad (14)$$

where Nu is Nusslt number, X is characteristic dimension of the system, λ_a is air thermal conductivity. In general form Nusslt number for free convection case can be write as (15)

$$Nu = C(Gr \cdot Pr)^n \cdot K \quad (15)$$

where Gr is Grashof number, Pr is Prandtl number, C & n are constants, K is dimensionless function. C, n and K depends from shape of cooling surface and orientation in space.

In conditions of our model Pr and Gr numbers calculated by (16, 17)

$$Pr = \nu_a \rho_a c_a/\lambda_a = 0.708 \quad (16)$$

$$Gr = g\beta\Delta T \cdot X^3/\nu_a^2 = 7.972 \cdot 10^3 \quad (17)$$

where ν_a =1.568·10^{-5} [m^2/s] is kinematic viscosity of air, ρ_a = 1.1774 [kg/m^3] is density, c_a = 1006 [J/kg·K] is air specific heat, λ_a = 0.02624 [W/m·K] is thermal conductivity, g=9.81 [m/s^2] is acceleration of gravity, β = 3.33·10^{-3} [K^{-1}] is coefficient of volumetric expansion, ΔT is temperature gradient. Composition $Pr \cdot Gr$=5.643·10^3 corresponds laminar convection [7]. Using a constants C, n and K for this case we obtain expressions (18, 19) for boundary conditions for top and bottom outer walls of chip respectively.

$$P_1 = (T_a - T_s) \cdot (\alpha_1 F_{s1}) \quad (18)$$

$$P_2 = (T_a - T_s) \cdot (\alpha_2 F_{s2}) \quad (19)$$

where α_1=12.281 [W/m^2·K], α_2= 6.14 [W/m^2·K].

2.3 Implementation

Program CFD_ACE+ used for numerical solving of equations (2, 3, 4). Finite volume method with Conjugate Gradient Squared + Precondition Solver was applied for Velocity and Enthalpy. 3D geometry model was developed for chip and microchannel. Transient simulation (until quasi stationary state) with adaptive time step was used for numerical simulation.

We simulated a process of etching in microchannel for 100 μm straight channel with constant flow rate of KOH solution R=10^{-12} [m^3/s]. Area of silicon wall was F_{ch}=10^{-6} [m^2]. Two calculations were carry out with different initial conditions: T=300 and T=380 [K]. As the approximation, to simplify a simulation, we assume a constant value of P_R = 6.17·10^{-4} [W] (that means the constant etching rate at temperature 380 [K]). In addition under the same conditions we simulate a microstructure in SOG chip with large etching surface of silicon (F_{ch}=10^{-4} [m^2]).

3. Results of numerical simulation

Figure 5 shows a flow velocity in microchannel. The maximum value is 1.9·10^{-4} [m/s].
Distribution of temperature in SOG chip is presented in figure 6. Figure 7 (A) illustrates the variation of the temperature over time for the model.

Journal of Physics: Conference Series **681** (2016) 012035 doi:10.1088/1742-6596/681/1/012035

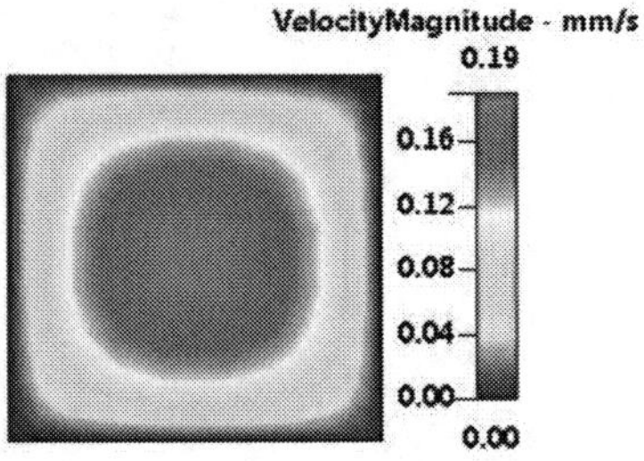

Figure 5. Velocity in microchannel ($R=10^{-12}$ [m^3/s])

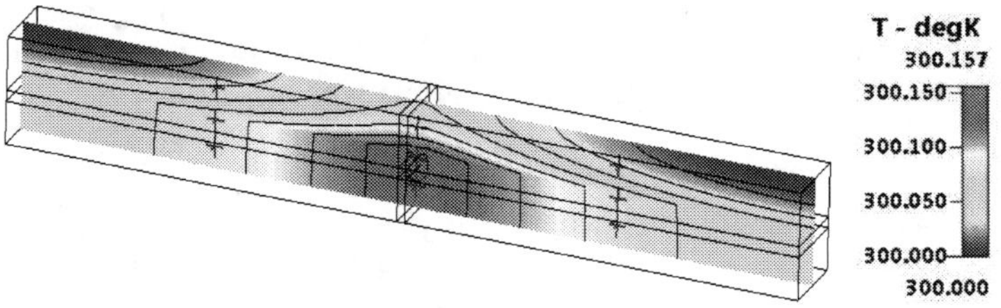

Figure 6. Temperature distribution in cross-section of chip ($F_{ch}=10^{-6}$ [m^2], $R=10^{-12}$ [m^3/s])

As we can see the maximum value of temperature is 300.157 [K]. It is means that insignificant heating during etching in microchannel with small area of opened silicon cannot lead to self-heating of SOG chip. So to provide a high rate of etching reaction we should supply heat to microstructure. But absence of self-heating is advantage when we need to have a low etching rate to control of microchannel geometry precision. We can see another situation for microstructure with large etching surface of silicon ($F_{ch}=10^{-4}$ [m^2]). Figure 7 (B) shows the change in temperature over time for the case. Large etching area leads to high power of inner heating that result to significant temperature increasing. We see maximum temperature at 351.483 [K]. It shows that such microstructure can have self-heating effects during the KOH etching in microchannel. We should take into account the phenomenon when developing the technology for structures with large open surface (membranes, microfluidics filters, mixers, etc). For some devices it is possible to etch without external heating.

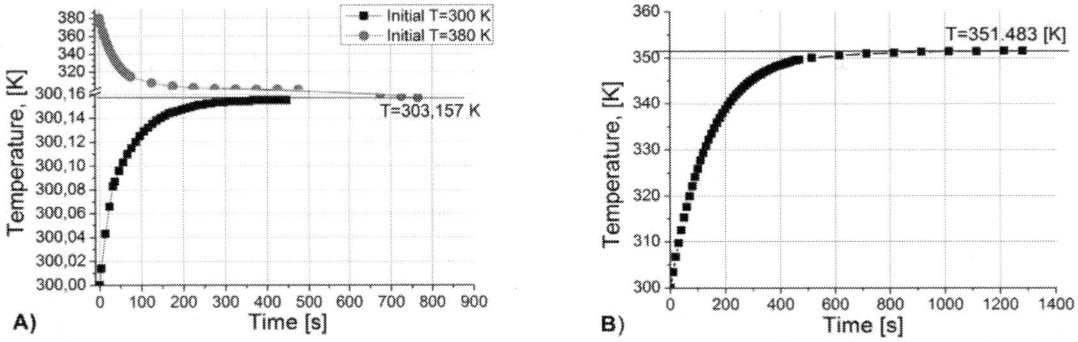

Figure 7. Maximum chip temperature change over time. (A) $F_{ch}=10^{-6}$ [m^2], (B) $F_{ch}=10^{-4}$ [m^2]

4. Analytical modeling

The second approach, which was used in the work, is analytical expression of equation (1). In this case we apply tree assumption: etching rate is constant; thermal conductivity of chip is infinite; temperature in the channel is equal temperature on the top and bottom surface (we denote it as T). With this assumption we can write heat balance equation (1) as (20):

$$F_{ch}\cdot\left(-\frac{\Delta H_R \rho_{Si} R_{etch}}{M_{Si}}\right)+c_e\rho_e R(T_i-T_{out})+(T_a+T_s)(\alpha_1 F_{s1}+\alpha_2 F_{s2})=0 \quad (20)$$

Express temperature from (20) we get final equation (21)

$$T=\frac{\left[F_{ch}\left(-\frac{\Delta H_R \rho_{Si} R_{etch}}{M_{Si}}\right)+c_e\rho_e R T_i+T_a(\alpha_1 F_{s1}+\alpha_2 F_{s2})\right]}{c_e\rho_e R+(\alpha_1 F_{s1}+\alpha_2 F_{s2})} \quad (21)$$

Using (21) we can fast calculate the equilibrium temperature of the SOG chip. Figure 8 shows plots of equilibrium temperature for different etching areas and flow rates of KOH solution. In table 1 we see a compilation between numerical simulation and analytical solution. It is possible to note that even under our bold assumptions the values of temperature do not have significant differences. That means the analytical model preferably to use for fast calculations in the first approximation.

Table 1. Comparison of two models

F_{ch} [m^2]	3D Simulation	Analytical modeling
10^{-6}	ΔT=0.157 [K]	ΔT=0.335 [K]
10^{-4}	ΔT=51.483 [K]	ΔT=51.385 [K]

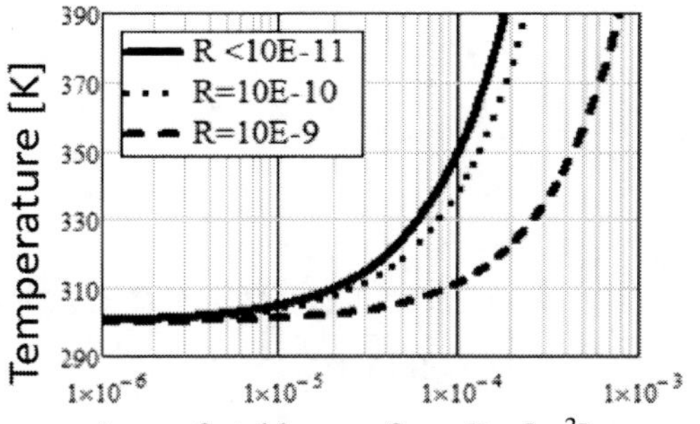

Figure 8. Equilibrium temperature of the chip. Analytical model

5. Conclusions and future plans

In the work we investigated a heat transfer problem for SOG chip during the process of etching in microchannel. Developed 3D computer model and analytical approach give the same results about equilibrium temperature of the chip. Analytical model may provide a fast estimation of temperature in the first approximation while numerical calculations allow obtain detailed information about etchant flow and temperature distribution in chip volume and their changes over time. Despite obtained results real experiments have many problems. The general issue is hydrogen generation during the etching process of silicon by KOH. Its bubbles have a strong influence on flow regime and local heat balance. Investigation of the problem requires advance multi scale 3D computer modeling, what we are planning to carry out in our further work. In spite of this the technology of etching-in-microchannel is very promising process for microfabrication of many complicated devices with unique features, such as microreactor for silicon dioxide deposition [8]. In future work we will continue our comprehensive research of the technology of etching-in-microchannel.

References

[1] Modrea A 2014 *Procedia Technology* ***12***, 283-288
[2] Araci I E & Brisk P 2014 *Current opinion in biotechnology* ***25***, 60-68
[3] Nguyen N T, Wereley S T 2002 *Fundamentals and applications of microfluidics*. Artech House
[4] De Boer et al. 2000 *Journal of Microelectromechanical Systems* **9.1**, 94-103
[5] Kohler M 2008 *Etching in microsystem technology*. John Wiley & Sons
[6] Nijdam A J 2001 Anisotropic wet-chemical etching of silicon pits, peaks, principles, pyramids and particles. University of Twente
[7] Wong H Y 1977 Handbook of essential formulae and data on heat transfer for engineers
[8] Konakov S A & Krzhizhanovskaya V V 2015 *Journal of Physics: Conference Series* **574** 012145

Journal of Physics: Conference Series **681** (2016) 012036 doi:10.1088/1742-6596/681/1/012036

Monte Carlo simulation of V/III flux ratio influence on GaAs island nucleation during MBE

O A Ageev, M S Solodovnik, S V Balakirev, I A Mikhaylin
Department of Nanotechnologies and Microsystems, Institute of Nanotechnologies, Electronics and Electronic Equipment Engineering, Southern Federal University, 2 Shevchenko Street, Taganrog, 347928, Russia

E-mail: ageev@sfedu.ru

Abstract. The kinetic Monte Carlo simulation of GaAs/GaAs(001) molecular beam epitaxial growth considering V/III flux ratio influence on nucleating island characteristics is presented. It is shown that the island density increases with the surface coverage increase and reaches saturation after deposition of ~0.1 monolayer of GaAs. The increase of V/III flux ratio from 3 to 40 leads to the increase of the island density from $1.9 \cdot 10^{12}$ to $2.6 \cdot 10^{12}$ cm^{-2}. At the same time the average size decreases from 4.4 to 4.1 nm. The island size distribution function narrows with V/III flux ratio increase. This is attributed to the shortage of gallium atoms in comparison with deposited arsenic molecules that prevents large island formation and leads to the dramatic growth of little island concentration. The simulation demonstrates good agreement with experimental results.

1. Introduction

Molecular beam epitaxy (MBE) is one of the most promising fabrication method for nano- and optoelectronic, photonic and quantum processing device structures. A more accurate understanding of the fundamental kinetic processes during epitaxial growth including submonolayer regime would aid the manufacturing process.

During MBE of III-V semiconductor structures their parameters are primarily controlled by substrate temperature and growth rate. At the same time V/III flux ratio is not usually considered to alter growing structure characteristics although experiments show that it has a significant influence on nucleating island size and surface density [1-3].

By now, many experimental and theoretical techniques are developed to reveal growth mechanisms during epitaxy of semiconductors. These methods have permitted to scrutinize a lot of parameters of growth process, such as activation energy [4-6], migration rate [7], diffusion coefficient [1,4,8], diffusion length [7,8]. However, very little attention is focused on macroscopic characteristics which are average island size, form, surface density and size distribution function.

In this paper, we report kinetic Monte Carlo simulation of GaAs epitaxial growth on the GaAs(001) surface considering V/III flux ratio effect on the growing island geometrical characteristics. Monte Carlo method enables calculations in a wide range of technological parameters and at large time and length scales what is unattainable with *ab initio* calculations.

Journal of Physics: Conference Series **681** (2016) 012036 doi:10.1088/1742-6596/681/1/012036

2. Description of the model

In this simulation, the GaAs(001) surface is assumed to have zinc blende crystal structure with $\beta2(2\times4)$ reconstruction. This means that there are arsenic dimer rows alternating with missing trenches on the surface [9]. We consider the growth from the fluxes of two principally different particles [10,11] on the surface with complicated structure, so it is necessary to take into account a lot of microscopic processes. Each process is implemented as a set of simple events depending on a particle type, location and environment. A basic parameter of an event i is an activation energy barrier E_i which defines the frequency f_i of occurrence of this event by the Arrhenius law:

$$f_i = \nu_0 \exp(E_i / kT), \tag{1}$$

where ν_0 is the attempt frequency set at 10^{13} s^{-1} [12]; k is the Boltzmann constant; and T is the growth temperature.

The following microscopic processes are involved in the simulation and predefine more than 40 elementary events:

- Ga adsorption and desorption. Ga atoms are deposited into random gallium sublattice sites with unity sticking probability and do not desorp any longer;
- Ga surface diffusion. Ga adatom migration is implemented as a sequence of site-by-site diffusion hops. Ga diffusion is anisotropic so that the activation energy of free Ga atom diffusion is 1.3 eV along the $[1\bar{1}0]$ direction [13] and 1.5 eV along the [110] direction [1];
- As_2 physisorption and surface diffusion. Arsenic is deposited on the surface from the As_2 molecule flux and occupy random doubled sites in the arsenic sublattice. As_2 molecules can only adsorb chemically (chemisorb) on Ga adatoms whereas interaction of As_2 with uncovered surface is via weakly bound (physisorbed) state [4]. The presence of this state enhances As_2 molecule migration. Since one arsenic molecule performs $\sim10^9$ hops during one gallium hop we assume that one diffusion step of As_2 molecule has random distribution:

  $$l = -\lambda_{\mathrm{As}} \ln \delta, \tag{2}$$

  where λ_{As} is arsenic molecule diffusion length depending on technological parameters and varying from 1 to 30 μm [14-16], δ is a random number having uniform distribution in the range (0,1]. Since site-by-site migration of arsenic molecules is neglected, the calculations have been accelerated essentially.

  Arsenic diffusion activation energy is 0.2 eV [17]. The contribution of atomic As is neglected since As_2 dimers do not dissociate under typical growth conditions [18].
- As_2 desorption. We assume that arsenic can re-evaporate from the surface only from the physisorbed state, and the activation barrier of this process is equal to 0.37 eV [19,20].
- As_2 chemisorption. As_2 molecules can only chemisorb on Ga adatoms. However, the activation energy of this process varies from 0.25 eV for the chemisorption in a trench site on two gallium dimers to 0.55 eV in case of the chemisorption on a dimer row [19].

  The reverse transition of a chemisorbed As_2 molecule to the physisorbed state is probable. The activation energy also depends on the location and environment of an arsenic dimer and varies from 1.7 to 2.6 eV [4].

As opposed to the traditional kinetic Monte Carlo method, the characteristic time interval in our simulation is deterministic, but the number of events n_{ev} occurring during one time step is distributed randomly:

$$n_{\mathrm{ev}} = -n_0 \ln \delta, \tag{3}$$

where n_0 is a nominal number of events.

3. Results and discussion

We reveal that islands preferentially nucleate in the trenches and have elongated form along the $[1\bar{1}0]$ direction (figure 1) what is connected with the diffusion anisotropy [1]. The (2×4) reconstruction remains on the surface during the entire submonolayer growth. The morphology of islands formed under the same technological parameters is in good agreement with the scanning tunnelling microscopy images [13] and other theoretical models [1,18].

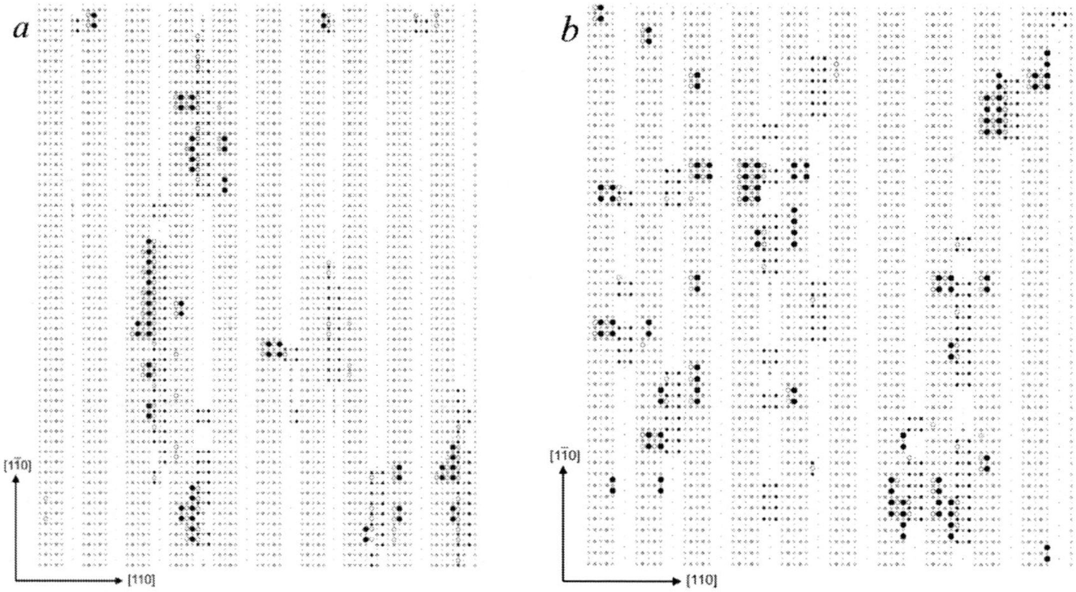

Figure 1. Island morphology in a simulation area of 160 Å × 200 Å after deposition of 0.1 ML GaAs at $T = 580$°C, $v = 0.1$ nm/s: *a*) $J_{As/Ga} = 3$ *b*) $J_{As/Ga} = 40$. The open circles represent Ga atoms and the filled circles As atoms. The substrate is marked in grey and deposited material in black.

Figure 1 shows that V/III flux ratio $J_{As/Ga}$ increase leads to larger surface scattering of islands and smaller anisotropy of their form. At higher V/III flux ratio we observe a large number of conglomerates consisting of several atoms and being nucleation centers of big islands.

In order to estimate gallium arsenide growing film characteristics quantitatively the island surface density (figure 2), average size (figure 3) and size distribution function (figure 4) have been calculated.

The island density rises with the coverage increase and reaches saturation after deposition of ~0.1 monolayer (ML) of GaAs (figure 2), in consistence with experiments [21]. The increase of V/III flux ratio leads to the island density growth. This is attributed to the suppression of Ga adatom diffusion caused by As_2 molecule concentration increase. Consequently, Ga adatoms become nuclei of new clusters rather than incorporate into stable islands. The saturation value of the island density at the substrate temperature $T = 580$°C, growth rate $v = 0.1$ nm/s and $J_{As/Ga} = 10$ is $2.2 \cdot 10^{12}$ cm^{-2}, in good agreement with experiments [22].

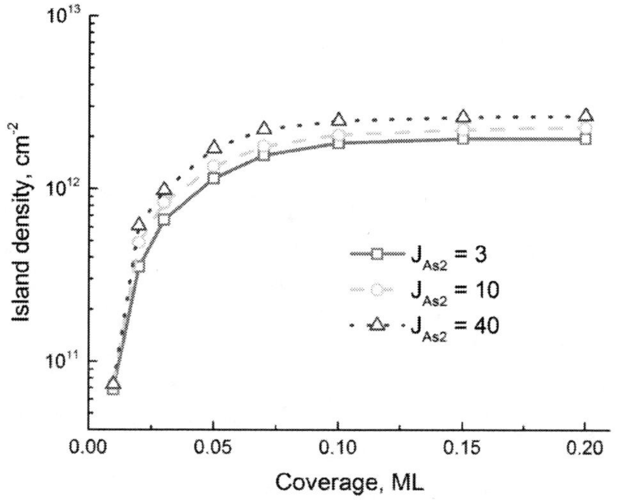

Figure 2. Island surface density as a function of coverage at $T = 580°C$, $v = 0.1$ nm/s and different V/III flux ratios.

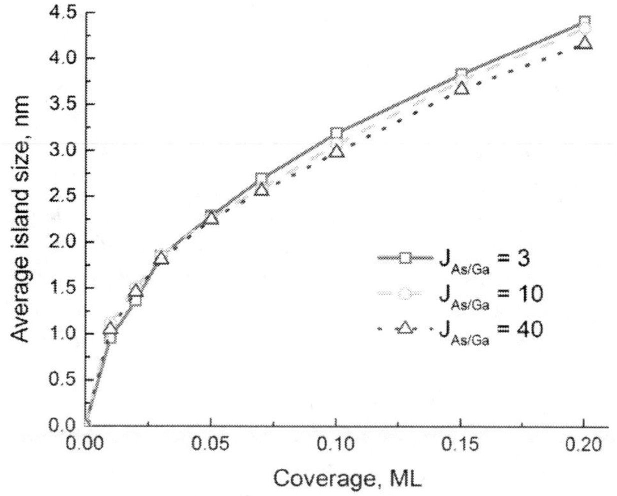

Figure 3. Island average size as a function of coverage at $T = 580°C$, $v = 0.1$ nm/s and different V/III flux ratios.

The average size of islands increases monotonically with the coverage increase (figure 3). However, V/III flux ratio increase leads to the slight decrease of island size due to above reasons. The alteration of the size is less significant than that of the density. It is connected with the consumption of deposited material for the island growth at any V/III flux ratio whereas the nuclei concentration remains the same value after a certain moment. The average island size at $T = 580°C$, $v = 0.1$ nm/s and coverage $\theta = 0.2$ ML is equal to ~4 nm that corresponds to experimental data [22].

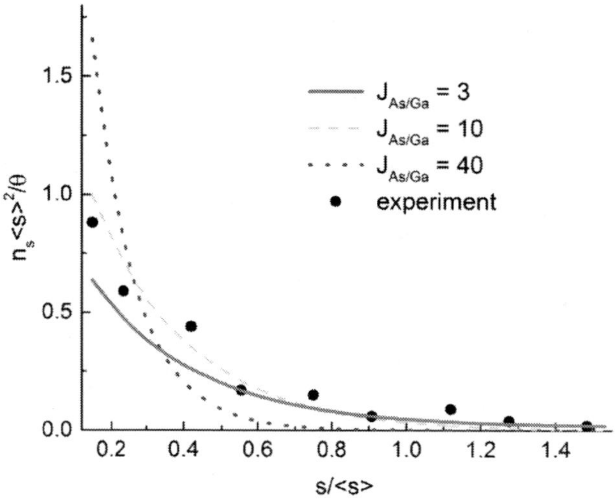

Figure 4. Island size distribution function at $T = 580°C$, $v = 0.1$ nm/s and different V/III flux ratios. Experimental data are taken from [13].

The simulation size distribution function is based on the concept [23] that the number density n_s of islands consisting of s atoms can be stated as:

$$n_s = \frac{\theta}{<s>^2} f(s/<s>), \tag{4}$$

where $<s>$ is the average size of islands formed at $J_{As/Ga} = 10$; f is a scaling function. The form of the distribution for the GaAs(001) surface shows that islands have a critical size of one atom and grow by gallium dimer capture [21] in contrast with (110) and (111)A surfaces [24]. The simulation size distribution is in good quantitative agreement with the experimental dependence for $J_{As/Ga} = 10$ [13]. Figure 4 shows that V/III flux ratio increase narrows the distribution with its shift to the small-size region. It is mainly due to the increase of arsenic molecule surface concentration and respective formation of small unstable clusters. At the same time islands are distributed quite uniformly along the size axis at small flux ratio $J_{As/Ga} = 3$ as adatoms can form quite many big islands along with little clusters.

4. Conclusion

In summary, the present simulation makes it possible to investigate GaAs epitaxial growth mechanisms thoroughly and take into account V/III flux ratio influence on the nucleating island characteristics. The flux ratio alteration does not affect the average island size significantly but can change the nucleation mechanisms and island morphology in whole. At large V/III flux ratios the density of little clusters increases whereas there is more homogeneous island size distribution on the surface at small V/III flux ratios. Being able to control nucleating island characteristics, V/III flux ratio is very important technological parameter of molecular beam epitaxy method.

This work was supported by the Russian Science Foundation Grant No. 15-19-10006. The results were obtained using the equipment of Common Use Center and Education and Research Center "Nano-technologies" of Southern Federal University.

5. References

[1] LaBella V P, Bullock D W, Ding Z, Emery C, Harter W G and Thibado P M 2000 *J. Vac. Sci. Tech.* A **18** 1526
[2] Ledentsov N N *et al.* 1996 *Sol. St. Electron.* **40** 785
[3] Riel B J, Hinzer K, Moisa S, Fraser J, Finnie P, Piercy P, Fafard S and Wasilewski Z R 2002 *J. Cryst. Growth* **236** 145
[4] Morgan C G, Kratzer P and Scheffler M 1999 *Phys. Rev. Lett.* **82** 4886
[5] Murdick D A, Wadley H N G and Zhou X W 2007 *Phys. Rev.* B **75** 125318
[6] Shiraishi K and Ito T 1998 *Phys. Rev.* B **57** 6301
[7] Amrani A, Djafari Rouhani M and Mraoufel A 2011 *Appl. Nanosci.* **1** 59
[8] Kangawa Y, Ito T, Taguchi A, Shiraishi K, Irisawa T and Ohachi T 2002 *Appl. Surf. Sci.* **190** 517
[9] Daweritz L and Ploog K 1994 *Semicond. Sci. Tech.* **9** 123
[10] Foxon C T and Joyce B A 1977 *Surf. Sci.* **64** 293
[11] Tok E S, Neave J H, Zhang J, Joyce B A and Jones T S 1997 *Surf. Sci.* **374** 397
[12] Kley A, Ruggerone P and Scheffler M 1997 *Phys. Rev. Lett.* **79** 5278
[13] Avery A R, Dobbs H T, Holmes D M, Joyce B A and Vvedensky D D 1997 *Phys. Rev. Lett.* **79** 3938
[14] Nishinaga T and Shen X Q 1994 *Appl. Surf. Sci.* **82-83** 141
[15] Nishinaga T 2004 *Prog. Cryst. Growth Charact. Mater.* **48-49** 104
[16] Higuchi Y, Uemura M, Masui Y, Kitada T, Shimomura S, Hiyamizu S 2003 *J. Cryst. Growth* **251** 80
[17] Tok E S, Neave J H, Allegretti F E, Zhang J, Jones T S and Joyce B A 1997 *Surf. Sci.* **371** 277
[18] Kratzer P and Scheffler M 2002 *Phys. Rev. Lett.* **88** 036102
[19] Garcia J C, Neri C and Massies J 1989 *J. Cryst. Growth* **98** 511
[20] Foxon C T and Joyce B A 1975 *Surf. Sci.* **50** 434
[21] Itoh M, Bell G R, Joyce B A and Vvedensky D D 200 *Surf. Sci.* **464** 200
[22] Itoh M, Bell G R, Avery A R, Jones T S, Joyce B A and Vvedensky D D 1998 *Phys. Rev. Lett.* **81** 633
[23] Evans J W and Bartelt M C 1994 *J. Vac. Sci. Technol.* A **12** 1800
[24] Ratsch C, Smilauer P, Zangwill A, Vvedensky D D 1995 *Surf. Sci.* **329** L599

Journal of Physics: Conference Series **681** (2016) 012037 doi:10.1088/1742-6596/681/1/012037

Structure and stability of silicon nanoclusters passivated by hydrogen and oxygen: evolutionary algorithm and first-principles study

V S Baturin[1,2], S V Lepeshkin[1,2], N L Matsko[1,2] and Yu A Uspenskii[1]

[1] P.N. Lebedev Physical Institute, Russian Academy of Sciences - 119991 Leninskii prosp. 53, Moscow, Russia
[2] Moscow Institute of Physics and Technology - Dolgoprudny, Moscow Region 141700, Russia

E-mail: lepeshkin@lpi.ru

Abstract. We investigate the structural and thermodynamical properties of small silicon clusters. Using the graph theory applied to previously obtained structures of $Si_{10}H_{2m}$ clusters we trace the connection between geometry and passivation degree. The existing data on these clusters and structures of $Si_{10}O_{4n}$ clusters obtained here using evolutionary calculations allowed to analyze the features of $Si_{10}H_{2m}$ clusters in hydrogen atmosphere and $Si_{10}O_{4n}$ clusters in oxygen atmosphere. We have shown the basic differences between structures and thermodynamical properties of silicon clusters, passivated by hydrogen and silicon oxide clusters.

1. Introduction

Silicon nanoclusters are in the center of intensive theoretical and experimental studies for over two decades. The reason of such a great attention comes from several different fields. These systems are the promising objects to perform the further minimization of silicon-based microelectronic devices via the use of single electron transistors on quantum dots [1]. Clusters of silicon with different doping atoms exhibit photoluminescence which differs them from bulk silicon in a fundamental way (see e.g. [2]). This phenomena, having numbers of possible applications, has no clear explanation yet. Though it is obvious that the surface atoms and, generally, size-induced rearrangements of atoms play crucial role here. Thus, knowing of the atomic structure is of great importance. Another problem relevant for applications is the determination of conditions that allow mass production of identical particles. The most of widely used cluster synthesis methods, such as laser ablation produce ensembles of atoms, which are non-uniform in general case [3]. However, experimental and theoretical data on the atomic structure of nanoclusters is limited. Here we present the study on structural and thermodynamical properties of small silicon clusters, passivated by hydrogen and oxygen, with formulas $Si_{10}H_{2m}$ (m=0–11) [4] and $Si_{10}O_{4n}$ (n=0–6).

2. Computational techniques

The scheme we used to find the most stable structures of clusters is based on the evolutionary algorithm and code USPEX [5,6] coupled with ab-initio total energy calculations and local structure

Journal of Physics: Conference Series **681** (2016) 012037 doi:10.1088/1742-6596/681/1/012037

optimization. The principle of genetic algorithm lies within the analogy with natural selection. First, the code generates the initial set of structures called the first generation. Each of these structures is made satisfying several conditions. First, they have one of the randomly chosen symmetries. The number of symmetries is infinite, but here we use all 32 crystallographic point groups plus pentagonal, decagonal and icosahedral groups. Second, the distances between the neighbouring atoms should be larger than some predefined threshold value. The third condition is connectivity, i.e. all the strong bonds in the clusters form only one, non-separated network. Besides these conditions, in all other respects, atoms are placed in random positions. As the first generation is created, all the structures are locally optimized and ranged by energy. Then the most energetically unfavourable ones are discarded and the rest (about 60%) are used in creation of the next generation. A few best clusters are left unchanged, while the rest undergo the action of evolutionary operators. The latter can be divided in two groups. The first one — mutations — produce an offspring structure from a single parent, by performing mutations along soft vibrational mode, permutations of atoms, random distortions, etc. The second group, heredity, makes the essence of the evolutionary algorithm and distinguishes it from other heuristic optimization schemes. It consists of operators which use a few (usually two) parent structures to make an offspring. The most common example is the cut-and-splice operation [7]. It is important to note, that despite the use of symmetry at the stage of initialization, the algorithm slightly breaks it right before the local minimization. This is done because the optimal structure can be asymmetrical and also because highly symmetrical structures are often correspond to a very shallow local minima or even maxima of total energy. As the next generation created and relaxed the procedure repeats and terminates when the lowest energy structure remains unchanged for sufficiently large number of generations. In our studies the number of generations per each cluster was 30-60 with 20-50 structures in each generation. The local minimization and energy calculation is performed within PAW-PBE approach in density functional theory [8, 9] as implemented in VASP code [10]. We used supercell with 10 Å vacuum distance between clusters and the cutoff energy of 350 eV.

3. Structural analysis of $Si_{10}H_{2m}$ clusters

Here we present the results of structure analysis and thermodynamical consideration performed on silicon clusters $Si_{10}H_{2m}$, (m=0–11). The ground-state structures of clusters, obtained from ab-initio USPEX calculations in our previous study [4] are presented on figure 1.

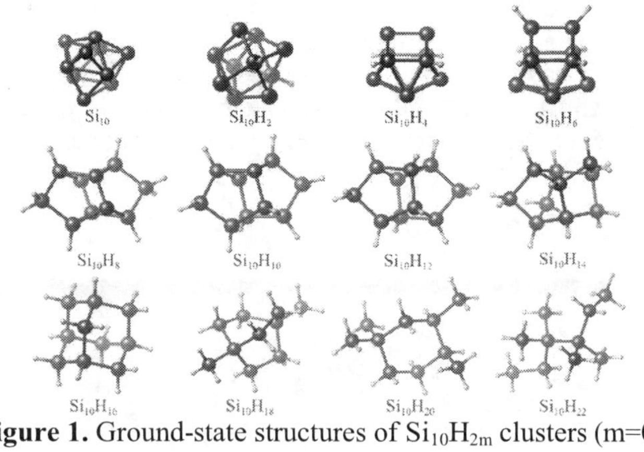

Figure 1. Ground-state structures of $Si_{10}H_{2m}$ clusters (m=0–11)

Looking at low-passivated structures, Si_{10} – $Si_{10}H_{10}$, we see that unlike in the usual silicon compounds, silicon atoms have coordination numbers (numbers of closest neighbors) different from the standard valence of silicon, i.e. four. This can be explained by means of the graph theory and the angles between adjacent bonds. On the one hand, atoms of silicon tend to have four neighbors, by the number of valent electrons. On the other hand, bonds of fourfold-coordinated silicon atoms, e.g. in bulk silicon, are oriented with tetrahedral angles, equal to 109°28'. We build the graph, corresponding to

Journal of Physics: Conference Series **681** (2016) 012037 doi:10.1088/1742-6596/681/1/012037

the cluster, in the following way. Hydrogen atoms are neglected, and the edges of graph represent the non-branching chains of Si-Si bonds, so the Si atoms participating in three or more Si-Si bonds are represented as graph's vertices (figure 2).

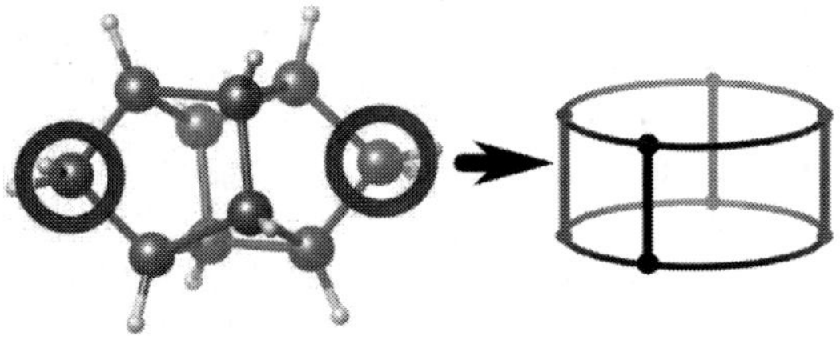

Figure 2. Cluster and corresponding graph. Atoms in circles do not form the graphs vertices.

It is clear from figure 2, that silicon atom forms the vertex if it is bound to one or zero hydrogen atoms. Assuming that graphs constructed in such way are planar, which is true for all $Si_{10}H_{2m}$ clusters under consideration, we can say that it satisfies cycle double cover problem [11], meaning that it can be drawn on a sphere as a superposition of minimal cycles, with each graph edge participating only in two cycles (all other further assumptions concerning graphs do not affect the generality of results). Such cycles are called faces (by analogy with convex polyhedra). If a graph has v vertices, e edges and f faces, it satisfies Euler formula

$$f=e-v+2$$

If we demand fourfold coordination of Si atoms in Si_nH_{2m} cluster, then its graph has $v=2(n-m)$ vertices, $e=(3/2)v=2(n-m)$ edges and $f=n-m+2$ faces. The average number of atoms per one face is $n_{av}=(4n-2m)/f=(4n-2m)/(n-m+2)$. The numerator here comes from the fact, that each vertex atom is related to three cycles, and each on-edge atom - to two cycles. In case of $Si_{10}H_{2m}$ we have $n_{av}=2+16/(12-m)$. The important consequence is that the average angle between adjacent atoms is $\alpha_{av}=180°-360°/n_{av}$. If a cycle has 5 or more atoms ($m>5$), the mean angle can be greater than 103° which is close to tetrahedral angle. If a cycle is big enough (i.e. $m>7$), it has degrees of freedom to bend, so the angles will be equal to energetically favourable tetrahedral angles. On the other hand, if $m<5$ (Si_{10} - $Si_{10}H_{10}$) average angle $\alpha_{av}<96°$, so the structure would experience so called 'ring strain'. The structure 'avoids' it by deviating the coordination number of silicon atoms from their valency or by following the delocalized bonding scheme instead of covalent one as in case of Si_{10}. This simple consideration is one of many examples when graph theory gives useful insights in understanding of motifs in structure formation, e.g. in automatic distinguishing of structural isomers or classification of crystal structures.

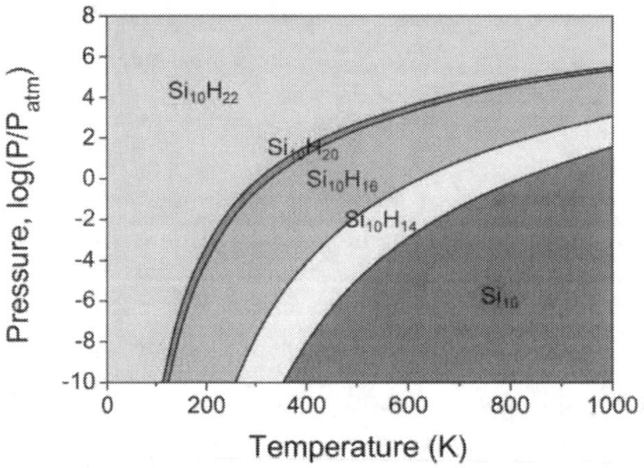

Figure 3. Phase diagram of $Si_{10}H_{2m}$ (m=0–11) clusters in hydrogen atmosphere.

Journal of Physics: Conference Series **681** (2016) 012037 doi:10.1088/1742-6596/681/1/012037

The thermodynamical approach allows to find the most abundant clusters in hydrogen atmosphere with given temperature and pressure. The results, obtained from minimization the corresponding Gibbs free energy of formation is presented on Figure 3. We can see that at normal pressure only Si_{10}, $Si_{10}H_{14}$, $Si_{10}H_{16}$, $Si_{10}H_{20}$ and $Si_{10}H_{22}$ clusters can exist at reasonable temperatures (T<1000 K).

4. $Si_{10}O_{4n}$ clusters

In this part we present the ground-state atomic structures of silicon nanoclusters $Si_{10}O_{4n}$ (n=0–6) which are calculated using the evolutionary algorithm within density functional theory (see figure 4).

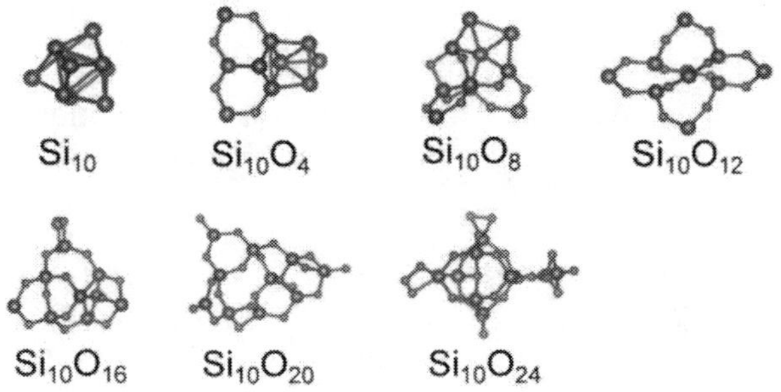

Figure 4. Ground state of $Si_{10}O_{4n}$ (n=0–6) clusters.

The atomic structures of $Si_{10}O_{4n}$ changes drastically with increasing of number of oxygen atoms. At low concentrations oxygen atoms go presumably to the cluster core forming there small regions of quartz-like structure. When oxygen concentration exceeds the SiO_2 composition, new added O atoms occupy the cluster surface and make a cluster more reactive.

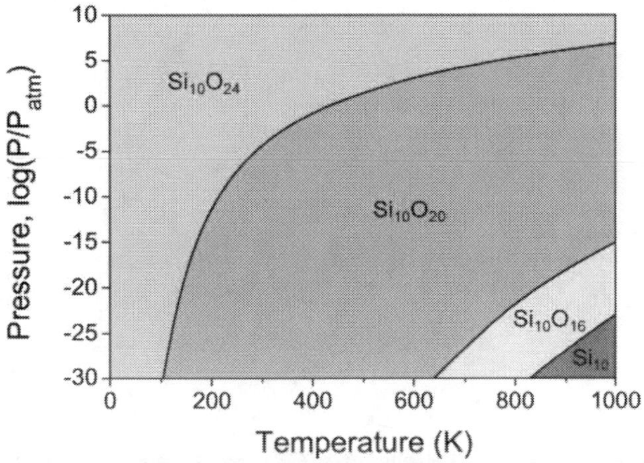

Figure 5. Phase diagram of $Si_{10}O_{4n}$ (n=0–6) clusters in oxygen atmosphere.

Figure 5 shows phase diagram of $Si_{10}O_{4n}$ clusters in oxygen atmosphere. In a wide range of pressures and temperatures the most stable clusters are $Si_{10}O_{20}$ and $Si_{10}O_{24}$. Clusters with low number of oxygen atoms appears only at very small pressures and high temperatures.

5. Conclusions

The structural analysis based on graph theory has shown that the unusual coordination numbers of silicon atoms in $Si_{10}H_{2m}$ clusters can be explained by the tendency of the system to form tetrahedral angles between adjacent bonds. We obtained a formula showing that tetrahedral angles can be achieved only in cases of m>5, which is in agreement with results of global optimization. Using evolutionary algorithm results we performed the thermodynamical consideration of silicon nanoclusters $Si_{10}H_{2m}$ (m=0–11) and $Si_{10}O_{4n}$ (n=0–6). We found that of all possible clusters with formula $Si_{10}H_{2m}$ only Si_{10}, $Si_{10}H_{14}$, $Si_{10}H_{16}$, $Si_{10}H_{20}$ and $Si_{10}H_{22}$ clusters can exist in thermodynamical equilibrium with hydrogen atmosphere in a wide range of pressures and temperatures. In oxygen atmosphere the most stable clusters are Si_{10}, $Si_{10}O_{16}$, $Si_{10}O_{20}$ and $Si_{10}O_{24}$ of all possible clusters with formula $Si_{10}O_{4n}$, but Si_{10} and $Si_{10}O_{16}$ clusters can exist only at very small pressures and high temperatures.

Acknowledgments

This work is supported by the Russian Foundation for Basic Research (grants 13-02-00655, 14-02-00583, 13-02-00913) and the Government of the Russian Federation (No. 14.A12.31.0003). Calculations were performed at the Joint Supercomputer Center (Russian Academy of Sciences, Moscow, Russia).

References

[1] Wilson W. L. Szajkowski P. F. Brus L. E. *Science* 262 1241 (1993)
[2] Belomoin G. et. al. *Appl. Phys. Lett.* 80 841 (2002)
[3] Akiko Okano, Kunio Takayanagi, *Applied Surface Science* 127–129 362-367 (1998)
[4] V S Baturin, S V Lepeshkin, N L Matsko, Artem R Oganov and Yu A Uspenskii *EPL* 106 37002 (2014)
[5] Oganov A. R. and Glass C. W. *J. Chem. Phys.* **124** 24474 (2006)
[6] Oganov A. R., Lyakhov A. O. and Valle M. *Acc. Chem. Res.* **44** 227 (2011)
[7] Deaven, D.M. and Ho, K.M. *Phys. Rev. Lett* **75** 288 (1996)
[8] Blöchl, P.E. *Physical Review B* **50** (24) 17953–17978
[9] Perdew J. P., Burke K. and Ernzerhof M. *Phys. Rev. Lett* **77** 3865 (1996)
[10] G. Kresse and J. Hafner *Phys. Rev. B* **47** 558 (1993).
[11] Huck A. *Discrete Appl. Math.* **99** 71-90 (2000)

Journal of Physics: Conference Series **681** (2016) 012038 doi:10.1088/1742-6596/681/1/012038

Impact of defects on electrical connectivity of monolayer of ideally aligned rods

Yu Yu Tarasevich[1], D O Dubinin[1], V V Laptev[1,2], N I Lebovka[3,4]
[1]Astrakhan State University, Astrakhan, Russia
[2]Astrakhan State Technical University, Astrakhan, Russia
[3]Institute of Biocolloidal Chemistry named after F.D. Ovcharenko, NAS of Ukraine, Kiev, Ukraine
[4]Taras Shevchenko Kiev National University, Department of Physics, Kiev, Ukraine

E-mail: tarasevich@asu.edu.ru

Abstract. The processes of formation of electrically conductive films filled by aligned elongated nanoparticles, i.e. nanotubes, nanowires or fibers attract great attention in nanotecnological applications. The alignment can be controlled by external electric fields, evaporation-driven self-assembly and assisted by different other techniques. This work studies the impact of defects on electrical connectivity of ideally aligned monolayer of rods (k-mers). By means of Monte Carlo simulation the problem of percolation for conductive rods on a discrete insulating substrate (square lattice) is analyzed. The aspect ratio of the particles changes within the interval 1–64, the insulating defects were distributed both on the lattice and on the particles. We found that even a very small amount of the insulating defects on the particles can destroy the electrical connectivity. The critical concentration of the defects decreases as the aspect ratio of the particles increases.

1. Introduction

Over the past decade, single-walled carbon nanotubes (SWCNTs), multi-walled carbon nanotubes (MWCNTs), heteronanotubes and similar objects attract the great attention due to their exceptional properties [1]. Hybrid and functionalized tubular structures have been constructed via several modification paths. These modifications include the presence of molecules, generation of defects, and partial or full replacement of the carbon atoms, always maintaining a nanotube structure [2]. The promising applications have motivated efforts to produce thin films filled with the ordered nanotubes. These films inherit anisotropy of the one-dimensional carbon nanotube structure. Various methods for aligning nanotubes have been explored [3]. Evaporation-driven self-assembly (EDSA) is one of these methods [4]. EDSA is based on the well-studied 'coffee-ring effect' [5]. When a particle-laden droplet desiccates on a horizontal hydrophilic substrate, the particles suspended in the droplet tend to aggregate at the edges of the droplet, forming a ring-like structure. EDSA provides an opportunity to fabricate large-area, electronically monodisperse, aligned single-walled carbon nanotube thin films [6].

During EDSA, droplets of nanorods dispersed in solvents are deposited on substrate surfaces and the solvent is allowed to evaporate. Strong contact line pinning in combination with evaporation induces an outward flow inside the droplet. This flow can produce dense nanorod deposition inside coffee stain fringes, where large scale lateral ordering of the nanorods can be

Journal of Physics: Conference Series **681** (2016) 012038 doi:10.1088/1742-6596/681/1/012038

observed. The long axis of the rods is oriented parallel to the contact line [7]. Self-assembly of single-walled carbon nanotubes into a sheet by drop drying has been reported [8].

A solution-based method produces highly oriented carbon nanotube (CNT) arrays. Inside these arrays, long-range nematic-like ordering of carbon nanotube stripes was observed [9]. Structures of aligned single-wall carbon nanotubes can be produced using a template-based fluidic assembly process [10]. Recently, a novel solvent evaporation method to align SWNTs in patterns having nanoscale width and micronscale length has been reported [11]. Novel means of building hybrid assemblies of crystalline C_{60} and carbon nanotubes (CNTs) has been presented, wherein aligned CNT films direct the crystallization and orientation of C_{60} rods from solution [12].

Another method to produce aligned single-wall carbon nanotube structures is based on application of the electric field. This method was used to form SWCNTs polymer composites [13]. Controlled placement of carbon nanotubes may be performed using alternating-current electric field [14].

Many efforts have been done to simulate the properties of layers of aligned rod-like objects. Thin film formation from a suspension of rod-like nanoparticles via solvent evaporation was studied by simulation [15]. When particles were strongly charged, they spontaneously aligned in suspension forming well-aligned thin film after complete drying.

Effect of nanotube alignment on percolation conductivity in carbon nanotube or polymer composites was investigated both experimentally and theoretically by Monte Carlo simulations [16]. The investigations showed that percolation conductivity of a stick network depends on alignment as well as concentration. Both dependencies exhibit critical (power-law) behavior. The highest conductivity occurs for slightly aligned, rather than isotropic, sticks.

Monte Carlo simulations has been used to study the effects of nanotube alignment and measurement direction on the resistivity in single-walled carbon nanotube multiple layers [17]. The minimum resistivity occurs for a partially aligned rather than a perfectly aligned nanotube film. The film resistivity becomes highly dependent on the measurement direction when nanotubes are strongly aligned.

A lattice gas model with Monte Carlo simulation has been used to study the morphological self-assembled patterns [18]. Moreover, percolation in two-dimensional nanoparticle films from colloidal self-assembly has been investigated numerically in this work.

Numerous both theoretical investigations and computer simulations have been devoted to the common physical behaviour of the systems of rod-like particles. Exact calculations and enumerations of systems of hard rigid rods distributed at closed packed density have been performed [19]. Monte Carlo simulations have been carried out for a system of straight rigid rods on a square lattice with only hard-core interactions [20], for the isotropic-nematic phase transition in a system of long straight rigid rods on two-dimensional square and triangular lattices [21, 22], for the isotropic-nematic phase transition in a system of self-assembled rigid rods on a lattice [23, 24], and for the self-assembled rigid rods on triangular lattices [25]. Particularly, it was demonstrated that a system of straight rigid rods on a square lattice with only hard-core interactions undergoes a phase transition from the low-density disordered phase to a nematic phase and then again undergoes a re-entrant phase transition from the nematic phase to a disordered phase [20]. Particularly, percolation of aligned rigid rods on two-dimensional square lattices has been investigated [26, 27].

Recently, impact of defects on the percolation of partially oriented rigid rods (k-mers) deposited of a square lattice has been reported [28, 29, 30]. Nevertheless, influence of the combined defects on percolation of perfectly aligned rods needs additional attention. This report presents the results of the Monte Carlo simulation plied to the problem of percolation for conductive rods on a discrete insulating substrate (square lattice). The insulating defects were distributed both on the lattice and on the particles. The impacts of the aspect ratio of the

Journal of Physics: Conference Series **681** (2016) 012038 doi:10.1088/1742-6596/681/1/012038

particles and concentration different types of insulating defects on the electrical connectivity are discussed.

2. Details of simulation

A problem on a discrete two-dimensional substrate (square lattice $L \times L$ sides) is analyzed. We applied periodic (toroidal) boundary conditions to eliminate the effect of the border. The substrate was assumed as insulating. The substrate was initially randomly filled with the insulating point defects at given concentration d_l. The defects prevent the deposition of the particles. The deposited objects were linear k-mers (particles occupying k adjacent sites of the square lattice). In fact, the value of k is the aspect ratio of the particles. The objects were oriented strongly along one direction, e.g., along vertical axis y. We used random sequential adsorption to deposit the particles [31]. Interparticle hard-core repulsion was assumed, i.e. overlapping of the particles was strictly prohibited, as a result a monolayer was formed. The particles were treated as conducting. For each given value of d_l, we filled the lattice with the particles to the jamming state. At the jamming state, there is no possibility to adsorb even one additional particle, because any empty space of the appropriate size and shape is absent. The jamming concentration is denoted as p. Then, we randomly replaced some sites of the particles by the insulating defects; the defect concentration, d_k, being defined as the fraction of insulating sites on the particles, i.e. the number of such defects is $d_k p L^2$, and checked whether a path over the conducting particles spanning entirely the substrate. If the such path can be found, the electrical connectivity is present and the monolayer conducts the electricity. This situation is defined as the percolation [32]. We checked the percolation in two perpendicular directions, x and y, and used two criteria: there is percolation in both directions (criterion AND), there is percolation at least along one direction (OR). In contrast to [33], we chose to treat spiral clusters as wrapping (percolating). We repeated the process 1000 times for different configurations of defects and particles and found the probability, $R(d_k)$, that percolation would occur at given concentrations of defects on the deposited particles. The abscissa of the inflection point of the curve was treated as the estimation of the critical concentration of defects for the given lattice size. We performed a scaling analysis to obtain the critical concentration at the thermodynamic limit

$$d_k(L) - d_k(\infty) \sim L^{-1/\nu_c},$$

where $\nu_c = 4/3$ is the critical exponent [32]. For $k \leq 32$, we used the lattice sizes $L = 100k, 200k$, and $400k$. For $k = 64$, we used only two lattice sizes $L = 100k$ and $200k$ but the scaling analyzes were performed using the two criteria (AND, OR).

The highly efficient algorithm, proposed by Frank and Lobb [34] was utilized for finding conductivity of a square lattice, filled with the k-mers. The Frank and Lobb algorithm utilises the repeated application of a sequence of series, parallel and star-triangle (Y–$\triangle$) transformations to the bonds of the lattice. The final result of this sequence of transformations is reduction of any finite portion of the lattice to a single bond that has the same conductance as the entire lattice. We used four equivalent resistors (conductors) with $\sigma_f = 10^6$ and $\sigma_i = 1$ for occupied and empty sites [35], respectively, instead of each cell. The details of the calculations can be found in [36].

3. Results and Discussion

Below the critical concentrations of the defects, the monolayer is conducting (Fig. 1a) since the conducting parts of the particles form a spanning cluster. Above the critical concentrations of the defects, the monolayer is insulating since despite the existence of large conducting regions, these regions are not connected with each other (Fig. 1b).

Fig. 2a presents a phase diagram on the plane $(d_l; d_k)$. The regions below the curves correspond to the conducting (percolation) states. Even a small concentration of defects inhibits

Figure 1. The fragment (80×80) of adsorbed monolayer 800×800 of elongated particles. Particle aspect ratio is 8. (a) The monolayer is conducting. $d_l = 0.05$, $d_k = 0.05$. (b) The monolayer is insulating. $d_l = 0.075$, $d_k = 0.075$. The conducting subsystem is shown in blue. Insulating substrate is shown in white. The insulating point defects on the substrate are shown in black. The insulating regions of the particles are shown in light blue. The isolated clusters of the conducting particles are shown in red; these clusters do not contribute in the conductivity of the monolayer.

the deposition of long objects. The curves separating percolating and non-percolating states are convex for short objects ($k \leq 8$) and concave for the larger objects. As the aspect ratio, k, increases, the area of the percolating region on the phase plane ($d_l; d_k$) decreases faster than the product of $0.5 d_l d_k$. We can expect that in the case of long objects, the area will be extremely narrow and located along the axes. The result suggests that for objects with a large aspect ratio, even a very small concentration of defects can inhibit the percolation when both kinds of defects are present. Fig. 2b suggests that the monolayers constructed from the aligned rod-like particles cannot conduct even any kind of defects is absent when the aspect ratio of such particles is large enough. Our rough estimation gives the critical aspect ratio $k \gtrsim 100$.

Figure Fig. 3 presents examples of electrical conductivity σ in the vertical (y) and horizontal (x) directions as a function of defects concentration, d_k, at fixed values of d_k for $L = 800$ and $k = 8$. Behaviour of electrical conductivity $\sigma(d_k)$ corresponds with direct estimations of the percolation threshold from the phase diagram at the Fig. 2a. There is a sharp transition of electrical conductivity near the percolation threshold. Moreover, the behaviours of $\sigma_x(d_k)$ and $\sigma_y(d_k)$ are slightly different. Data evidence (see, inset to Fig. 3) that at fixed value of d_l the percolation concentration d_k) for x direction was higher that for y direction. It reflects the presence of anisotropy of electrical conductivity for strictly ordered systems.

4. Conclusion

Using the model of random sequential adsorption, we studied the electrical conductivity of the monolayers of perfectly aligned elongated particles with aspect ratio up to 64. We considered deposition of the particles onto a substrate with preliminary placed insulating point defects. Some regions of the particles may be insulating. It should be kept in mind that no real interaction between the particles is taken into account in the model; the anisotropic monolayer is formed 'by hand'. We found that even a very small amount of the insulating regions on the particles can destroy the conductivity of the monolayer. The critical concentration of the defects decreases as

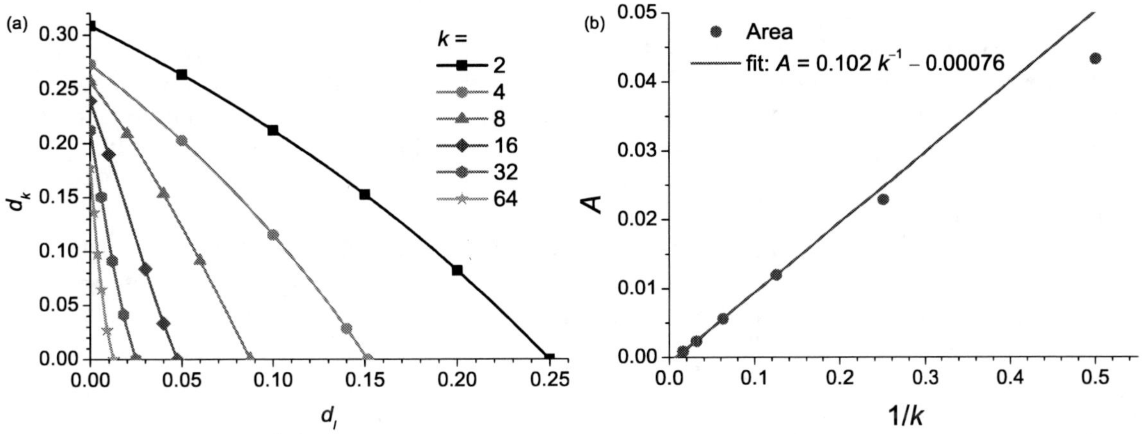

Figure 2. (a) Phase diagram presented as the concentration of the defects on the particles, d_k, versus the concentration of the defects on the substrate, d_l. The areas below the curves correspond to the conducting (percolation) states. (b) Area of the conducting state on the phase diagram as a function of the inverse of the aspect ratio of the particles.

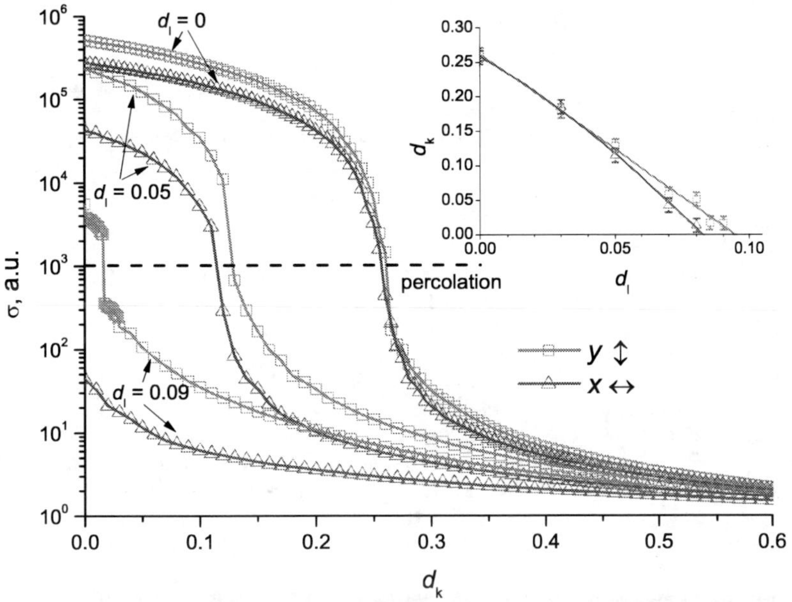

Figure 3. Examples of electrical conductivity σ in the vertical (y) and horizontal (x) directions as a function of defects concentration, d_k, at fixed values of d_k for $L = 800$ and $k = 8$. The k-mers are perfectly oriented dimers along the vertical direction. Inset shows the percolation concentration of the defects on the particles, d_k, versus the concentration of the defects on the substrate, d_l.

the aspect ratio of the particles increases. The behaviour of the monolayers of aligned particles is very similar to the behaviour of the isotropic monolayers [29, 30].

The obvious shortcoming of our model is that the model shows only the fact of presence or absence of the conductivity, rather than the numerical value of the resistivity of the layer.

The future efforts should be focused on the calculations of the resistivity values for different defect concentrations and aspect ratios. Moreover, computer simulations of monolayers formed by particles with the larger aspect ratio ($k > 64$) is of interest because the real nanotubes have the large aspect ratio, e.g. the MWCNTs have the aspect ratio ≥ 100 [37].

Acknowledgements
The reported research is supported by the Ministry of Education and Science of the Russian Federation, Project No. 643, the Russian Foundation for Basic Research, Grant No. 15-02-90402 Ukr_a, and the National Academy of Sciences of Ukraine, Project No. 43-02-15(U).

References

[1] Charlier J C, Blase X and Roche S 2007 *Rev. Mod. Phys.* **79**(2) 677–732
[2] Ayala P, Arenal R, Loiseau A, Rubio A and Pichler T 2010 *Rev. Mod. Phys.* **82**(2) 1843–1885
[3] Goh P, Ismail A and Ng B 2014 *Compos. Part A: Appl. S.* **56** 103–126
[4] Lin Z (ed) 2010 *Evaporative Self-assembly of Ordered Complex Structures* (World Scientific Publishing Company) ISBN 9814304689
[5] Deegan R D, Bakajin O, Dupont T F, Huber G, Nagel S R and Witten T A 1997 *Nature* **389** 827–829
[6] Shastry T A, Seo J W T, Lopez J J, Arnold H N, Kelter J Z, Sangwan V K, Lauhon L J, Marks T J and Hersam M C 2013 *Small* **9** 45–51
[7] Nobile C, Carbone L, Fiore A, Cingolani R, Manna L and Krahne R 2009 *J. Phys. Condens. Matt.* **21** 264013
[8] Duggal R, Hussain F and Pasquali M 2006 *Adv. Mater.* **18** 29–34
[9] Ko H, and Tsukruk V V 2006 *Nano Lett.* **6** 1443–1448
[10] Kim Y L, Li B, An X, Hahm M G, Chen L, Washington M, Ajayan P M, Nayak S K, Busnaina A, Kar S and Jung Y J 2009 *ACS Nano* **3** 2818–2826
[11] Yang D, Wang L, Zhang X, Wang D, Shen Z and Li S 2011 *Nano-Micro Lett.* **3** 146–152
[12] Meshot E R, Patel K D, Tawfick S, Juggernauth K A, Bedewy M, Verploegen E A, De Volder M F L and Hart A J 2012 *Adv. Funct. Mater.* **22** 577–584
[13] Park C, Wilkinson J, Banda S, Ounaies Z, Wise K E, Sauti G, Lillehei P T and Harrison J S 2006 *J. Polym. Sci. Part B* **44** 1751–1762
[14] Chen Z, Hu W, Guo J and Saito K 2004 *J. Vac. Sci. Technol. B* **22** 776–780
[15] Ohta S, Inasawa S, Koike O, Fujita M and Yamaguchi Y 2009 *Appl. Phys. Express* **2** 065002
[16] Du F, Fischer J E and Winey K I 2005 *Phys. Rev. B* **72**(12) 121404
[17] Behnam A, Guo J and Ural A 2007 *J. Appl. Phys.* **102** 044313
[18] Hsu C L, Chu S M, Wood K and Yang Y R 2007 *phys. status solidi a* **204** 1856–1862
[19] Baumgärtner A 1985 *J. Physique Lett.* **46** 659–666
[20] Ghosh A and Dhar D 2007 *Europhys. Lett.* **78** 20003
[21] Matoz-Fernandez D A, Linares D H and Ramirez-Pastor A J 2008 *Europhys. Lett.* **82** 50007
[22] Matoz-Fernandez D, Linares D and Ramirez-Pastor A 2012 *Eur. Phys. J. B* **85** 296
[23] López L G, Linares D H and Ramirez-Pastor A J 2009 *Phys. Rev. E* **80**(4) 040105
[24] Tavares J M, Holder B and Telo da Gama M M 2009 *Phys. Rev. E* **79**(2) 021505
[25] Almarza N G, Tavares J M and Telo da Gama M M 2011 *J. Chem. Phys.* **134** 071101
[26] Longone P, Centres P M and Ramirez-Pastor A J 2012 *Phys. Rev. E* **85**(1) 011108
[27] Romiszowski P and Sikorski A 2013 *CMST* **19** 115–121
[28] Tarasevich Y Y, Laptev V V, Vygornitskii N V and Lebovka N I 2015 *Phys. Rev. E* **91**(1) 012109
[29] Lebovka N I, Tarasevich Y Y, Dubinin D O, Laptev V V and Vygornitskii N V 2015 *Phys. Rev. E* **92**(6) 062116
[30] Tarasevich Y Y, Laptev V V, Burmistrov A S and Shinyaeva T S 2015 *J. Phys. Conf. Ser.* **633** 012064
[31] Evans J W 1993 *Rev. Mod. Phys.* **65**(4) 1281–1329
[32] Stauffer D and Aharony A 1992 *Introduction to Percolation Theory* (London: Taylor & Francis)
[33] Newman M E J and Ziff R M 2001 *Phys. Rev. E* **64**(1) 016706
[34] Frank D J and Lobb C J 1988 *Phys. Rev. B* **37**(1) 302–307
[35] Yuge Y and Onizuka K 1978 *J. Phys. C Solid State* **11** L763–L766
[36] Cherkasova V A, Tarasevich Y Y, Lebovka N I and Vygornitskii N V 2010 *Eur. Phys. J. B* **74** 205–209
[37] Guo J, Liu Y, Prada-Silvy R, Tan Y, Azad S, Krause B, Pötschke P and Grady B P 2014 *J. Polym. Sci. Pol. Phys.* **52** 73–83

Journal of Physics: Conference Series **681** (2016) 012039 doi:10.1088/1742-6596/681/1/012039

Computational investigation of stable formation condition for Fe_xNi_{1-x} alloy films on paramagnetic substrate

Marina V. Mamonova, Vladimir V. Prudnikov, Dmitrii V. Pilipenko

Department of Theoretical Physics, Omsk State University, Omsk 644077, Russia

E-mail: mamonovamv@omsu.ru, prudnikv@univer.omsk.su, virgo54@mail.ru

Abstract. The adsorption process of ultrathin ferromagnetic Fe and Ni alloy film on a nonmagnetic metal substrate is studied by the variational spin-density functional method. The conditions for the stable formation of ferromagnetic films with respect to island adsorption are discovered with a change of values of the coverage parameter and alloy component concentration. The equilibrium values of vacuum gap and film thickness are determined from the minimum of the total interfacial energy. The energy characteristics of films Fe_xNi_{1-x} on close-packed surfaces W, Cu and Ag with effects of magnetic ordering are calculated for different temperatures.

1. Introduction

The amazing properties of thin magnetic alloy films are the subject of great research interests [1–3], which are determined by many their applications in microelectronics and computer technology. For example, permalloy $Fe_{20}Ni_{80}$ is used in transformer laminations and magnetic recording heads. In this work, the spin density functional method was used for description of the temperature and ferromagnetic ordering influence on the adsorption properties of Fe_xNi_{1-x} alloy films onto the surface of nonmagnetic materials. For theoretical description we applied the method of multi-parameter test functions [4] the parameters of which are determined by the numerical minimization procedure of the interfacial energy functional in the subsurface region.

The experimental investigations of ultrathin magnetic films [5] show that the magnetic properties of system Fe/Cu are characterized by two-dimensional Ising model type, and Fe/Au by two-dimensional XY model type. Therefore, in this work we use for description of temperature dependence for the relative magnetization $m(T)$ the results of renormalization-group finite-size consideration for XY type systems.

2. Model and methods

Consider a semi-infinite metal with an average charge density n_1, which is limited by an infinite flat surface and occupies the region $z < -D$. The adsorbate film with a charge density n_2 occupies the region $D < z < D + h$, where h is the thickness of the film. Between the adsorbent and adsorbate in this model there is a vacuum gap of width $2D$.

Define the interfacial interaction energy per unit area of contact as an integral over z from the bulk density of the free energy of the electron gas:

Table 1. The values of input parameters.

Me	$T_c(\Theta = 1)$	r_c	d	$\sigma(\infty)$	Z
Fe	521	0.95	3.82	2170	4
Ni	209	1.27	2.35	1820	2
W		0.92	4.21	2800	6
Cu		1.10	3.92	1750	2
Ag		1.15	6.30	1205	1

$$\sigma = \int_{-\infty}^{+\infty} \{f[n(z,\beta)] - f[n_0(z)]\}\, dz. \tag{1}$$

The volume density of the free energy of the inhomogeneous electron gas can be represented as a gradient expansion:

$$f[n(z)] = w_0[n(z)] + w_2[n(z), |\nabla n(z)|^2] + w_4[n(z), |\nabla n(z)|^4] - T(s_{id} + s_{order}), \tag{2}$$

where

$$w_0[n_{\uparrow\downarrow}(z)] = w_{kin} + w_{cul} + w_{ex} + w_{cor} \tag{3}$$

is the energy density of the homogeneous electron gas, including successively the kinetic, Coulomb, exchange and correlation energy, s_{id}, s_{order} are the entropic contributions to the free energy taking into account the temperature changes in the entropy for the electron gas and the effects of magnetic ordering in the electronic subsystem, respectively. Detailed expression for all of this contributions given in [4]. The densities n_+ and n_- can be expressed in terms of relative magnetization $m = M(T)/M(T = 0)$ of electrons as follows:

$$n_{\uparrow\downarrow}(z) = \frac{n(z)(1 \pm m(z))}{2} \tag{4}$$

We are taking into consideration the inhomogeneous spatial distribution of magnetization in surficial region in the following view:

$$m(z) = \begin{cases} 0.5m_0 e^{\beta(z-D)}[1 - e^{-\beta h}], & z < D \\ 0.5m_0[1 - 0.5e^{-\beta(z-D)} - 0.5e^{\beta(z-D-h)}], & D < z < D + h \\ 0.5m_0 e^{-\beta(z-D)}[e^{\beta h} - 1], & z > D + h \end{cases} \tag{5}$$

where m_0 which is determined by two-dimensional XY model type.

$$m_0(T) \simeq \left[\frac{T_c^{(s)}(x) - T}{T_c^{(s)}(x)}\right]^{0.23} \tag{6}$$

The critical temperature $T_c^{(s)}$ of magnetic ordering of the monatomic film depends on the parameter Θ:

$$T_c^{(s)}(\Theta) \cong \Theta T_c^{(b)} \frac{z_{surf}}{z_{bulk}}, \tag{7}$$

Journal of Physics: Conference Series **681** (2016) 012039 doi:10.1088/1742-6596/681/1/012039

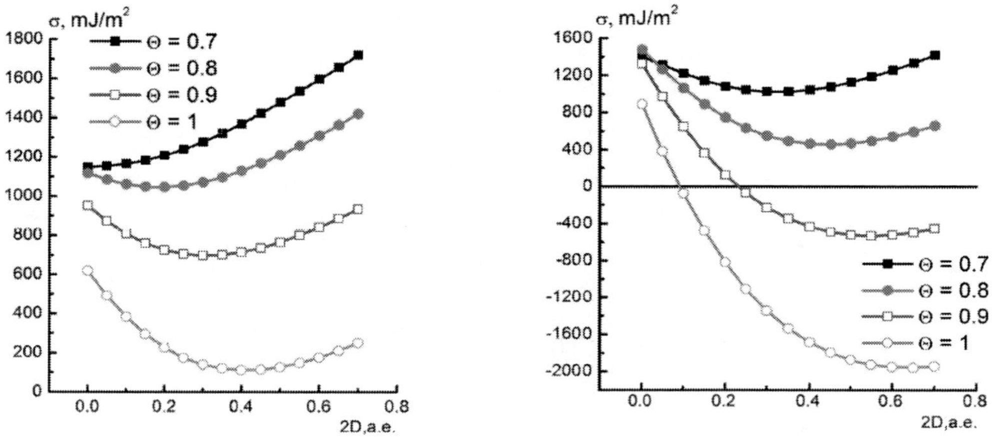

Figure 1. The dependence of total interfacial energy for $Fe_xNi_{1-x}/Cu(111)$ and $Fe_xNi_{1-x}/W(110)$ systems on vacuum gap at T=100K and $x = 80\%$ for different coverage parameter Θ.

where z_{surf} is the number of nearest neighbors in the ferromagnetic film, and z_{bulk} in the bulk ferromagnet. In this case, for the critical temperature of magnetic ordering of the bulk ferromagnet alloys we used its experimental dependence on iron concentration [5].

The inclusion of the discreteness in the ion distribution leads to the corrections to the electrostatic interaction energy due to the ion–ion and electron–ion interactions. As a result, the interfacial interaction energy can be written in the form

$$\sigma = \sigma_0 + \sigma_{ei} + \sigma_{ii}, \qquad (8)$$

where σ_0 is the contribution of the electronic system in the framework of the jellium model, σ_{ii} is the correction to the ion–ion electrostatic interaction energy, and σ_{ei} is the correction to the energy associated with the difference in electrostatic interactions of electrons with discrete ions and the uniform "jelly" background. The parameters of the discreteness in the ion distribution presented in table 1 as Ashcroft pseudopotential cutoff radius r_c, charge of the ions Z and interplanar distance d also depend on the concentration of alloy components:

$$r_c = \left(\frac{xZ_{Fe}(r_{cFe})^3 + (1-x)Z_{Ni}(r_{cNi})^3}{xZ_{Fe} + (1-x)Z_{Ni}} \right)^{1/3}, \qquad (9)$$
$$Z = xZ_{Fe} + (1-x)Z_{Ni}, \qquad d = xd_{Fe} + (1-x)d_{Ni}.$$

We have developed a multiparameter version of the functional of electron density with test functions which take into account, along with the heterogeneity of the distribution of electron density, the existence of the equilibrium distance between the adatoms and the substrate determined from the condition of the minimum interfacial energy of interaction [3]. Knowing the interfacial energy of the metal contact with the film, as well as the surface energy of each component of a bimetallic system, it is easy to find the energy of adhesion of the system as a work that needs to be carried out to remove the substrate and the film from each other to infinity. We determine the adsorption energy as the specific adhesion energy per adsorbed atom

$$E_{ads}(\Theta) = \min_{\beta,h,D} \left[\frac{\sigma(\infty) - \sigma(\beta,h,D,T,\Theta)}{n_s} \right], \qquad (10)$$

where n_s is the surface atomic concentration in the film.

Table 2. Conditions for formation of stable $Fe_xNi_{1-x}/Cu(111)$ film on Ag substrate.

T, K	Θ	$X, \%$	T, K	Θ	$X, \%$
0	0.9	[40]	300	0.8	[80-100]
100	0.7	[100]		0.9	[60-80]
	0.8	[60-100]		1	[40]
	0.9	[0-40]			
200	0.8	[60-100]	400	0.9	[80-100]
	0.9	[20-60]		1	[80]
	1	[0-20]			

Table 3. Conditions for formation of stable $Fe_xNi_{1-x}/Cu(111)$ film on Cu substrate.

T, K	Θ	$X, \%$
0	0.7	[30-100]
	[0.8-1]	All concentrations
200	0.8	[50-100]
	[0.9-1]	All concentrations
600	0.9	[17-100]
	1	All concentrations

Table 4. Conditions for formation of stable $Fe_xNi_{1-x}/Cu(111)$ film on W substrate.

T, K	Θ	$X, \%$
0	[0.5-0,7]	All concentrations
	0.8	[0-60]
	0.9	[0-20]
	1	[0]
100	[0.7,0.8]	All concentrations
	0.9	[0-40]
	1	[0-20]
200	0.7	[40-100]
	0.8	[20-100]
	0.9	[0-60]

3. Results of calculations

The equilibrium vacuum gap $D = D_{min}$ which simulates the effective influence of the atomic roughness of the substrate and the equilibrium film thickness $h = h_{min}$ are determined from the minimum of the total interfacial energy $\sigma(\beta, h, D, T, \Theta)$. As shown in Figure 1, minimum of $\sigma(\beta, h, D, T, \Theta)$ is disappears for some values of the parameter Θ or $\sigma(\beta_{min}, h_{min}, D_{min}T, \Theta)$ takes negative values. This approach makes it possible to predict the formation of stable Fe_xNi_{1-x} coatings or to reveal their absence (the case of island adsorption) in depending on concentration x. Conditions for formation of stable $Fe_xNi_{1-x}/Cu(111)$ film on different substrates, obtained from our calculations, are given in Tables 2-4.

We present the results of calculations of adsorption energy for alloy film on Cu, W and Ag substrates in Figures 2, 3. Calculations show that for all substrate the adsorption energy E_{ads} of film decreases with increasing Θ in the range of stability of film and with increase of temperature. Also, E_{ads} significantly increases with growth of iron concentrations x in alloy Fe_xNi_{1-x}.

Results of calculations, given in Figure 4, demonstrate that the film thickness h_{min} decreases with an increase in the coverage parameter Θ in the range of stability of film and keeps a constant value for different iron concentrations.

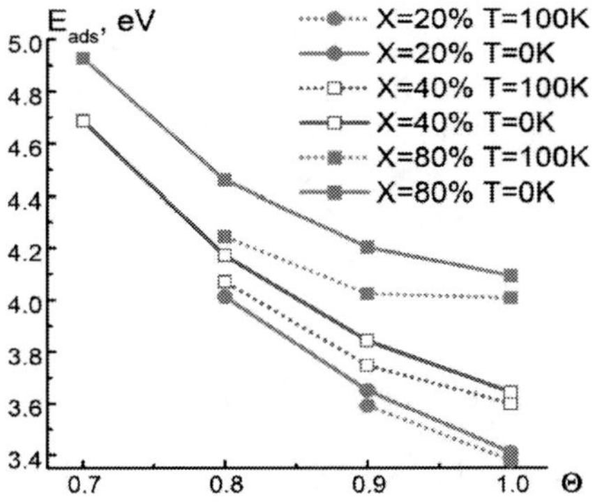

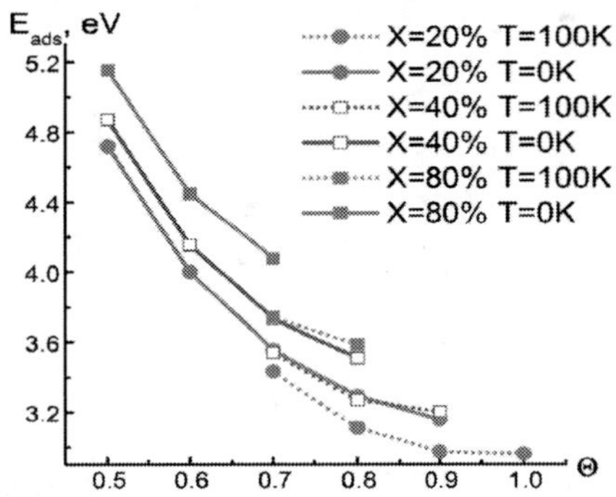

Figure 2. The dependence of the adsorption energy E_{ads} on the coating parameter Θ for $Fe_xNi_{1-x}/Cu(111)$ and $Fe_xNi_{1-x}/W(110)$ for different temperatures and iron concentrations x.

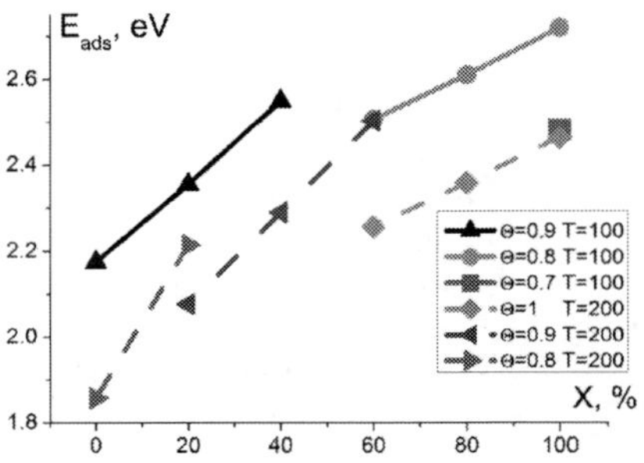

Figure 3. The dependence of the adsorption energy E_{ads} on iron concentrations x for $Fe_xNi_{1-x}/Ag(111)$ system for different temperatures and the coating parameter Θ.

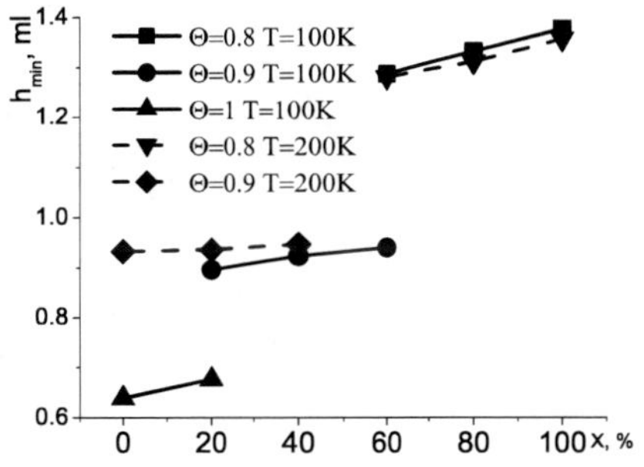

Figure 4. The dependence of equilibrium thickness h_{min} on iron concentration x for different parameter Θ for $Fe_xNi_{1-x}/Ag(111)$.

4. Conclusions

The calculation of the adsorption energy of Fe_xNi_{1-x} alloy film on the close-packed surface faces Cu, W and Ag substrates are carried out in dependence on parameter of coverage Θ by the spin-density functional method for different temperatures. The stable monolayer film on Cu(111) substrate is formed only for values of $\Theta \geq 0.7$ at $T = 0K$ and for $\Theta \geq 0.8$ at $T \geq 100K$. For samples with threshold values of Θ, the stable films are realized for high concentration of Fe. Stable film Fe_xNi_{1-x} on W(110) substrate is realized for $\Theta > 0.7(T > 0K)$, but for $\Theta = 0.9-1.0$ it is formed stable state for low concentrations of Fe only. The behavior of a interfacial energy for Ag is similar to W, that cause losing stability for large values of Θ.

Acknowledgments

Support by Russian Science Foundation through project No. 14-12-00562 is acknowledged.

References

[1] Tryputen L, Guo F, Liu F, *et al* 2015 *Phys. Rev* B **91** 014407
[2] Ustinov V V, Milyaev M A and Naumova L I 2014 *SPIN* **04** 1440001
[3] Mamonova M V, Morozov N S , Prudnikov V V 2009 *Physics of the Solid State.* **51** 2169
[4] Mamonova M V, Prudnikov V V, and Prudnikova I A 2013 *Surface Physics: Theoretical models and experimental methods.* (CRC Press: Tailor and Francis Group) p. 384
[5] Schumann F O , Wu S Z , Mankey G J and Willis R F 1997 *Phys. Rev.* B **56** 2668

Dynamics of a liquid drop in porous medium saturated by another liquid under gravity

A.O. Ivantsov[1,2] **and T.P. Lyubimova**[1,2]

[1] Institute of Continuous Media Mechanics UB RAS, Perm, Russia
[2] Perm State University, Perm, Russia

E-mail: aivantsov@icmm.ru, lyubimovat@mail.ru

Abstract. The work deals with numerical simulations of settling or ascension process of a liquid drop in porous media saturated by another liquid. The calculations were carried out using the Darcy model by Level set method with adaptive mesh refinement algorithm that dynamically refines computational mesh near interface. It is shown that the drop is unstable and the finger instability develops at the forefront of moving drop for any ratio of the viscosities of liquids. Under modulated pressure gradient small-scale perturbations of interface are suppressed and in the case of modulation with large enough intensity drop becomes stable.

1. Introduction

In the framework of the conventional Darcy model a planar displacement front in a porous medium is known to be unstable if the displacing fluid has lower viscosity than the displaced one [1]. The instability is associated with development of penetrating fingers of the less viscous fluid. They grow due to lower friction with porous matrix. The process is virtually unimpeded by viscous momentum transfer, which is negligible in porous media. This mechanism explains the short-wave nature of the instability.

In practice the displaced fluid often forms compact inclusions surrounded by the displacing fluid. In this case one should consider displacement fronts on both sides of the inclusion. In the framework of the linear theory of stability the dynamics of the steady settling of spherical inclusion of one liquid in a porous medium filled with another liquid is studied in [2]. It is shown that monotonically increasing perturbations are localized at the forefront of moving droplets, in agreement with the Rayleigh-Taylor mechanism. On the other hand, a two-parameter family of eigenmodes was found, with the spectrum of eigenvalues filling the entire complex plane.

In the present paper the numerical simulation of a liquid drop behaviour in a porous medium under gravity is performed and the effect of time-modulation of external pressure gradient is analyzed. The problem is studied under assumption of thin front between the phases, interpenetration of phases typical for Buckley-Leverett model [3] is not considered. Calculations are performed using the Level set method. The method was developed in [4–6] where interface between fluids of different properties is represented as transition layer of finite thickness so fluids are considered as one medium with sharp change of properties in the transition region. In [7] a new enhanced treatment of the Level set method is introduced that allows to perform simulation for fluids having large density and viscosity ratios.

Journal of Physics: Conference Series **681** (2016) 012040 doi:10.1088/1742-6596/681/1/012040

The paper is organized as follows. In Sec. II we formulate the problem in terms of nondimensional governing equations and interface boundary conditions and introduce the nondimensional parameters. The method of computation of drop dynamics is described in Sec. III. In Sec. IV and Sec. V, we present the results of calculations. Section VI contains the discussion and concluding remarks.

2. Problem statement

We consider a uniform isotropic porous medium filled by the fluid wherein single drop of another fluid with different density and viscosity moves within the first fluid in gravity field, Fig. 1. The drop settles or ascends depending on density ratio in a cylindrical channel with rigid boundary. It was shown in [2] that this problem can be transformed into the problem of a drop driven by an imposed pressure gradient.

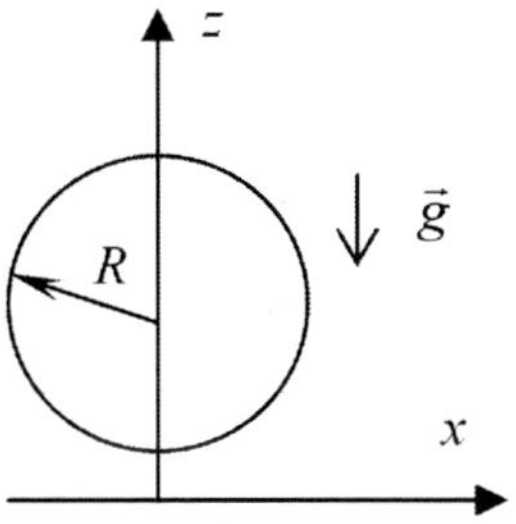

Figure 1. Problem geometry

The drop is initially spherical and the interface thickness between liquids is supposed to be small, much less when the drop radius. We neglect the effect of capillary forces. Let us introduce the coordinate system such that z-axis is directed vertically upward (see fig. 1).

The dynamics of system under consideration is described by Darcy model and the continuity equation in each liquid:

$$\nabla p_j + \frac{\varepsilon \eta_j}{k} \boldsymbol{v}_j = \rho_j g \boldsymbol{\gamma} \tag{1}$$

$$\operatorname{div} \boldsymbol{v}_j = 0 \tag{2}$$

where p_j is pressure, $\boldsymbol{v}_j$ is hydrodynamic velocity of liquid (not the filtration rate), k and ε are permeability and porosity coefficients, η_j is dynamic viscosity, ρ_j is liquid density, g is acceleration of gravity, $\boldsymbol{\gamma}$ is unit vector directed downwards, j denotes inner 1 or outer 2 fluids.

At the interface the kinematic condition and the condition of continuity of the normal component of velocity are held:

$$\frac{\partial F}{\partial t} + (\vec{v}\nabla)F = 0, \quad [v_n] = 0. \tag{3}$$

where brackets $[f]$ denotes the jump of function f across the interface, so $[f] = f_1 - f_2$; $F(x, y, z, t) = 0$ is the equation determining the shape of the interface. At rigid boundary the condition of impermeability is imposed.

3. Numerical method

Numerical modeling of the liquid droplet motion is performed using the Level set method [7]. The basic idea of the method is the introduction of the distance function - an auxiliary function, which has different sign in different phases, and zero value of this function corresponds to the interface location. At the same time the interface is treated as the transition layer with sharply varying parameters. As a result, a two-phase system can be described as a single medium with parameters (density, viscosity) depending from the distance function. The calculations were performed for a cylindrical cavity with height H_0 and radius R_0 bounded by solid boundaries. We considered axisymmetric dynamics of the drop.

In accordance with the idea of the method, the equations (1) should be supplemented with the transport equation for the distance function and equations that determine the distribution of the density and dynamic viscosity:

$$\frac{\mathrm{d}\phi}{\mathrm{d}t} = \frac{\partial \phi}{\partial t} + \boldsymbol{v}\nabla\phi = 0 \tag{4a}$$

$$\rho(\phi) = \rho_1 + (\rho_2 - \rho_1) H(\phi) \tag{4b}$$

$$\eta(\phi) = \eta_1 + (\eta_2 - \eta_1) H(\phi) \tag{4c}$$

where $H(\phi)$ is Heaviside function defined by the expression

$$H(\phi) = \begin{cases} 0 & \text{if } \phi < -\varepsilon \\ \frac{1}{2}\left[1 + \frac{\phi}{\varepsilon} + \frac{1}{\pi}\sin(\pi\phi/\varepsilon)\right] & \\ 1 & \text{if } \phi > \varepsilon \end{cases} \tag{5}$$

where ε is half-thickness of the transition layer. We assumed that $\varepsilon = n\Delta x$, Δx is the distance between the nodes of computational mesh, $n = 3$ is the number of nodes located in the half-thickness of the transition layer.

As the typical scales of the problem, we chose R_0 for length, $kg(\rho_2 - \rho_1)/\varepsilon(\eta_2 + 0.5\eta_1)$ for velocity, $\rho_1 g R_0$ for pressure, ρ_2 for density and η_2 for viscosity. Note, that selected dimension of velocity corresponds to known analytical formula for the velocity of steady motion of spherical droplet in porous medium under vibrations [2]. Conducting the standard nondimensionalization of the problem (1-3) and equations (4) we obtain the following nondimensional equations:

$$\eta\boldsymbol{v} = A(\rho\gamma - \nabla p) \tag{6a}$$

$$\operatorname{div}\boldsymbol{v}_j = 0 \tag{6b}$$

$$\frac{\partial\phi}{\partial t} + \boldsymbol{v}\nabla\phi = 0 \tag{6c}$$

$$\rho(\phi) = \lambda + (1 - \lambda) H \tag{6d}$$

$$\eta(\phi) = \mu + (1 - \mu) H \tag{6e}$$

The problem contains the following dimensionless parameters: $\lambda = \rho_1/\rho_2$ is ratio of the liquids densities, $\mu = \eta_1/\eta_2$ is the ratio of viscosities, $A = \mu + 0.5/(\lambda - 1)$ is the dimensionless viscosity, $\varsigma = H_0/R_0$ is the height of the cavity, and $r_c = r_0/R_0$ is the drop radius.

The level set method requires high accuracy treatment near the interface of the drop where parameters of medium sharply changes. Therefore we implement adaptive mesh refinement (AMR) algorithm that allows maintaining fine enough mesh near the interface during all time of calculations. The AMR algorithm is realized using Paramesh libraries [8] that supports parallel computations based on Message-Passing Interface and dynamic load balancing procedure for distributing the workloads across multiple computing resources. Thus, the mesh is automatically changed to minimize diffusion of the grid near the interface. In addition, at each time step, the corrections of the position of the interface is performed such that the thickness of the transition layer is kept constant [7]. As result, variation of the drop mass during calculations did not exceeded 1%.

We introduce the flow function to describe incompressible flow with axial symmetry. Poisson equation for the stream function is solved implicitly by Generalized minimal residual method [9]. To implement this method the library Aztec is used [10], which can effectively solve the system of linear equations with a sparse matrix distributed among computational nodes. The time derivative is approximated by second-order scheme. An approximation of the spatial derivatives is carried out by finite volume method based on the integral form of the conservation equations.

4. Droplet dynamics in gravity field

Calculations were performed for cavity with $\varsigma = 4$ and the droplet radius $r_c = 0.4$. The initial coordinate of the drop mass center is $z_0 = 3\varsigma/4$. At top and bottom of the cavity the condition of periodicity is imposed.

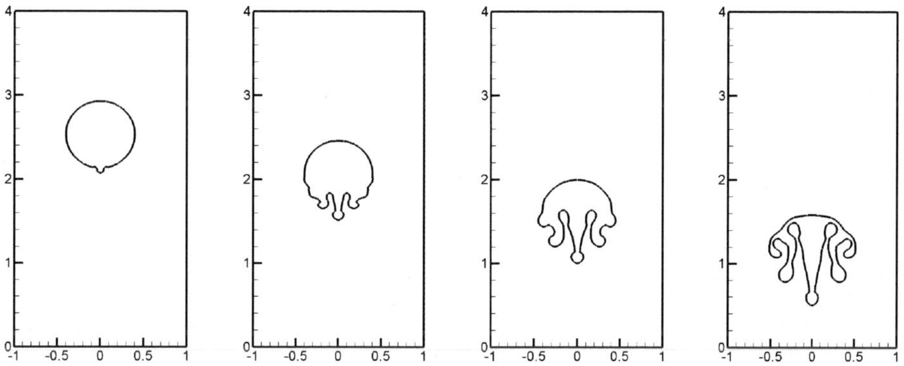

a) $t = 0.5$ b) $t = 1$ c) $t = 1.5$ d) $t = 2$

Figure 2. Drop settling for $A = 12.5$, $\lambda = 1.2$, $\mu = 2$

Numerical modeling of the droplet dynamics subjected to gravity showed that the droplet is unstable and instability develops at the forefront of moving droplets regardless of the ratio of fluids viscosities (see Figs. 2, 3). The results are in good agreement with the conclusions made in the paper [2], where in the frame work of linear stability analysis it is shown that the spectrum decrements fills the entire complex plane and monotonically increasing perturbations are localized at the forefront of moving droplets.

Due to the fact that the drop is unstable with respect to perturbations of an arbitrarily large wavenumber, the result of numerical simulation depends on parameters of the computational mesh. With the increase of the mesh resolution the perturbations of smaller scale become capable to be resolved. The rate of their growth is larger then growth rate of other perturbation so the evolution of the drop shape changes significantly. The effect is illustrated in Fig. 3.

5. Drop stability under modulated pressure gradient

Consider the drop stability in the gravity under the influence of time-modulated vertical pressure gradient. Effect of linearly polarized vibration on the stability of plane displacement front in

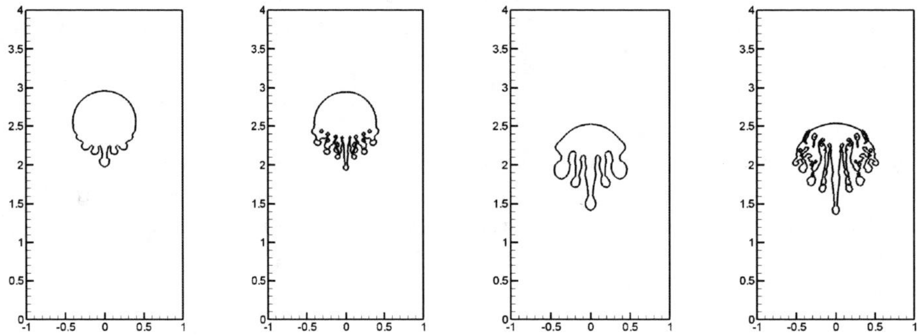

a) $t = 0.5$ b) $t = 0.5$ c) $t = 1$ d) $t = 1$

Figure 3. Evolution of the drop shape for $A = 5$, $\lambda = 1.2$, $\mu = 0.5$. Number of mesh nodes per unit of length near the interface equals 128 for (a), (c) and 256 for (b), (c)

a porous medium is investigated in [11]. It was shown that vibrations suspended short-wave perturbations of the displacement front. One could expect that the modulated pressure gradient is able to stabilize the drop.

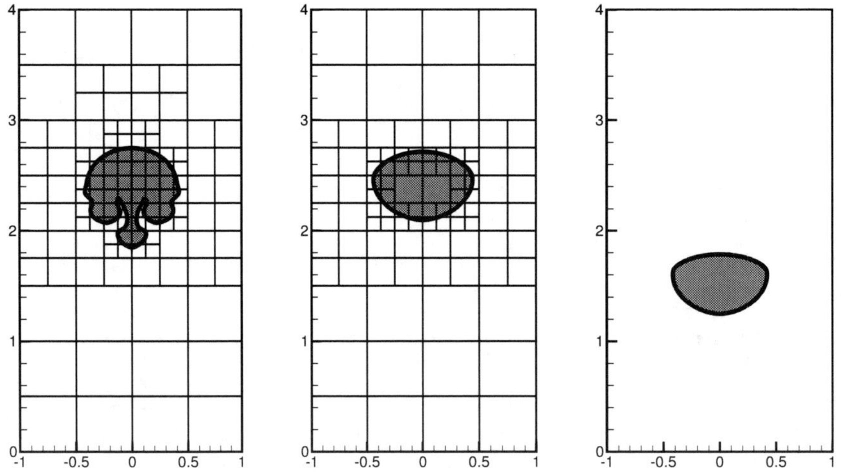

a) $t = 0.7$, $U = 10$ b) $t = 0.7$, $U = 20$ b) $t = 2$, $U = 20$

Figure 4. Drop shape under modulated pressure gradient for $W = 1000$, $A = 5$, $\lambda = 1.2$, $\mu = 0.5$

The influence of pressure gradient was taken into account by changing the boundary condition for the velocity field. According to (1) the modulated pressure gradient leads to vertical oscillating flow of fluid far from the drop. Therefore at the upper and lower boundaries of the computational domain vertical velocity component should be

$$v_z = U cos(Wt) \tag{7}$$

where U is the dimensionless amplitude of the velocity and W is the dimensionless frequency of modulation.

Calculations show that small-scale perturbations of interface are suppressed by modulated pressure gradient and only large-scale perturbations are developed (see Fig. 3 and Fig. 4a). Increasing of modulation amplitude and frequency leads to intensification of stabilizing effect. For large enough intensity of modulation drop becomes stable (Figs.4b, 4c). It is worth noting that for the case of stable drop results of calculation do not depend on calculation mesh.

Fig. 4 also illustrates the structure of calculation mesh. As it is mentioned above we implement AMR algorithm. Computational mesh consists of a number of blocks of equal size, near the interface additional blocks are created to maintain fine enough mesh. The size of each block shown in Fig. 4 is 16x16.

6. Conclusion

Numerical modeling of the liquid drop dynamics in porous medium saturated with another liquid under gravity is carried out. Simulations showed that the droplet is unstable and regardless of the ratio of liquid viscosities, the finger instability develops at the forefront of moving drop.

The results of modeling performed in the framework of Darcy filtration model depends on computational mesh parameters due to the fact that the drop is unstable to perturbations of an arbitrarily large wave number. If the density of mesh nodes is increased, then perturbations of smaller scale can be realized.

Under modulated pressure gradient small-scale perturbations of interface are suppressed and in the case of modulation with large enough intensity drop is stable.

Acknowledgments

The work was carried out under financial support of Russian Science Foundation (grant No. 14-01-00090).

References

[1] Barenblatt G, Entov V and Ryzhik V 1990 *Theory of Fluid Flows Through Natural Rocks* (Springer Netherlands)
[2] Lyubimov D and Shklyaev S 2005 *Fluid Dynnamics* **40** 777–784
[3] Kaasschieter E 1999 *Computational Geosciences* **3** 23–48
[4] Lyubimov D and Lyubimova T 1990 *Modeling in Mechanics* **1** 136–140
[5] Osher S and Sethian J 1988 *J. Comp. Phys.* **79** 12–49
[6] Brackbill J, Kothe D and Zemach C 1992 *Journal of Computational Physics* **100** 335–354
[7] Sussman M, Smereka P and Osher S 1994 *Journal of Computational Physics* **114** 146–159
[8] MacNeice P, Olson K, C Mobarry R d and Packer C 2000 *Computer Physics Communications* **126** 330–354
[9] Saad Y and Schultz M 1986 *J. Sci. Stat. Comput.* **7** 856–869
[10] Tuminaro R S, Heroux M, Hutchinson S A and Shadid J N 1999 Aztec user guide URL `http://acts.nersc.gov/aztec/`
[11] Lyubimov D and Sedelnikov G 2006 *Fluid Dynamics* **41** 3–11

Vibrational convection of ternary mixtures in rectangular cavities in zero gravity conditions

T.P. Lyubimova[1,2], N.A. Zubova[1], V.M. Shevtsova[1]
[1] Laboratory of Computational Hydrodynamics, Institute of Continuous Media Mechanics UB RAS, Russia, 614013, Perm, Academika Koroleva str., 1,
[2] The Department of Theoretical Physics, Perm State University, Russia, 614990, Perm, Bukireva str., 15

E-mail: lubimova@psu.ru

Abstract. The present paper is devoted to the investigation of the onset and nonlinear regimes of the vibrational convection of ternary mixtures in rectangular cavities with rigid boundaries in zero gravity conditions. We consider translational linearly polarized vibrations of finite amplitude and frequency. The axis of vibrations is perpendicular to the temperature gradient. The problem is solved by finite difference method in the framework of 2D unsteady approach. Numerical data on the time evolution of instantaneous and average fields and characteristics of convective flows and concentration fields are obtained for different vibration intensities and aspect ratios.

1. Introduction

In [1] the Soret induced convection of the binary mixtures in a horizontal layer subjected to the vertical temperature gradient and longitudinal high-frequency vibration was investigated. It was found that the vibrations exert destabilizing effect at positive values of the separation ratio and stabilizing effect at negative separation ratios. The influence of high-frequency vibrations of different orientations on the Soret-induced convection of the methane-butane mixture in square cavity in low gravity conditions was studied in [2]. As in [1], the study was performed in the framework of the averaged approach. In [3] the effect of vibrations along the axis perpendicular to the temperature gradient on three-dimensional regimes of the Soret-induced convection of water-isopropanol mixture in cubic cavity in low gravity conditions was investigated. The onset and nonlinear regimes of the Soret-induced convection of ternary mixtures in square cavity at different levels of gravity was studied in [4, 5]. The effect of longitudinal high-frequency vibrations on the onset of Soret-induced convection of ternary mixtures in horizontal layers subjected to the vertical temperature gradient was investigated in [6].

The effect of vibrations is used to control the flows in the melt, which in general is a multicomponent system, during crystal growth [7, 8]. The present work is devoted to the investigation of the onset and nonlinear regimes of the vibrational convection of ternary mixtures in rectangular cavities with rigid boundaries subjected to vertical temperature gradient in zero gravity conditions. The interest to this problem is associated with the experiments on transport phenomena in ternary mixtures conducted on board of the International Space Station.

Journal of Physics: Conference Series **681** (2016) 012041 doi:10.1088/1742-6596/681/1/012041

2. Formulation of the problem

Let us consider vibrational convection of multicomponent mixture, consisting of n components, in a rectangular cavity with rigid impermeable boundaries in zero gravity conditions. We introduce Cartesian coordinate system as follows: x-axis is parallel to the long sides of the rectangle, y-axis is parallel to the short sides, the origin of coordinate system is located in lower left corner. The cavity performs longitudinal (parallel to x-axis) harmonic vibrations with amplitude a and angular frequency ω. The short sides of the rectangle are thermally insulated, at the long sides the constant different temperatures are maintained.

Unsteady equations of thermosolutal convection of ternary mixture subjected to the translational vibrations of finite frequency and amplitude in zero gravity conditions, in the Boussinesq approximation in the reference system of oscillating cavity are [9]:

$$\frac{\partial \vec{u}}{\partial t}+\vec{u}\nabla\vec{u}=-\nabla p+\nabla^2\vec{u}+\frac{\mathrm{Ra_v}}{\mathrm{Pr}}cos(\Omega t)(T+\boldsymbol{I}\cdot\boldsymbol{C})\vec{j}, \quad (1)$$

$$\frac{\partial \boldsymbol{C}}{\partial t}+\vec{u}\nabla\boldsymbol{C}=\frac{1}{\mathbf{SC}}\left(\nabla^2\boldsymbol{C}-\boldsymbol{\varepsilon}\nabla^2 T\right), \quad (2)$$

$$\frac{\partial T}{\partial t}+\vec{u}\nabla T=\frac{1}{\mathrm{Pr}}\nabla^2 T, \quad (3)$$

$$\nabla\cdot\vec{u}=0. \quad (4)$$

Here $\vec{u}$ is the velocity vector, p is the pressure, $\boldsymbol{C}=(C_1,\ldots,C_{n-1})^T$ is the transpose vector of concentrations, $\boldsymbol{I}=(1,\ldots,1)$ is the unit vector and $\vec{j}$ is the unit vector in the direction of vibrations.

The equations are written in the dimensionless form using the following quantities as the scales: the height of the cavity H for the length, ν/H for the velocity, H^2/ν for the time, $\rho_0\nu/H$ for the pressure, temperature difference between the lower and upper boundaries ΔT for the temperature, and $\beta_T\Delta T\mathbf{B}^{-1}$ for the vector of concentrations. Here ν is the viscosity of the mixture, $\mathbf{B}$ is the diagonal matrix of the coefficients of concentration dependence on the density and β_T is the thermal expansion coefficient. Equations (1)-(4) contain the following dimensionless parameters: $\mathrm{Pr}=\nu/\chi$ is the Prandtl number, χ is the thermal diffusivity, $\boldsymbol{\varepsilon}=-\beta_T^{-1}\mathbf{BD}^{-1}\boldsymbol{D}_T$ is the vector, consisting of $(n-1)$ separation ratios, characterizing the thermal diffusion separation in mixture for each solutes. $\mathbf{SC}=\nu^{-1}\mathbf{BDB}^{-1}$ is the matrix of parameters of the dimension $(n-1)\times(n-1)$, where $\mathbf{D}$ – is the matrix of the molecular diffusion coefficients, $\boldsymbol{D}_T$ is the vector of thermal diffusion coefficients, $\mathrm{Ra_v}=a\omega^2\beta_T\Delta TH^3/(\nu\chi)$ is the vibrational analogue of the Rayleigh number, $\Omega=\omega H^2/\nu$ is the dimensionless frequency.

The boundary conditions are:

$$x=0,L:\vec{u}=0,\frac{\partial T}{\partial x}=\frac{\partial \boldsymbol{C}}{\partial x}=0, \quad (5)$$

$$y=0,H:\vec{u}=0,T=\pm 1/2,\frac{\partial}{\partial y}(C-\boldsymbol{\varepsilon}T)=0. \quad (6)$$

3. Numerical method

In the case of binary mixture the matrix of molecular diffusion coefficients is reduced to a single number. For ternary mixtures the matrix $\mathbf{D}$ has a dimension (2×2), that leads to the appearance of additional terms in the equation for component concentration responsible for cross-diffusion, and complicates the calculations. Reduction of the molecular diffusion coefficient matrix to the diagonal form allows to eliminate the cross diffusion terms in the equations, decreasing the number of governing parameters of the problem, and the equations for the concentrations of the components become independent of each other. Such transformation applied to the problem (1)-(6) in dimensionless form can be written as [4, 10]:

$$\boldsymbol{C} = \mathbf{BV}(\mathbf{BQ})^{-1}\widehat{\boldsymbol{C}},\ \boldsymbol{\varepsilon} = \mathbf{BV}(\mathbf{BQ})^{-1}\widehat{\boldsymbol{\varepsilon}},$$

where $\mathbf{V}$ is the matrix whose columns are the eigenvectors $v_i = (v_{i1}, \ldots, v_{in-1})^T$ of the matrix $\mathbf{D}$; $\mathbf{Q} = diag\{q_1, \ldots, q_{n-1}\}$, $q_i = \beta_i^{-1}\sum_{j=1}^{n-1}\beta_j v_{ij}$.

The problem (1)-(6), written in terms of stream function and vorticity, was solved numerically by finite difference method. The spatial derivatives were approximated by central differences. Unsteady equations were solved using explicit finite difference scheme with a constant time step $h^2/8$, where h is constant spatial step. The Poisson equation for the stream function was solved by successive over-relaxation method. The vorticity values at the boundaries of the cavity were calculated using the Thom formula [11].

The calculations were carried out for rectangular cavities with aspect ratios 1:1 and 5:1, on the grids with the numbers of nodes 41×41 and 201×41 respectively. The initial conditions corresponded to small perturbations of stream function and uniform vertical gradients of temperature and solutes concentrations:

$$\psi = 10^{-8} sin(2\pi x) sin(\pi y),\ T = 1/2 - y \text{ and } \boldsymbol{C} = \boldsymbol{\varepsilon}(1/2 - y).$$

The calculations were carried out for two ternary mixtures with different combinations of the separation ratios of solutes: for Mixture 1 both solutes had positive separation ratios $\varepsilon_1 = 0.3$, $\varepsilon_2 = 0.1$, and for Mixture 2 – negative separation ratios $\varepsilon_1 = -0.3$, $\varepsilon_2 = -0.1$. The molecules of solutes with positive separation ratios concentrate in the warm area of cavity, and the molecules of solutes with negative separation ratios – in cold area. The values of the Prandtl number and Schmidt numbers were taken the same for both mixtures: $\mathrm{Pr} = 10$, $\mathrm{Sc}_1 = 100$, $\mathrm{Sc}_2 = 1000$. The chosen parameters are typical for the liquid mixtures. The value of the dimensionless frequency of vibrations was also fixed $\Omega = 445$. For the cavity of 1 cm height filled by ternary liquid mixture with the viscosity $\nu \sim 10^{-6}$ m^2/s this value corresponds to the dimensional vibration frequency approximately equal to 1 Hz. Vibrational Rayleigh number was varied in the range $\mathrm{Ra_v} = 10^3 \div 2\cdot10^5$.

4. Results

Figures 1-2 show the numerical data on the time evolution of the average stream function and the difference in the solute concentration values at the centers of the upper and lower cavity boundaries for Mixtures 1 and 2 for the cavity with aspect ratio 5:1. The average values were obtained by averaging of the instantaneous values over the period of vibrations.

We consider the physical situation in which the static gravity field is absent and the initial state corresponds to the established linear distribution of temperature and solute concentrations. In the case of infinite horizontal layer, in such situation at low values of the vibrational Rayleigh number the average flow is absent, convection arises in a threshold manner at a certain critical value of $\mathrm{Ra_v}$. In the case of rectangular cavity with aspect ratio 1:5, as follows from our calculations (Figs. 1, 2) at low intensities of vibrations (low values of vibrational Rayleigh number) in both mixtures weak average flow is observed and the differences in the component concentrations values at the centers of upper and lower boundaries are not very different from the initial values, corresponding to the established linear distributions in the absence of convection. At large $\mathrm{Ra_v}$ for the Mixture 1 (Fig. 1) in the first stage of time evolution the average flow of high intensity arises, then oscillatory decrease of flow intensity with time is observed. For the Mixture 2, significant vibration intensity is required for the development of marked average flow: the curves corresponding to $\mathrm{Ra_v} = 5\cdot 10^4$ and $\mathrm{Ra_v} = 10^5$ in Figure 2a are close to the line $\overline{\psi} = 0$ and only at $\mathrm{Ra_v} = 2\cdot 10^5$ the value of average stream function markedly different from zero is established. The difference in the solute concentrations values for both mixtures significantly decreases with the vibration intensity growth (Figs. 1bc, 2bc) resulting in the amplification of the fluids mixing.

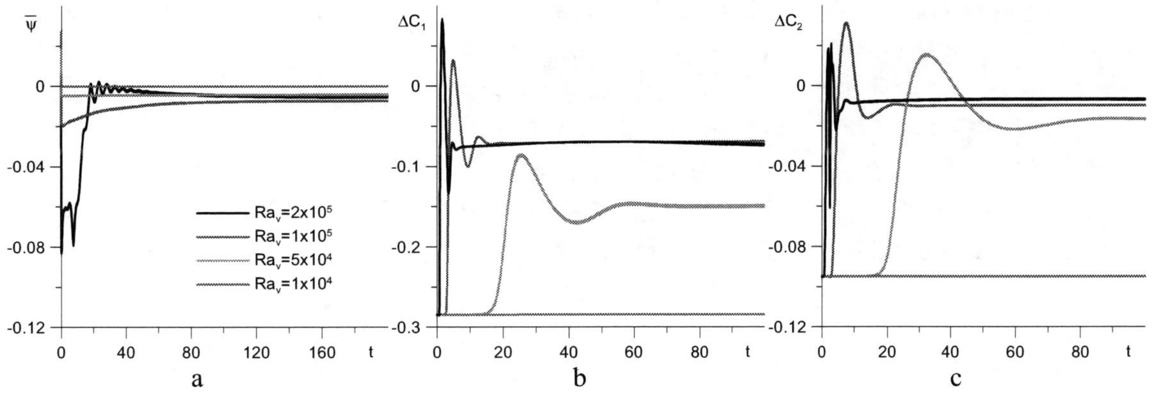

Figure 1. Time evolution of the average stream function at point $(0.275;0.275)$ (a) and difference in the solute concentrations at the centers of lower and upper boundaries (b,c) for Mixture 1 in the cavity with aspect ratio 5:1

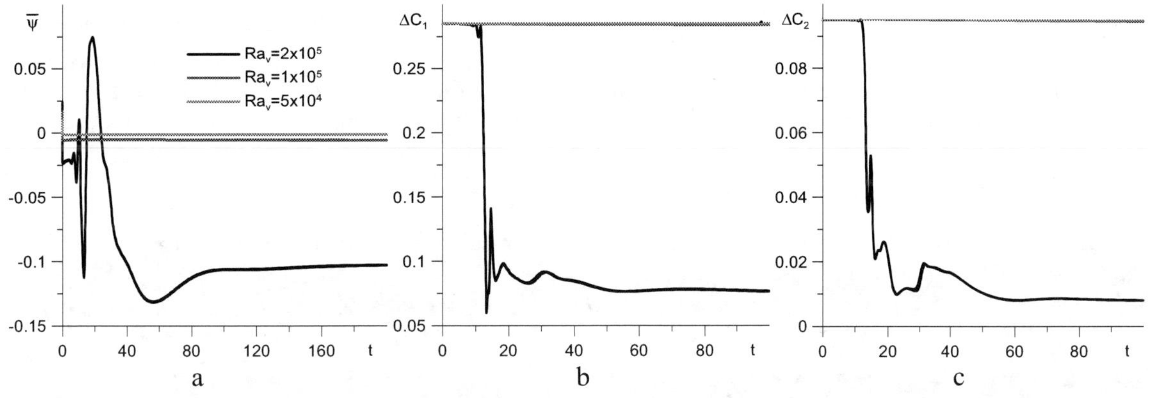

Figure 2. Time evolution of the average stream function at point $(0.275;0.275)$ (a) and difference in the solute concentrations at the centers of lower and upper boundaries (b,c) for Mixture 2 in the cavity with aspect ratio 5:1

The transformation of the flow structure during the period of oscillations, which coincides with the period of external vibrations, for Mixture 1 at $\mathrm{Ra_v} = 2\cdot 10^5$ is shown in Figure 3a. Time moments for the corresponding flow structures are marked in the plot of the temporal evolution of the stream function (Fig. 3b). As one can see, during the period the change of the direction and inclination of vortices occurs.

Fig. 4 shows the fields of steady average flow and solutes concentrations for Mixtures 1 and 2 at different values of the vibrational Rayleigh number. It is seen that at low values of the vibrational Rayleigh number (Fig. 4a, b) for both mixtures the flow in the central part of the cavity is absent, there are only four weak vortices near the side walls. The isolines of the solute concentrations are also deformed only near the side walls. At large values of Ra_v the average flow has multivortex structure, the fields of solute concentrations are strongly deformed by the flow (Fig. 4b, d). The isolines of the second solute concentration (the lower fields in Fig. 4a-d) are deformed more significantly than that of the first solute, due to the larger Schmidt number.

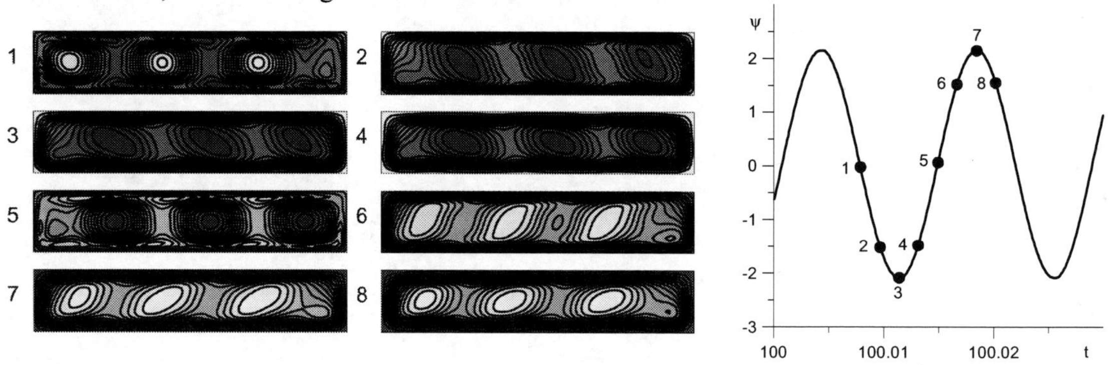

a b

Figure 3. Transformation of instantaneous stream function fields (a) during the period of stream function at point $(0.275;0.275)$ oscillations (b) for Mixture 1 at $Ra_v = 2\cdot10^5$

In the case of square cavity at low values of the vibrational Rayleigh number for both mixtures the average four-vortex flow is observed (Fig. 5b). At large Ra_v for Mixture 1 the regime with large diagonal vortex and two smaller vortices in the corners of the cavity is realized (Fig. 5b). For Mixture 2 such regime was not found in the considered range of Ra_v. Four-vortex structure of average flow in square cavity is the limit case of four-vortex structure of average flow in a rectangular cavity when the distance between side walls becomes smaller. The isolines of the first solute as well as in the case of rectangular cavity with aspect ratio of 5:1, are less deformed than the isolines of the second solute. Temporal evolution of the flow characteristics in square cavity is qualitatively similar to that for the rectangular cavity with aspect ratio 5:1 (Figs. 1, 2).

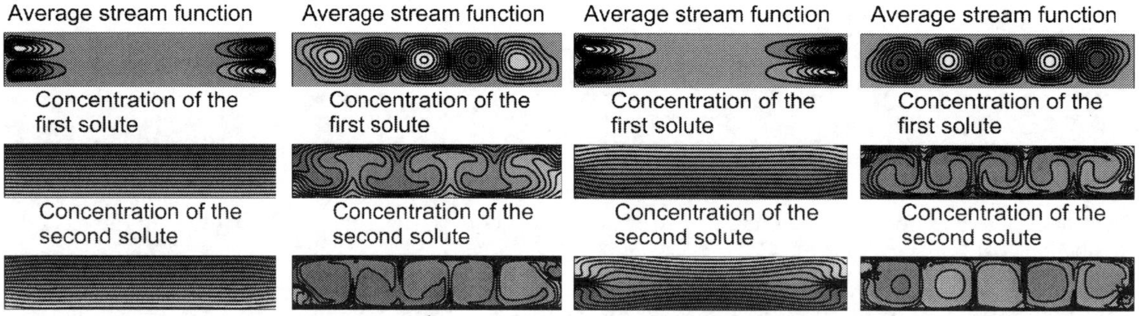

a b c d

Figure 4. Streamlines of average flow and solute concentration distributions for Mixture 1 at $Ra_v = 1\cdot10^4$ (a) and $Ra_v = 1\cdot10^5$ (b) and for Mixture 2 at $Ra_v = 1\cdot10^5$ (c) and $Ra_v = 2\cdot10^5$ (d)

Figure 6 shows the dependence of the Nusselt number per unit length on the vibrational Rayleigh number for cavities with aspect ratios 1:1 and 1:5. In both cases, the change of the regime in Mixture 1 occurs at $Ra_v = 5\cdot10^4$. As one can see, the transformation of flow structure is accompanied by the change of slope of the curve $Nu/L(Ra_v)$, as well as it occurs in single-component fluid [9].

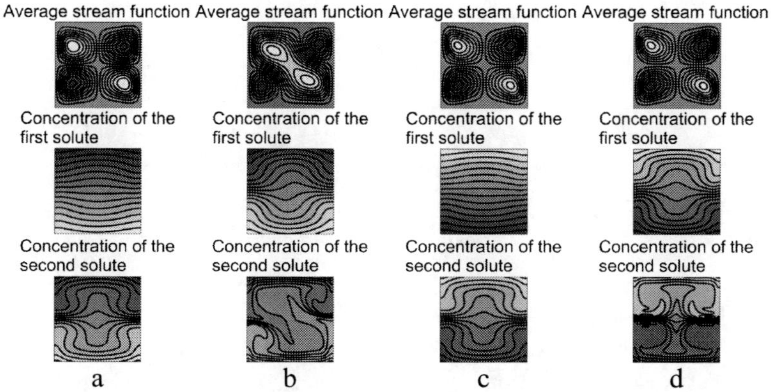

Figure 5. Streamlines of average flow and solute concentration distributions for Mixture 1 at $Ra_v = 5 \cdot 10^4$ (а) and $Ra_v = 1 \cdot 10^5$ (б) and for Mixture 2 at $Ra_v = 1 \cdot 10^5$ (а) and $Ra_v = 2 \cdot 10^5$ (г)

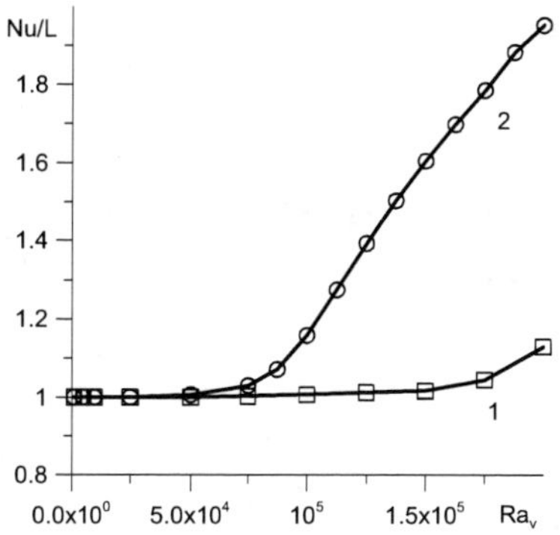

Figure 6. Dimensionless heat flux per unit length versus Ra_v for Mixture 1: 1 – aspect ratio 1:1, 2 – aspect ratio 5:1

5. Conclusions

Vibrational convection of ternary mixture in square and horizontally-elongated rectangular cavities is studied numerically on the basis of 2D unsteady approach. It is found, that in the case of square cavity, at small vibrational Rayleigh numbers four-vortex average flow of low intensity is generated. With the increase of Ra_v the transformation of average flow structure occurs – large diagonal vortex and two smaller vortices in the cavity corners are formed.

In the case of rectangular cavity elongated in the horizontal direction, at small vibrational Rayleigh numbers the average flow consists of two pairs of low intensive vortices located near the sidewalls whereas in the main part of the cavity the average flow is nearly absent. With the increase of Ra_v the transformation of average flow structure occurs – the multivortex average flow occupying the whole cavity is formed.

The transformation of average flow structure in both cases is accompanied by the change of the slope of the curve $Nu/L(Ra_v)$.

Acknowledgments

This work was supported by Russian Science Foundation (grant No. 14-01-00090).

References

[1] Gershuni G Z, Kolesnikov A K, Legros J C, Myznikova B I 1997 *Journal of Fluid Mechanics* **330** 251-269
[2] Lyubimova T, Shklyaeva E, Legros J C, Shevtsova V, Roux B 2005 *Advances in Space Research* **36** 70-74
[3] Shevtsova V, Melnikov D, Legros J C, Yan Y, Saghir Z, Lyubimova T, Sedelnikov G, Roux B 2007 *Physics of Fluids* **19** 017111
[4] Lyubimova T P, Zubova N A 2014 *Microgravity Science and Technology* **26(4)** 241-247
[5] Lyubimova T P, Zubova N A 2015 *The European Physical Journal E* **38** 19
[6] Lyubimova T P 2015 *The European Physical Journal E* **38** 43
[7] Lyubimova T P, Lyubimov D V, Ivantsov A O 2014 *Journal of Crystal Growth* **385** 77-81
[8] Lyubimova T P, Parshakova Ya N 2014 *Journal of Crystal Growth* **385** 82-87
[9] Gershuni G Z, Lyubimov D V 1998 *Thermal Vibrational Convection* (New York: Wiley)
[10] Ryzhkov I I, Shevtsova V M 2009 *Physics of Fluids* **21** 014102
[11] Thom A, Apelt C J 1961 *Field computations in engineering and physics* (London: Van Nostrand)

The oscillations of cylindrical drop under the influence of a nonuniform alternating electric field

A A Alabuzhev[1,2] and M A Kashina[2]
[1]Institute of Continuous Media Mechanics UB RAS, Perm 614013, Russia
[2]Perm State University, Perm 614990, Russia

E-mail: alabuzhev@mail.ru

Abstract. The forced oscillations of incompressible fluid drop under the alternating electric field are considered. In equilibrium, the drop has the form of a cylinder bounded axially parallel solid planes and contact angle is right. The drop is surrounded by an incompressible fluid with another density. The external nonuniform electric field acts as an external force that causes motion of the contact line. In order to describe this contact line motion the modified Hocking boundary condition is applied: the velocity of the contact line is proportional to the deviation of the contact angle and the speed of the fast relaxation processes, which frequency is proportional to twice the frequency of the electric field. It is shown that under uniform electric field the oscillation amplitude at the resonance frequency increases with the increase of the Hocking parameter. In the case of nonuniform electric field the oscillation amplitude is always limited.

1. Introduction

The stability and dynamics of fluid drops subjected to external electrical forces continue to attract research efforts [1-4]. One of the most important factor influencing the dynamics and manipulation processes is electrowetting (EW) [4,5]. This branch of science has been actively developed recently when electrowetting-on-dielectric (EWOD) was proposed [6-8]. Now EWOD has been put into applications in various fields, such as digital (droplet) microfluidic devices for bioanalysis (lab-on-a-chip) [9, 10], variable-focus liquid lenses [11,12], electronic display technology [13,14] etc.

The typical electrowetting devices are shown in figure 1. When a direct current (DC) voltage is applied between the drop and the electrode, the varying contact angle ϑ of the drop can be expressed by Young–Lippmann equation [6,7]

$$\cos\vartheta = \cos\vartheta_0 + \frac{\varepsilon\varepsilon_0}{2d\sigma}V^2, \quad (1)$$

where V is the value of the applied voltage, ϑ_0 is the contact angle without applied voltage – equilibrium contact angle, which is defined by well-known Young's equation, σ is the interfacial tension between the droplet and the surrounding fluid, d is the thickness of the dielectric film, ε_0 and ε are the permittivity of vacuum and the dielectric layer, respectively. The last term in the equation (1) is the modified term of Lippmann equation [16]. Note, that $C = \varepsilon\varepsilon_0 / d$ is capacitance per unit area [5] and $E_w = CV^2 / 2\sigma$ is EW number (it represents the ratio of the electrostatic energy to the liquid – surrounding fluid interfacial energy [8]). For alternating current (AC) instead of V^2 in equation (1) the square value of the voltage, or effective voltage, U^2 is used [5].

[1] To whom any correspondence should be addressed.

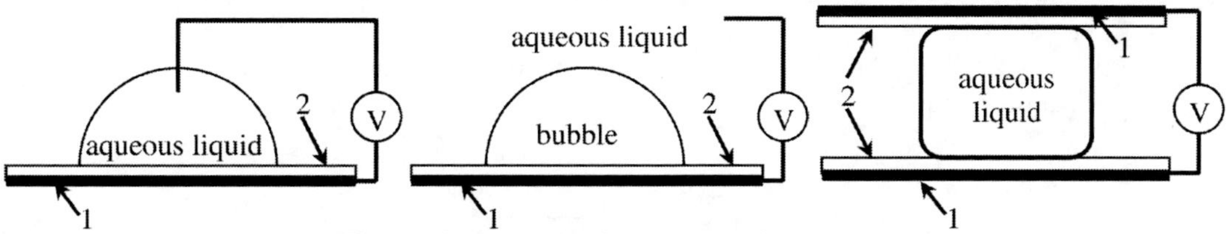

Figure 1. Typical schematic configurations.
1 – electrode, 2 – dielectric layer.

The Young–Lippmann equation (1) can be derived by several approaches [4,5,8]. However this equation is adequate for the voltage V which is smaller than the critical voltage V_* [4,5,7,15]. When $V \geq V_*$ the electrowetting contact angle does not increase with increasing of applied voltage, this effect is known as contact angle saturation. Also article [17] states that the Young–Lippmann equation surprisingly well agrees with the experimental data for the cases when local contact angle ϑ_0 (without electric field) is varied between 38° and 120°.

The contact angle ϑ depends on the time when AC is used. We propose to change the equation (1), which describes the dynamics of the contact angle, because it is not always correct. The most frequently used Hocking condition (by virtue of its simplicity) is the one employed in a study [18] of the damping of standing waves between two vertical walls. This condition assumes a linear dependence between the velocity of motion of the contact line and the contact angle (in the case of right equilibrium contact angle):

$$\frac{\partial \zeta^*}{\partial t} = \Lambda \vec{k} \cdot \vec{\nabla} \zeta^*, \tag{2}$$

where ζ^* is the deviation of the interface from the equilibrium position, $\vec{k}$ is the external normal to the solid surface, Λ is a phenomenological constant (the so-called wetting parameter or Hocking parameter), having the dimension of velocity. The special cases of boundary condition (2) are the requirement of the fixed contact line ($\zeta^* = 0$, the so-called pinned-end edge condition) and the constant contact angle ($\vec{k} \cdot \vec{\nabla} \zeta^* = 0$). For example, the equation (2) one can find in the studies of oscillations of a sessile drop and bubble [19-21], cylindrical drop and bubble [22,23], "sandwiched" drop [24] and capillary bridge [25].

In the present paper, we consider the behavior of a droplet sandwiched between two flat plates under AC. In order to describe the motion of the contact line the modified boundary condition (2) is used: the velocity of the contact line is proportional to the deviation of the contact angle and the speed of the fast relaxation processes, whose frequency is proportional to twice of the electric field frequency. In future we plan to compare the results with experimental data [26].

2. Problem formulation

We consider the dynamic behavior of an incompressible liquid drop of density ρ_i^* surrounded by other liquid of density ρ_e^* (here and in the following, quantities with subscript i refer to the drop, and those with subscript e to the surrounding liquid). The system is bounded by two parallel solid surfaces (see figure 2) separated by a distance h^*. The equilibrium contact angle ϑ_0 between the side surface of the drop and the solid surface is equal to $\pi/2$. The external nonuniform alternating electric field acts as an external force that causes the contact line motion.

According to the problem symmetry it is convenient to introduce the cylindrical coordinates r^*, α, z^*. The azimuthal angle α is reckoned from the x-axis. Let the surface of the droplet be described by the equation $r^* = R_0^* + \zeta^*(\alpha, z^*, t^*)$. Assuming potential liquid motion, we introduce the velocity potential $\vec{v}^* = \vec{\nabla}\varphi^*$. By choosing the length R_0^*, height h^*, the density $\rho_e^* + \rho_i^*$, the time $\sigma^{-1/2}\sqrt{(\rho_e^* + \rho_i^*) R_0^{*3}}$, the velocity potential $A^*\sqrt{\sigma}\left((\rho_e^* + \rho_i^*) R_0^{*3}\right)^{-1/2}$, the pressure $A^*\sigma (R_0^*)^{-2}$, and the deviation of the surface A^* as characteristic quantities, we pass to dimensionless variables and obtain the following linear problem

$$p_j = -\rho_j \varphi_{jt}, \ \Delta\varphi_j = 0, \ j = i, e, \tag{3}$$

$$\Delta = \frac{1}{r}\frac{\partial}{\partial r}\left(r\frac{\partial}{\partial r}\right) + \frac{1}{r^2}\frac{\partial^2}{\partial \alpha^2} + b^2\frac{\partial^2}{\partial z^2},$$

$$r = 1: \ [\varphi_r] = 0, \ \zeta_t = \varphi_r, \ [p] = \zeta + \zeta_{\alpha\alpha} + b^2\zeta_{zz}, \tag{4}$$

$$z = \pm\frac{1}{2}: \ \varphi_z = 0, \tag{5}$$

$$r = 1, \ z = \pm\frac{1}{2}: \ \zeta_t = \pm\lambda\left(\zeta_z + af(\alpha)\cos(2\omega t)\right), \tag{6}$$

where p is the fluid pressure, $f(\alpha)$ is the function of nonuniform electric field, the square brackets denote the jump in the quantity at the interface between the external liquid and the drop. The second term in boundary condition (6) is external action and written similarly to the Young–Lippmann equation (1): $\cos\vartheta = E_w = \tilde{E}_w V^2$ as $\cos\vartheta_0 = 0$ in our problem, therefore $\zeta_z \sim \operatorname{ctg}\vartheta = E_w \big/ \sqrt{1 - E_w^2} = \tilde{E}_w V^2 \big/ \sqrt{1 - \tilde{E}_w^2 V^4} \approx \tilde{E}_w V^2 + O\left(V^6\right) \approx \tilde{E}_w V^2 \sim V^2$. The boundary-value problem (3)–(6) involves six parameters: the aspect ratio, the dimensionless density, the wetting parameter, the AC frequency and amplitude

$$b = R_0 h^{-1}, \ \rho_i = \rho_i^*\left(\rho_e^* + \rho_i^*\right)^{-1}, \ \rho_e = \rho_e^*\left(\rho_e^* + \rho_i^*\right)^{-1}, \ \lambda = \Lambda^*\sigma^{-1/2}\sqrt{\left(\rho_e^* + \rho_i^*\right)R_0^*},$$

$$\omega = \omega^*\sigma^{-1/2}\sqrt{\left(\rho_e^* + \rho_i^*\right)R_0^{*3}}, \ a = A^* C\sigma^{-3/2}\sqrt{\left(\rho_e^* + \rho_i^*\right)R_0^{*3}} \big/ 2.$$

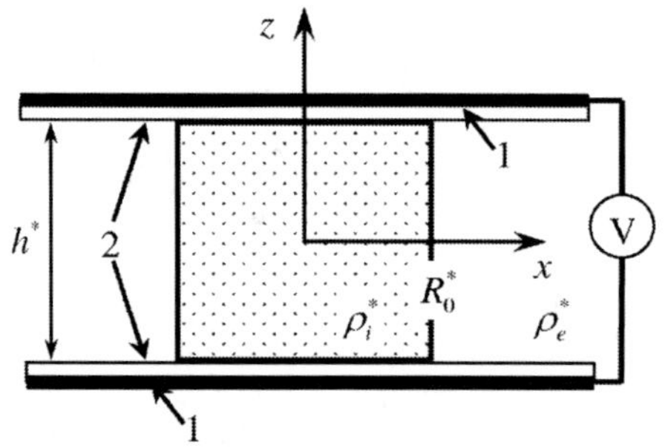

Figure 2. Problem geometry (1 – electrode, 2 – dielectric layer).

3. Uniform electric field

In order to investigate the problem it is convenient to begin with a consideration of the uniform electric field, i.e. $f(\alpha) = 1$. Note that the natural oscillations of a cylindrical drop and those eigen frequencies were investigated in [22]. Our interest deals with the axisymmetric oscillation modes, therefore system of the equations and boundary conditions (3)–(6) are axially symmetric at $f(\alpha) = 1$.

In view of axial symmetry, the solution of the Laplace equation (6) is written as

$$\varphi_i(r, z, t) = \operatorname{Re}\left(\sum_{k=0}^{\infty} a_k \operatorname{I}_0\left((2k+1)\pi b r\right)\sin\left((2k+1)\pi z\right)e^{i2\omega t}\right), \tag{7}$$

$$\varphi_e(r, z, t) = \operatorname{Re}\left(\sum_{k=0}^{\infty} b_k \operatorname{K}_0\left((2k+1)\pi b r\right)\sin\left((2k+1)\pi z\right)e^{i2\omega t}\right), \tag{8}$$

where I_0 and K_0 are modified Bessel functions. The kinematic condition on the free surface (the second condition in (4)) gives the expression for the surface deviation

$$\zeta(z, t) = \operatorname{Re}\left(\left(\sum_{k=0}^{\infty} c_k \cos(2k\pi z) + d\cos\left(\frac{z}{b}\right)\right)e^{i2\omega t}\right). \tag{9}$$

Substituting solutions (7)–(9) into (3)–(6) we obtain expressions for the unknown amplitudes a_k, b_k, c_k and d. These expressions are not given due to their cumbersome forms.

The dependence of the amplitude of the surface oscillations and the contact angle at the upper plate on the frequency of the driving force is given in figure 3 for different values of the Hocking parameter and of the aspect ratio. As it was mentioned in [20,22-24], at certain values of eigen frequencies the drop motion is independent of the wetting parameter: at any λ the contact line remains motionless. At such points the amplitude of the contact line goes to zero. This is so-called "antiresonance" frequencies. Values of these frequencies do not depend on λ and depend on the parameters of the drops (aspect ratio b, density ρ_i etc.).

Note one feature of the behavior of the drop side surface. The study of the contact line dynamics applying Hocking boundary condition (2) have shown that this condition leads to the energy dissipation. Non-dissipative cases are only $\zeta = 0$ and $\zeta_z = 0$. For example, the study of forced oscillations under vibrations [18–25] has shown that the oscillations amplitude of the surface tends to infinity in the limiting cases, but bounded at finite values. It can be seen from figure 3 that the amplitude of contact line oscillations increases with the increase of λ. However the amplitude decreases with increasing aspect ratio b and vibration frequency ω. The cause of the first phenomenon is obvious: the larger drop, the harder it moves. Apparently the second reason is related to the fact that dissipation for high-frequency oscillations is greater than for low-frequency oscillations [19,20,22].

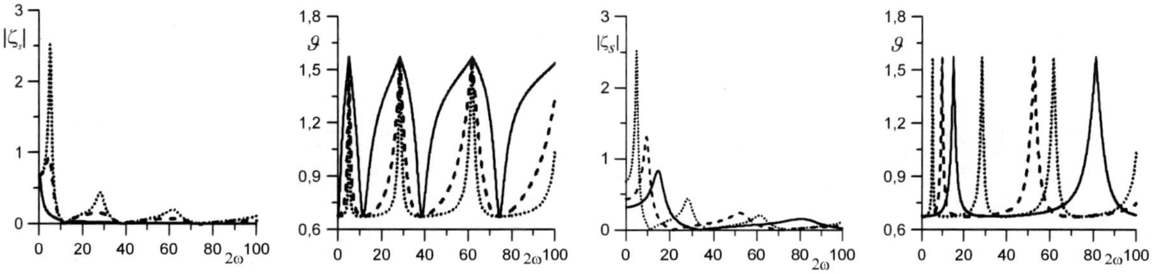

(a) (b) (c) (d)

Figure 3. The maximum deviation of the lateral surface ζ_s (a,c) and the contact angle ϑ (b,d) for three different values of the Hocking parameter λ (a,b) and of the aspect ratio b (c,d). (a,b) $b=1$, $a=2.5$, $\rho_i=0.7$, $\lambda=0.5$ (solid line), $\lambda=3.5$ (dashed line), $\lambda=10$ (dotted line). (c,d) $\lambda=10$, $a=2.5$, $\rho_i=0.7$, $b=2$ (solid line), $b=1.5$ (dashed line), $b=1$ (dotted line).

4. Nonuniform electric field

The function $f(\alpha)$ is expanded into the Fourier series in eigen functions of the Laplace operator. Let us considered the particular case of nonuniform electric field: $f(\alpha)=\sin(\kappa x)\big|_{r=1}=\sin(\kappa\cos(\alpha))$, where κ is wave number. The solutions of the velocity potential and the surface deviation are written as

$$\varphi_i(r,z,t)=\mathrm{Re}\left(\sum_{m=0}^{\infty}\sum_{k=0}^{\infty}a_{mk}R_{mk}^{i}(r)\sin((2k+1)\pi z)\cos((2m+1)\alpha)e^{i2\omega t}\right), \quad (10)$$

$$\varphi_e(r,z,t)=\mathrm{Re}\left(\sum_{m=0}^{\infty}\sum_{k=0}^{\infty}b_{mk}R_{mk}^{e}(r)\sin((2k+1)\pi z)\cos((2m+1)\alpha)e^{i2\omega t}\right), \quad (11)$$

$$\zeta(z,t)=\mathrm{Re}\left(\left(\sum_{m=0}^{\infty}\sum_{k=0}^{\infty}c_{mk}\sin((2k+1)\pi z)\cos((2m+1)\alpha)+d_0 z\cos(\alpha)+\sum_{m=1}^{\infty}d_m sh\left(\frac{\sqrt{(2m+1)^2-1}}{b}z\right)\cos((2m+1)\alpha)\right)e^{i2\omega t}\right), \quad (12)$$

Journal of Physics: Conference Series **681** (2016) 012042 doi:10.1088/1742-6596/681/1/012042

where I_m and K_m are modified Bessel functions of m order. Substituting solutions (10)–(12) into (3)–(6), we obtain expressions for the unknown amplitudes a_{mk}, b_{mk}, c_{mk}, d_0 and d_m.

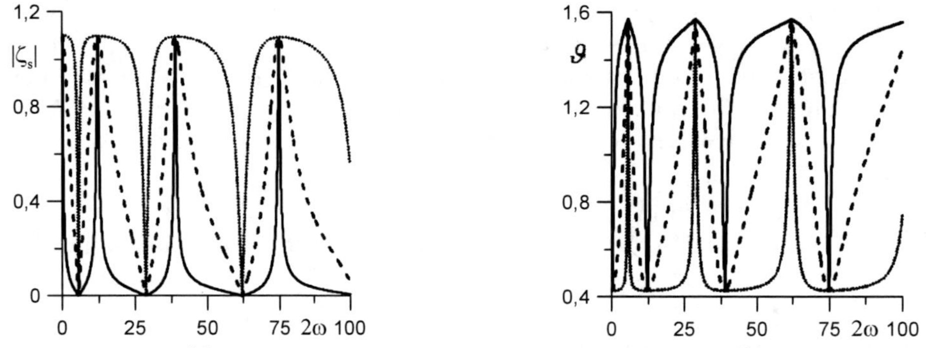

(a) (b)

Figure 4. The maximum deviation of the lateral surface ζ_s (a) and the contact angle ϑ (b) for three different values of Hocking's constant λ (a,b)
$b = 1$, $a = 2.5$, $\rho_i = 0.7$, $\kappa = 1$, $\lambda = 0.1$ (solid line), $\lambda = 1$ (dashed line), $\lambda = 10$ (dotted line)

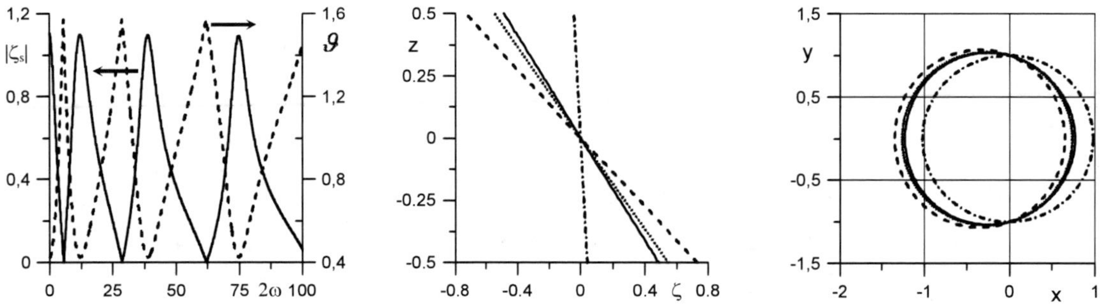

(a) (b) (c)

Figure 5. The maximum deviation of the lateral surface ζ_s and the contact angle ϑ (a), form of later surface (b) and form of contact line (c) ($b = 1$, $a = 2.5$, $\rho_i = 0.7$, $\kappa = 1$, $\lambda = 1$).
(b,c) $\omega = 2$, $T = 2\pi\omega^{-1}$, $dT = 0.125T$, $t = 0$ (solid line), $t = dT$ (dashed line), $t = 2dT$ (dotted line), $t = 3dT$ (dash-dotted line).

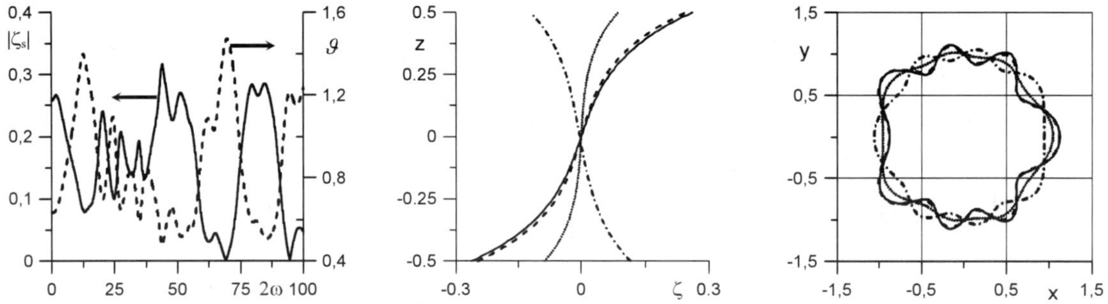

(a) (b) (c)

Figure 6. The maximum deviation of the lateral surface ζ_s and the contact angle ϑ (a), form of later surface (b) and form of contact line (c) ($b = 1$, $a = 2.5$, $\rho_i = 0.7$, $\kappa = 10$, $\lambda = 1$).
(b,c) $\omega = 2$, $T = 2\pi\omega^{-1}$, $dT = 0.125T$, $t = 0$ (solid line), $t = dT$ (dashed line), $t = 2dT$ (dotted line), $t = 3dT$ (dash-dotted line).

The dependence of the amplitude of the surface oscillations and the contact angle at the upper plate on the frequency of the driving force is given in figure 4 for different values of the Hocking parameter and $\kappa = 1$. It is obvious that the translational mode of oscillations is excited. The amplitude of the side surface oscillations has identical local maximum unlike to the uniform electric field. It is obvious that the minimum value of the contact angle achieved when the amplitude deflection is maximum (see figure 4 and 5). In addition, the minimum value of the contact angle is different from the zero and independent on parameter λ. There is no traveling wave along the lateral surface (see figure 5), the section of the lateral surface is a straight line in the plane xz (fig. 5b).

Azimuthal modes excited by a larger wave number $\kappa = 10$ (see figure 6). The traveling wave propagates along the lateral surface of the drop in this case.

5. Conclusions
The data about deviation of frequency and surface characteristics depending on Hocking parameter, frequency and amplitude of an external nonuniform electric field and the geometric parameters of the system are obtained. It is shown that under uniform electric field the oscillation amplitude at the resonance frequency increases with the increase of the Hocking parameter. It is similar to forced oscillations of the drop under the mechanical axisymmetric vibration [20]. There are also «anti-resonant» frequencies (i.e. the frequencies at which the amplitude of contact line deviation is zero) under the alternating electric field as under the mechanical vibrations. In the case of the nonuniform electric field the oscillations amplitude is always limited similarly to the forced translational vibrations of a drop [19,24]. Note that we consider the weakly nonuniform electric field which leads to translational vibrations, while a strongly nonuniform electric field leads to azimuthal vibrations.

Acknowledgments
This work was supported by the Russian Science Foundation (project 14-21-00090).

6. References
[1] Melcher J R and Taylor G I 1969 *Annual Review of Fluid Mechanics* **1** 111
[2] Pelekasis N A, Economou K and Tsamopoulos J A 2001 *Phys. of Fluids* **13** 3564
[3] Shutov A A 2002 *Technical Physics* **47** 1501
[4] Mugele F and Baret J-C 2005 *J. Phys.: Condens. Matter.* **17** 705
[5] Chen L and Bonaccurso E 2014 *Adv. Colloid Interface Sci.* **210** 2
[6] Berge B 1993 *Comptes Rendus Acad. Sci. II* **317** 157
[7] Quilliet C and Berge B 2001 *Curr. Opin. Colloid Interface Sci.* **6** 34
[8] Zhao Y-P and Wang Y 2013 *Rev. Adhesion Adhesives* **1** 114
[9] Hua Z, Rouse J L, Eckhardt A E, Srinivasan V, Pamula V K, Schell W A, Benton J L, Mitchell T G and Pollack M G 2010 *Anal Chem* **82** 2310-16
[10] Li J, Wang Y,Chen H and Wan J 2014 *Lab Chip* **14** 4334-37
[11] Kuiper S and Hendriks B H W 2004 *Appl. Phys. Lett.* **85**, 1128–30
[12] Li C and Jiang H 2014 *Micromachines* **5** 432-41
[13] Hayes R A and Feenstra B J 2003 *Nature* **425** 383–5
[14] Roques-Carmes T, Hayes R A, Feenstra B J and Schlangen L J M 2004 *J. Appl. Phys* **95** 4389–96
[15] Mugele F 2009 *Soft Matter* **5** 3377–84
[16] Lippmann G 1875 *Ann. Chim. Phys.* **5** 494–549
[17] Adamiak K 2006 *Microfluid Nanofluid* **2** 471-80
[18] Hocking L M 1987 *J. Fluid Mech.* **179** 253-66
[19] Lyubimov D V, Lyubimova T P and Shklyaev S V 2004 *Fluid Dynamics* **39** 851–62
[20] Lyubimov D V, Lyubimova T P and Shklyaev S V 2006 *Phys. Fluids* **18** 012101
[21] Shklyaev S and Straube A V 2008 *Phys. Fluids* **20** 052102
[22] Alabuzhev A A and Lyubimov D V 2007 *J. Appl. Mech. Tech. Phys.* **48** 686–93
[23] Alabuzhev A A 2014 *Computational Continuum Mechanics* **7** 151–61 (in Russian)
[24] Alabuzhev A A and Lyubimov D V 2012 *J. Appl. Mech. Tech. Phys.* **53** 9–19
[25] Borkar A and Tsamopoulus J 1991 *Phys. Fluids A* **3** 2866–74
[26] Mampallil D, Eral H B, Staicu A, Mugele F and van den Ende D 2013 *Phys. Rev. E* **88** 053015

The translational oscillations of a cylindrical bubble in a bounded volume of a liquid with free deformable interface

A A Alabuzhev[1,2] and M I Kaysina[2]
[1]Institute of Continuous Media Mechanics UB RAS, Perm 614013, Russia
[2]Perm State University, Perm 614990, Russia

E-mail: alabuzhev@mail.ru

Abstract. We consider the eigen and forced translation oscillations of a cylindrical gas bubble surrounded by an incompressible fluid with free deformable interface. The bubble has an equilibrium cylindrical shape and is bounded axially by two parallel solid surfaces. Dynamics of contact lines is taken into account by an effective boundary condition: velocity of the contact line is assumed to be proportional to contact angle deviation from the equilibrium value. The equilibrium contact angle is right. Eigen frequency decreases with liquid outer free surface radius decreasing and increases with the radius-to-height ratio increasing. We found that the main translational frequency of eigen oscillation vanishes at a certain value of the Hocking parameter (the so-called wetting parameter). The eigen frequencies of the bubble are higher than the eigen frequencies of fluid volume.

1. Introduction

Problems with the contact-line dynamics taken into account have been examined in various formulations [1-4]. In the study of the dynamic contact angle Hocking condition is widely used (by virtue of its simplicity), and it is the one employed in a study [5] of the damping of standing waves between two vertical walls. This condition assumes a linear relationship between the velocity of the contact line motion and the contact angle (for the case of right equilibrium contact angle)

$$\frac{\partial \zeta^*}{\partial t^*} = \Lambda \vec{k} \cdot \vec{\nabla} \zeta^* , \qquad (1)$$

where ζ^* is the deviation of the interface from the equilibrium position, $\vec{k}$ is the external normal to the solid surface, Λ is a phenomenological constant (the so-called wetting parameter or Hocking parameter) having the dimension of the velocity. The special cases of the boundary condition (1) are the requirement of the fixed contact line ($\zeta^* = 0$, the so-called pinned-end edge condition) and the constant contact angle ($\vec{k} \cdot \vec{\nabla} \zeta^* = 0$). Article [5] also presents a qualitative comparison with the experimental studies of the other authors. The equation (1) was used in the study of oscillations of the sessile drop and bubble [6-8], cylindrical drop and bubble [9,10], capillary bridge [11] and "sandwiched" drop [12]. The particular case of pinned-end edge was used for capillary bridge in [13,14] and sessile drop [15]; the case of constant contact angle was used for cylindrical drop in [16] and for sessile drop in [17].

The more difficult condition was proposed in [18]. This condition involves an ambiguous dependence of the contact angle on the contact line velocity (for simplicity, the equilibrium contact angle is generally assumed to be a right angle)

[1] To whom any correspondence should be addressed.

$$\frac{\partial\zeta^*}{\partial t^*}=\begin{cases}\Lambda(\gamma-\gamma_c),\gamma>\gamma_c\\ 0,\ |\gamma|\le\gamma_c\\ \Lambda(\gamma+\gamma_c),\gamma<-\gamma_c\end{cases},\qquad(2)$$

where $\gamma=\partial\zeta^*/\partial z^*$ is the contact angle deviation from the equilibrium value, z^* is the coordinate normal to the solid wall and extending deep into the fluid. Results of physical experiments [19,20] are described well by formula (2) for small deviations of the contact angle. The equation (2) was used in the study of oscillations of the sessile drop [21], hemispherical bubble [22] on solid plate and cylindrical drop [23].

In the present study, we consider the translational oscillations of cylindrical bubble which surrounded by a liquid with free deformable interface. We apply condition (1) and continue the study of the axisymmetric oscillations [10,24,25]. It is important to note that the external liquid has the free nondeformable interface in [10,24] and the rigid wall in [25].

2. Problem formulation

Consider a gaseous bubble surrounded by a liquid with a free external surface (see figure 1). The system is bounded by two parallel solid surfaces separated by a distance h^*. The bubble and liquid volume are assumed to be cylindrical, with a radius r_0^* and R_0^* in equilibrium, respectively, implying that the every equilibrium contact angle is 90^0. The system is acted upon by an external vibration force with amplitude A^* and oscillation frequency ω^*. The force directed parallel to the solid surfaces.

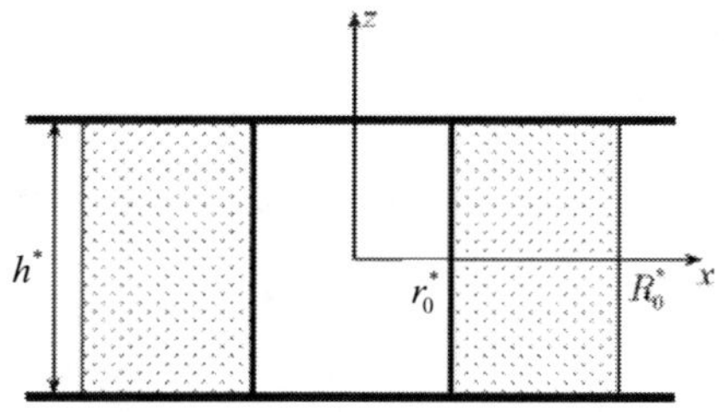

Figure 1. Problem geometry.

The characteristic amplitude of oscillations A^* is small compared to the equilibrium radius r_0^*. We assume that, on the one hand, the fundamental oscillation frequency ω^* is large enough for the viscosity could be ignored, and, on the other hand, the oscillation frequency is small enough, so that we can use the incompressibility conditions $\delta=\sqrt{\nu/\omega^*}<<r_0^*$ and $\omega^* r_0^*<<c$ (δ is the boundary-layer thickness, c is the sound velocity and ν is the kinematic viscosity of liquid).

Since the problem is symmetric, it is convenient to introduce cylindrical coordinates (r^*, α, z^*). The azimuthal angle α is reckoned from the x axis. Let the lateral surface of the bubble and the external surface of liquid volume be described by the following equations, respectively

$$r^*=r_0^*+\zeta^*(\alpha,z^*,t^*),\ r^*=R_0^*+\xi^*(\alpha,z^*,t^*).$$

Following [10,24], we use $\sqrt{\rho_e^* r_0^{*3}/\sigma^*}$, r_0^*, h^*, A^*, $A^*\sigma^*/r_0^{*2}$, $A^*\sqrt{\sigma^*/\rho_e^* r_0^*}$ as the scales for the time, length, height, deviation of bubble surface and free surface from its equilibrium position, pressure, and velocity potential, respectively. Here, σ^* is the surface tension and ρ_e^* is the liquid density. Thus, the dimensionless boundary value problem is determined by (intermediate steps can be found in [10,24])

$$p_e=-\left(\varphi_t+\omega^2 r\cos(\alpha)e^{i\omega t}\right),\ \Delta\varphi=0,\ p_i=-2n_p P_g^* r_0^*/\sigma^*\langle\zeta\rangle\equiv-P_0\langle\zeta\rangle,\qquad(3)$$

$$\Delta=\frac{1}{r}\frac{\partial}{\partial r}\left(r\frac{\partial}{\partial r}\right)+\frac{1}{r^2}\frac{\partial^2}{\partial\alpha^2}+b^2\frac{\partial^2}{\partial z^2},$$

$$r=1:\ \zeta_t=\varphi_r,\ [p]=\zeta+\zeta_{\alpha\alpha}+b^2\zeta_{zz},\tag{4}$$

$$r=R_0:\ \xi_t=\varphi_r,\ [p]=\xi+\xi_{\alpha\alpha}+b^2\xi_{zz},\tag{5}$$

$$z=\pm\frac{1}{2}:\ \varphi_z=0,\tag{6}$$

$$r=1,\ z=\pm\frac{1}{2}:\ \zeta_t=\pm\lambda\zeta_z,\tag{7}$$

$$r=R_0,\ z=\pm\frac{1}{2}:\ \xi_t=\pm\lambda\xi_z,\tag{8}$$

where p_e is the liquid pressure, φ is potential of liquid velocity, p_i is the gas pressure in the bubble, n_p is polytropic (e.g., adiabatic) exponent, P_g^* is dimension gas pressure in the bubble. The boundary-value problem (3)–(7) involves five parameters: the aspect ratio, the radius of free surface, the wetting parameter, the frequency and amplitude

$$b=r_0^*/h^*,\ R_0=R_0^*/r_0^*,\ \lambda=\Lambda^*\Big/\sqrt{\rho_e^* r_0^*/\sigma^*},\ \omega=\omega^*\sqrt{\rho_e^* r_0^{*3}/\sigma^*},\ \varepsilon=A^*/r_0^*.$$

3. Natural translational oscillations of the bubble

In order to investigate the problem, it is convenient to begin with a consideration of the natural oscillations of a cylindrical bubble. By the evenness of the natural oscillation modes is meant the evenness of the functions under a change of sign of the axial coordinate z. In view of translational symmetry, the solution of the boundary value problem (3)–(8) without external force is written as

$$\varphi(r,z,t,\alpha)=i\left(\sum_{n=0}^{\infty}\left(a_n^{(1)}R_{1n}^i(r)+b_n^{(1)}R_{1mn}^e(r)\right)\sin((2n+1)\pi z)+\right.$$
$$\left.+\sum_{n=0}^{\infty}\left(a_n^{(2)}R_{2n}^i(r)+b_n^{(2)}R_{2n}^e(r)\right)\cos(2\pi nz)\right)e^{i\alpha}e^{i\Omega t},\tag{9}$$

$$\zeta(\alpha,z,t)=\left(d_1z+d_2z^2+\sum_{n=0}^{\infty}h_n\sin((2n+1)\pi z)+\sum_{n=0}^{\infty}c_n\cos(2n\pi z)\right)e^{i\alpha}e^{i\Omega t},\tag{10}$$

$$\xi(\alpha,z,t)=\left(D_1z+D_2z^2+\sum_{n=0}^{\infty}H_n\sin((2n+1)\pi z)+\sum_{n=0}^{\infty}C_n\cos(2n\pi z)\right)e^{i\alpha}e^{i\Omega t},\tag{11}$$

where Ω is eigen frequency, $R_{1n}^i(r)=\mathrm{I}_1((2n+1)\pi br)$, $R_{1n}^e(r)=\mathrm{K}_1((2n+1)\pi br)$, $R_{20}^i(r)=r$, $R_{2n}^i(r)=\mathrm{I}_1(2\pi nbr)$, $R_{2n}^e(r)=r^{-1}$, $R_{2n}^e(r)=\mathrm{K}_1(2\pi nbr)$, I_1 and K_1 are modified Bessel functions of the first order.

Substituting solutions (9)–(11) into (3)–(8), we obtain a spectral-amplitude problem which eigenvalues are the values of the natural oscillation frequency Ω. From the solution of this problem it follows that the eigenvalues are found from the equations:

even modes
$$\sum_{n=1}^{\infty}\frac{1}{\pi^2n^2}\frac{\Omega_{2n}^2(R_0)}{\omega^2-\Omega_{2n}^2(R_0)}=-\frac{\lambda}{i\omega},\tag{12}$$

$$\frac{\displaystyle\sum_{n=1}^{\infty}\frac{1}{\pi^2n^2}\frac{\Omega_{2n}^2(1)\omega^2S_8}{-\Omega^4+\Omega^2S_2-S_3}+\frac{S_5(S_7-S_6)}{\Omega^2}}{\displaystyle\sum_{n=1}^{\infty}\frac{1}{\pi^2n^2}\frac{\Omega_{2n}^2(1)-\Omega^2}{-\Omega^4+\Omega^2S_2-S_3}-\frac{1}{6}-\frac{S_5S_4}{\Omega^2}-\frac{\lambda}{i\Omega}}\frac{S_4}{\Omega^2}-\frac{S_6}{\Omega^2}+\frac{1}{6}+\sum_{n=1}^{\infty}\frac{1}{\pi^2n^2}\frac{\Omega^2(\Omega^2-\Omega_{2n}^2(R_0))}{-\Omega^4+\Omega^2S_2-S_3}=-\frac{\lambda}{i\Omega},\tag{13}$$

$$\Omega_{2n}^2(1)=4\pi^2n^2b^2\frac{R_{2n}'^e(R_0)R_{2n}'^i(1)-R_{2n}'^e(1)R_{2n}'^i(R_0)}{R_{2n}'^i(R_0)R_{2n}^e(1)-R_{2n}^i(1)R_{2n}'^e(R_0)},\ \Omega_{2n}^2(R_0)=4\pi^2n^2b^2\frac{R_{2n}'^e(1)R_{2n}'^i(R_0)-R_{2n}'^e(R_0)R_{2n}'^i(1)}{R_{2n}'^e(1)R_{2n}^i(R_0)-R_{2n}^e(R_0)R_{2n}'^i(1)},$$

$$S_2=\Omega_{11n}^2(1)+\Omega_{11n}^2(R_0),\ S_3=\Omega_{11n}^2(1)\cdot\Omega_{11n}^2(R_0),\ S_4=-\frac{4b^2R_0}{(R_0-1)^2},\ S_5=\frac{R_0+1}{2R_0^2},\ S_6=\frac{2b^2(R_0^2+1)}{(R_0-1)^2},$$

$$S_7=\frac{2b^2(R_0^2-1)}{R_0^2+1}.$$

Odd modes
$$\sum_{n=0}^{\infty}\frac{4}{\pi^2(2n+1)^2}\frac{\Omega^2}{\Omega_{1n}^2(1)-\Omega^2}+\frac{1}{2}=-\frac{\lambda}{i\Omega},\tag{14}$$

$$\sum_{n=0}^{\infty}\frac{4}{\pi^2(2n+1)^2}\frac{\Omega_{1n}^2(R_0)}{\Omega^2-\Omega_{1n}^2(R_0)}=-\frac{\lambda}{i\Omega},\tag{15}$$

$$\Omega_{1n}^2(1)=(2n+1)^2\pi^2b^2\frac{R_{1n}'^e(1)R_{1n}'^i(R_0)-R_{1n}'^e(R_0)R_{1n}'^i(1)}{R_{1n}^i(1)R_{1n}'^e(R_0)-R_{1n}'^i(R_0)R_{1n}^e(1)},$$

$$\Omega_{1n}^2(R_0)=\frac{(2n+1)^2\pi^2b^2(R_{1n}'^e(1)R_{1n}'^i(R_0)-R_{1n}'^e(R_0)R_{1n}'^i(1))}{R_{1n}'^e(1)R_{1n}^i(R_0)-R_{1n}^e(R_0)R_{1n}'^i(1)}.$$

Here, $\Omega_{1n}^2(1)$, $\Omega_{1n}^2(R_0)$, $\Omega_{2n}^2(1)$, $\Omega_{2n}^2(R_0)$ are the natural oscillation frequencies of the freely moving contact line. The complex algebraic equations (12)–(15) have complex solutions, this leads to oscillation damping due to the dissipation on the contact line.

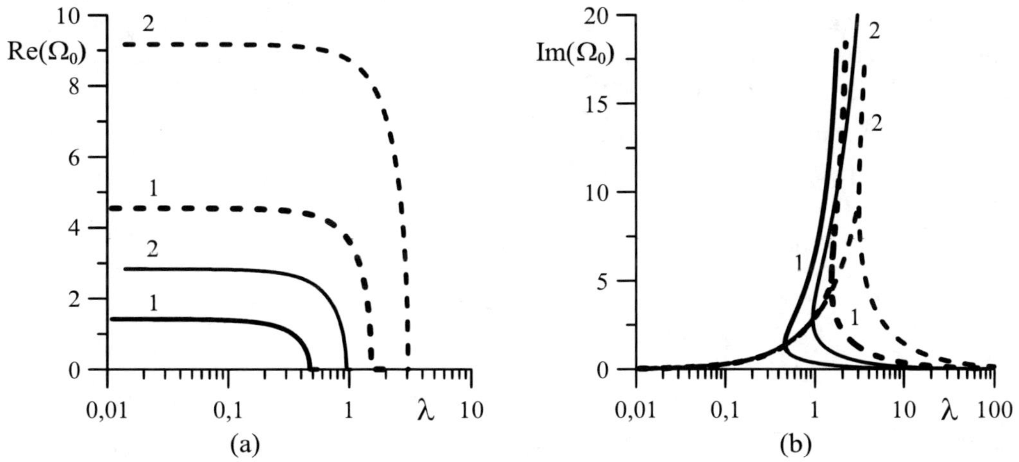

Figure 2. Frequency (a) and damping ratio (b) of natural oscillations vs wetting parameter λ for Ω_0 ($R_0=5$, $P_0=5$).
$b=1$ – line 1, $b=2$ – line 2, liquid – solid line, bubble – dashed line.

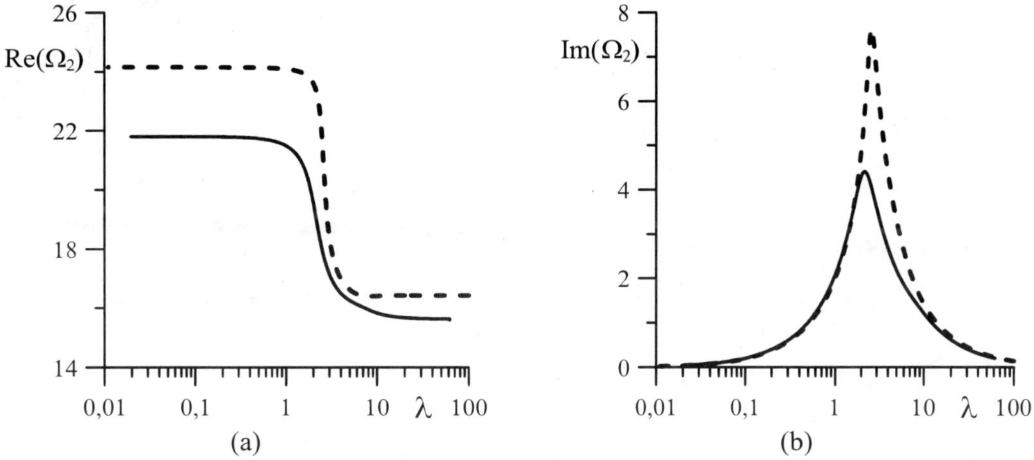

Figure 3. Frequency (a) and damping ratio (b) of natural oscillations vs wetting parameter λ for Ω_2 ($R_0=5$, $P_0=5$, $b=1$).
liquid – solid line, bubble – dashed line.

Equations (12)–(15) were solved numerically with the usage of the two-dimensional secant method. Figures 2 and 3 show the real part of $\text{Re}(\Omega)$ (oscillation frequency) and imaginary part

$\mathrm{Im}(\Omega)$ (damping ratio) of the complex natural frequency Ω for the oscillation even modes Ω_0 (i.e., $k=0$ is the wavenumber) and Ω_2. Here and below for the frequency indices, consecutive numbering of the wavenumber is used: even values of k correspond to the even modes (the solution of Eqs. (12) ,(13)) and odd values of k correspond to the odd modes (the solution of Eqs. (14),(15)). The zero translational mode $k=0$ describes the displacement of the liquid volume as a whole. In the considered case (with the contact line dynamics taken into account) the displacement is larger in the central part of the column than near the ends. Elastic forces cause the liquid volume to take the original shape, resulting in the return motion of its center of mass. As the capillary parameter increases, the shift between the center and periphery of the free surface decreases. For a certain value of the capillary parameter, the difference of the value of the shift disappears and the zero eigen frequency vanishes (figure 2a). For large values of λ, the damping ratio takes two values (figure 2b).

From figure 3 it follows that as λ increases, the frequency decreases monotonically, the damping ratio has a maximum for a finite value of the capillary parameter and tends to zero as $\lambda \to 0$ and $\lambda \to \infty$. Note, that the eigen frequencies of the bubble are higher than the eigen frequencies of the fluid volume. It is the result of the influence of the surface tension force on external interface.

Figure 4 shows the real part of $\mathrm{Re}(\Omega)$ (oscillation frequency) and imaginary part $\mathrm{Im}(\Omega)$ (damping ratio) of the complex natural frequency Ω for the oscillation odd mode Ω_1. From figures 2 and 4 it follows that the natural oscillation frequency increases as the parameter b increases (i.e., as the bubble equilibrium radius increases or as the bubble height decreases).

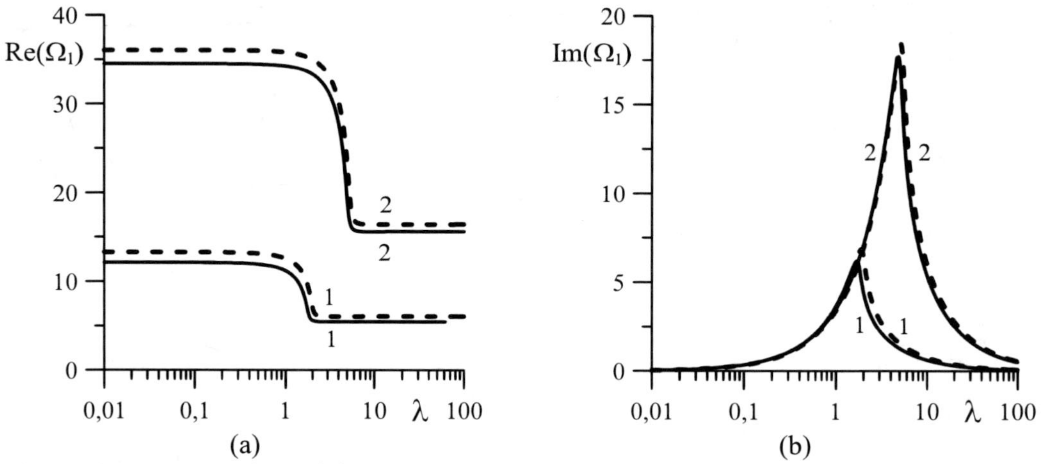

Figure 4. Frequency (a) and damping ratio (b) of natural oscillations vs wetting parameter λ for Ω_2 ($R_0=5$, $P_0=5$).
$b=1$ – line 1, $b=2$ – line 2, liquid – solid line, bubble – dashed line.

4. Forced translation oscillations of a bubble

Here we consider the problem of forced oscillations of a cylindrical bubble. The fields of velocity potential and surfaces deflections are defined by

$$\varphi(r,z,t,\alpha)=\mathrm{Re}\left(i\sum_{n=0}^{\infty}\left(a_n R_{2n}^{i}(r)+b_n R_{2n}^{e}(r)\right)\cos(2\pi nz)\cos(\alpha)e^{i\omega t}\right), \tag{16}$$

$$\zeta(\alpha,z,t)=\mathrm{Re}\left(\left(z^2 d+\sum_{n=0}^{\infty}c_n\cos(2n\pi z)\right)\cos(\alpha)e^{i\omega t}\right), \tag{17}$$

$$\xi(\alpha,z,t)=\mathrm{Re}\left(\left(z^2 D+\sum_{n=0}^{\infty}C_n\cos(2n\pi z)\right)\cos(\alpha)e^{i\omega t}\right). \tag{18}$$

Substituting solutions (16)–(18) into (3)–(8), we obtain expressions for the unknown amplitudes a_n, b_n, c_n, C_n, d and D. These expressions are not given due to their cumbersome forms.

5. Conclusions

Eigen frequency decreases with the decrease of the liquid outer free surface radius and increase with the bubble radius-to-height ratio increasing. We found that the main translational frequency of the natural oscillations vanishes at a certain value of the Hocking parameter. The eigen frequencies of the bubble are higher than the eigen frequencies of the fluid volume. It is the result of the influence of the surface tension force on external interface. Also note that the frequencies of the translational modes are independent of the gas pressure inside the bubble.

Well-marked resonance effects are found in the study of the forced oscillations. Thus, one can choose the bubble radius-to-height ratio such that the characteristic frequency of any mode is equal to zero, in order to ultimately, to determine the wetting parameter. There are not the «anti-resonant» frequencies under the translational vibrations, i.e. the vibration frequencies at which the amplitude of contact line deviation is zero [7,10].

Acknowledgments
This work was supported by the Russian Foundation for Basic Research (project 14-01-96017-r-ural).

6. References

[1] De Gennes P G 1985 *Rev. Mod. Phys.* **57** 827
[2] Voinov O V 1976 *Fluid Dyn* **11** 714–721
[3] Bonn D, Eggers J, Indekeu J, Meunier J and Rolley E 2009 *Rev. Mod. Phys.* **81** 739–805
[4] Zhang L and Thiessen D B 2013 *J. Fluid Mech.* **719** 295–313
[5] Hocking L M 1987 *J. Fluid Mech.* **179** 253-66
[6] Lyubimov D V, Lyubimova T P and Shklyaev S V 2004 *Fluid Dynamics* **39** 851–62
[7] Lyubimov D V, Lyubimova T P and Shklyaev S V 2006 *Phys. Fluids* **18** 012101
[8] Shklyaev S and Straube A V 2008 *Phys. Fluids* **20** 052102
[9] Alabuzhev A A and Lyubimov D V 2007 *J. Appl. Mech. Tech. Phys.* **48** 686–93
[10] Alabuzhev A A 2014 *Computational Continuum Mechanics* **7** 151–61 (in Russian)
[11] Borkar A and Tsamopoulus J 1991 *Phys. Fluids A* **3** 2866–74
[12] Alabuzhev A A and Lyubimov D V 2012 *J. Appl. Mech. Tech. Phys.* **53** 9–19
[13] Tsamopoulos J, Chen T and Borkar A 1992 *J. Fluid Mech.* **235** 579–609
[14] Demin V A 2008 *Fluid Dyn* **43** 524–532
[15] Kartavih N N and Shklyaev S V 2007 *Bulletin of Perm University. Series: Physics* no 1(6) 23–28 (in Russian)
[16] Alabuzhev A A and Lyubimov D V 2005 *Fluid Dyn* **40** 183-192
[17] Ivantsov A O 2012 *Bulletin of Perm University. Series: Physics* no 3(21) 16–23 (in Russian)
[18] Hocking L M 1987 *J. Fluid Mech.* **179** 267
[19] Ablett R 1923 *Philos. Mag.* **46** 244
[20] Dussan V E B 1979 *Annu. Rev. Fluid Mech.* **11** 371
[21] Fayzrakhmanova I S and Straube A V 2009 *J. Fluid Mech.* **21** 072104
[22] Fayzrakhmanova I S, Straube A V and Shklyaev S 2011 *J. Fluid Mech.* **23** 102105
[23] Alabuzhev A A 2012 *Bulletin of Perm University. Series: Physics* no 1(29) 35-41 (in Russian)
[24] Alabuzhev A A and Kaysina M I 2015 *Bulletin of Perm University. Series: Physics* no 1(29) 35-41 (in Russian)
[25] Alabuzhev A A and Kaysina M I 2015 *Bulletin of Perm University. Series: Physics* no 2(30) 56-68 (in Russian)

Journal of Physics: Conference Series **681** (2016) 012044 doi:10.1088/1742-6596/681/1/012044

Elastic properties of static charge-stabilized colloidal crystal with two-dimensional hexagonal lattice

Y V Alexandrov, A A Batanova, E V Gladkova, P E Dyshlovenko, A N Nagatkin and A F Nizametdinov

Department of Applied Mathematics and Informatics, Ulyanovsk State Technical University, 32 Severny Venets Street, Ulyanovsk 432027, Russia

E-mail: p.dyshlovenko@mail.ru

Abstract. Two-dimensional charge-stabilized colloidal crystal with a hexagonal crystal lattice and constant electric potential on the particles is studied numerically. The properties of the crystal are described by the theory based on the Poisson-Boltzmann non-linear differential equation. The force constants and the equilibrium pressure in the crystals are calculated for a broad range of the lattice parameters. The numerical procedures of the force constant and pressure determination is briefly described. Elastic constants of the first and second order are also calculated on the base of the force constants to obtain information about the many-body effective interactions in the system. It was shown that the Cauchy relation between the elastic constants breaks down in the whole range of the lattice parameter especially at higher densities. This can be interpreted as an effect of the many-body effective interaction between the particles in the crystal.

1. Introduction

Colloidal crystals are dispersions of macroscopic colloidal particles arranged into a regular crystalline lattice. They play an increasingly important role as a basis for the development of novel materials, in particular, in photonic crystal manufacture [1, 2, 3]. They are also a well-defined model systems resembling behavior of conventional crystals, whose properties, however, are much easier to control and observe. In charge-stabilized colloidal crystal, the particles are electrically charged and immersed into a liquid electrolyte.

In the present paper, we study numerically the two-dimensional charge-stabilized colloidal crystal obeying the general non-linear Poisson-Boltzmann (PB) equation making no a priory assumptions about the properties of the inter-particle interaction. Within the framework of the theory based on the PB equation, we calculate the force constants and equilibrium pressure in the system as well as elastic constants. Analysis of the elastic constants provides some information about the possible presence of the many-body interactions in the system. If each pair of the particles in a regular crystal lattice interacts via a central pair potential, then the elastic constants must satisfy the Cauchy relations. Violation of the Cauchy relations means that the total energy of the crystal cannot be represented as a sum of pair contributions of such kind and that the many-body effective interactions can play some role in the system. It is shown in the paper that this is the case for the crystal studied.

Journal of Physics: Conference Series **681** (2016) 012044 doi:10.1088/1742-6596/681/1/012044

2. Model and governing equations

Colloidal crystal under consideration consists of circular colloidal particles of radius R arranged in a two-dimensional hexagonal lattice with the lattice constant a. Alternatively, the particles can be treated as infinitely long uniform cylinders in three dimensions. The system of the particles is immersed in a binary symmetrical univalent electrolyte (1:1 electrolyte). The particles are electrically charged rigid dielectric bodies. Electric potential ϕ_0 at the surface of the particles is kept constant (the so-called constant-potential or *cp* model).

We assume that the charge-stabilized colloidal system obeys the non-linear PB equation [4, 5] for the electric potential ϕ. The PB equation for the 1:1 electrolyte has the form

$$\nabla^2 \phi = -\frac{q_e n}{\epsilon \epsilon_0} \left[exp\left(-\frac{q_e \phi}{k_B T} \right) - exp\left(\frac{q_e \phi}{k_B T} \right) \right] \tag{1}$$

in the rationalized SI unit system. This equation is valid in the electrolyte domain outside the particles. Here n is the bulk number concentration of either species of the electrolyte, q_e is the elementary charge, ϵ is the relative dielectric permittivity of the electrolyte, ϵ_0 is the electric constant, k_B is the Boltzmann constant, T is the absolute temperature.

To benefit from the dimensionless form of the governing equations and boundary conditions we introduce units $\kappa^{-1} = (2nq_e^2/\epsilon\epsilon_0 k_B T)^{-1/2}$ (Debye length) and $k_B T/q_e$ for length and electrostatic potential respectively. From here on all the variables are thought to be dimensionless. The dimensionless equation for our system can then be written as

$$\nabla^2 \phi = \sinh \phi \tag{2}$$

Electric field at the particle-electrolyte interface meets simple Dirichlet boundary condition

$$\phi = \phi_0, \tag{3}$$

where ϕ_0=*const* is a potential at the particles. Electric field inside the particles is equal to zero.

Domain of the problem for numerical solution consists of the Wigner-Seitz cells of the arbitrarily chosen central particle 0 and its 18 nearest neighbours of order 1 to 3. Interior of the particles is excluded from the domain. The central particle can be shifted slightly from its equilibrium position in different spatial configurations during the numerical experiments. Equilibrium position of a particle is given by vector $\mathbf{M} = m_1\mathbf{a}_1 + m_2\mathbf{a}_2$ where m_1 and m_2 are integer numbers and $\mathbf{a}_1$ and $\mathbf{a}_2$ are the primitive vectors of the lattice ($\mathbf{a}_1$ is parallel to the x-axis and $\mathbf{a}_2$ is directed at angle $\pi/3$ relative to $\mathbf{a}_1$).

The homogeneous Neumann boundary condition

$$\nabla\phi \cdot \mathbf{n} = 0, \tag{4}$$

where $\mathbf{n}$ is an outer unit normal, holds at the outer boundary of the domain. This means the normal component of the electric field is equal to zero at the external boundary.

Differential equation (2) with boundary conditions (3) and (4) constitute a boundary problem on the domain described above. Solution $\phi = \phi(x, y)$ of this problem fully determine the properties of the system. Calculation of both the forces and pressure is based on the stress tensor T of the PB equation:

$$T = \nabla\phi \otimes \nabla\phi - \left(\frac{1}{2}|\nabla\phi|^2 + \cosh\phi - 1 \right) I, \tag{5}$$

where I is an identity matrix.

Numerical solution was carried out by the finite-element method. Calculations were partly supported by the Supercomputing Center of Lomonosov Moscow State University [6]

3. Force constants

Force constants $C^{\mathbf{M}}_{\alpha\beta}$ of a two-dimensional monatomic crystal are defined by the following expression [7]:

$$C^{\mathbf{M}}_{\alpha\beta} = \frac{\partial^2 U}{\partial Z_{\alpha,\mathbf{N}} \partial Z_{\beta,\mathbf{N}+\mathbf{M}}}, \tag{6}$$

where U is the (free) energy of the system, $Z_{\alpha,\mathbf{N}}$ is an α-component of the displacement $\mathbf{Z}$ of the particle pointed by vector $\mathbf{N}$ from its equilibrium position, $\alpha = x, y$, $\beta = x, y$, $\mathbf{N}$ and $\mathbf{M}$ are vectors of the Bravais lattice. Since force constants do not depend on vector $\mathbf{N}$ due to the translational symmetry of potential energy U, one can set $\mathbf{N} = 0$ without loss of generality. Force constants for any particular $\mathbf{M}$ can be arranged into a square 2×2 matrix $C^{\mathbf{M}} = \left\| C^{\mathbf{M}}_{\alpha\beta} \right\|$.

Derivative

$$F_{\alpha,\mathbf{M}} = -\frac{\partial U}{\partial Z_{\alpha,\mathbf{M}}}, \tag{7}$$

is the α-component $F_{\alpha,\mathbf{M}}$ of the force on the particle $\mathbf{M}$. Then the force constant

$$C^{\mathbf{M}}_{\alpha\beta} = -\frac{\partial F_{\alpha,\mathbf{M}}}{\partial Z_{\beta,\mathbf{0}}}. \tag{8}$$

The force component $F_{\alpha,\mathbf{M}}$ can be calculated directly, without differentiation, by integrating the stress tensor (5):

$$F_{\alpha,\mathbf{N}} = \oint\limits_{\Gamma} \mathbf{e}_\alpha \cdot T \cdot \mathbf{n}\, dl, \quad \alpha = x, y, \tag{9}$$

where L is the length of the Wigner-Seitz cell's boundary Γ, $\mathbf{n}$ is an outer unit normal to the element length dl, and $\mathbf{e}_\alpha$ is a corresponding unit basis vector of the Cartesian coordinate system.

The functional dependences $F_{\alpha,\mathbf{M}}(Z_{\beta,\mathbf{0}})$, $\beta = x, y$ for each α and $\mathbf{M}$ in the domain were determined for a set of 21 equally spaced values of $Z_{\beta,\mathbf{0}}$ comprising of zero value, ten steps in the positive β-direction, and ten of the same steps in the negative β-direction. The largest shift of particle 0 from its equilibrium position amounted to 10% of the separation distance $(a - 2R)$ between the neatest particles. Numerical calculations were actually carried out only for positive displacements. The data for negative displacements were obtained from the data for positive ones with the help of inversion symmetry.

To find the force constants, the functions $F_{\alpha,\mathbf{M}}(X)$ for all vectors $\mathbf{M}$ and $\alpha = x, y$ with $X = Z_{\beta,\mathbf{0}}$, $\beta = x, y$, were fitted with polynomials:

$$F_{\alpha,\mathbf{M}}(\mathbf{X}) = a_0 + a_1 X + a_2 X^2 + \ldots + a_n X^n, \tag{10}$$

The best fitting was obtained with $n = 3$. Then $C^{\mathbf{M}}_{\alpha\beta} = -a_1$.

The amount of calculation can be significantly reduced taking into account the point symmetry of a particular crystal. For the hexagonal lattice, it is sufficient [8] to make shift of the central particle only in positive x-direction, while the y-directional shifts can be avoided. Moreover, only an (upper) half of the domain can be used since the crystal retains the mirror symmetry relative the x-axis during the x-shifts. If the force constant matrix $C^{\mathbf{M}}$ of particle $\mathbf{M}$ is known, then the matrix $C^{\mathbf{M}'}$ of nearest neighbour $\mathbf{M}'$ of the same order can be found from the following relation:

$$C^{\mathbf{M}'} = R^{\mathrm{T}}(\theta) C^{\mathbf{M}} R(\theta). \tag{11}$$

Here θ is the angle of rotation from $\mathbf{M}$ to $\mathbf{M}'$. Matrix $R(\theta)$ is a matrix of rotation through the angle θ, and superscript T means matrix transposition. For that reason, the minimal set of

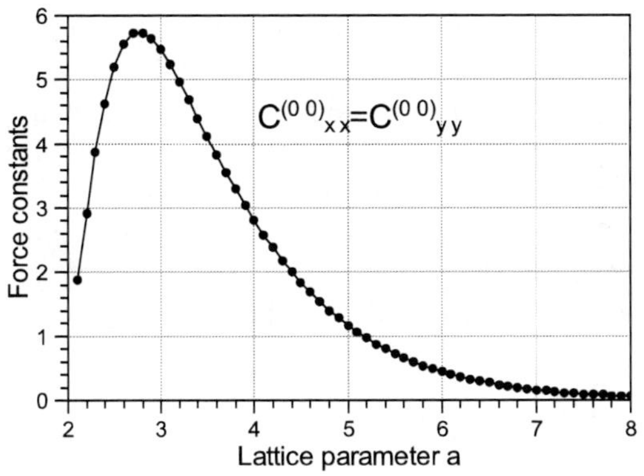

Figure 1. Force constants $C^{(0,0)}$.

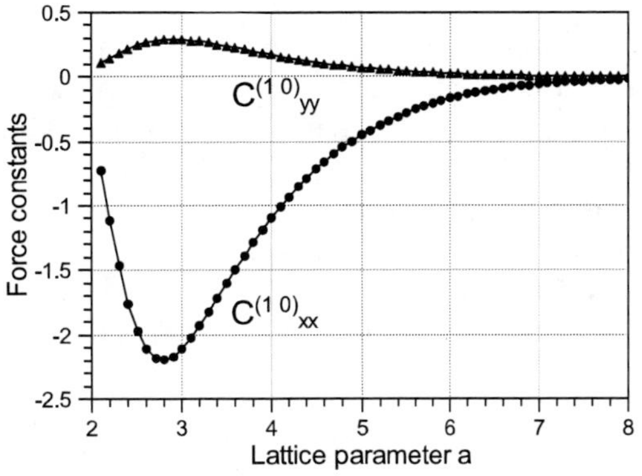

Figure 2. Force constants $C^{(1,0)}$ for the 1st-order nearest neighbours.

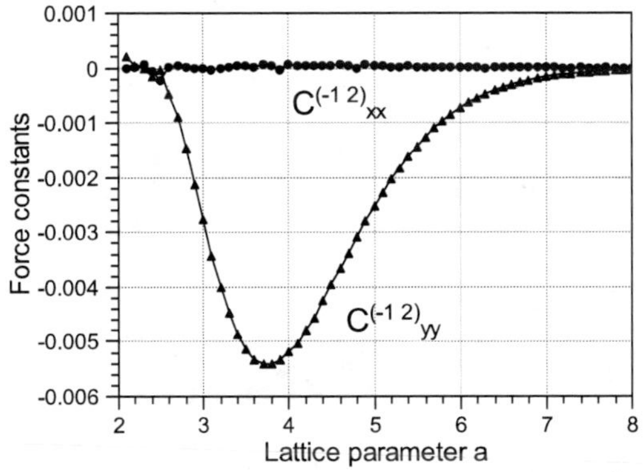

Figure 3. Force constants $C^{(-1,2)}$ for the 2nd-order nearest neighbours.

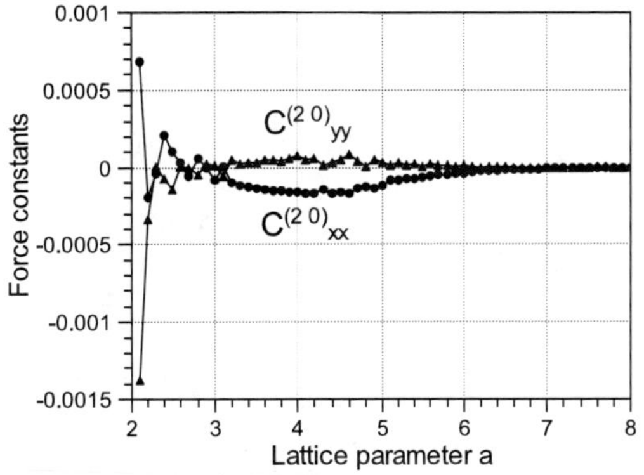

Figure 4. Force constants $C^{(2,0)}$ for the 3rd-order nearest neighbours.

data for the force constants can include only matrices $C^{(1,0)}$, $C^{(-1,2)}$ and $C^{(2,0)}$ which are all diagonal.

Numerical results for the force constants at $R = 1$ and $\phi_0 = 2$ are shown in Figs. 1 to 4. The scales of the vertical axes are very different in the figures. The curves for nearest neighbours of higher orders are rather noisy. For the neighbours of the 3rd order, the force constants are hardly resolved in our calculations and the results give only the order of magnitude. Behavior of the force constants for the large lattice parameters is close to exponential. In contrast to the crystal with constant charge density on the particles [8], the curves do not rise monotonically while the lattice parameter decreases. There are definite extrema on the curves so that the rigidity of the crystal decreases at high densities.

4. Elastic constants

Elastic constants describe macroscopic elasticity of the whole piece of matter. Elastic constants of the second order $C_{\alpha\beta\sigma\tau}$ appear in a series expansion of the strain-energy density with respect to the Lagrangian finite strain tensor components $\eta_{\alpha\beta}$ [9]. An alternative set of elastic constants

$B_{\alpha\beta\sigma\tau}$ arises from expansion with respect to infinitesimal strain tensor components $\epsilon_{\alpha\beta}$. There are only three nontrivial second-order elastic constants $C_{\alpha\beta\sigma\tau}$ for the hexagonal lattice, C_{xxxx}, C_{xxyy}, C_{xyxy}. The two sets of elastic constants are related to each other by the following expressions: $C_{xxxx} = B_{xxxx} + p$, $C_{xxyy} = B_{xxyy} - p$, $C_{xyxy} = B_{xyxy} + p$, where p is the pressure (more precisely, the excess of the equilibrium osmotic pressure over the pressure in the bulk).

Elastic constants can be expressed via the force constants according to the theory of elasticity of the solid under stress [10]. Pressure p is calculated through the stress tensor (5) by averaging the normal component of the force on the boundary of the Wigner-Seitz cell [11]:

$$p = -\frac{1}{L}\oint_{\Gamma} \mathbf{n} \cdot T \cdot \mathbf{n}\, dl, \tag{12}$$

where L is the length of the Wigner-Seitz cell's boundary Γ. The integration in (12) is made for the colloidal particle located in its equilibrium position. Numerical results for the elastic constants for $R = 1$ and $\phi_0 = 2$ are shown in Fig. 5 and Fig. 6.

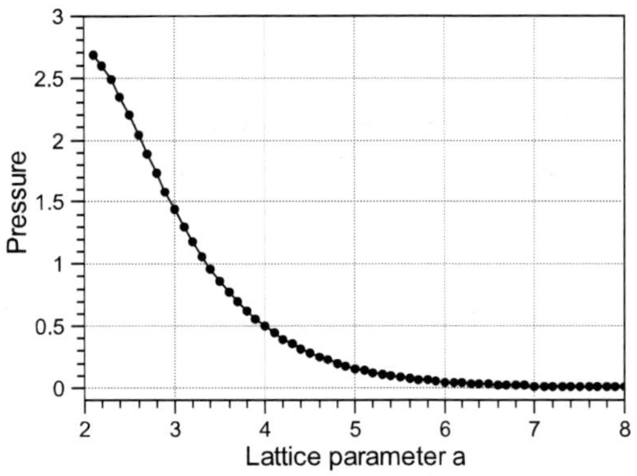

Figure 5. Equilibrium pressure (elastic constant of the 1st order).

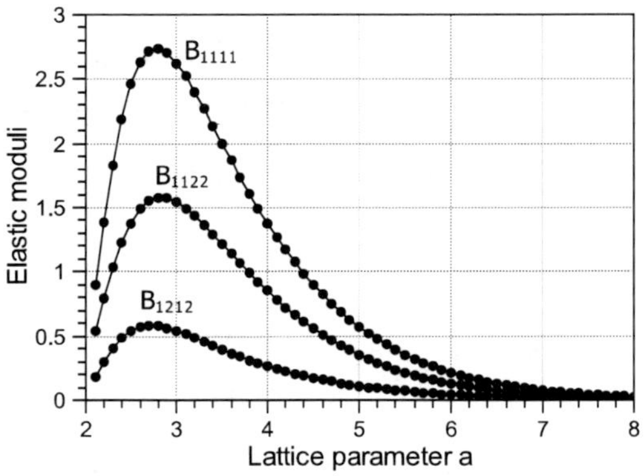

Figure 6. Elastic constants of the 2nd order B_{ijkl}.

Within the one-component model [12] the interaction in a colloidal system is reduced to the effective interaction between the colloidal particles only. If this interaction is central pairwise then Cauchy relations between the elastic constants holds additionally [9]. In the case of two-dimensional hexagonal lattice, there is only one such relation:

$$C_{1122} = C_{1212}. \tag{13}$$

The ratio C_{1122}/C_{1212} of the two constants is shown in Fig. 7

It is seen from Fig. 7 that there is a noticeable deviation of ratio C_{1122}/C_{1212} from unit in the whole region of the lattice parameter except possibly for very large separation distances. Effective interaction is close to be pairwise only at a low density of the system where the interaction itself is very weak. The deviation from the Cauchy relation grows monotonically while the density rises up to the almost contact between the particles. Relation $C_{1122} < C_{1212}$ holds for any values of the lattice parameter. That means that many-body interactions play some role in the system and proper elastic constants cannot be obtained with pair central potential of any kind.

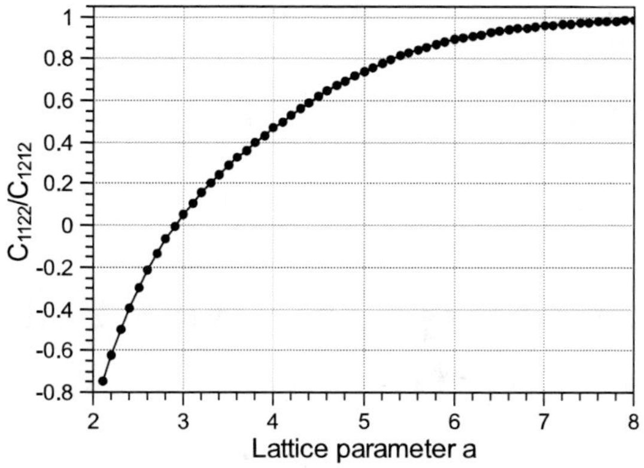

Figure 7. Verification of the Cauchy relation $C_{1122} = C_{1212}$.

5. Conclusions

Two-dimensional charge-stabilized colloidal crystal with hexagonal crystal lattice and constant potential on the particles was studied numerically within the static lattice approximation. The system obeys the general non-linear Poisson-Boltzmann equation. Numerical procedure based on light disturbance of the system by shifting a single particle from its equilibrium position was proposed for determination of the force constants. Complete set of the force constants of the nearest neighbours up to the third order and equilibrium pressure were calculated for a broad range of lattice parameter of the crystal.

Elastic constants of the first and second order were also calculated and used to verify the validity of the Cauchy relation $C_{1122} = C_{1212}$. Significant deviation from the Cauchy relation was observed that proves the presence of many-body interactions in the system. Further calculations are needed for quantitative estimation of many-body contributions into the total interparticle interaction.

References

[1] Tarhan I I and Watson G H 1996 *Phys. Rev. Lett.* **76** 315
[2] van Blaaderen A, Ruel R and Wiltzius P 1997 *Nature* **385** 321
[3] Joannopoulos J D, Villeneuve P R and Fan S H 1997 *Nature* **386** 143
[4] Israelachvili J N 1991 Intermolecular and Surface Forces (Academic Press) chapter 12
[5] Derjaguin B V, Churaev N V and Muller V M 1987 *Poverkhnostnye Sily (Surface Forces)* (Moscow: Nauka) (in Russian) chapter 1
[6] Sadovnichy V, Tikhonravov A, Voevodin V and Opanasenko V 2013 *"Lomonosov": Supercomputing at Moscow State University.* In Contemporary High Performance Computing: From Petascale toward Exascale (Chapman & Hall/CRC Computational Science) (Boca Raton, USA: CRC Press) pp 283-307
[7] Feynman R P 1972 *Statistical Mechanics* (Massachusetts: W. A. Benjamin) chapter 1
[8] Dyshlovenko P E 2007 *Colloid J.* **69** 13
Dyshlovenko P E 2007 *Kolloidny Zhurnal* **69** 18 (Russian original)
[9] Barron T H K and Klein M L 1965 *Proc. Phys. Soc.* **85** 523
[10] Wallace D C 1965 *Rev. Mod. Phys.* **37** 57
[11] Trizac E and Hansen P 1997 *Phys. Rev. E* **56** 3137
[12] Belloni L 2000 *J. Phys.: Condens. Matter.* **12** R549

Coherence of Noisy Oscillators with Delayed Feedback Inducing Multistability

Anastasiya V Pimenova[1] **and Denis S Goldobin**[1,2,3]

[1]Institute of Continuous Media Mechanics, UB RAS, Perm 614013, Russia
[2]Department of Theoretical Physics, Perm State University, Perm 614990, Russia
[3]Department of Mathematics, University of Leicester, Leicester LE1 7RH, UK

E-mail: `Anastasiya.Pimenova@gmail.com`, `Denis.Goldobin@gmail.com`

Abstract. For self-sustained oscillators subject to noise the coherence, understood as a constancy of the instantaneous oscillation frequency, is one of the primary characteristics. The delayed feedback has been previously revealed to be an efficient tool for controlling coherence of noise-driven self-sustained oscillator. The effect of the delayed feedback control on coherence is stronger for a longer delay time. Meanwhile, the instantaneous frequency of a noise-free oscillator can exhibit multistability for long delay time. The impact of the delay-feedback-induced multistability on the oscillation coherence, measured by the phase diffusion constant, of a noisy oscillator is studied in this work both numerically and analytically.

1. Introduction

Delayed feedback was found to be a highly efficient tool for controlling the coherence of noisy oscillators [1, 2, 3, 4]. Even weak feedback with sufficiently long delay time can diminish or enhance the phase diffusion constant, quantitative measure of coherence, by one order of magnitude and even more. The theory of the delay feedback control of the phase diffusion was developed in [1, 2, 4].

However, for long delay times and vanishing noise, the multistability of the mean frequency can occur. In the presence of multistability, noise results in intermittent switchings between states with different mean frequencies and, thus, the phase diffusion is contributed not only by fluctuations around the mean linear growth of the phase but also by the alternation of 'local' mean growth rates. For weak noise—which is a physically relevant situation—it turns-out to be possible to find natural variable in terms of which the switching between two stable phase growth rates becomes a perfect telegraph process. In [5], frequency multistability and noise-induced switchings were studied for extremely long delay times, although without consideration of coherence and phase diffusion.

In this work, on the basis of the 'telegraphness' property, we construct the analytical theory of the effect of delayed feedback on the phase diffusion in the presence of multistability. In agreement with the results of numerical simulation we analytically derive that the phase diffusion constant has giant peaks close to the points where the residence times in two states are equal.

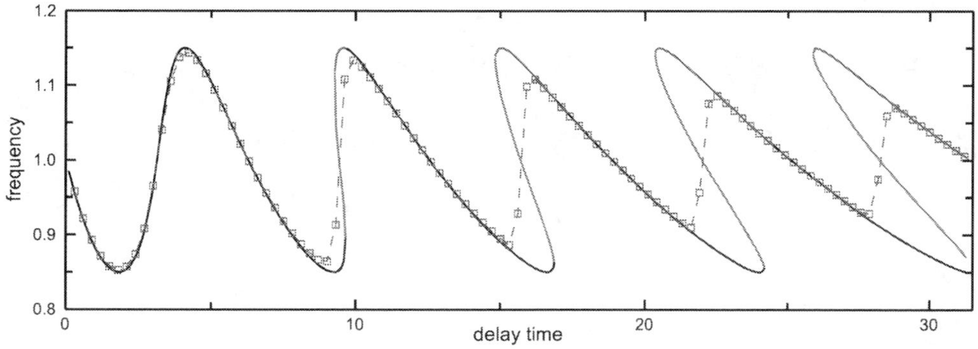

Figure 1. Mean oscillation frequency Ω *vs* delay time τ for $\Omega_0 = 1$ and $a = 0.15$. Solid line: the noise-free case (solution to equation (2)). Squares: numerical simulation for $\varepsilon^2 = 0.02$.

2. Phase Reduction

The dynamics of a limit-cycle oscillator subject to weak action can be described within the framework of the phase reduction, where the system state in determined solely by the oscillation phase [6, 7]. The phase reduction can be used as well for the case of weak noise, including a δ-correlated one [8, 9]. For nearly harmonic oscillators subject to white Gaussian noise and linear delayed feedback the phase equation reads [2, 7]

$$\dot{\varphi} = \Omega_0 + a\sin[\varphi(t-\tau) - \varphi(t)] + \varepsilon\xi(t)\,, \tag{1}$$

where Ω_0 is the natural frequency of the oscillator, a and τ are the strength and delay time of the feedback, respectively, ε is the noise strength, $\xi(t)$ is the normalised δ-correlated Gaussian noise:

$$\langle\xi(t)\xi(t+t')\rangle = 2\delta(t'), \qquad \langle\xi\rangle = 0.$$

Recently, a strong impact of weak anharmonicity for recursive delay feedback has been reported for the effects under our consideration [10]; however, for a 'simple' delay the harmonic approximation and model reduction (1) remain well justified for many physical, chemical and biological systems (*e.g.*, see [7, 11]).

Noise disturbs the linear growth of the phase, resulting in its diffusion according to the law $\langle[\varphi(t) - \langle\varphi(t)\rangle]^2\rangle = Dt$ for $t \to \infty$. In the control-free case, $D = D_0 = 2\varepsilon^2$. The diffusion constant D measures the coherence of oscillations; the smaller D the better coherence. Within the framework of the linear-in-noise approximation, delayed feedback has been found to change the mean oscillation frequency Ω and the diffusion constant D as follows [1, 2, 10];

$$\Omega = \Omega_0 - a\sin\Omega\tau\,, \tag{2}$$

$$D = \frac{D_0}{(1 + a\tau\cos\Omega\tau)^2}\,. \tag{3}$$

3. 'Telegraphness' of the instantaneous frequency in the presence of weak noise

3.1. Multistability of oscillation frequency for vanishing noise

Let us first discuss the system dynamics for vanishing noise. In figure 1 the solution of equation (2) is plotted for the noise-free case with the solid line. One can notice that for large τ the frequency Ω is not always unique, which has already been noticed in [2, 12]. The linear stability analysis for this system reveals that the linear growth solution $\varphi(t) = \Omega t$ is unstable when $a\tau\cos\Omega\tau < -1$. In [2], the stability analysis can be found for the monotonous

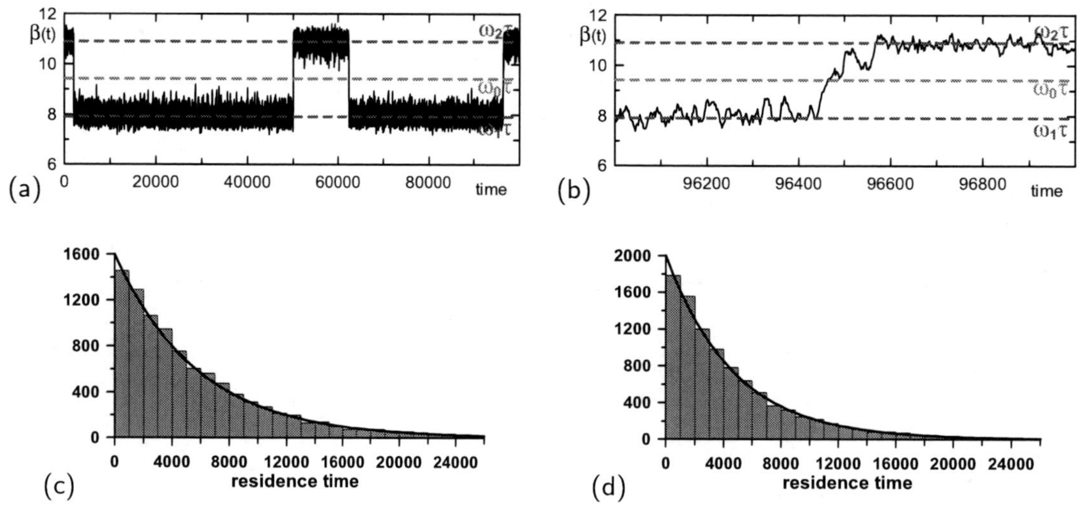

Figure 2. Sample of the dynamics of the system with multistability in the presence of weak noise for $\varepsilon^2 = 0.02$, $\Omega_0 = 1$, $\tau = 9.5$. (a) and (b): dynamics of $\beta(t) = \varphi(t) - \varphi(t-\tau)$ in time. (c) and (d): the distribution of residence times for ω_1 and ω_2, respectively, for 9551 switching events; the solid lines are exponential fitting of these distributions.

perturbations, the analysis for the oscillatory perturbations (to be published elsewhere) does not change this condition. If one moves along the curve in figure 1 from the origin, this condition will be fulfilled on the segments of the line going from the minimal values of Ω towards the maximal ones between points with vertical inclination. These segments, say ω_0, are plotted with gray and separate the branches of stable solutions which are plotted with blue, ω_1, and red, ω_2, in the domains of multistability.

Equation system (2)–(3) is derived within the framework of the linear-in-noise approximation, which considers only small fluctuations around a state with constant phase growth rate Ω; it does not handle hopping between several states with $\Omega = \omega_i$. Equation (3) has to be considered as a 'local' phase diffusion at the state with corresponding Ω. The system dynamics is dominantly controlled by the delay term, which makes the events of transition between two stable states of the noise-free system non-local in time; for instance, $\dot{\varphi}$ can be kicked-out by noise from ω_1, reach ω_2 and then again return to ω_1. Thus, the noise-driven dynamics of the system in the presence of multistability is a non-trivial problem requiring subtle consideration. For longer delay time τ there are domains with more than two locally stable regimes. As a first step, in this work we restrict our consideration to the case of multistability between two stable regimes; for $\Omega_0 = 1$ and $a = 0.15$ (figure 1) this is the case as long as $\tau \lesssim 31.5$.

3.2. Dynamics of the system subject to weak noise

Let us rewrite equation (1) in the form

$$\omega(t) \equiv \dot{\varphi} = \Omega_0 - a\sin\beta(t) + \varepsilon\xi(t)\,, \qquad \dot{\beta}(t) = \omega(t) - \omega(t-\tau)\,,$$

where $\beta \equiv \varphi(t) - \varphi(t-\tau)$ is the phase growth per delay time. This integral form of our dynamic system suggests that $\beta(t)$ can be a suitable 'natural' variable to trace the switchings between two states. Indeed, in figure 2 one can see that in terms of β the system state is a 'telegraph' signal; transitions between two states occur without immediate reversals—each reverse transition occurs as an event independent of the previous transition—the residence time distributions for

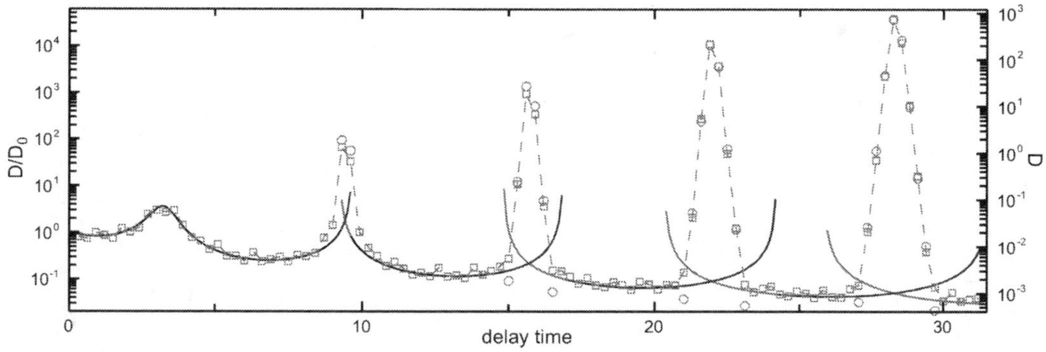

Figure 3. Phase diffusion constant D *vs* delay time τ for $\Omega_0 = 1$, $a = 0.15$ and $\varepsilon^2 = 0.02$. Solid lines: 'local' diffusion at stable states. Green squares: numerical simulation. Orange circles: diffusion constant (equation (4)) of a telegraph process with given $\omega_{1,2}$ and mean residence times $T_{1,2}$ taken from the results of numerical simulation.

two states are exponential. This 'telegraphness' can be used to develop analytical theory of the dynamics of a system with multistability.

4. Phase diffusion in the presence of multistability

4.1. Analytical theory

The diffusion constant of an asymmetric telegraph process altering between two states with growth rates ω_1 and ω_2 and mean residence times T_1 and T_2, respectively, can be calculated;

$$D_{\text{tel}} = \frac{2(\omega_2 - \omega_1)^2 T_1^2 T_2^2}{(T_1 + T_2)^3}\,. \tag{4}$$

The net phase diffusion is contributed by 'telegraph' diffusion and local diffusions at states $\omega_{1,2}$;

$$D = D_{\text{tel}} + \frac{T_1}{T_1 + T_2} D_1 + \frac{T_2}{T_1 + T_2} D_2\,, \tag{5}$$

where $D_{1,2} = D_0/(1 + a\tau\cos\omega_{1,2}\tau)^2$ (equation (3)).

4.2. Phase diffusion on the edges of the multistability domains

Singular behaviour is observed for the local phase diffusion at the edges of the multistability domains (figure 3). In figure 3 the local phase diffusions D_i are plotted with the solid line. The line is black for the domains with no multistability and coloured otherwise; the same colours mark corresponding regimes in figure 1. One can notice that at the edges of the multistability domains for the regime experiencing the tangential bifurcation at this edge the local phase diffusion (3) tends to infinity. This divergence of local diffusion rises the question how the net diffusion behaves on the edges of the multistability domains. To answer this question, one can perform a normal form analysis of the system near the tangential bifurcation in the presence of weak noise and derive scaling properties for quantifiers of interest.

Let us consider a point of the tangential bifurcation where regimes ω_0 and ω_2 appear (annihilate), $\tau = \tau_{2*}$, and describe the noise-induced behaviour of the system for $0 < \tau - \tau_{2*} \ll 1$. The normal form of the system dynamics near the bifurcation point reads $\dot{x} = -dU/dx + \varepsilon\xi(t)$ with $U(x) = -\alpha x + x^3/3$ and $\alpha = k_2(\tau - \tau_{2*})$, where k_2 is a positive number of order of magnitude

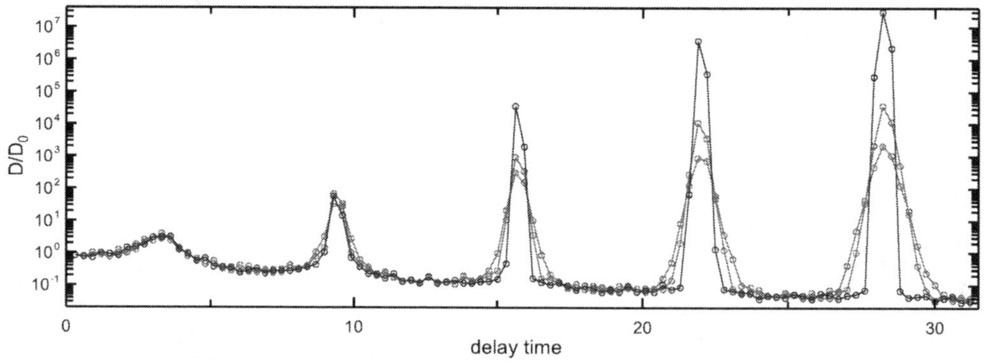

Figure 4. Phase diffusion constant D *vs* delay time τ: numerical simulation for $\Omega_0 = 1$, $a = 0.15$ and $\varepsilon^2 = 0.01$ (blue), 0.02 (red), 0.03 (green).

of 1. For this normal form the mean first passage time from the potential well can be calculated for weak noise;

$$T_2(\tau \approx \tau_{2*}) \approx \frac{\pi}{2}[k_2(\tau - \tau_{2*})]^{-\frac{1}{2}} \exp\left(\frac{4}{3\varepsilon^2}[k_2(\tau - \tau_{2*})]^{\frac{3}{2}}\right). \tag{6}$$

Since $T_2 \ll T_1$, the telegraph diffusion (4) becomes $D_{\rm tel}(\tau \approx \tau_{2*}) \approx 2(\omega_2 - \omega_1)^2(T_2^2/T_1^3)$, which tends to zero for $\tau \to \tau_{2*}$. Subtle analysis of equations (2)–(3) yields the scaling law for the local diffusion $D_2 \propto D_0/(\tau - \tau_{2*})$; it will be convenient to introduce constant K_2 as follows $D_2 = (D_0/2)K_2\sqrt{k_2}/(\tau - \tau_{2*})$. The behaviour of D_1 is regular around $\tau = \tau_{2*}$. Hence, equation (4) takes the form

$$D = D_{\rm tel} + D_1 + \frac{K_2\varepsilon^2}{T_1(\tau - \tau_{2*})^{3/2}} e^{\frac{4}{3\varepsilon^2}[k_2(\tau - \tau_{2*})]^{3/2}}. \tag{7}$$

One can see that for $\tau - \tau_{2*} > \varepsilon$, where the weak noise approximation and our derivations are valid, the last term in (7) rapidly decreases as τ tends to τ_{2*}. Thus, both the telegraph diffusion and the contribution of the diverging local diffusion into the net diffusion vanish at the edge of the multistability domains, and the behaviour of the net diffusion is regular, as it can be also seen from the results of numerical simulation (see figure 3).

4.3. Behaviour of phase diffusion near the centres of the multistability domains

Within a multistability domain the mean residence time T_1 decreases as τ increases and tends to zero at the right edge of the multistability domain, while T_2 increases starting from zero at the left edge of the domain. At certain value τ_* these times become equal, $T_1 = T_2 = T_*$. Let us consider the vicinity of this point. For small deviations $\delta\tau = \tau - \tau_*$, one can write $T_{1,2} \approx T_* \exp[(d\ln T_{1,2}/d\tau)\delta\tau] \equiv T_* \exp[\mp(q_{1,2}/\varepsilon^2)\delta\tau]$. Notice, since the mean residence times depend on τ exponentially strong with small number ε^2 in the denominator of the argument of the exponential, we linearise not T_i but $\ln T_i$ next to τ_*. Coefficients $q_{1,2}$ are positive number of order of magnitude of 1. Hence,

$$D(\tau \approx \tau_*) = \frac{(\omega_2 - \omega_1)^2 T_*}{4} \frac{\exp\left(\dfrac{q_2 - q_1}{2\varepsilon^2}(\tau - \tau_*)\right)}{\cosh^3\left(\dfrac{q_1 + q_2}{2\varepsilon^2}(\tau - \tau_*)\right)} + \frac{D_1 + D_2}{2}. \tag{8}$$

The exponential in the nominator is a small correction to the cube of hyperbolic cosine in the denominator. One has to expect a strong peak near $T_1 = T_2$ with height somewhat above $(\omega_2 - \omega_1)^2 T_*/4$. Notice, for the weak noise, the latter expression is exponentially large, as T_* is exponentially large. Moreover, for vanishing noise the width of the peak $\propto \varepsilon^2$, while its height is exponentially large in $1/\varepsilon^2$, *i.e.*, the integral of this peak over τ diverges exponentially fast.

In figure 3 one can see these peaks in the results of numerical simulation. Noteworthy, these peaks have a well pronounced triangular shape in the linear–log scale, meaning equation (8) to represent the behaviour of the phase diffusion remarkably well for the major part of each multistability domain, not only at its centre. In figure 4 one can see how the noise strength influences peaks in multistability domains. The numerical simulation data in all the figures are calculated with time series of length $2.5 \cdot 10^9$.

5. Conclusion

In this paper we have developed the theory of the effect of delayed feedback on coherence of noisy phase oscillators in the presence of frequency multistability induced by this time delay. The coherence has been quantified by the phase diffusion constant D. The process of alternation between two states has been demonstrated to be well representable as an asymmetric Markovian 'telegraph' process. Employment of this 'telegraphness' property allows constructing the framework for analytical study on the problem. The behaviour of the phase diffusion constant has been revealed to be smooth on the edges of the multistability domains. Giant peaks in the dependence of the phase diffusion constant on delay time have been discovered near points where the mean residence times in two states are equal (see equation (8) and figures 3 and 4). Remarkably, their 'integral strength' increases for vanishing noise, as their width $\propto \varepsilon^2$ while height is exponentially large in $1/\varepsilon^2$. For the longer delay time the peaks become taller.

Acknowledgments

The work has been supported by the Russian Science Foundation (Grant No. 14-21-00090).

References

[1] Goldobin D, Rosenblum M and Pikovsky A 2003 Coherence of noisy oscillators with delayed feedback *Physica* A **327** 124–128
[2] Goldobin D, Rosenblum M and Pikovsky A 2003 Controlling oscillator coherence by delayed feedback *Phys. Rev.* E **67** 061119
[3] Boccaletti S, Allaria E and Meucci R 2004 Experimental control of coherence of a chaotic oscillator *Phys. Rev.* E **69** 066211
[4] Goldobin D S 2014 Uncertainty Principle for Control of Ensembles of Oscillators Driven by Common Noise *Eur. Phys. J.* ST **223**(4) 677–685
[5] D'Huys O, Jungling Th and Kinzel W 2014 Stochastic switching in delay-coupled oscillators *Phys. Rev.* E **90** 032918
[6] Winfree A T 1967 Biological Rhythms and the Behavior of Populations of Coupled Oscillators *J. Theoret. Biol.* **16** 15–42
[7] Kuramoto Y 2003 *Chemical Oscillations, Waves and Turbulence* (New York: Dover)
[8] Yoshimura K and Arai K 2008 Phase Reduction of Stochastic Limit Cycle Oscillators *Phys. Rev. Lett.* **101** 154101
[9] Goldobin D S, Teramae J-N, Nakao H and Ermentrout G B 2010 Dynamics of Limit-Cycle Oscillators Subject to General Noise *Phys. Rev. Lett.* **105** 154101
[10] Goldobin D S 2011 Anharmonic resonances with recursive delay feedback *Phys. Lett.* A **375** 3410–3414
[11] Masoller C 2002 Noise-Induced Resonance in Delayed Feedback Systems *Phys. Rev. Lett.* **88** 034102
[12] Niebur E, Schuster H G and Kammen D M 1991 Collective frequencies and metastability in networks of limit-cycle oscillators with time delay *Phys. Rev. Lett.* **67** 2753–2756

Detailed numerical simulation of shock-body interaction in 3D multicomponent flow using the RKDG numerical method and "DiamondTorre" GPU algorithm of implementation

Boris Korneev[1], Vadim Levchenko[2]

[1] Moscow Institute of Physics and Technology, Dolgoprudny, Russia
[2] Keldysh Institute of Applied Mathematics, Moscow, Russia

E-mail: boris.korneev@phystech.edu

Abstract. Interaction between a shock wave and an inhomogeneity in fluid has complicated behavior, including vortex and turbulence generating, mixing, shock wave scattering and reflection. In the present paper we deal with the numerical simulation of the considered process. The Euler equations of unsteady inviscid compressible three-dimensional flow are used into the four-equation model of multicomponent flow. These equations are discretized using the RKDG numerical method. It is implemented with the help of the DiamondTorre algorithm, so the effective GPGPU solver is obtained having outstanding computing properties. With its use we carry out several sets of numerical experiments of shock-bubble interaction problem. The bubble deformation and mixture formation is observed.

1. Introduction

This paper studies the problem of interaction between a shock wave and an inhomogeneity in a gas. Being one of the basic topics in fluid dynamics, considered process has a lot of applications in industry, medicine, cosmology and other fields, and represents a complicated task both for analytical and numerical research.

The mathematical model of the dynamics of the mixture of fluids is the system of Euler equations. The issue of multicomponent flow is resolved in this work by using so called 4-equation model [1], which means that we introduce extra advection equation of volume fraction of mixture components. It is applied in the quasi-conservative form and does not require special treatment of the interface.

We use the RKDG method (Runge-Kutta discontinuous Galerkin method) for numerical solution of considered equations [2]. RKDG method is the high-order, non-oscillating and explicit method, based on the finite volume approach in symbiosis with the Galerkin method as the technique of increasing the local order of approximation. For the numerical fluxes construction, the exact or approximate Riemann solvers are used. The method is high-order, free from spurious oscillations, local and has clear logical structure simplifying its coding.

For the GPU implementation, a modified version of the algorithm called DiamondTorre developed by authors is used [3]. It uses the technique of special space-temporal tiling inside a 'window of computations', increasing the computational intensity of the algorithm. Specially adjusted for large scale problems, the algorithm shows outstanding performance compared to the maximum possible, which finally allows to obtain results for sufficiently fine grids using only a single PC with GPU [4, 5]. Performance results for our solver are shown.

Numerical validation of obtained solver is made. First of all, we examine one-dimensional tests based on the Riemann problem. Due to existence of analytical solutions of such tests, it is easy to test solver accuracy and other properties of obtained numerical results, comparing them with exact solutions and numerical solutions made by other solvers [1]. Results for some commonly used test problems are presented, showing good quality of our solver.

Three-dimensional problem of interaction between a steady spherical bubble of gas put inside some gaseous medium, particularly, air and a planar shock wave is considered. The sphere is chosen as a simplest model of a body in the wider problem of shock-body interaction. For the numerical point of view, it can be treated as a hard-level test for the solver, as it is resource-consuming fully three-dimensional problem. The pattern of the solution behavior is known for the problem at early times. In this paper we show the numerical results for some set of initial parameters of the problem and compare them with some well-known works [1, 6, 7].

2. Governing equations

The dynamics of fluid mixture in space when viscous effects are negligible is described by the system of Euler equations shown below

$$\frac{\partial \mathbf{U}}{\partial t} + \frac{\partial \mathbf{F}}{\partial x} + \frac{\partial \mathbf{G}}{\partial y} + \frac{\partial \mathbf{H}}{\partial z} = 0, \tag{1}$$

We use Cartesian coordinates (x, y, z) here. $\mathbf{U} = (\rho, \rho u, \rho v, \rho w, E)^{\mathbb{T}}$ is a vector of conservative variables, $E = \rho(\varepsilon + \frac{1}{2}(u^2 + v^2 + w^2))$ is full energy, $\mathbf{F}(\mathbf{U}) = (\rho u, \rho u^2 + p, \rho uv, \rho uw, (E+p)u)^{\mathbb{T}}$, $\mathbf{G}(\mathbf{U}) = (\rho v, \rho vu, \rho v^2 + p, \rho wv, (E+p)v)^{\mathbb{T}}$, $\mathbf{H}(\mathbf{U}) = (\rho w, \rho uw, \rho vw, \rho w^2 + p, (E+p)w)^{\mathbb{T}}$ are Eulerian fluxes.

2.1. Model of multicomponent flow

There are a lot of ways how to mathematically describe the multicomponent fluid flow. They can be defined into two groups. First are so called interface-tracking techniques, where the boundaries between two different fluids are taken into the account. Second is called volume-of-fluid (VOF) technique and it means that we do not pay special attention to the interfaces between different components, observing dynamics of every single component instead. We believe that for our case of gases the second approach is more appropriate, as it easily describes mixing and does not require numerically difficult interface tracing operations. In this paper the four-equation model of multicomponent flow is used [1]. Each component of the flow in every point of space is represented by the volume fraction α_i, $0 \geq \alpha_i \leq 1$, $\sum\limits_i \alpha_i = 1$. The advection equation for each volume fraction component is considered

$$\frac{\partial \alpha_i}{\partial t} + u\frac{\partial \alpha_i}{\partial x} + v\frac{\partial \alpha_i}{\partial y} + w\frac{\partial \alpha_i}{\partial z} = 0, \tag{2}$$

$p = p(\rho, \varepsilon, \alpha_i)$ is equation of state which completes the system of equations. Further the EoS of the ideal gases mixture is used given by the following equations

$$\begin{aligned} p &= \rho\varepsilon(\gamma - 1), \\ \frac{1}{\gamma - 1} &= \sum_i \frac{\alpha_i}{\gamma_i - 1} \end{aligned} \tag{3}$$

γ_i is the adiabatic index of the i component of the mixture.

Journal of Physics: Conference Series **681** (2016) 012046 doi:10.1088/1742-6596/681/1/012046

2.2. Quasi-conservative form of the system of equations

We transform the equations (1) and (2) to obtain the quasi-conservative form of the governing system. As (2) is in non-divergent form, rewrite it the following way

$$\frac{\partial \alpha_i}{\partial t} + \frac{\partial (v_j \alpha_i)}{\partial x_j} = \alpha_i \frac{\partial v_j}{\partial x_j}$$

Now left-hand side is in the divergent form, and the right-hand expression can be treated as a source term.

Introducing $\mathfrak{U} = [\mathbf{U}, \alpha_i]^{\mathbb{T}}$, $\mathfrak{F}_j = [\mathbf{F}_j, v_j \alpha_i]^{\mathbb{T}}$, $\mathfrak{S} = \left[\mathbf{0}, \alpha_i \frac{\partial v_j}{\partial x_j}\right]^{\mathbb{T}}$ we obtain the full system of equations written in the divergent form with the source term in the right-hand side

$$\frac{\partial \mathfrak{U}}{\partial t} + \frac{\partial \mathfrak{F}_j}{\partial x_j} = \mathfrak{S} \tag{4}$$

3. RKDG method

3.1. DG spacial discretization

Let the domain is discretized with the cells L_j.

Approximate solution $\mathfrak{U}_h$ at every cell L_j is written in form of the expansion in the specified basis functions $\{\varphi_i(x)\}_{i=1}^k$

$$\mathfrak{U}_h = \sum_{i=1}^{k} \mathbf{U}_i(t) \varphi_i(x).$$

Inserting it into the equation (4), requesting the result to be orthogonal to all the basis elements and using the Divergence theorem we obtain

$$\sum_{i=1}^{k} \frac{d\mathbf{U}_i(t)}{dt} \int\limits_{L_j} \varphi_i \varphi_l dV + \int\limits_{\delta L_j} \varphi_l \overline{\mathfrak{F}_j} n^j dS = \int\limits_{L_j} \left(\frac{\partial \varphi_l}{\partial x_j} \mathfrak{F}_j + \varphi_l \mathfrak{S} \right) dV, \tag{5}$$

where $\overline{\mathfrak{F}_j} = \left[\overline{\mathbf{F}_j}, \overline{v_j \alpha_i}\right]^{\mathbb{T}}$ is the numerical flux. We use HLLC flux [8] flux for constructing $\overline{\mathbf{F}_j}$ and Lax-Friedrichs flux [9] for $\overline{v_j \alpha_i}$.

3.2. Time discretization and limiting

In the previous paragraph we obtained spatially discretized ODE system (5), where the highest derivative is resolved. In terms of the RKDG method the explicit Runge-Kutta method is supposed to apply with limiting the solution in a special way after each RK stage. It is necessary to use the limiter for suppressing the spurious oscillations near big gradients. Various types of limiters have been developed (minmod, ENO and WENO-type etc) [2, 10]. Reconstruction of the coefficients is a local procedure, depending only on the values of the considered cell and the cells which are adjacent to it.

3.3. Concluding remarks and specifications

The method has several significant features, including the explicit stencil, high-order accuracy in time and space, and non-oscillation (TVDM) of the solutions.

Further in this work we use 3D Cartesian grid, second order in space and time RKDG method on piecewise linear basis in every cell, HLLC numerical flux for the fluid dynamics terms and Lax-Friedrichs flux for the volume fraction terms and the *minmod* limiter.

We would like to point out that the method combines both quite simplicity and high-order accuracy. That makes it a suitable candidate for building on it the efficient solver for simulating big three-dimensional problems. In the next section we consider the numerical algorithm of the implementation on GPU.

Journal of Physics: Conference Series **681** (2016) 012046 doi:10.1088/1742-6596/681/1/012046

4. DiamondTorre algorithm of the implementation

Here we suppose the approach called *LRnLA* to build the effective solver algorithm. LRnLA stands for "locally recursive non-locally asynchronous algorithms". Developed since 1990s, firstly published in [11], now this technique is used to construct various algorithms, including ones for GPU, for various physical codes [3, 4, 5, 12]. Stencil topology properties are used as a basis of the algorithm. Its structure is well-compatible with the computer architecture, having effective interaction between different levels of hierarchical computer systems (i.e GPU-CPU), resulting in the increased performance of the algorithm.

In this work the GPU algorithm called 'DiamondTorre' is used [3, 12]. The main points of the algorithm are following: the DiamondTorre algorithm's parts are associated with CUDA tools, while whole domain can be larger than the GPU memory amount because of 'computational window' using. The algorithm has the economic CPU-GPU exchange and is highly efficient.

4.1. Performance results

We use the domain with $512 \times 512 \times 1024$ cells (20 GB) for the performance tests. The solver is run on several GPUs with varying properties. The results of the testing are shown in figure 1. Performance gained is up to $5 \cdot 10^7$ cells per second. We note that using 32GB CPU DDR the

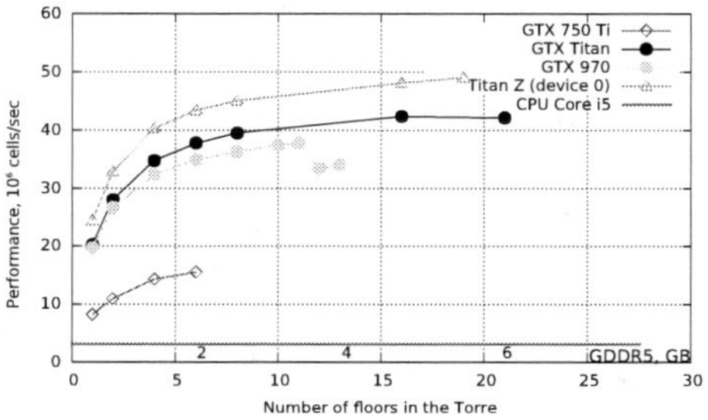

Figure 1. Speed of calculations depending on the window size

task with the domain of dramatically big size of about $4 \cdot 10^8$ cells can be calculated.

Dealing with two-component problems, we get only one extra equation in addition to homogeneous fluid [3], so the performance stays practically the same in this case. Further we consider only one- or two- component flows.

5. One-dimensional tests

We provide a series of one-dimensional tests for validation of the solver. They basically represent the Riemann problem having known analytical solution. We show below the results of two well-known tests: Sod problem and Lax problem both in the two-component setting. The tests are taken from [1]. Initial conditions are given in the table 1, the discontinuity was located in the middle of the domain.

Test #	ρ_L	v_L	p_L	α_L	ρ_R	v_R	p_R	α_R	$\gamma(\alpha)\|_{\alpha=0}$	$\gamma(\alpha)\|_{\alpha=1}$
1	0.445	0.699	3.527	0.9999	0.5	0	0.8565	0.0001	1.4	1.6
2	1	0	1	0.9999	0.125	0	0.1	0.0001	1.4	1.6

Table 1. Initial conditions for the tests

We show solutions for grids of 100 and 1000 cells in one graph. Results show quite good quality of numerical solutions expressing in very small spurious oscillations, excellent shock and good contact capturing.

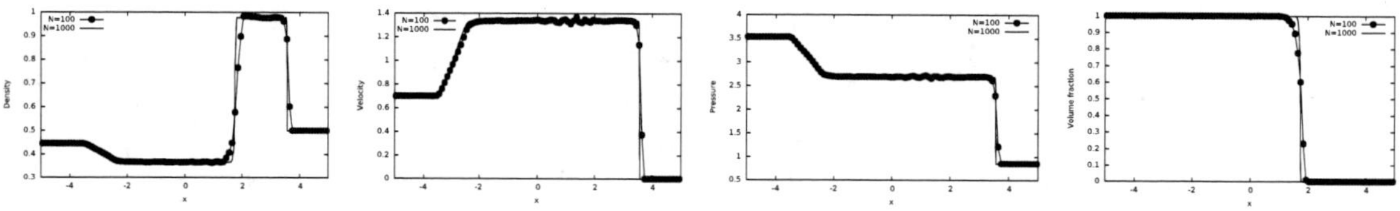

Figure 2. Results of the test 1 (Lax problem) solution at time $t = 1.3$

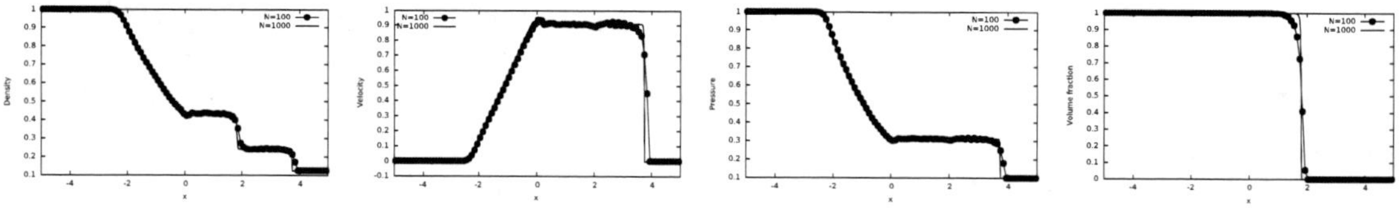

Figure 3. Results of the test 2 (Sod problem) solution at time $t = 2.0$

6. Bubble-shock interaction problem

6.1. Problem statement

The bubble is a spheric region where density ρ_B is different from the background one ρ_0 and pressures are equal. Shock wave moves and hits the bubble.

The dimensionless units of measure are chosen connected with the bubble size and the background gas condition as $x = R_0\tilde{x}$, $y = R_0\tilde{y}$, $z = R_0\tilde{z}$; $u = a_0\tilde{u}$, $v = a_0\tilde{v}$, $w = a_0\tilde{w}$; $t = \tilde{t}R_0/a_0$; $\rho = \rho_0\tilde{\rho}$; $p = \rho_0 a_0^2\tilde{p} = \tilde{p}p_0/\gamma$. The shock wave is characterized by the Mach number M of its front moving velocity. The Atwood number $At = \frac{\rho_B - \rho_0}{\rho_B + \rho_0}$ is used for the characteristic of the density inside the bubble.

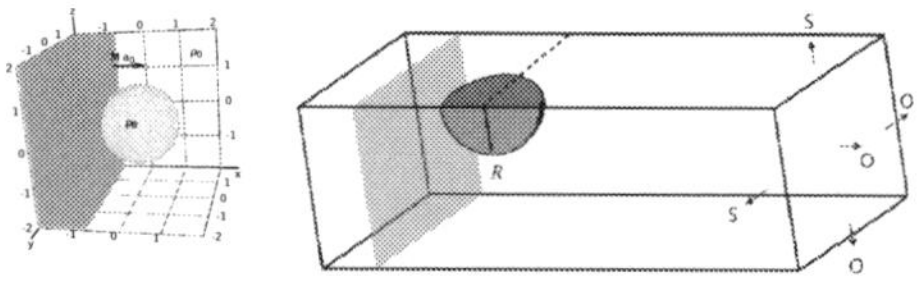

Figure 4. View of the numerical domain at the initial time. On the right figure the setting using the one-fourth of the domain is shown

One quarter of the whole domain can be considered, introducing the boundary condition of symmetry on the diametrical cross-sections.

6.2. Results for $At < 0$

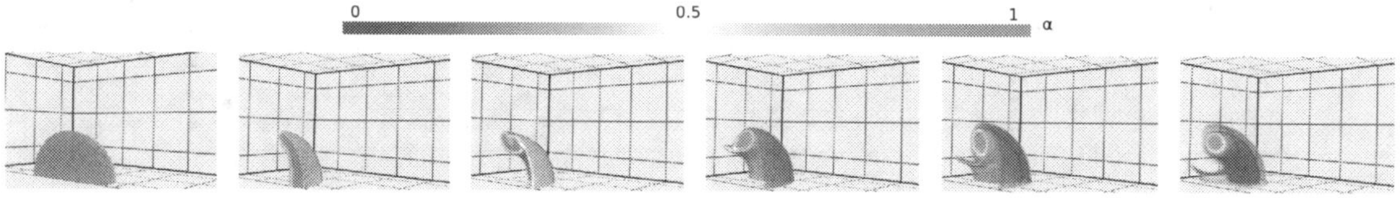

(a) (b) (c) (d) (e) (f)

Figure 5. Rendering of He component initially representing the bubble at different time moments: (a) $t = 0.0$, (b) $t = 0.6$, ... (e) $t = 3.0$, (f) $t = 3.6$; grid size is $600 \times 300 \times 300$

First case is considered when the bubble of He ($At = -0.757$) is hit by a shock at $M = 3$. This setting is known for long-living He vortex ring formation and is considered in various works including [6, 7]. Figure 5 shows how He component (red) is distributed in the domain at different time moments. One can see the vortex ring appearance in 5(c)–5(f).

6.3. Results for $At > 0$

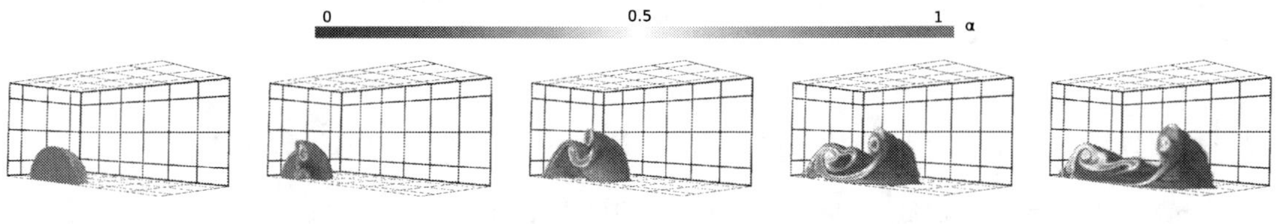

(a) (b) (c) (d) (e)

Figure 6. Rendering of the freon component initially representing the bubble at different time moments: (a) $t = 0.0$, (b) $t = 1.0$, (c) $t = 2.0$, (d) $t = 3.0$, (e) $t = 4.0$; grid size is $600 \times 300 \times 300$

Second case considered in the paper is when the bubble from heavy freon with $At = 0.613$ is hit with highly intensive shock at $M = 5$. The process is turbulent and a lot of vortices are created, tearing the bubble. We observe intense mixing of the bubble gas and air, which mainly has turbulent nature. The results have good coincidence with [6] and other known data.

7. Conclusion

In the paper the approach of studying the dynamics of multicomponent fluid using the Euler equations and 4-equation model is successfully applied. High-order numerical scheme is developed by the RKDG method and then validated. The solver having outstanding computing properties is constructed using the DiamondTorre algorithm of LRnLA class. The bubble-shock interaction problem with various initial conditions is resolved.

Acknowledgments

The work is supported by the "UMNIK" grant of the Foundation for Assistance to Small Innovative Enterprises. Also MIPT 5—100 Program is gratefully acknowledged.

References

[1] Gryngarten L D and Menon S 2011 *49th AIAA Aerospace Sciences Meeting including the New Horizons Forum and Aerospace Exposition, no. AIAA-2011-294, Orlando, FL*
[2] Cockburn B and Shu C 2000 Runge-Kutta Discontinuous Galerkin Methods for Convection-dominated Problems Tech. rep. NASA
[3] Korneev B A and Levchenko V D 2015 *Procedia Computer Science* **51** 1292 – 1302 ISSN 1877-0509 URL `http://www.sciencedirect.com/science/article/pii/S1877050915011229`
[4] Zakirov A and Levchenko V 2009 *Session 3AP* 169
[5] Perepelkina A, Levchenko V and Goryachev I 2014 *Journal of Physics: Conference Series* vol 510 (IOP Publishing)
[6] Niederhaus J H, Greenough J, Oakley J, Ranjan D, Anderson M and Bonazza R 2008 *Journal of Fluid Mechanics* **594** 85–124
[7] Hejazialhosseini B, Rossinelli D and Koumoutsakos P 2012 *arXiv preprint arXiv:1210.3822* URL `http://arxiv.org/abs/1210.3822`
[8] Toro E 2009 *Riemann Solvers And Numerical Methods for Fluid Dynamics: A Practical Introduction* (Springer)
[9] Lax P D 2005 *Selected Papers Volume I* 198–232
[10] Zhu J, Qiu J, Liu T and Khoo B C 2011 *Appl. Numer. Math.* **61** 554–580 ISSN 0168-9274
[11] Levchenko V D 2005 *Information technology and computing systems* **1**
[12] Perepelkina A, Levchenko V and Goryachev I 2014 *41st EPS Conference on Plasma Physics*

Model for the evolution of the time profile in optimistic parallel discrete event simulations

L Ziganurova[1,2], M A Novotny[3], L N Shchur[1,2]

[1]Science Center in Chernogolovka, 142432, Russia
[2]National Research University Higher School of Economics, 101000, Moscow
[3]Department of Physics and Astronomy, and Center for Computational Sciences, Mississippi State University, Mississippi State, MS 39762-5167, USA

E-mail: shchur@itp.ac.ru

Abstract. We investigate synchronisation aspects of an optimistic algorithm for parallel discrete event simulations (PDES). We present a model for the time evolution in optimistic PDES. This model evaluates the local virtual time profile of the processing elements. We argue that the evolution of the time profile is reminiscent of the surface profile in the directed percolation problem and in unrestricted surface growth. We present results of the simulation of the model and emphasise predictive features of our approach.

1. Introduction

The Parallel Discrete Event Simulation (PDES) approach [1] is thought to be very promising for large scale simulations. The main attractive feature of the approach is that it is scalable by construction: i) there are limitations on the problem size [2], and ii) it is proven for some simple cases it is deadlock free [3]. Practical implementations for using a PDES approach in computational physics is limited, primarily due to the folklore that de-synchronisation between processes grows with the system size. We have to mention that this fact is proved for the conservative algorithm of PDES [3], and not adequately investigated for the optimistic algorithm. At the same time, the most powerful implementation of the PDES approach is the optimistic algorithm. It is widely used for simulations in areas outside of computational physics [4]. In this paper we investigate the problem of synchronisation in optimistic PDES using a simple model of the optimistic algorithm. This model has some similarity with the Korniss et al model [3] for the conservative algorithm, and captures essential features of the optimistic algorithm. The dynamical behaviour of the optimistic PDES model is quite different from the conservative PDES mode. The two models belong to different universality classes. The Korniss et al model belongs to the universality class of Kardar-Parisi-Zhang equation [KPZ], while our model demonstrates features of the universality class of a roughening transition [5].

In the second section we formulate our model. In the third section we present results and analysis. In the final section we discuss how predictive properties of our model can be mapped onto real simulations in computational physics.

Journal of Physics: Conference Series **681** (2016) 012047 doi:10.1088/1742-6596/681/1/012047

2. Model

PDES considers the evolution in time of the ensemble of N objects. One can think on the object as an Ising spin in Glauber dynamics [6] or as a particle that forms clusters [7]. Each object i from $(1, 2, ...N)$ may change its state A_i at some moment of time, this change is named an *event*. In the examples, the event is the spin flip or the particle sticking to the cluster. The event happens at some definite moment of time, so the time associated with the event is a discrete variable. The idea of PDES is to replace the interaction of objects, which is usually realised in simulations through the access to the data held in the hardware memory, with the mechanism of message sending [1, 2]. Object i while changing its state A_i sends a message $S(A_i; \tau_i)$ with the time stamp τ_i, which is the value of the *local virtual time* (LVT$_i$) of the object i. This message is received by all objects which may depend on the state A_i. The ensemble of the LVT$_i$ forms the LVT profile over the objects.

Let us map the ensemble of objects onto the ensemble of processing elements (PEs). A PE may be the computing node, or CPU, or core, or thread, depending on the appropriate hardware/software combination available for the simulation [8]. Accordingly, the LVT is now associated with the PEs. Therefore, the problem of PE synchronisation is connected with the problem of keeping causality in the object evolution. Possible algorithms can be classified into three groups [8] and named as conservative, optimistic, and FaS. We do not discuss here details of the classification and of the associated algorithms, and refer the reader to a recent paper with a short review on the subject [9]. We concentrate here rather on the optimistic algorithm.

The optimistic algorithm name comes from the idea to give each PE the possibility to proceed, with the assumption that causality is not violated. Each PE$_i$ holds messages ordered by the time stamp in the list LR$_i$ of received messages. In the process of performing the simulation, the PE uses information written in the message, which is the state A_j of the PE$_j$ changed at LVT=τ_j. The process of simulation and the process of receiving messages are performed in parallel. Each PE$_i$ holds also a list of sent messages, LS$_i$, which is also ordered with the time stamps.

If it happens that incoming message A_j contains the stamp τ_j with the time smaller than the time of the message on the top of the LS$_i$, and if the state A_i of that message has been simulated using the old state of processing element j, it means that causality has been violated. The problem of causality violation is solved using the mechanism of a rollback in virtual times. In this case, PE$_i$ should return back to the state A_i contained in the message on the top of LS$_i$. If it happens that the time stamp of this message is still larger than the time of the message on the top of LR$_i$, than one more rollback is required to be performed, and it is necessary to inform other PEs that this message is not correct. This is done through the mechanism of sending anti-messages. PE$_i$ takes message A_i from the top of LS$_i$ and sends anti-message $\tilde{A}_i$ with the same information as in A_i. PEs receiving anti-message $\tilde{A}_i$ annihilate it with the corresponding A_i. It is clear that the rollback process may create an avalanche of rollbacks of some length.

We propose the following model for the evolution of the time horizon in optimistic PDES algorithms. We abbreviate in the following our model as OA, optimistic algorithm model. For simplicity, let us consider that all PEs may be ordered in a one-dimensional chain, and only nearest neighbours depend on the information contained in the messages. These assumptions were used previously in a model of the conservative algorithm [3] and of the FaS algorithm [8]. The first step of the model consists of the forward evaluation of LVT$_i$. We start with the flat profile of local virtual times of L processing elements, $\tau_i = 0$ with $i = 1, 2, ..., N$. We advance LVT$_i$ of each PE$_i$ with a random time step taken from the exponential distribution with mean F, this is the forward step of the model. Then we have to relax the LVT time profile under the rollback process. We assume that the length of the avalanche follows an exponential distribution with mean B, and each (randomly chosen) PE$_i$ may rollback B times to the right or left value of LVT, τ_{i-1} or τ_{i+1}.

3. Simulations

We performed numerical simulations for different values of parameter F ranging from 1 to 10, and for values of the parameter B from 0.2 to 4.4. We measured the speed of propagation of the local virtual time profile, as well as the width of its spreading. We can look for the evolution at clock time t of the average value of the LVT profile $\tau_{av}(t) = (1/N)\sum_{i=1}^{N} \tau_i(t)$, and of the evolution of the minimal value of the LVT profile $\tau_{min} = min\{\tau_i(t) : i = 1, 2, ..., N\}$, which is the value of the Global Virtual Time [1].

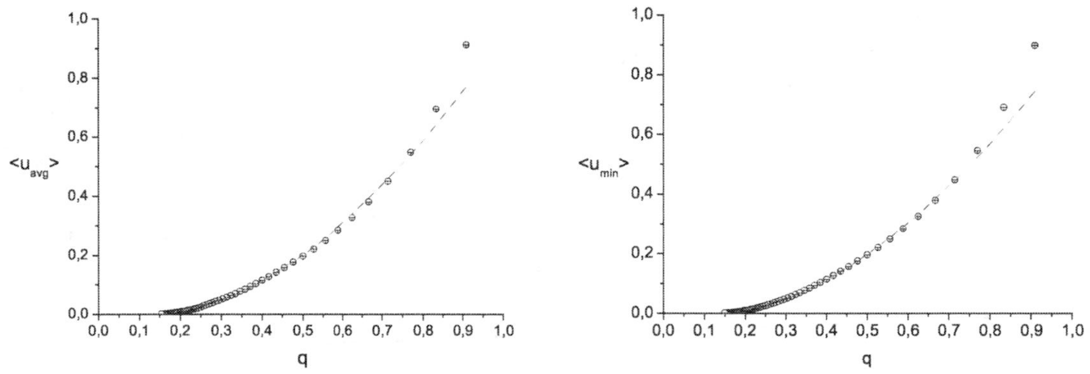

Figure 1. Dependence of the average LVT profile speed u_{av} (left) and GVT speed u_{min} on the parameter q for $F = 1$ (points with error bars were computed as the variation over 100 independent runs). The dashed line is a fit to the data.

Simulation shows that both the average LVT $\tau_{av}(t)$ and the GVT $\tau_{min}(t)$ grow linearly with time, and the slope of the curves define the corresponding speeds of the time profile, u_{av} and u_{min}. We found it useful to introduce a variable $q = 1/(1 + B)$. Figures 1 shows variation of these quantities as a function of q for the parameter value F=1. A fit to the data with the form $u = u_0(q - q_c)^{\nu}$ gives $u_0 \approx 1$, $q_c \approx .115$, and $\nu \approx 1.73$. The value of the exponent ν is very close to the one of directed percolation [10]. It is interesting to compare our model to the model of unrestricted surface growth (USD) [5]; they have some similarities and some differences. The similarity is in the existence of two main steps in both models: adsorption of atoms in the USD model and the time increment in the OA model, and the desorption of atoms in the USD model and rollbacks in the OA model. The differences are in the probabilities, which is q for adsorption and $(1 - q)/2$ for desorption in the USD model, while the probabilities in the OA model are independent parameters. The value of $q_c \approx 0.233$ in the USD model is about two times larger than in the OA model.

Figures 2 shows that curves $u(q)$ calculated for different values of the parameter F can be collapsed on one curve dividing u_{av} or u_{min} by the value of F.

The width of the LVT time profile grows with an exponent of approximately 0.53. This is illustrated in Figures 3 for values of parameters $F = 1$, $B = 0.182$ (left) and $F = 1$, $B = 0.53$ (right). Note that the slopes are about the same while the amplitudes are quite different.

There is the possibility to compare the time evolution in our OA model with the conservative algorithm model discussed in [3]. We can calculate the number of local minima $N(t)$ of the time profile. This is the number of PEs which may successfully evaluate in the conservative algorithm, and number of PEs which will not rollback on the next step in the OA. Figure 4 shows the density of local minima $N(t)/N$ of the time profile for a number of values of the parameter q, as well as for the conservative algorithm. We notice that formally this picture

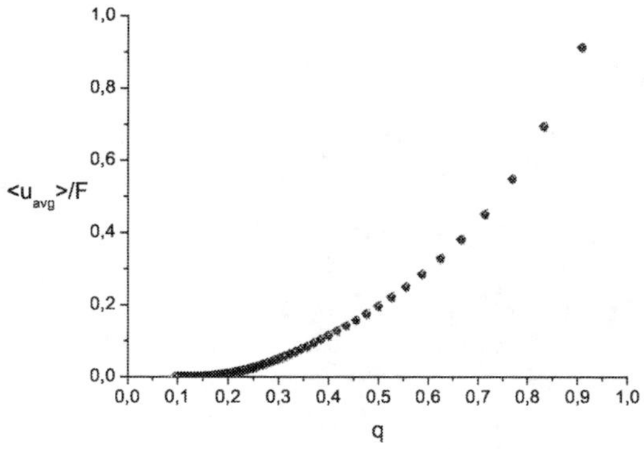

Figure 2. Data collapse of u_{av} (left) and u_{min} (right). Values of F vary from 1 to 10.

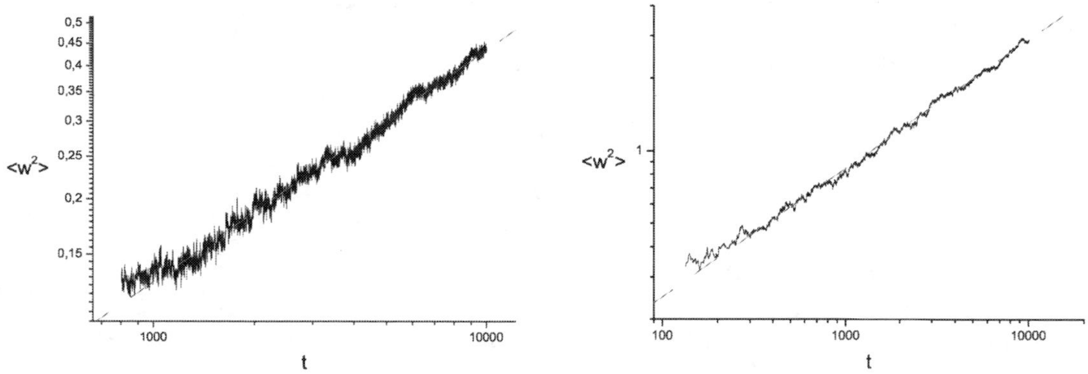

Figure 3. Time dependence of the LVT profile width: $F = 1$, $q = 0.182$ (left) and $q = 0.53$ (right).

demonstrates that the optimistic algorithm is more efficient as the number of active sites is larger than in the conservative algorithm. At the same time, the speed of the time profile evolution is growing from the top of figure 4 from the bottom (starting from $q = 0.185$ and progressing toward $q = 0.833$). The speed of OA model is equal to the speed of the optimistic algorithm model at the value of $q \approx 0.6$.

4. Discussion

We proposed a model for the evolution of the time horizon in optimistic PDES simulations. Simulations of the OA model demonstrate that some properties of OA is reminiscent of those of unrestricted surface growth [5], which belongs to the universality class of directed percolation[11]. This observation may be used for the prediction of the behaviour of real optimistic PDES simulations. Predictions may be performed by matching parameters of the OA model to those of the PDES simulations. For example, optimistic Time Warp simulations show indications of self-organised criticality[12]. We plan to perform case studies of optimistic PDES, and analyse possible mappings of PDES onto the OA model. Probably, some extensions of the present OA model should also be investigated.

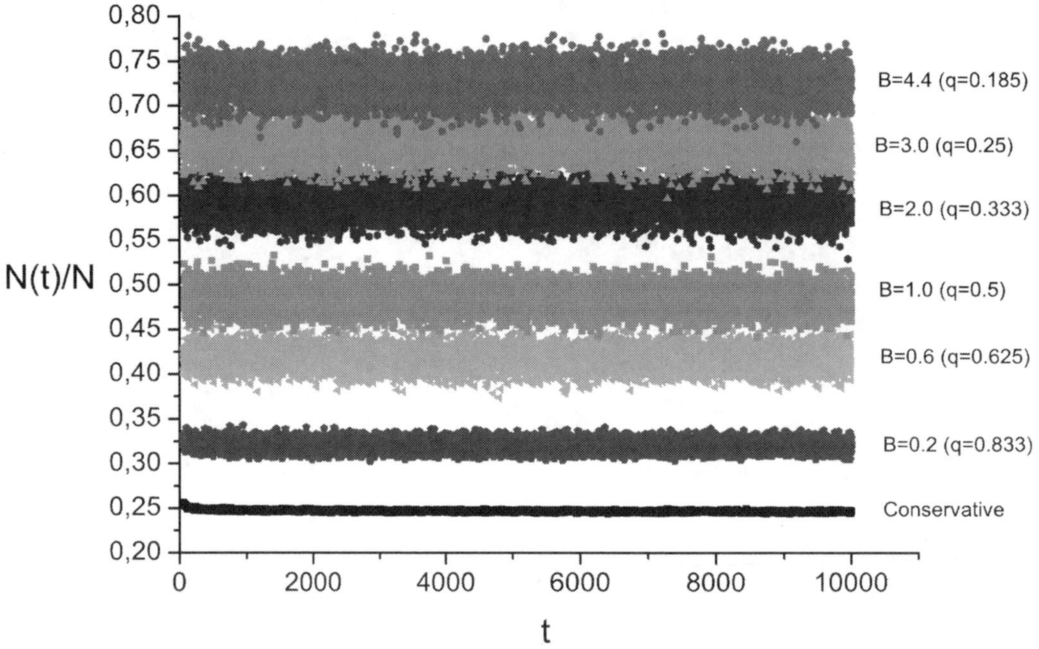

Figure 4. Density of the number of local minima as function of the time t. Black line = conservative algorithm model; coloured lines = OA model for a number of values of q.

5. Acknowledgments

This work by LZ and LNS was supported by the Russian Science Foundation (project nos. 14-21-00158). MAN was supported in part by a grant from the US National Science Foundation (DMR-1151387).

References

[1] Jefferson D R 1985 *ACM Transactions on Programming Languages and Systems* **7**: 404
[2] Fujimoto R M 1990 *Communication of ACM* **33**:31
[3] Korniss G, Toroczkai Z, Novotny M A, Rikvold P A 2000 *Physical Review Letters* **84**:1351
[4] Barnes PD Jr, Jefferson D R, Schordan M, Quinlan M D, Carothers C D, Kale L V 2014 *Proceedings of the 2014 Winter Simulation Conference, Savannah, GA, USA - December 07-10, 2014*
[5] Alon U, Evans H, Hinrichsen H, Mikamel D 1996 *Physical Review Letters* **76**:2746
[6] Lubachevsky B D 1988 *Journal of Computational Physics* **75**: 103
[7] Gonzalés S, Thornton A R, Luding S 2011 *Computer Physics Communications* **182**: 1842
[8] Shchur L N, Novotny M A 2004 *Physical Review E* **70**: 026703
[9] Shchur L N, Shchur L V 2014 *Journal of Physics: Conference Series* **640**: 012065
[10] Odor G 2004 *Reviews of Modern Physics* **76**: 663
[11] Domany E, Kinzel W 1984 *Physical Review Letters* **53**: 311
[12] Sloot P M A, Overeinder B J, and Schoneveld A 2001 *Computer Physics Communications* **142**: 76

Integrating GPGPU computations with CPU coroutines in C++

Pavel A Lebedev

Subdepartment of Computer Security, National Research University Higher School of Economics, 34 Tallinskaya Ulitsa, Moscow, Russia

E-mail: cygnus@michiru.ru

Abstract. We present results on integration of two major GPGPU APIs with reactor-based event processing model in C++ that utilizes coroutines. With current lack of universally usable GPGPU programming interface that gives optimal performance and debates about the style of implementing asynchronous computing in C++, we present a working implementation that allows a uniform and seamless approach to writing C++ code with continuations that allow processing on CPUs or CUDA/OpenCL accelerators. Performance results are provided that show, if corner cases are avoided, this approach has negligible performance cost on latency.

Keywords: C++, CUDA, OpenCL, coroutines.

1. Introduction

Despite various efforts on standardization, both open and proprietary, there is currently no universally available programming interface to access compute capabilities of heterogeneous systems. The two main competitors — NVIDIA CUDA [1] and OpenCL [2] — are being pushed by NVIDIA and AMD respectively. While CUDA is only supported on NVIDIA devices, OpenCL support is more widespread, but its implementations frequently have inconsistent behavior or performance. We will discover an example of such difference later in this paper. Automatic offloading tools are also in their infancy, which also contributes to the need to manually target these APIs to exploit full capabilities of the hardware.

C++ is a proven language for high-performance computing, but it lags in standardization of concurrent and parallel processing facilities. This includes recently rediscovered solution to elegant asynchronous processing in the form of coroutines. Coroutines are a generalization of subroutines ("functions" in C++) that may suspend execution and return control to the caller to be resumed later [3]. Different types of coroutines are useful to clearly express generators, state machines, cooperating pipeline stages and other types of control flow.

Traditional asynchronous code in C++ contains lots of relatively small functions that specify each other as a point at which the algorithm should continue once an invoked asynchronous call completes. This quickly complicates program structure in comparison with blocking synchronous code that doesn't need to split algorithms with multiple asynchronous calls into pieces. Coroutines can be used as the primary means of reverting tangled and piecewise asynchronous code back to serial and readable form.

Currently there are several proposals that cover coroutines, which differ in the type of coroutines supported (stackless or stackful) and, implicit or explicit suspension syntax and memory requirements and allocation strategies. A comparison of these approaches is given in [4]. While none of these approaches have been accepted into C++ language standard so far, many of them have working prototypes.

The first proposal to add stackful coroutine support to the language was a purely library-based solution based on Boost.Context library [5]. On the other hand, resumable functions [6] have been implemented by Microsoft as part of the recent release of Visual C++ compiler and are an example of leveraging compiler support to reduce memory requirements for saving stackless coroutine state. There are also hybrid proposals, for example [7]. Obviously, solutions based on language extensions have potential to be more efficient than pure library-based implementations.

CUDA and OpenCL both provide blocking and nonblocking API calls. Blocking calls are not usable in scalable code because they force to dedicate a separate operating system thread to every independent logical flow that works in parallel. This results in losses due to excessive context switch overhead and is known in networking applications as C10K problem [8]. Asynchronous APIs allow to avoid it, but the program's readability suffers, as we've explained earlier.

In this work we'll discuss how to seamlessly integrate computations on CPU and I/O with GPU operations that allow to specify any combinations of asynchronous operations in a unified form. While we're waiting for coroutines to become an integral part of the C++ language, we may already obtain an estimate of abstraction penalty that coroutines impose in comparison with traditional code that uses GPU APIs.

2. Test implementation

In addition to using GPUs, programs perform computations on the CPU and input/output with storage, network and other devices. Computations directly consume CPU time, but waiting for I/O doesn't. According to this scheme, operations on GPUs should be treated as another type of I/O as they consume no CPU time while waiting for results. This includes both data exchange with GPU and computations that run on it.

Operating system threads that are waiting for I/O requests to complete can be reused to perform other computations meanwhile. The reactor-based completion event handling [9] allows to exploit these opportunities, but decomposes algorithms to chains of continuations that are more difficult to read and reason about than traditional blocking code.

Coroutines allow to combine efficiency and scalability of this asynchronous approach while maintaining readability of the implementation. In this case coroutines suspend immediately after an asynchronous call and specify a small thunk that resumes them as their continuation. This construct integrates properly with existing dispatch and scheduling facility that is already present in asynchronous code.

While C++ already has future/promise as a way to perform asynchronous computations, a more rich set of facilities is required to specify policies for execution. Again, there are several competing proposals that tackle this problem. In this work we'll be using a reference implementation of executors and asynchronous operations [10] that underpins current C++ Netowrking Library proposal, that is itself a derivative of asio library [11], a proven solution for writing scalable network I/O software in C++. The coroutines will be provided by the aforementioned Boost.Context library [12]. We choose these libraries because they are readily available, do not depend on particular compiler or operating system, and represent the least efficient library-based approach. This allows us to treat our performance as a guaranteed lower bound, which can only be improved on.

Reactor-based event processors allow easy integration with foreign event notification methods. Integration with asynchronous programming framework can allow GPU operations to have the same invocation syntax as other asynchronous operations. Even more important is that this will allow these operations to be composable with asynchronous operations of other types: different GPU API, network or disk I/O etc. This greatly simplifies design of complex and general applications in heterogeneous environments. In fact, current design of C++ futures is not generally useful exactly because it has poor composition support, but this is being worked on [13]. It is important to note that both GPU APIs support multiple command queues, therefore composition may be required even in scenarios that work with a single GPU.

Fortunately, both CUDA and OpenCL are capable of invoking callbacks as notifications for events. The relevant functions are `cuStreamAddCallback` (CUDA) and `clSetEventCallback` (OpenCL). Both memory copy and code execution operations can be integrated in this way. Thanks to the use of coroutines, the code remains visually clear as if it was using blocking operations both for I/O and GPU calls, while actually retaining its asynchronous nature.

An example of how the presented solution influences code design is given below. Consider a scenario of receiving a fixed width data block from network into host memory that is mapped into GPU address space, processing it in place and sending the result back. The following code assumes different GPU APIs have already been wrapped for use with asio reactor and employs generic programming to take advantage of this fact.

```
template<typename Stream,typename Kernel>
void sync_code(Stream& stream,array_view view,Kernel& kernel)
{
    while(read(stream,view)==view.size()){
        kernel.invoke(view.data(),view.size());
        write(stream,view);
    }
}

template<typename Stream,typename Kernel,typename CompletionToken>
void async_code(Stream& stream,array_view view,Kernel& kernel,CompletionToken&& done)
{
    struct io_op
    {
        struct data
        {
            Stream& stream;
            array_view view;
            Kernel& kernel;
            CompletionToken&& done;
        }

        shared_ptr<data> d;

        io_op(Stream& stream,array_view view,Kernel& kernel,CompletionToken&& done)
            : d(make_shared<data>(
                stream,view,kernel,std::forward<CompletionToken>(done)))
        {
            write_done(0,{});
        }

        void write_done(size_t bytes_written,error_code ec)
        {
            async_read(d->stream,d->view,bind(&io_op::read_done,*this));
        }

        void read_done(size_t bytes_read,error_code ec)
        {
            if(bytes_read==d->view.size())
```

Journal of Physics: Conference Series **681** (2016) 012048 doi:10.1088/1742-6596/681/1/012048

```
                kernel.async_invoke(d->view.data(),d->view.size(),
                                    bind(&io_op::compute_done,*this));
            else
                std::forward<CompletionToken>(d->done)();
        }

        void compute_done(error_code ec)
        {
            async_write(d->stream,d->view,bind(&io_op::write_done,*this));
        }
    } op(stream,view,kernel,std::forward<CompletionToken>(done));
}

template<typename Stream,typename Kernel>
void async_coro_code(Stream& stream,array_view view,Kernel& kernel)
{
    while(async_read(stream,view,yield)==view.size()){
        kernel.async_invoke(view.data(),view.size(),yield);
        async_write(stream,view,yield);
    }
}
```

`sync_code` function represents the most simple blocking solution: it is concise, but consumes an OS thread for its execution. It returns when there is no more data to process.

`async_code` function shows the complexity of traditional asynchronous approach that requires splitting the logical loop into multiple intermediate operation completion handlers. These examples do not contain any error checking for clarity of presentation, but in real code its presence would complicate code even further. This function also contains shared ownership memory allocation to satisfy completion handler technical requirements. As is customary in asynchronous code, this function returns after initiating its operation and calls the provided continuation when its job is done.

The last function `async_coro_code` is the result of approach presented in this work. It mixes GPU operations with I/O on an arbitrary stream abstraction (network, disk, etc) and is as simple as the blocking code while retaining benefits of being asynchronous. `yield` is a global function object that suspends the coroutine when it is passed as continuation parameter using boost.coroutine. It also changes the invoked operation's semantics back to synchronous case, where it returns the result of the operation directly. The implementation of this object is based on the one provided by asio library, but our version adds additional features, such as support for coroutine-specific local storage, that extends the use of thread-specific local storage to coroutines.

Our experience shows that for most tasks overhead of coroutines against pure callback-based code following reactor pattern is insignificant, but even straightforward integration of foreign APIs involves additional overhead of system calls with unknown latency. Both APIs do not document the exact set of threads that user callbacks may be invoked on. Since we follow the executors model with strict guarantees for this kind of behavior, and given the restriction of operations that may be invoked in callbacks, we have to repost the notification to the proper thread from the callback. This involves at least one context switch and is another source of latencies.

3. Performance comparison

We have tested our implementation on two machines with different operating systems, both equipped with NVIDIA and AMD GPUs:

(i) Core i7-920 CPU, NVIDIA GTX 580 GPU, AMD Radeon HD 7570 GPU, Linux OS.

Journal of Physics: Conference Series **681** (2016) 012048 doi:10.1088/1742-6596/681/1/012048

Table 1. Performance results for Linux OS.

API	Variant	Time, μs Synchronous code	Coroutines
CUDA	spin/yield	24 ± 1	75 ± 10
	block	75 ± 15	75 ± 10
OpenCL	NVIDIA	26 ± 3	600 ± 400
	AMD	145 ± 60	145 ± 60
	AMD(poll)	85 ± 20	N/A

(ii) Core 2 Quad Q6600 CPU, NVIDIA GTX 460 CPU, AMD Radeon R7 265 GPU, Windows OS.

Our goal is to measure relative overhead of using coroutines with GPUs. In our tests we measure execution time for the common scenario of sending data to the GPU, performing computations there and sending results back to CPU. To make coroutine overhead more visible, we use the least possible amount of data (1 machine word) and trivial computation (single GPU thread with a single increment instruction) to process it. For this test GPU performance difference between all 4 GPUs is irrelevant, instead we compare performance of different APIs on different operating systems.

We've tested CUDA in different context modes that determine the way CPU waits for GPUs (see flags of `cuCtxCreate`). Spin mode corresponds to a busy loop waiting for completion. Yield mode also uses a wait loop, but it gives up its thread time slice on every iteration. In all our scenarios and different system loads we did not find any measurable difference between spin and yield options, so they are presented as a single entry. Block mode uses unspecified thread synchronization primitive to wait for GPU completion without consuming any CPU time.

First we consider results for Linux OS presented in table 1. For NVIDIA CUDA the overhead is about $50\mu s$ for the cases where no kernel synchronization is used. While huge for this test case, computations that are worth offloading to the GPU take orders of magnitude more time, so in real applications it won't contribute noticeably. For example, this consumes 0.3% of time alloted to a single frame of a real-time 60 frames per second visualization. When using blocking synchronization, the overhead is unmeasurable. Blocking synchronization is preferable for long-running computations, on low power devices and when trying to utilize all computation power of a system for high-performance computations.

The OpenCL implementation from NVIDIA is strangely a lot slower when coroutines are utilized. Even then, this should be acceptable for long-running kernels. We believe this to be a driver or runtime quality of implementation issue. OpenCL for AMD GPUs is slightly slower for the non-coroutine baseline, but the overhead is comparable to CUDA blocking case when coroutines are used. Unlike CUDA, OpenCL has no standard or other way to specify whether to use synchronization primitives, spin loops or other ways to detect GPU events. Since performance of AMD OpenCL implementation is more close to the blocking case, we also tested a manual spin loop implementation ("poll" entry in the table). Surprisingly, this allows for better latency, which seems to confirm that AMD implementation does use blocking. We consider it another case in support of OpenCL not fulfilling its role of the universal API.

Unfortunately, we've discovered incompatibility between implementation of CPU context switching in Boost.Context library on Windows and NVIDIA runtime. This issue is known and is a result of lack of documentation of Windows OS internals. In anticipation of such problems Boost.Context on Windows can use existing Fiber API to achieve compatibility at the price of

Journal of Physics: Conference Series **681** (2016) 012048 doi:10.1088/1742-6596/681/1/012048

Table 2. Performance results for Windows OS (preliminary).

API	Variant	Time, μs Synchronous code	Coroutines
CUDA	spin/yield	120 ± 20	250 ± 25
	block	230 ± 20	260 ± 25
OpenCL	NVIDIA	160 ± 20	$*$
	AMD	200 ± 30	205 ± 30
	AMD(poll)	155 ± 15	N/A

performance. While it allowed us to run tests, it doesn't work reliably, therefore we consider results for the Windows OS to be preliminary.

Most of the results follow the same pattern as on Linux. The AMD OpenCL implementation on Windows is slower than NVIDIA, but remains competitive. The main problem is the test of coroutines with NVIDIA OpenCL implementation, that is unacceptably slow (about 0.036 *seconds*, not included in the table). In some runs this test hanged and failed to complete at all. We believe this to be another incompatibility with the use of fibers. While we've run into several problems on Windows, we believe these can be overcome with future updates to drivers and libraries. When the C++ language evolves to include coroutine support in the standard with compiler-assisted optimizations, our solution can be adapted to use them, which will allow us to eschew complete CPU context switches and corresponding problems.

This shows that existing libraries and APIs already allow unifying of asynchronous programming in heterogeneous environments with acceptable overheads and good readability. Our results serve as a worst-case estimate that is already acceptable barring corner cases and will be improved in the future.

References

[1] NVIDIA CUDA
`http://www.nvidia.com/object/cuda_home_new.html`
[2] Khronos OpenCL
`https://www.khronos.org/opencl/`
[3] Knuth D 1997 *The Art of Computer Programming* 3rd ed, vol 1 (Amsterdam: Addison-Wesley) pp. 193-200
[4] Riegel T 2015 P0073R0: On unifying the coroutines and resumable functions proposals
`http://www.open-std.org/jtc1/sc22/wg21/docs/papers/2015/p0073r0.pdf`
[5] Kowalke O 2015 N4397: A low-level API for stackful coroutines
`http://www.open-std.org/jtc1/sc22/wg21/docs/papers/2015/n4397.pdf`
[6] Nishanov G and Vandevoorde D 2015 N4499: Draft wording for Coroutines (Revision 2)
`http://www.open-std.org/jtc1/sc22/wg21/docs/papers/2015/n4499.pdf`
[7] Kohlhoff C 2015 P0114R0: Resumable expressions (revision 1)
`http://www.open-std.org/jtc1/sc22/wg21/docs/papers/2015/p0114r0.pdf`
[8] Kegel D The C10K Problem
`http://www.kegel.com/c10k.html`
[9] Schmidt D et al. 2000 *Pattern-Oriented Software Architecture* vol 2 (New York: Wiley)
[10] Kohloff C 2015 P0113R0: Executors and Asynchronous Operations, Revision 2
`http://www.open-std.org/jtc1/sc22/wg21/docs/papers/2015/p0113r0.html`
[11] asio C++ library
`http://think-async.com`
[12] Boost.Context
`http://www.boost.org/doc/libs/develop/libs/context/doc/html/index.html`
[13] N4399: Working Draft, Technical Specification for C++ Extensions for Concurrency 2015 ed Laksberg A
`http://www.open-std.org/jtc1/sc22/wg21/docs/papers/2015/n4399.html`

Journal of Physics: Conference Series **681** (2016) 012049 doi:10.1088/1742-6596/681/1/012049

Floating-point performance of ARM cores and their efficiency in classical molecular dynamics

V Nikolskiy[1,2], V Stegailov[2,1]

[1]National Research University Higher School of Economics, 20 Myasnitskaya ulitsa, Moscow 101000, Russia

[2]Joint Institute for High Temperatures of Russian Academy of Sciences (JIHT RAS), Izhorskaya st. 13 Bd. 2, Moscow 125412, Russia

Abstract. Supercomputing of the exascale era is going to be inevitably limited by power efficiency. Nowadays different possible variants of CPU architectures are considered. Recently the development of ARM processors has come to the point when their floating point performance can be seriously considered for a range of scientific applications. In this work we present the analysis of the floating point performance of the latest ARM cores and their efficiency for the algorithms of classical molecular dynamics.

1. Introduction

The development of computer hardware of the last decades followed the Moore's law that the number of transistors in a dense integrated circuit doubled every one and a half or every two years. Nowadays the development of the circuit technology has nearly reached its physical limits and the further increase of performance is not based on the growth of integrated circuits, but on the supercomputers that combine huge numbers of processing elements. It is estimated that supercomputers of the exaflops era are going to consist of millions of computing elements. Energy efficiency is essential for creation of such systems. Priorities in hardware development are shifted from creating faster computing elements to creating more energy efficient elements that can serve as building blocks for large systems and to creating an appropriate interconnect. A typical example is the development of IBM Blue Gene systems, the first of which was based on a notoriously weak processor with a clock speed of only 700 MHz [1].

GPU accelerators are an important current hardware trend in high performance computing (HPC). Initially this technology was focused on the mass market (video games accelerators). An attractive "price-to-performance" ratio ensured the adaptation of such hardware to HPC.

Another similar trend is a mass use of ARM processors in small-scale computing machinery and especially in smartphones and tablets [2]. Market demands determine the growth of performance of ARM CPUs while maintaining the high level of energy efficiency. In the latest family of ARM processors it became possible to use a special module for floating point calculations that opened a way to HPC related tasks [3]. In this paper, we analyze the performance of several examples of ARM Cortex-A processors family both in the terms of peak and Linpack performance and in a test case of classical molecular dynamics (MD).

Nowadays the MD method has an important place among the high-performance computing applications [4]. For example, the projects connected with classical MD correspond to about

Journal of Physics: Conference Series **681** (2016) 012049 doi:10.1088/1742-6596/681/1/012049

1/5 of the total number of projects and of the allotted computational time in the Innovative and Novel Computational Impact on Theory and Experiment (INCITE) program of the US Department of Energy [1].

Along with the development of theoretical foundations of the MD method [5], there is a rapid growth of the number of its applications in modern multiscale models based on the description of processes at atomistic level in physics, chemistry, biology, materials science and other areas. Examples of new parallel programming approaches for MD appear constantly (e.g. [6]). However the modern record model size of trillions of particles [7] corresponds, for example, to only a few μm^3 of metal at normal density. The increase of the maximum MD simulation time is even a more complicated problem [8]. Expanding the limitations of MD methods is inextricably connected with advancement in HPC technology.

The plan of this paper is as follows. The second section describes the hardware and the software used to test the performance of ARM systems. In the third section we consider the Linpack-like performance using the RGbenchMM test and the peak performance of ARM cores. The fourth section brings together the results of a test MD problem for different CPU types, including our results for ARM Cortex-A5 CPU. In the fifth section we compare power consumption of Intel-based and ARM-based systems on the example of an MD test.

2. Hardware and software details

The ARM architecture is a microprocessor architecture with a reduced instructions set (Advanced RISC Machine). This architecture is subdivided into different types of cores. We consider the Cortex-A family and Cortex-A5, Cortex-A9 and Cortex-A15 cores. These names do not defined CPUs types strictly, but they are licensed patterns used by different vendors for (often customized) CPU designs (e.g., Scorpion and Krait cores). All types of cores considered in this work include the floating point module (VFPv3 or VFPv4).

The main results of the work are obtained with ODROID-C1. This is a portable minicomputer made by Hardkernel co. Ltd., which contains a quad core processor Amlogic S805 Cortex-A5 with VFPv4 modules on each core and a graphics accelerator Mali-450 MP2 that we do not use. The processor operates at 1.5 GHz with 1 GB of DDR3 SDRAM. The device has eMMC and microSD slots, 4 USB ports and 10/100/1000 Mbps Ethernet with a RJ-45 port. Power is connected via a special adapter. The monitor is connected via microHDMI port. We use ODROID-C1 either with Linux Ubuntu 14.05.1 LTS OS (odroid 3.10.67-55 #1 SMP) with the LXDE lightweight graphical environment, or with Android 4.4.2 OS. Several different smartphones running Android OS are used to test the performance of other types of ARM Cortex cores. Analysis of energy consumption of the minicomputer ODROID-C1 is carried out with a digital watt-meter “Smart Power” made by Hardkernel co. Ltd. The previously benchmarks and energy consumption tests are used for the analysis as well.

The tests of the performance of microprocessors have a long history of development [9]. The *de facto* standard unit for describing the performance in the field of scientific computing and mathematical modeling is the number of floating-point operations per second (flop per second or flops). The measurement of this property is usually associated with the Linpack test. However the ARM architecture was originally focused on integer operations, and the peak performance of ARM cores in terms of flops is not declared by vendors (however, as shown below, it can be estimated by specific tests).

LAMMPS [10] and the Lennard-Jones fluid model are used for CPU performance tests in classical MD algorithms (32K atoms at the density 0.8442 σ^{-3}, the potential cut-off radius 2.5 σ that gives about 55 neighbors per atom, 100 time steps with the NVE integration scheme).

The GCC compiler ver. 4.9 is used for the compilation of the LAMMPS under Linux OS. The `-mcpu-cortexa5 -mfpu=vfpv4-d16 -mfloat-abi=softfp` compiler options are used (performance differences from `-mfloat-abi=hard` are negligible). The serial compilation without

Journal of Physics: Conference Series **681** (2016) 012049 doi:10.1088/1742-6596/681/1/012049

multithreading is used for the execution on a single Cortex-A5 core.

Table 1. The results of the RGBenchMM test on different devices with ARM processors

ARM-core	R_{max}, MFlops/core	Flop/cycle	Source
Cortex-A5	284	0.2	ODROID-C1 (this work)
Cortex-A8	66	0.1	[11] (Google Nexus S)
Cortex-A9	372	0.3	Google Nexus 7
Cortex-A9	460, 614	0.4	Samsung Galaxy S II X, Note II
Cortex-A9	352 - 655	0.2 - 0.4	[11] (8 devices)
Scorpion	575, 588	0.4	Samsung Galaxy S Plus, S II
Scorpion	446 - 772	0.4 - 0.5	[11] (4 devices)
Krait	613 - 896	0.4 - 0.5	[11] (14 devices)
Krait 400	1073	0.5	Sony Xperia Z2
Krait 400	1038 - 1197	0.5	[11] (11 devices)
Cortex-A15	1125	0.7	Samsung Nexus 10
Cortex-A15	1164 - 1502	0.6 - 0.7	[11] (4 devices)

3. Estimation of peak performance for ARM cores

Unlike Intel, AMD and IBM, the developers and vendors of ARM processors do not publish the information about the peak performance in flops. According to recent studies [12] it takes 1-2 cycles for the Cortex-A9 core to perform one flop, while Cortex-A15 cores can make in 1 cycle all operations including fused multiply-add or fused multiply-accumulate (FMA) operations.

The RGBenchMM test application [13], based on the implementation of the DGEMM-type procedures with the possibility of parallelizing into 2 or 4 threads is available for Android OS. This application was written in C++ in the Android NDK and is devoid of the understated performance inherent to Java programs running Android OS. In particular, the executable code uses special FMA instructions in the key operations [14]. The data on the absolute performance of R_{max} in MFlops at a single core and on the relative performance in flop units per one processor cycle is presented in the table 1 (with respect to the CPU clock frequency).

The table shows that, unlike the data mentioned above [12] obtained for peak performance R_{peak} in specialized tests, the RGBenchMM test performance R_{max} of Cortex-A9 and Cortex-A15 cores is about half lower than R_{peak}. This is due to the fact that about one half of the cycles in the RGBenchMM test is used for data transfers. We see that these effects are even more pronounced for "weaker" cores like Cortex-A5 and A8. Based on these data, it seems reasonable to assume that the peak performance of the Cortex-A5 core equals 2-3 cycles per flop (taking into account the FMA operations) or 4-6 cycles per flop (excluding FMA).

4. Comparison of CPU cores using the LAMMPS MD test

From the perspective of HPC hardware end-users processors differ in peak performance R_{peak} measured in flops. Broadly speaking, the consumer expects a time-to-solution to be inversely proportional to R_{peak}:

$$t \sim \frac{1}{R_{peak}}. \tag{1}$$

In parallel computing the possibility of decreasing the calculation time is limited by many factors (algorithm parallelization, communication etc.) However, even for a sequential algorithm the

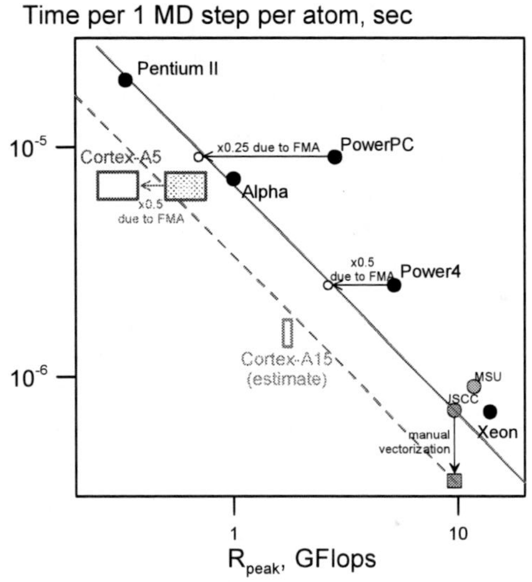

Figure 1. The comparison of LAMMPS performance for the Lennard-Jones liquid model on different CPUs.

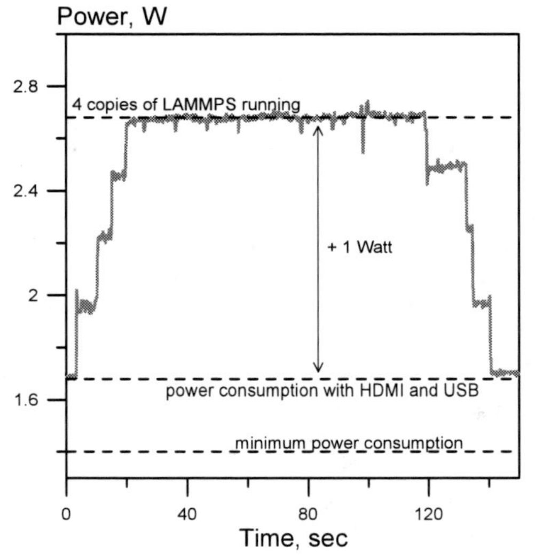

Figure 2. The ODROID-C1 power consumption when running 4 copies of LAMMPS with 4 identical MD tasks.

relation (1) may not hold for a particular task and different CPU types because of different data transfer rates and other factors (cache, compiler, operating system) as well as the use of a special kind of operations like FMA that make the peak performance value "task specific" [4].

A reasonable "common denominator" is needed for the comparison of different types of CPUs (as well as heterogeneous computing elements). This role can be naturally played by the total peak performance R_{peak} [4]. Figure 1 shows the times for 1 MD time step per 1 atom of the Lennard-Jones fluid model.

The benchmark results for one core (black circles) from the LAMMPS website [10] are shown for the Intel Pentium II Over-Drive 333 MHz, DEC Alpha 500 MHz, PowerPC 440 700 MHz, Power4 1.3 GHz and Intel Xeon 3.47 GHz processors. In this case, the calculation time is determined by the structure of the microprocessor core and the ability of the compiler to create an efficient executable code. All the points (except for a square) correspond to the same LAMMPS code written in C++, but to different compilers. The results for the Intel and DEC processors are in a good agreement (1). If we scale the peak performance of IBM Power CPUs to exclude the FMA operations that are not used in the classical MD algorithm then the IBM Power points move to the common dependence $6.84 \cdot 10^3$ Flop/R_{peak} pretty well (the solid line at figure 1).

The results of the tests [4] made on the supercomputer "Lomonosov" of the Lomonosov Moscow State University without vectorization of the code (a purple circle) and on the supercomputer MVS-10P of JSCC RAS without (an orange circle) and with (an orange square) vectorization of the code are presented. In this case LAMMPS is compiled by Intel C++ compiler with manual vectorization, which is oriented at the Intel Xeon architecture (the USER-INTEL module in LAMMPS). This kind of code optimization results in acceleration of calculations in 2 times corresponding to $3.36 \cdot 10^3$ Flop/R_{peak} (the dotted line in figure 1).

As mentioned above the ARM Cortex-A5 R_{peak} value is not declared by the manufacturer. Rectangles at figure 1 show the results of the LAMMPS test with one Cortex-A5 core. Two rectangles illustrate the uncertainty of its peak performance (either corrected for FMA operations or not), and the uncertainty of time values (the 4-core Amlogic S805 CPU with 1 copy of

Journal of Physics: Conference Series **681** (2016) 012049 doi:10.1088/1742-6596/681/1/012049

LAMMPS running gives 5.95 μs/atom/step and 7.8 μs/atom/step with 4 copies).

It is evident that the Cortex-A5 architecture together with the compiler provides a highly efficient use of hardware resources for floating-point operations. The calculation time values at the given peak performance follow the same trend as Intel Xeon CPUs with manual vectorization (the USER-INTEL module in LAMMPS). However, in the case of Cortex-A5 no manual tuning of the LAMMPS code is required!

The value of the Cortex-A15 peak performance is known better [12]. An estimate of the calculation time for the LAMMPS test considered and a system similar to Samsung Nexus 10 can be performed by rescaling the results for ODROID-C1 using the R_{max} from table 1. The figure 1 shows that we can expect a slightly higher performance for Cortex-A15 in comparison with IBM Power4.

Table 2. The energy consumption in the Lennard-Jones liquid test calculations (for 1 atom and 1 MD integration step) at the full load of 4 cores of a server/minicomputer. The benchmark data for Intel Xeon servers are used (for Ivy Bridge and Haswell).

System	**Power, W**	**Time per atom per step, μs**	**Energy per atom per step, μJ**
Ivy Bridge Server	316 [15]	0.053	16.8
Ivy Bridge Server	451.8 [16]	0.029	13.1
Haswell Server	462.9 [16]	0.027	12.5
ODROID-C1	2.4	1.75	4.2
Equivalent of Samsung Nexus 10	$\sim$ 8 [17]	0.4	3.2

5. Comparison of energy efficiency

Comparison of different computer systems by energy efficiency is a significant challenge. Usually the power consumption of a processor itself is under consideration. It seems to be more reasonable to consider a power consumption of an entire system (a server, a minicomputer, a smartphone). In this work we compare the energy consumption of several servers under conditions on the full CPU load by the Lennard-Jones liquid MD test.

Figure 2 displays power consumption changes during sequential executions of 4 independent MD test calculations. Results show that the ODROID-C1 power consumption at the full load can be estimated as 2.4 W (subtracting the power consumed by microHDMI and USB ports).

According to the data of Abdurachmanov et al. [17] it is possible to determine the power consumption at the full load of the ODROID XU+E minicomputer based on four Cortex-A15 cores (and probably overestimated due to the presence of idle Cortex-A7 cores on the same CPU chip). At the same time there are the data on the energy consumption of the servers based on Intel Xeon processors with the Ivy Bridge and Haswell architectures [15, 16]. It allows us to estimate the amount of energy required to calculate 1 MD step for 1 atom for the MD test considered (table 2). The calculation times are estimated using a universal dependence (figure 1) of the peak performance without consideration of FMA operations (128 GFlops for E5-2650 v2, 240 GFlops for E5-2690 v2 and 249.6 GFlops for E5-2690 v3). The calculation time for ODROID-C1 with the use of all 4 cores is 7 μs (in correspondence to figure 1). Table 2 shows an estimate for a system with a quad-core Cortex-A15 processor as well.

We see that the calculations of MD tests on systems with ARM Cortex-A processors require approximately 5 times less energy than the calculations on the latest models of Intel processors.

6. Conclusions

We have analyzed the results of the RGBenchMM test (similar to Linpack) for a variety of systems based on ARM Cortex-A CPUs including smartphones and the minicomputer ODROID-C1. Floating point peak performance of Cortex-A5 cores in ODROID-C1 has been estimated.

We have compiled the LAMMPS for ODROID-C1 and performed classical MD benchmarks. We have used the metric "calculation time vs. peak performance". By comparison with other CPUs we have shown the efficient use of floating point hardware capabilities of the Cortex-A architecture in combination with the GCC 4.9 compiler. The Cortex-A5 architecture provides 2 times faster calculations with LAMMPS than others CPU architectures at the same level of peak performance and without the need for manual tuning of the code.

The power consumption of ODROID-C1 has been measured. ARM Cortex-A cores have been found to be 3-5 times more energy efficient for the MD benchmark considered than Intel Xeon Ivy Bridge and Haswell cores.

Acknowledgments

The research (of Stegailov) was supported by the Basic Research Program at the National Research University HSE, Russia in 2015.

References

[1] Stegailov V V and Norman G E 2014 *Program Systems: Theory and Applications* **5** 111 – 152 ISSN 2079-3316 URL `http://psta.psiras.ru/read/psta2014_1_111-152.pdf`
[2] An interview with Steve Furber URL `http://cacm.acm.org/magazines/2011/5/107684-an-interview-with-steve-furber/fulltext`
[3] Mitra G, Johnston B, Rendell A, McCreath E and Zhou J 2013 Use of simd vector operations to accelerate application code performance on low-powered arm and intel platforms *Parallel and Distributed Processing Symposium Workshops PhD Forum (IPDPSW), 2013 IEEE 27th International* pp 1107–1116
[4] Stegailov V V, Orekhov N D and Smirnov G S 2015 Hpc hardware efficiency for quantum and classical molecular dynamics *Parallel Computing Technologies* (*Lecture Notes in Computer Science* vol 9251) ed Malyshkin V (Springer International Publishing) pp 469–473 ISBN 978-3-319-21908-0 URL `http://dx.doi.org/10.1007/978-3-319-21909-7_45`
[5] Norman G E and Stegailov V V 2013 *Mathematical Models and Computer Simulations* **5** 305–333 ISSN 2070-0482 URL `http://dx.doi.org/10.1134/S2070048213040108`
[6] Podryga V O and Polyakov S V 2015 *Mathematical Models and Computer Simulations* **7** 456–466 ISSN 2070-0482 URL `http://dx.doi.org/10.1134/S2070048215050105`
[7] Eckhardt W, Heinecke A, Bader R, Brehm M, Hammer N, Huber H, Kleinhenz H G, Vrabec J, Hasse H, Horsch M, Bernreuther M, Glass C, Niethammer C, Bode A and Bungartz H J 2013 591 TFLOPS multi-trillion particles simulation on SuperMUC *Supercomputing* (*Lecture Notes in Computer Science* vol 7905) ed Kunkel J, Ludwig T and Meuer H (Springer Berlin Heidelberg) pp 1–12 ISBN 978-3-642-38749-4
[8] Piana S, Klepeis J L and Shaw D E 2014 *Current Opinion in Structural Biology* **24** 98 – 105 ISSN 0959-440X
[9] Curnow H J and Wichmann B A 1976 *The Computer Journal* **19** 43 – 49
[10] LAMMPS web site URL `http://lammps.sandia.org`
[11] Swedroid — Nordens största Android-community URL `http://www.swedroid.se`
[12] Garg R Exploring the floating point performance of modern arm processors URL `http://www.anandtech.com/show/6971/exploring-the-floating-point-performance-of-modern-arm-processors`
[13] Garg R Rgbenchmm - android apps on google play URL `https://play.google.com/store/apps/details?id=org.codedivine.rgbench&hl=en`
[14] Garg R Prelim analysis of RGBenchMM URL `http://codedivine.org/2012/09/25/prelim-analysis-rgbenchmm/`
[15] Supermicro — Ivy bridge based DCO superserver power & cost savings URL `http://www.supermicro.com/white_paper/white_paper_Ivy-Bridge-Power.pdf`
[16] Intel Xeon E5-2600 v3 review: Haswell-EP redefines fast URL `http://www.tomshardware.com/reviews/intel-xeon-e5-2600-v3-haswell-ep,3932-9.html`
[17] Abdurachmanov D, Arya K, Bendavid J, Boccali T, Cooperman G, Dotti A, Elmer P, Eulisse G, Giacomini F, Jones C D, Manzali M and Muzaffar S 2014 *Journal of Physics: Conference Series* **513** 052008 URL `http://stacks.iop.org/1742-6596/513/i=5/a=052008`

Improving the efficiency of solving discrete optimization problems: The case of VRP

A Belov and S Slastnikov

National Research University "Higher School of Economics", Moscow, Russia

E-mail: sslastnikov@gmail.com

Abstract. Paper is devoted constructing efficient metaheuristics algorithms for discrete optimization problems. Particularly, we consider vehicle routing problem applying original ant colony optimization method to solve it. Besides, some parts of algorithm are separated for parallel computing. Some experimental results are performed to compare the efficiency of these methods.

1. Introduction

Optimization algorithms are known to be useful in many types of problems in mathematics, physics, chemistry, biology, statistics and so on. Particularly, many physical systems are governed by minimization principles. For example, in thermodynamics, a system coupled to a heat bath always takes the state with minimal free energy. Besides, some other applications are well known: determination of the self affine properties of polymers in random media, solution of the protein folding problem, analysis of X-ray data, optimization of lasers and optical fibers, etc. [1].

In general discrete optimization is a branch of mathematical programming, which includes the extremal problems with variables defined on discrete sets. Almost all discrete optimization problems have application in practice. A large number of discrete optimization problems in one form or another can be reduced to a vehicle routing problem. Therefore, the algorithms and principles for solving vehicle routing problems are widely used for solving other discrete optimization problems.

Consider vehicle routing problem with capacity constraints (CVRP). Given a graph $G = (V, A, d)$, where $V =\{ v_0, v_1,..,v_n\}$ is the set of vertices (v_0 is a depot and other vertices are clients). A is a set of edges connecting vertices of the graph G. For each arc (i,j) a non-negative distance d_{ij} between clients v_i and v_j is given . Positive demand c_i is set for client i, capacity of each of the m vehicles is bounded by C_k $(k=1,...,m)$. Also the following restrictions are given:

- each client should be visited exactly once;
- the depot is the start and the end point of all the routes.

The aim of the task is the construction of route with minimal total distance which satisfies the demands of all customers and does not violate the restrictions described above.

In section 2 mathematical setting of CVRP is performed. Section 3 is devoted to the original modification of ant colony optimization algorithm for CVRP. The algorithm parallelization scheme and some computational results of its application to the model tasks are presented in section 4. Finally, in section 5 these results are discussed.

Journal of Physics: Conference Series **681** (2016) 012050 doi:10.1088/1742-6596/681/1/012050

2. Mathematical setting

The mathematical formalization of CVRP is the following: to minimize the objective function

$$F = \sum_{k=1}^{m} \sum_{i=0}^{n} \sum_{j=0}^{n} d_{ij} X_{ij}^{k} \to \min \tag{1}$$

under constraints

$$\sum_{i=1}^{n} c_i \sum_{j=0}^{n} X_{ij}^{k} \le C_k, \quad \forall k = 1,...,m, \tag{2}$$

$$\sum_{j=1}^{n} X_{0j}^{k} \le 1, \quad \forall k = 1,...,m, \tag{3}$$

$$\sum_{i=1}^{n} X_{i0}^{k} \le 1, \quad \forall k = 1,...,m, \tag{4}$$

$$\sum_{k=1}^{m} \sum_{j=0}^{n} X_{ij}^{k} = 1, \quad \forall i = 1,...,n, \tag{5}$$

$$\sum_{i=0}^{n} X_{ih}^{k} - \sum_{j=0}^{n} X_{hj}^{k} = 0, \quad \forall h = 1,...,n; \ \forall k = 1,...,m, \tag{6}$$

$$X_{ij}^{k} \in \{0,1\}, \quad \forall i = 0,...n; \ j = 0,...,n; \ \forall k = 1,...,m, \tag{7}$$

$$X_{ii}^{k} = 0 \quad \forall i = 0,...,n; \ \forall k = 1,...,m, \tag{8}$$

if $S_k = \{(i,j) : X_{ij}^{k} = 1\}$, then

$$\forall k = 1,...,m \quad \forall (i,j) \in S_k \neq \varnothing \ \exists (j_p, j_{p+1}) \in S_k \ (p = 0,...,l): \ j_0 = j_{l+1} = 0, \ (i,j) = (j_r, j_{r+1}), \ r \le l. \tag{9}$$

X_{ij}^{k} equals 1 if the vehicle k follows the customer i to customer j, and 0 otherwise.

The objective function (1) defines the total distance value passed by all vehicles. Conditions (2)–(9) specify the problem constraints and are discussed by many authors (see, for example, [2]).

CVRP is NP-hard problem, so only metaheuristics are suitable to solve it for practical applications. In the next section we consider an effective optimization algorithm based on the modification of ant colony method.

3. Modified ant colony optimization method

Ant colony optimization (ACO) is a swarm intelligence method for solving different hard discrete optimization problems. The approach is to use the model of food searching in colonies of ants that mark the path by a special chemical substance called pheromone. Left traces attract other ants, which, passing along marked paths, in turn enhances the smell of the pheromone. Over the time pheromone evaporates, so the ants could adapt their behavior to the external environment changes. Classical ACO for CVRP could be found in papers [2, 3].

Here we consider the following modification of ACO for CVRP. Initially, each vehicle k starts from the depot, and the set M_k of customers, included in its route, is empty. The next customer j to visit is chosen by the probabilistic criterion:

$$j = \begin{cases} \arg\max\limits_{u \notin M_k} [\tau_{iu} (d_{iu})^{-\beta}], & \text{with probability } p_0, \\ S, & \text{with probability } 1 - p_0, \end{cases} \tag{10}$$

where i is the current client (or depot on the first step); τ_{iu} is pheromone amount on the route between clients i and u; β is the parameter establishing the importance of distance in comparison to pheromone quantity; p_0 is the parameter. S is a discrete random variable with probability distribution $p_k(i,j)$, where $p_k(i,j)$ is the probability for the vehicle k to move from client i to client j. First part of formula (10) we call the "determined rule" for client choosing and the second – "stochastic rule". On the first step (iteration)

$$p_k(i,j) = \begin{cases} \tau_{ij} \dfrac{\tau_{iu}}{d_{iu}^{\ \beta}} \Big/ \sum\limits_{v \notin M_k} \tau_{iv} \dfrac{\tau_{iu}}{d_{iu}^{\ \beta}}, & \text{if } j \notin M_k, \\ 0, & \text{if } j \in M_k, \end{cases} \tag{11}$$

and on the next steps is proposed to remember the total distances of best and worst routes of one vehicle (L and R, respectively). In the case, when moving from customer i to customer j is contained in the worst route, then the probability

$$p_k(i,j) = \frac{L}{R} \cdot \frac{\tau_{ij}}{d_{ij}^{\ \beta}} \Bigg/ \left(\sum_{u \neq j, u \notin M_k} \frac{\tau_{iu}}{d_{iu}^{\ \beta}} + \frac{L}{R} \cdot \frac{\tau_{ij}}{d_{ij}^{\ \beta}} \right) \tag{12}$$

and in other cases

$$p_k(i,j) = \begin{cases} \dfrac{\tau_{ij}}{d_{ij}^{\ \beta}} \Bigg/ \left(\sum\limits_{u \neq j, u \notin M_k} \dfrac{\tau_{iu}}{d_{iu}^{\ \beta}} + \dfrac{L}{R} \cdot \dfrac{\tau_{ij}}{d_{ij}^{\ \beta}} \right), & \text{if } j \notin M_k, \\ 0, & \text{if } j \in M_k. \end{cases} \tag{13}$$

Route for vehicle is completed when its load capacity is over or all clients are visited, then vehicle returns to the depot. Thus routes for all vehicles are consecutively constructed.

In order to improve future solutions, the pheromone trails should be updated to reflect the quality of the solutions found. Local pheromone update models its natural evaporation and ensures that no route becomes too prevalent. This update is made after route for current vehicle is done:

$$\tau_{ij}^{\ new} = (1-\alpha)\tau_{ij}^{\ old} + \alpha\tau_0, \tag{14}$$

where α is a parameter that controls the speed of evaporation, and τ_0 is the initial pheromone value. We propose to calculate the initial pheromone value as follows:

$$\tau_0 = \frac{1}{(n+1) \cdot \min\limits_{i \neq j} d_{ij}}. \tag{15}$$

Global trail updating is performed by adding pheromone to all of the arcs included in the best route of one vehicle and encourages the use of shorter routes in future solutions:

$$\tau_{ij}^{\ new} = (1-\alpha)\tau_{ij}^{\ old} + \frac{\alpha}{L}. \tag{16}$$

This process is repeated until the stopping criterion occurs. The best solution over all iterations is a 'good' approximation of the optimal objective function value.

The efficiency of this modified ACO algorithm was studied and discussed in [2].

4. Algorithm parallelization

In order to reduce the ACO execution time we divided some parts of algorithm into several parallel processes for calculation. On the block diagram presented on the Figure 1 these parts are highlighted by dashed lines.

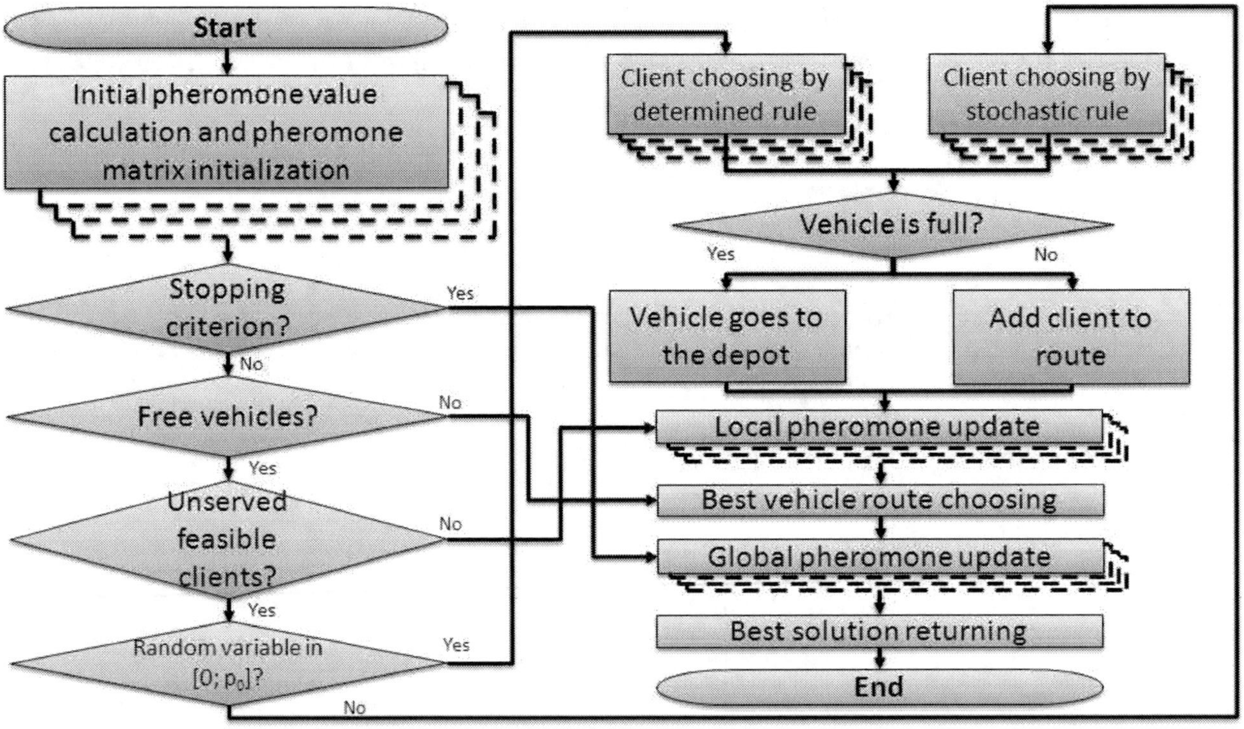

Figure 1. Block diagram of ACO algorithm with parallelization.

The modified ACO algorithms with and without parallel sections were implemented in C++ as dynamic shared object library. The library was written using the precise calculations library GNU Multi-Precision. Therefore, it becomes possible to reduce computing inaccuracies to a minimum without the computation time increasing.

As tools of parallel programming we used *OpenMP* and *BoostThread.* While the main advantage of *OpenMP* is in easy way of development and using, the main lack is in the absence of general support for working with arbitrary containers and custom data types. *BoostThread* is part of libraries designed specifically for the C++ language and provides opportunities for threads creation, management and synchronization. However, from the practical implementation side *BoostThread* has more complicated structure compared with *OpenMP* and is more complex in general.

Table 1 presents the solution results of CVRP by proposed modified ACO algorithm without the use of parallel computing and using the proposed scheme of parallelization. The experiments were conducted on the set of model tasks (benchmarks) from [4] with dimensions (number of clients) from 22 to 101. The table shows the solution time (in seconds) for some tasks using both methods. Improvements are presented both in absolute values (seconds) and relative values (percentage). The least relative improvement was obtained for the problem with 51 clients and amounted to 4.4%. The best result was obtained for the task with 101 clients, which gives relative improvement in 13.7%. Uneven improvements can be justified by the stochastic nature of the considered algorithm.

Table 1. Efficiency of parallel calculations in modified ACO algorithm.

Client amount	Execution time without parallel calculations, *seconds*	Execution time with parallel calculations, *seconds*	Absolute improvement, *seconds*	Relative improvement, %

22	5.4	5	0.4	7.4
30	20.4	19	1.4	6.9
51	118.6	113.4	5.2	4.4
76	337	317.8	19.2	5.7
101	770	664.6	105.4	13.7

5. Discussion and conclusions

First of all, note that the research results described in this paper are used for developing Transportation Management System for Russian Oil Company "Rosneft". After implementation of proposed algorithms the efficiency of transport logistics increased by 10% and decision making time for vehicle routing decreased by 20%.

When we begin to carry out computational experiments with parallel calculations of the algorithm we expected to get better time reducing results, but didn't reach the major improvement in execution time. Analyzing the reasons of such minor enhancement in time we can made the following conclusions.

The main shortcoming of an implementation of parallel calculations into ACO algorithm is the following. The biggest gain in time can be obtained by paralleling directly the process of the route building for each vehicle, i.e. while client choosing by determined and stochastic rules. But the arising problem is that the set of available clients to choose should be the same for all parallel processes. Otherwise there could be cases when the same client is chosen for several routes. Thus, if we follow this way of parallelization, it is necessary to block the set of available clients until the current process finishes client selection.

So the authors' future plans are to develop ACO method parallelization scheme to reduce execution time for big dimensions tasks.

References

[1] Hartmann A K and Rieger H 2002 *Optimization algorithms in physics* (Berlin: Wiley-VCH Verlag Berlin GmbH)
[2] Belov A and Slastnikov S 2014 A Metaheuristic Approach for the Problem of Motor Fuel Distribution *Proc. Comp. Sc.* **31** pp 143–150
[3] Bell J and McMullen P 2004 Ant Colony Optimization Techniques for the Vehicle Routing Problem *Adv. Eng. Inform.* **18** pp 41–48
[4] Augerat P *et al* 1995 Computational results with a branch and cut code for the capacitated vehicle routing problem *Technical report RR 949-M* (Grenoble: University Joseph Fourier)

New security infrastructure model for distributed computing systems

J Dubenskaya[1], A Kryukov[1], A Demichev[1] and N Prikhodko[2]

[1]Skobeltsyn Institute of Nuclear Physics, Lomonosov State University, Moscow, Russia
[2]Yaroslav-the-Wise Novgorod State University, Veliky Novgorod, Russia

E-mail: dubenskaya@theory.sinp.msu.ru

Abstract. At the paper we propose a new approach to setting up a user-friendly and yet secure authentication and authorization procedure in a distributed computing system. The security concept of the most heterogeneous distributed computing systems is based on the public key infrastructure along with proxy certificates which are used for rights delegation. In practice a contradiction between the limited lifetime of the proxy certificates and the unpredictable time of the request processing is a big issue for the end users of the system. We propose to use unlimited in time hashes which are individual for each request instead of proxy certificate. Our approach allows to avoid using of the proxy certificates. Thus the security infrastructure of distributed computing system becomes easier for development, support and use.

1. Introduction

Distributed computing systems (DCSs) are widely used by the researchers to solve different computational problems in various fields of natural science. DCS's are especially popular in the computer simulation in physics. Perhaps one of the most significant examples of DCS is the European GRID infrastructure (EGI) [1], including the Worldwide LHC Computing GRID (WLCG) [2], which is used for processing and simulation of experimental data from the Large Hadron Collider (LHC). The brilliant result of the LHC experiments is the discovery of Higgs boson, which could not be reached without the WLCG.

For the purpose of this paper we will consider a DCS as a set of interacting Web services which send requests to each other. Access points for such a system are user Web interfaces or command line interfaces installed on users' personal computers.

One of the most important parts of the DCS is a security infrastructure that provides: authentication and authorization, data integrity, encryption and so on. Information security is particularly important in such areas of science and technology as medicine, biological research, and engineering development. On the one hand, researchers need to be sure that the results of data processing and simulation are protected from unauthorized access; on the other hand, owners of the computational resources that

Journal of Physics: Conference Series **681** (2016) 012051 doi:10.1088/1742-6596/681/1/012051

make up the DCS want to have guarantees that only the authorized users will be able to submit computational requests to the system.

The security infrastructure for the DCS is to provide users with a comfortable and secure access to the remote resources. Currently in most DCSs security is based on the public key infrastructure (PKI) [3] in conjunction with proxy certificates [4]. Proxy certificate is a special short time living certificate used for the purpose of providing restricted rights delegation within a PKI based authentication system. The short lifetime of the proxy certificate is due to security reasons. In DCSs proxies are used in several cases:

- To grant the rights between users and services to access to computing resources.
- While processing a composite computational task. A composite task is a regular task that needs to be processed by several services successively (i.e. the second service receives requests from the first service directly). In this case rights delegation between services takes place.

Along with the incontestable benefits of strong security the mentioned approach has also some disadvantages. First of all it is about the usability issues. The fact is that requesting and management of the X.509 certificates and proxies requires deep understanding of the basic concepts of the PKI that not all users have. The second problem with proxies is a short lifetime, while one cannot predict how much time would take request processing. There are special services to support prolongation of proxy lifetime [5], and all this make the security infrastructure overcomplicated and difficult to interact with. For example. the security infrastructure of EGI/WLCG is shown on figure 1.

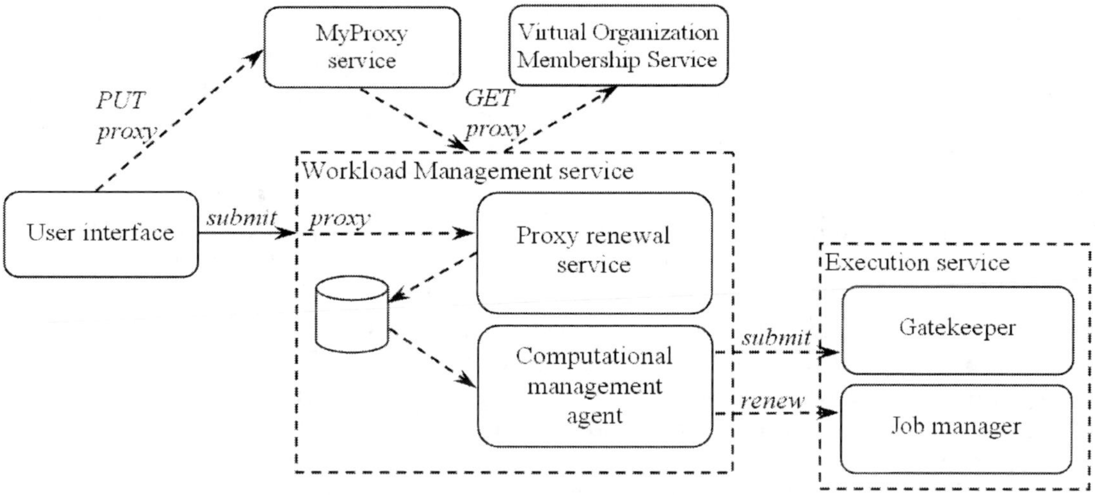

Figure 1. The security infrastructure of EGI/WLCG with proxy renewal procedure.

In this paper we propose an alternative approach to building of the security infrastructure for the DCSs with no use of the proxy certificates with short lifetime. The main idea is to generate a separate individual hash for each request and to register the request along with the corresponding hash in the new specially introduced infrastructure entity called 'validation service'. This service should be trusted by the other services of the DCS and should act on-line. Upon getting computational request each service checks its hash against the validation service and continues only if the hash is correct and is not yet used. Unlike proxy certificate, such a hash is not limited in time. This feature is not a big issue as the unlimited lifetime of the hash is compensated by the fact that it can be used only once and only for a specific request.

In the next section we present the detailed description of the proposed approach including step-by-step authentication, authorization, and delegation algorithms. In Discussion we consider possible weak points of the proposed security infrastructure and some solutions recommended to avoid them. In Conclusion the advantages of the proposed approach are analyzed in brief.

2. Proposed approach to security infrastructure building

Providing of the user-friendly, intuitive and yet secure interface can attract to the DCS lots of new users who are not experts in the field of computer science. To improve usability and facilitate access to the DCS we propose an alternative approach to building of the security infrastructure for the DCSs.

The main idea is in using login/password for user authentication while well tested PKI-based solutions are used for service-to-service interaction only. The authorization is implemented by using just in time approving of the rights via a special trusted validation service. The consequent security infrastructure is shown in figure 2.

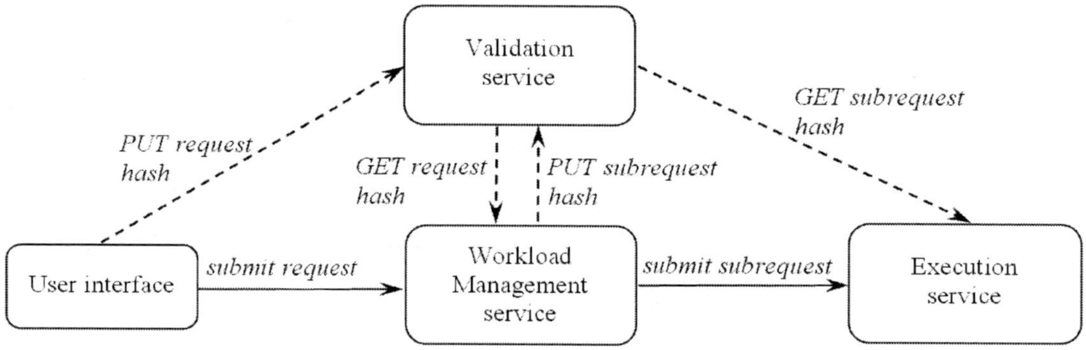

Figure 2. The proposed security infrastructure with validation service.

The main points of the proposed authentication and authorization algorithm are listed as follows:

- On the first request a user have to enter a valid pair of login and password. For strong security the multi-factor authentication can be implemented.
- After successful login a user obtains a special session key for further work with no need to enter login/password pair over and over again.
- Each next user request is implicitly supplied by the previously obtained session key that is used for password-free access to the validation service.
- Each request should be protected by the individual hash which is not limited in time. This request hash is generated on the user's side for every single computational request with respect to request generation time. Due to this approach the hashes of the two completely identical successively generated requests would be different.
- These request hashes are registered by the validation service in the special database and states of these hashes are tracked on real time.
- Having received a computational request every computational service checks against the validation service if the request hash is valid, correct and is not yet used, and if the user is authorized to pass the request, and if the validation service returns OK the service executes the request. This approach ensures impossibility of request changing during its passing and processing.

Rights delegation between computational services is also implemented via request hash in the following manner:

- The first service generates a new subrequest from the initial request.
- For the new subrequest a subrequest hash is generated.
- Then the first service registers the subrequest hash in the validation service (the same as for

Journal of Physics: Conference Series **681** (2016) 012051 doi:10.1088/1742-6596/681/1/012051

user's request).

- The first service passes the subrequest to the second service for processing. The second service examines the received subrequest hash via validation service.
- If the validation service says OK the second service continues processing.
- For a composite computational task a validation service collects a chain of the hashes which allow user to monitor the processing of the composite task and the status of each operation.

Thanks to this approach a user has even no need to know what a X.509 certificate and/or proxy certificate is. Session keys and request hashes are generated implicitly and are hidden from the user as well as details of the service interaction. Thus both high level of security along with seamless and easy access to the computational resources for the users are achieved thanks to intentional complication of the security infrastructure by adding a special validation service. So users are provided with seamless and easy access to the resources of the DCS and can concentrate on their computational needs.

3. Discussion

One of the possible weak points of the proposed approach is the requirement to have on-line access to the validation service for all other services of the DCS. The simulation using our prototype shows that such an infrastructure is quite stable and works fine at least for the systems with 20 user requests per second. For the critical high-availability systems it is possible to introduce two parallel validation services with on-line database replication. At this case one of the services acts as a master service that processes requests and another is a slave (an inactive full copy of the master). If the master service crashes it would be easy to switch to the slave service immediately with almost no loss of information.

An important benefit of the proposed security infrastructure is that during request processing the validation service collects all the information concerning each request in the DCS. This information can be used for monitoring purposes as well as for request revocation at any stage of processing.

4. Conclusion

The proposed approach improves user's operational performance and greatly enhances the competitive advantage for research organizations using DCSs. Scientists and engineers, as well as teachers and students can accelerate the practical results through the use of a simplified remote access to computational resources for data processing and modeling of processes in various fields of science and technology.

At the moment a prototype of the proposed security infrastructure has been developed. It includes one validation service and three other services that interact with users and with each other. The prototype is used to demonstrate the capabilities of the model and for heavy load tests.

Acknowledgments

This work is supported by the Ministry of Science and Education of Russian Federation, agreement No.14.604.21.0146.

References

[1] Kranzlmueller D, Lucas de J M, and Oster P 2010 The European Grid Initiative (EGI) *Remote Instrumentation and Virtual Laboratories* eds. Davoli F, Meyer N, Pugliese R, and Zappatore S (Boston Springer US) 61-66.

[2] Sciaba A, Andreeva J, Campana S, Donno F, Litmaath M, Magini N, Moscicki J T, and Renshall H 2010 Computing at the Petabyte scale with the WLCG. Worldwide LHC Computing Grid Tech. Rep. *CERN-IT-Note-2010-006* (Geneva CERN).

[3] Buchmann J A, Karatsiolis E, and Wiesmaier A 2013 *Introduction to Public Key Infrastructures* (Springer-Verlag Berlin Heidelberg).

[4] Tuecke S, Welch V, Engert D, Pearlman L, and Thompson M 2004 Internet X.509 Public Key Infrastructure Proxy Certificate Profile Tech. Rep. *RFC 3820.*

Journal of Physics: Conference Series **681** (2016) 012051 doi:10.1088/1742-6596/681/1/012051

[5] Kouril D and Basney J 2005 A credential renewal service for long-running jobs *IEEE/ACM International Workshop on Grid Computing* **13-14** 2-13.

A Web Tool for Research in Nonlinear Optics

Nikolay V Prikhod'ko[1], **Viktor A Abramovsky**[1], **Natalia V Abramovskaya**[1], **Andrey P Demichev**[1;2], **Alexandr P Kryukov**[1;2] **and Stanislav P Polyakov**[1]

[1] Yaroslav-the-Wise Novgorod State University, Velikiy Novgorod, Russia
[2] Skobeltsyn Institute of Nuclear Physics, Lomonosov Moscow State University

E-mail: `nikolai.prikhodko@novsu.ru`, `victor.abramovsky@novsu.ru`, `natalya.radchenko@novsu.ru`, `demichev@theory.sinp.msu.ru`, `kryukov@theory.sinp.msu.ru`, `s.p.polyakov@gmail.com`

Abstract. This paper presents a project of developing the web platform called WebNLO for computer modeling of nonlinear optics phenomena. We discuss a general scheme of the platform and a model for interaction between the platform modules. The platform is built as a set of interacting RESTful web services (SaaS approach). Users can interact with the platform through a web browser or command line interface. Such a resource has no analogues in the field of nonlinear optics and will be created for the first time therefore allowing researchers to access high-performance computing resources that will significantly reduce the cost of the research and development process.

1. Introduction

Successful development of optical devices requires complicated modeling of physical processes that occur in components of the devices and comprehensive database for optical properties of used materials. In many cases researchers need to run a large number of similar computing tasks to solve some problem in the field of nonlinear optics (NLO). Quite often, these tasks are performed by using already existing specialized application packages, the tasks being identical in form and differing only in the values of the input data (parameters). In this case, pre-installation and configuration of commonly used application packages on a computing resource and the availability of specialized web services as well as appropriate user web interfaces (SaaS model) allow researchers to increase efficiency of use of computing resources. The set of specialized web services and web application interfaces is called web platform for remote access to computing resources. Each individual pre-arranged task (a launch of an application package, an access to a data storage, and so on) is called "a tool". With the access to the tools provided by a platform, a user only has to specify particular values of the input parameters or input files, and the rest of the task description is generated automatically. Thus the user can carry out the generating of the task, its submission, monitoring, and getting the results via a standard web browser.

This approach proves to be particularly effective and convenient when the full research requires a set of application packages rather than just one of them. Moreover, often the output of some package is the basis for the generating of the task for the next application package in a series of stages of the research. Such series of tasks are called workflows. Besides the launch of computing packages, workflow may include other operations such as a query to relational database, visualization of the results or obtaining online data from an operating scientific facility. Carrying out of these operations can be made as simple as possible for end users by pre-configuring all the technical details (such as the Internet address of the resources, the specific formats of the query to a particular database or facility); in other words, creating a complete "tool". The user then only needs to formulate the essence of a specific request in a natural language (more details on the use of the web platforms in scientific research see, e.g. in [1]).

At present, there are many sources containing various mathematical models and large amount of data on the properties of nonlinear optical crystals in the form of a software [2], [3], [4]. Currently, however, there are no a single web based tool (web platform) which allows access to data and computational resources required for research and engineering tasks in nonlinear optics.

We present an on-line environment, namely the web platform WebNLO, intended for supporting research in nonlinear optics. The platform provides users with an access to a database for optical properties of crystal materials and to computational modules installed on remote resources which implement algorithms required for development of scientific models and devices in nonlinear optics.

2. WebNLO general scheme and design principles

The general scheme of the proposed web platform and interactions of its modules with each other are depicted in figure 1. Since the visible to user frontend to SaaS application cannot directly access internal systems such as databases and computation modules, the components of a SaaS application enable integration protocols and application programming interfaces (APIs) that operate over a wide area network. Typically these are protocols based on HTTP, SOAP, XML-RPC, REST. But the modern implementations are most commonly based on the REST architectural style combined with JavaScript Object Notation (JSON) as a message format. In most cases REST/JSON combination provides flexible protocol for message passing which is lightweight in terms of memory and processing power.

In this model everything is presented as a resource. All operations with these resources are strictly restricted and standardized. An essential advantage of RESTful-services is that the HTTP protocol is used directly as the message exchange protocol while in the other approaches HTTP is used only as a transport protocol for the SOAP serving as the messages exchange protocol. Moreover the REST/JSON has native support in all the modern web browsers (which usually serves as a client to SaaS applications). These properties make the use of the REST architectural style is very attractive for the development of web platforms. Further details and advantages of the REST architectural style can be found in [5],[6].

Since REST/JSON message passing scheme is quite simple, there are many popular frameworks which can be use for the platform construction, in particular, Django, Pyramid, Zope, Flask, etc. However since some services wrap an access to a database into REST/JSON, such a microframework like Flask should be used to implement REST interface for platform services. The microframework has minimal functionality, but it is sufficient for the operation of the WebNLO platform.

Considering that the main objective of the WebNLO project is to provide access to data and computational resources for researchers in nonlinear optics, we can identify the following required web platform modules (see figure 1):

- user management, authentication and authorization module;
- module of access to the NLO database;
- task management module;
- job execution module;

Journal of Physics: Conference Series **681** (2016) 012052 doi:10.1088/1742-6596/681/1/012052

- visualization module.

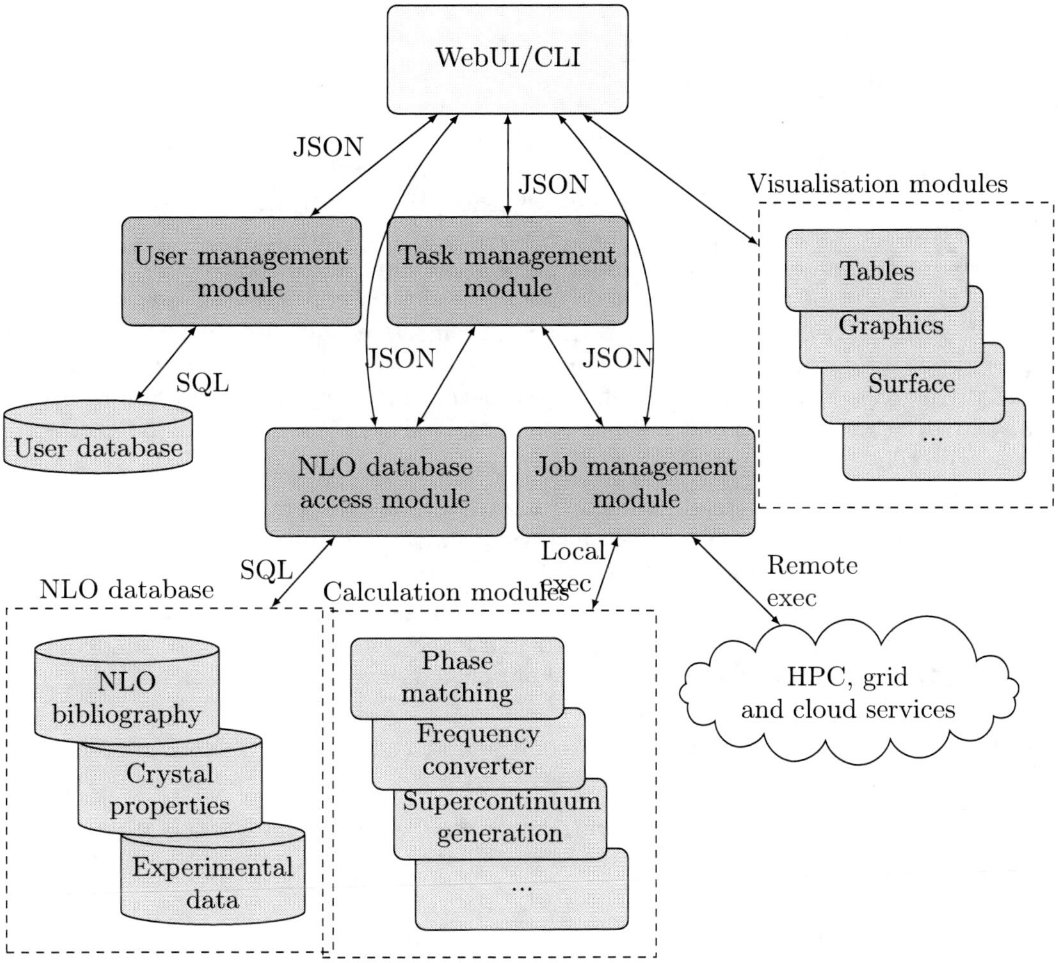

Figure 1. The general scheme of the WebNLO web platform

Since WebNLO has to interact with HPC, grid and cloud services, it is important to separate the stages of preparation and execution of computational tasks. Therefore WebNLO separates functionality of the task management module and the job management module:

- task management module
  - stores and selects local copies of computational modules for calculation of NLO processes.
  - prepares input data for jobs by means of the NLO database;
  - prepares jobs for job management module, for example, prepares parametric jobs for a parameter search;
- job management module
  - executs jobs provided by task management module on local servers, HPC, grid and cloud systems;
  - manages started jobs that includes monitoring, job termination, etc.

Journal of Physics: Conference Series **681** (2016) 012052 doi:10.1088/1742-6596/681/1/012052

An essential part of the platform is the NLO database which contains three main parts: bibliography, crystal parameters and experimental data. The number of relevant bibliographic entries is currently estimated as 10000-15000. The crystal parameters is separated into two categories: general parameters (e.g., name, symmetry and point group, etc.) and optical parameters (e.g., transparency range, refraction indices, etc.). These data are highly structured with very structured access from other modules, therefore an implementation on the top of the generic SQL server such as MySQL or PostgreSQL is preferable.

The module of user management as well as of authentication and authorization provides secure access of users to the web platform resources with the appropriate rights. The visualization module allows to present results of the computing jobs in a graphical form.

Since task management and job management modules essentially provide only management function they do not have special implementation requirements to software stack. Similarly, since the input and output data in the field of nonlinear optics does not typically exceed the scale of megabytes there are no special requirements to the performance of data exchange protocols.

At first stage of the project, the computational modules for simulation of NLO processes include the following calculations:

(i) calculation of the phase matching for the second harmonic generation [7, 8] that allows to study the second harmonic generation in crystals with a regular domain structure [9] by means of He-Ne laser with wavelength of 3.39 μm.

(ii) calculation of sum and difference frequency generation parameters [10] including the calculation of a phase-matching angle dependence on the azimuthal angle, tuning characteristics of pump wavelengths, crystal optical axis orientation and on the temperature of NLO crystals;

(iii) modeling the NLO processes in a supercontinuum generation by using microstructured optical fibers [11] and ultrashort-pulsed lasers [12].

All the modules receive information about the properties of optical crystals from the NLO database. Thus the typical research (workflow) in the WebNLO environment comprises of the following steps

- submitting query to the database and obtaining its result;
- creation of the input files for a computational job on the basis of the preceding step;
- launching the computational job and obtaining its result;
- visualization of the result.

Another example of typical workflows is a parametric search when the same computational job is repeatedly launched with different values of some input parameters for an optimization of properties of an optical device.

3. Conclusion

Currently the development of scientific web applications is booming around the world. In particular, development of web platforms for scientific research in different areas are widely underway around the world. Examples of such developments are nanoHUB (http://www.nanohub.org) for researches in nanotechnology and Yabi (https://ccg.murdoch.edu.au/yabi) for researches in the area of bioinformatics (other examples and discussions see in [1]). Though all the developments are based on the basic principles of service oriented architecture (SOA), their implementations are adapted to the specific application area or functional purpose (e.g., nanoHUB is intended mainly for educational purposes).

In this paper we presented the design of a web platform, entitled WebNLO, for computer modeling of nonlinear optics phenomena.The platform provides users with an access to database of optical properties of crystal materials and with a capability to execute computational tasks in the field of nonlinear optics. It includes computational modules which implement computational algorithms required for development of scientific models and devices in nonlinear optics. The platform uses the SaaS model and is built as a set of interacting RESTful web services. Users interact with the platform by means of a web browser or the command line interface. We have discussed the general scheme of WebNLO and the model of interactions between the platform modules. Such a resource which currently has no analogues will allow researchers to access high-performance computing resources that will significantly reduce the cost of the research and development process.

Acknowledgments

This work is supported under RSCF grant No.15-11-10023.

References

[1] Polyakov S P, Demichev A P, and Kryukov A P 2015 *Proceedings of the 4th International Young Scientists Conference on Computational Science, Procedia Computer Science* **66** (to be published)
[2] Smith A V 2015 *Crystal nonlinear optics with SNLO examples* (AS-Photonics)
[3] Grechin S G, Dmitriev V G, Sharandin E A and Kiiko V V 1998 *Proc. of the IX Conference on Laser Optics* (Technical Digest. TuA3-11) p 16
[4] LASCAD 2014 *LASCAD 3.6 manual*; https://www.las-cad.com.
[5] Richardson L and Ruby S 2008 *RESTful web service* (O'Reilly Media)
[6] Demichev A, Kryukov A, Shamardin L 2009 *Int. J. Software and Systems* **4** p 172 (in Russian)
[7] Stoleru R, Pascu M L and Dumitras A M 1996 *Pure and Applied Optics: Journal of the European Optical Society Part A* **5** p 119
[8] Kumar S C, Samanta G K, Devi K and Ebrahim-Zadeh M 2011 *Opt. Express* **19** p 11152
[9] Lu Y, Mao L, Cheng S, Ming N and Lu Y 1991 *Applied Physics Letters* **59** p 516
[10] Fradkin K, Arie A, Skliar A and Rosenman G 1999 *Applied Physics Letters* **74** p 914
[11] Russell P 2003 *Science* **299** p 358
[12] Krylov A A, Sazonkin S G, Lazarev V A, Dvoretskiy D A, Leonov S O, Pnev A B, Karasik V E, Grebenyukov V V, Pozharov A S, Obraztsova E D and Dianov E M 2015 *Laser Physics Letters* **12** p 065001